Peter Unger (Ed.)
Injection Molds

Peter Unger (Ed.)

Gastrow Injection Molds

130 Proven Designs

4th Edition

HANSER

Hanser Publishers, Munich • Hanser Gardner Publications, Cincinnati

The Editor:
Dr. Peter Unger, Rosengasse 1, 69469 Weinheim, Germany

Distributed in the USA and in Canada by
Hanser Gardner Publications, Inc.
6915 Valley Avenue, Cincinnati, Ohio 45244-3029, USA
Fax: (513) 527-8801
Phone: (513) 527-8977 or 1-800-950-8977
www.hansergardner.com

Distributed in all other countries by
Carl Hanser Verlag
Postfach 86 04 20, 81631 München, Germany
Fax: +49 (89) 98 48 09
www.hanser.de

Library of Congress Cataloging-in-Publication Data

Gastrow, Hans.
[Spritzgiess-Werkzeugbau in Beispielen. English]
Injection molds : 130 proven designs / Gastrow. -- 4th ed. / Peter Unger (ed.).
p. cm.
Includes bibliographical references.
ISBN-13: 978-1-56990-402-2
ISBN-10: 1-56990-402-2
1. Injection molding of plastics--Equipment and supplies. 2. Injection molding of plastics--Design. 3. Injection molding of plastics--Methodology. I. Unger, P. (Peter) II. Title. III. Title: Gastrow injection molds.
TP1150.G2713 2006
668.4'12--dc22
2006012030

Bibliografische Information Der Deutschen Bibliothek
Die Deutsche Bibliothek verzeichnet diese Publikation in der Deutschen Nationalbibliografie; detaillierte bibliografische Daten sind im Internet über <http://dnb.ddb.de> abrufbar.

ISBN-10: 3-446-40592-5
ISBN-13: 978-3-446-40592-9

Production Management: Oswald Immel
Typeset by Techset Ldt. UK
Coverdesign: MCP • Susanne Kraus GbR, Holzkirchen, Germany
Coverconcept: Marc Müller-Bremer, Rebranding, München, Germany
Printed and bound by Druckhaus Kösel, Krugzell, Germany

Preface to the Fourth Edition

For over 35 years, ever since the first edition of this explanatory collection of tried and proven examples of molds, the "Gastrow" has been serving two generations of mold designers and builders as reference work and as a problem solver for design tasks. That is how the new "Gastrow" should be regarded: as a book by the practical and for the practical, containing solutions for problems and design details for configuring injection molds.

For this fourth edition, changes and supplements were once again undertaken with the aim of representing the state of the art. At the same time, our aim was to preserve the knowledge that is "Gastrow's" lasting heritage. One new feature is a table of contents of all the molded parts produced in the molds described here. Again it was difficult, and not achieved to the degree desired, to obtain new examples of innovative mold building from actual practice for this new edition. The world has changed since "Gastrow". That may be regrettable, but it is certainly one of the results of "globalization": more or less justified worries about piracy of brands and products, as well as the worldwide trade in obvious plagiarisms advise us to be cautious. Although almost every mold is one of a kind, many people are shy about revealing their design ideas: copycats pay no development costs.

Nonetheless, the competition for and the education to excellence presuppose the presentation of capabilities. In one way or another, anything that can be purchased can ultimately be duplicated. The know-how embodied in every product often cannot be reconstructed without limitations, or at least not before many attempts to do so have failed. *Creative* entrepreneurs (mold builders) can obtain a competitive edge in this way.

We owe thanks to those who have given their kind assistance in this task along the way.

Spring 2006 The Editor

Preface to the Third Edition

The third English edition of "Gastrow" has been extensively revised to reflect the state of the art, and has also been considerably expanded. Outdated designs have been eliminated to make way for others, and the overall number has now been increased to 130. At the same time, the successful format of the previous editions remains unchanged. Modern technologies such as gas-assisted injection molding have been retained and augmented with others, such as three-component and metal injection molding. The literature references have been expanded, where appropriate, to include major publications.

The new "Gastrow" is the fruit of the labor of a great many authors. The editors would like to take this opportunity to extend their especial thanks to them.

The Editors

Preface to the Second Edition

The second English edition of Gastrow is now here. Since the appearance of the first (German) edition of this interpretative collection of tested and proven mold designs more than twenty-five years ago, this book has served two generations of designers and mold makers as a reference work and problem solver. This is also the intent of this new edition of Gastrow. It was not supposed to be a text book either then or now.
This new edition has been revised extensively. A large number of new molds representing the state of the art have been included. The computational methods given in earlier editions have been eliminated completely, since these are treated in a more up-to-date fashion and in greater detail in other literature (e.g. in Menges, Mohren "How to Make Injection Molds", 2nd edition, Carl Hanser Publishers). Whenever possible, the particular tool steels used have been listed with the respective examples. Accordingly, it appeared necessary to add a new chapter on material selection and surface treatment methods.
The second edition is easier to use: an overview (p. 17) with references to the particular design employed for a given mold simplifies the use of the book. Following the previous tradition, the spectrum of molds presented extends from the simplest design to those exhibiting the highest degree of difficulty. Nevertheless, all molds have one thing in common: each contains some special know-how, and they demonstrate the high technical standards moldmaking has reached today. The editors wish to thank all authors for their contributions to this new "Gastrow" and especially the translator Dr. Kurt Alex who prepared this English edition.

Fall 1992 The Editor

Preface to the First Edition

Hans Gastrow has been publishing examples of mold construction for injection molding since the mid-fifties. These were collected and published in 1966 in the first German edition of this book, which was widely acclaimed because there had been, until then, no other collection of its kind. The injection molding industry stood at the beginning of its great upturn and ideas for constructing good and economically feasible molds were received with great interest. Shortly after the publication of the first edition, Gastrow died. The second edition, published in 1975, kept the objectives set by the first. It does not aim to be a textbook but illustrates selected problems of injection mold construction with interesting and commercially tested solutions. Some of the examples from the original *Gastrow* were retained; others, from younger specialists, were added. The present English translation of the third German edition remains true to this principle. Along with a large number of new examples, principles of construction are also treated. At the time of the second edition's publication, some of them did not possess their present topicality, as for example, hot-runner molds. The solutions to the problems illustrated include molds from the simplest technology to the most complex multi-stage molds.

Summer 1983 The Editor

Contents

1 Principles of Mold Design

General Remarks

In an article reporting on the Ninth Euromold Fair, we read, [1] "Mold and die making is alive and well in Germany." The innovative strength of the field speaks for this claim. Even if production, and the know-how that goes with it, are being shifted out of the country, the truth is, "Much more significant for securing long-term perspectives are: continued technological progress with respect to production-cost cutting and product functionality, as well as unbending and far-sighted training to motivate the next generation." [2] From its very inception, the "Gastrow", being a reference work and source of ideas, has been dedicated to the goal of disseminating knowledge. This new edition aims to do so more as a collection of examples to help find design solutions. Computer methods, i.e., CAD, can at best supplement and optimize a design concept with, for example, rheological, thermal, and mechanical mold configuration, but, as all experience shows, cannot replace it. Moreover, it remains the case that the results of CAD have to be critically evaluated – a task that requires sophistication and practical experience. Thus it remains common practice in the production of precision-made injection molded parts to build a test mold, or at least a test cavity, in order to optimize dimensional stability, for example, and adapt to requirements (in several steps). CAD results often indicate only the determination for shrinkage (warping), a characteristic of molded parts, especially those made from semi-crystalline polymers, that is quite diffcult to quantify. Even so, development time and costs can undoubtedly be reduced by suitable computer methods. For information on applying computer methods, the reader should consult the relevant literature.

There may be no *objective* rule dictating *the* right way to classify anything, but there is *a* right way, namely to organize the subject matter so thoroughly that all phenomena are covered and so clearly that the mind receives a distinct overview of the total. Of course, time and experience cause us to see the phenomena differently, expand and alter the things to be classified and, in so doing, provide an additional pathway of understanding that does not always sit well with a classification system rooted in the past. In this respect, injection molds are no different from anything else: some of the terminology is theoretically clear, some does not become clear unless one knows when and where it came from. Since engineering is the practical offspring of science, historical example is a major source of knowledge as inspiration for the engineer, helping to bridge the gap between theory and practice. For the mold designer working on a problem, consulting previous practice can save time and locate the areas that require real work, i.e., innovation. He can see how others have faced and solved similar problems, while he can evaluate their results and create something even better – instead of "reinventing the typewriter". One basic requirement to be met by every mold intended to run on an automatic injection molding machine is this: the molded part has to be ejected automatically and not require subsequent finishing (degating, machining to final dimensions, etc.)

For practical reasons, injection molds are best classified according to both the major design features of the molds themselves and the molding-operational features of the molded parts. These include the

- type of gating/runner system and means of separation
- type of ejection system for molded parts
- presence or absence of external or internal undercuts on the part to be molded
- the manner in which the molded part is to be released.

The final mold design cannot be prepared until the part design has been specified and all requirements affecting the design of the mold have been clarified.

1.1 Types of Injection Molds

The DIN ISO standard 12165, "Components for Compression, Injection, and Compression-Injection Molds" classifies molds on the basis of the following criteria:

- standard molds (two-plate molds)
- split-cavity molds (split-follower molds)
- stripper plate molds
- three-plate molds
- stack molds
- hot runner molds

Generally, injection molds are used for processing

- thermoplastics
- thermosets
- elastomers

There are also cold runner molds for runnerless processing of thermosetting resins in analogy to the hot runner molds used for processing thermoplastic compounds and elastomers.

Sometimes runners cannot be located in the mold parting plane, or each part in a multi-cavity mold has to be center-gated. In such cases, either a second parting line (three-plate mold) is required to remove the solidified runner, or the melt has to be fed through a hot runner system. In stack molds, two or more molds are mounted back-to-back in the line of closing, but without multiplying the required holding force. The prerequisite for such solutions is large numbers of relatively simple, e.g., flat molded parts, and their attractiveness comes from reduced production costs. Today's stack molds are exclusively equipped with hot runner systems that have

to meet strict requirements, especially those involving thermal homogeneity.
For ejecting molded parts, mainly ejector pins are used. These often serve, in addition, to transfer heat and vent the cavity. Venting has become a major problem since electrical discharge machining (EDM) has become state-of-the-art. Whereas cavities used to be "built up" from several components, thus providing for effective venting at the respective parting planes, EDM has, in many cases, enabled the production of cavities from a single massive block. Special care must be taken to ensure that the melt displaces all air, and that no air remains trapped in the molded part – an especially sensitive issue. Poor ventilation can lead to deposits on cavity surfaces, and to the formation of burn spots (so-called "diesel effect") and even to corrosion problems. The size of venting gaps is essentially determined by the melt viscosity. They are generally on the order of 1/100 mm to approx. 2/100 mm wide. When extremely easy flowing melts are to be processed, vents have to measure in thousandths of a millimeter to ensure that no flash is generated. It must be noted that effective heat control is generally not possible in regions where a vent is provided. As for venting elements – such as venting inserts made from sinter metal – they require regular servicing due to time-factored pore-clogging that varies with the material being processed. Care must be taken when positioning venting elements in the cavity.
Moving mold components have to be guided and centered. The guidance provided by the tiebars for the moving platen of an injection molding machine can be considered as rough alignment at best. "Internal alignment" within the injection mold is necessary in every instance.
Tool steels are the preferred material for injection molds. The selection of materials should be very careful and based on the resins to be processed. Some of the properties required of tool steels are

- high wear resistance
- high corrosion resistance
- good dimensional stability (see also Section 1.9)

Molds made from aluminum alloys are also gaining in popularity, see also Section 1.10.3.1.

1.2 Types of Runners and Gates

1.2.1 Solidifying Systems

According to DIN 24450, a distinction is made between the terms

- 'runner' (also termed 'sprue') meaning that part of the (injection molding) shot that is removed from the molded part
- 'runner' meaning the channel that plasticated melt passes through from its point of entry into the mold up the gate and
- 'gate' meaning the cross-section of the runner system at the point where it feeds into the mold cavity.

The flow path of the melt into the cavity should be as short as possible in order to minimize pressure and heat losses. The type and location of runner/gate are important for:

- economical production
- properties of the molded part
- tolerances
- weld lines
- magnitude of molded-in stresses, etc.

The following list provides an overview of the most commonly encountered types of solidifying runner systems and gates.

- Sprus (Fig. 1.1)

are generally used when the parts have relatively thick walls or when highly viscous melts require gentle processing. The sprue has to be removed mechanically from the molded part after ejection. Appropriate sprue bushes are available as standard units in various versions, for example, with twist locks, temperature control, etc., see also ISO 10072. Due to their large flow diameters, conventional sprues exhibit minimal pressure loss. However, it must be taken into consideration that a too-large sprue can determine the cycle time. Thus maximum diameter ought not to exceed part wall-thickness plus approx. 1.5 mm. If temperature-controlled (cooled) sprue bushes are used, this value may be exceeded. Conventional sprues offer optimum holding time in the injection molding process. To prevent sink marks or non-uniform gloss, suffcient (separate) cooling power should be provided at a distance from the gate.

- Pinpoint (Fig. 1.2)

In contrast to the sprue, the pinpoint gate is generally separated from the molded part automatically. If gate vestige presents a problem, the gate d_1 can be located in a lens-shaped depression on the surface of the molded part. Commercially available pneumatic nozzles are also used for automatic ejection of a runner with pinpoint gate. Pinpoint gating has been especially successful in applications for small

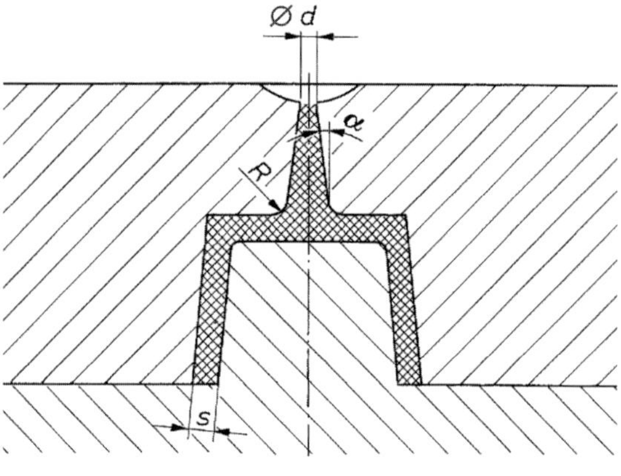

Figure 1.1 Conventional sprue
α = draft, s = wallthickness, d = sprue(diameter), $d \leq 1.5 + 5$ [mm]; $d \geq 0.5$ mm; 15 [mm]

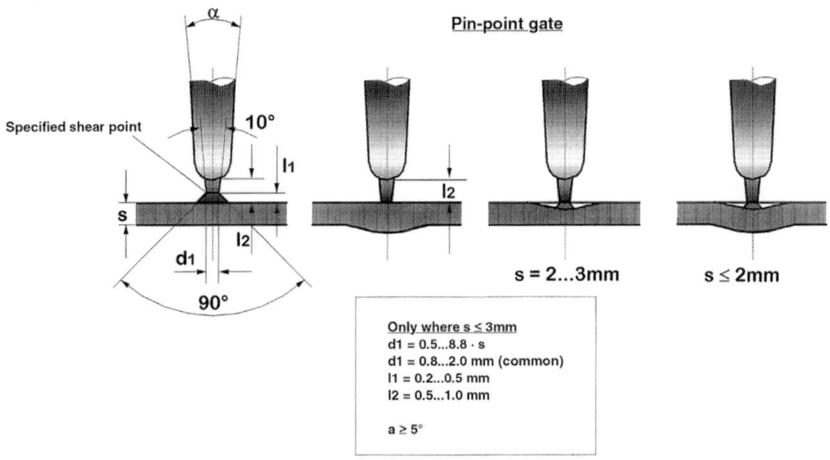

Figure 1.2 Pinpoint gate
(Courtesy: Ticona)

and/or thin-walled molded parts. At separation, however, drool has been a problem with certain polymers and premature solidification of the pin gate may make it diffcult to optimize holding time.

- Diaphragm gate (Fig. 1.3a)

The diaphragm is useful for producing, for instance, bearing bushings with the highest possible degree of concentricity and avoidance of weld lines. Having to remove the gate by means of subsequent machining is a disadvantage, as is one-sided support for the core. The diaphragm, Fig. 1.3, encourages jetting which, however, can be controlled by varying the injection rate so as to create a swelling material flow. Weld lines can be avoided with this type of gating.

- Disk gate (Fig. 1.3b)

This is used preferably for internal gating of cylindrical parts in order to eliminate disturbing weld lines. With fibrous reinforcements such as glass fibers, for instance, the disk gate can aggravate the tendency for distortion. The disk gate also must be removed subsequent to part ejection.

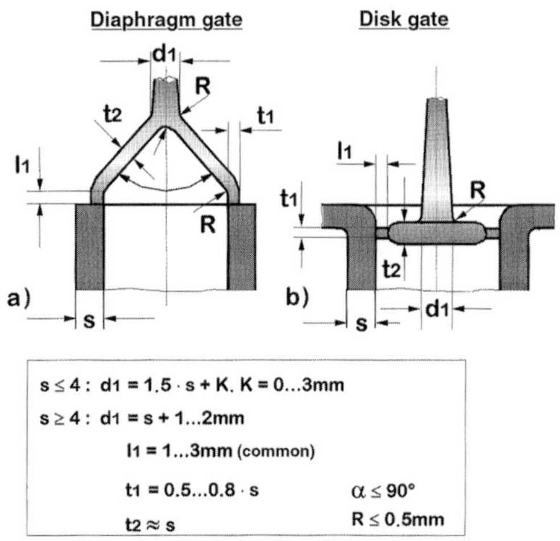

Figure 1.3 Diaphragm (a) and disk (b) gate
(Courtesy: Ticona)

- Film gate (Fig. 1.4)

To obtain flat molded parts with few molded-in stresses and little tendency to warp, a film gate over the entire width of the molded part is useful in providing a uniform flow front. A certain tendency of the melt to advance faster in the vicinity of the sprue can be offset by correcting the cross-section of the gate. In single-cavity molds, however, the offset gate location can cause the mold to open on one side, with subsequent formation of flash. The film gate is usually trimmed off the part after ejection, but this generally does not impair automatic operation. Immediately following removal, i.e., in the "first heat", the film gate should be separated mechanically, in order to ensure that the molded part does not warp in the gate area (since the gate's wall thickness is less than that of the molded part, greater and smaller differences in shrinkage may arise and encourage warping).

- Submarine gate (Fig. 1.5)

Depending on the arrangement, this type of gate is trimmed off the molded part during mold opening or directly on ejection at a specified cutting edge. The submarine gate is especially useful when gating parts laterally. Aside from potential problems due to premature solidification, submarine gates can have very small cross sections, leaving virtually no trace on the molded part. With abrasive molding compounds, increased wear of the cutting edge in particular is to be expected. This may lead to problems with automatic degating.

Runner systems should be designed to provide the shortest possible flow paths, avoiding unnecessary changes in direction, while achieving simultaneous and uniform cavity filling regardless of position in multi-cavity molds (assuming identical cavities) and ensuring identical duration of holding pressure for each cavity.

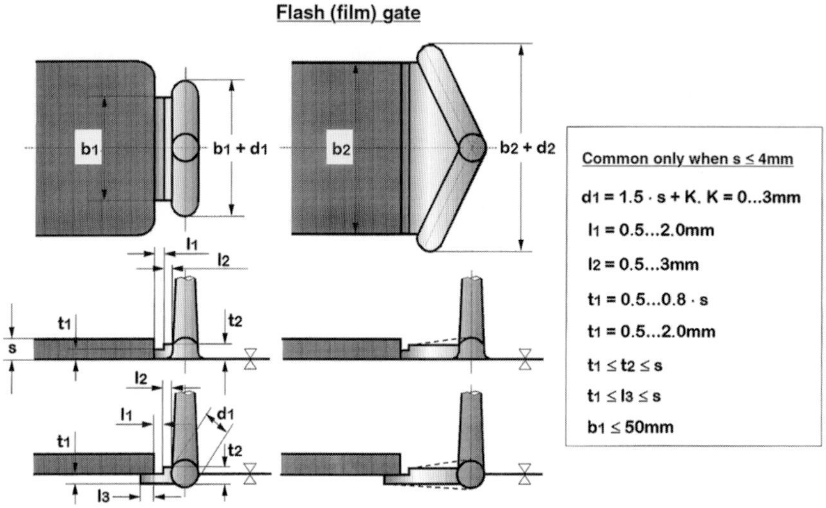

Figure 1.4 Flash or film gate
(Courtesy: Ticona)

For thermoplastics with a high modulus of elasticity (brittle-hard demolding behavior), the angle on the cutting edge has to be relatively small, e.g., $\alpha = 30°$. For thermoplastics with a low modulus of elasticity (viscoplastic removal behavior), *curved submarine gates* have proven successful, Figs. 1.6 and 1.7. In such molds, the gate is separated at a specified point, as with pinpoint gating. Using this type of gating, several submarine gates with short distances in between can produce approximately the same flow pattern as when a film gate is used, but with the considerable advantage that the gate is separated automatically from the molded part, Fig. 1.6. Certain peculiarities of this type of gate have to be kept in mind. For example, the runner must have a lengthened guide and, if necessary, a specified shear point, Fig. 1.6 (right segment), in order to ensure trouble-free separation and removal of the sprue. Replaceable runner inserts are available commercially. One-piece inserts manufactured by the MIM process, e.g., made from Catamold (BASF), are regularly available in round or angular versions with gate diameters between 0.5 and 3 mm [3]. An interesting new development is the swirl-flow insert, since it can be used to gate molded parts "around corners", Fig. 1.8. It is a good idea to provide for separate temperature control as close to the gate inserts as possible.

- Rectangular gate (Fig. 1.9)

Thanks to lower pressure losses and, in consequence, improved pressure transfer, the rectangular gate is sometimes an attractive alternative to point

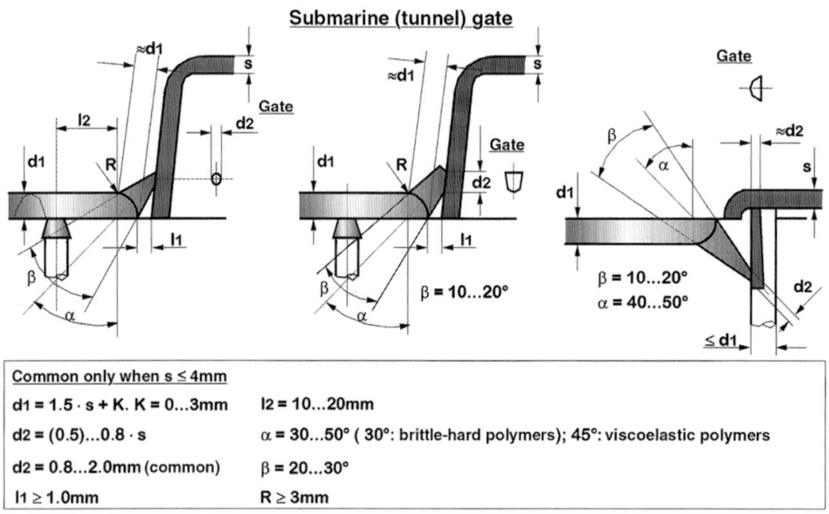

Figure 1.5 Submarine gate
(Courtesy: Ticona)

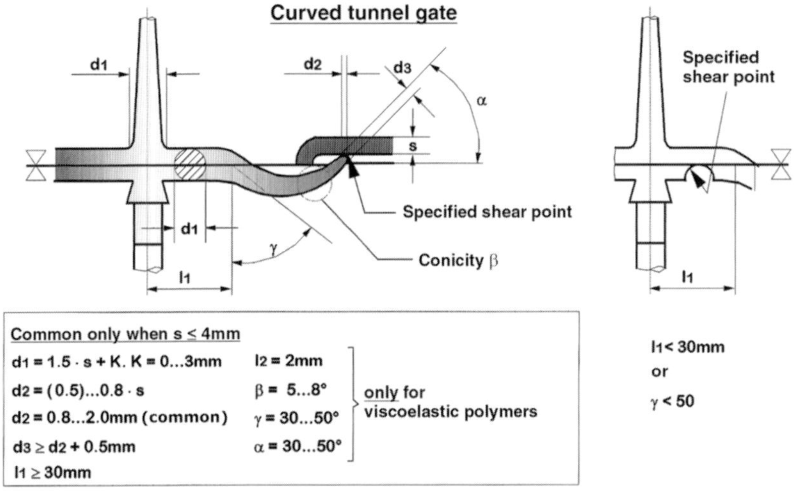

Figure 1.6 Curved submarine gate for viscoplastic polymers
(Courtesy: Ticona)

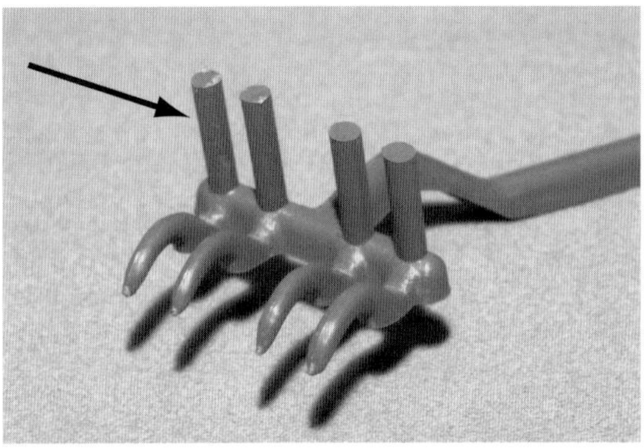

Figure 1.7 Curved submarine gate with lengthened guide ⟶

Figure 1.8 Curved submarine gate manufactured with swirl-flow insert (Source: Exaflow)

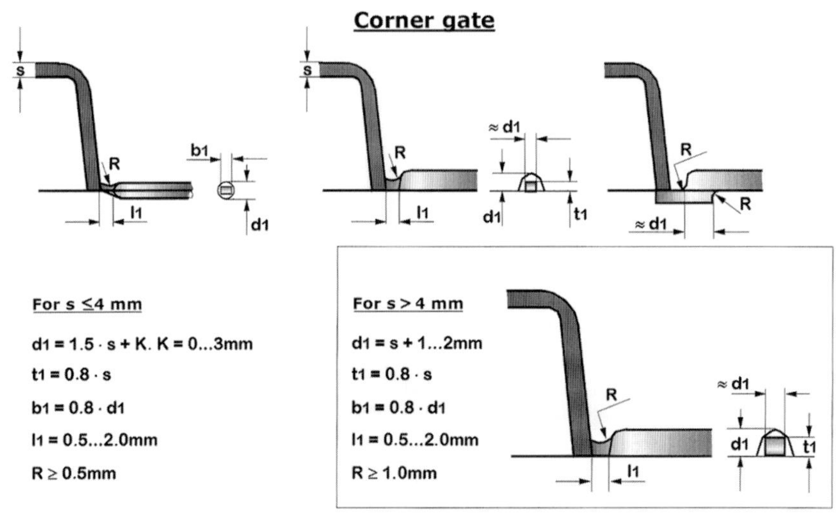

Figure 1.9 Rectangular gate
(Courtesy: Ticona)

gating. Thus rectangular gates are a good choice for molded parts requiring high reliability in operation. However, such gates have to be separated mechanically subsequent to removal. Runner systems should be designed to provide the shortest possible flow paths, avoiding unnecessary changes in direction, while achieving simultaneous and uniform cavity filling regardless of position in multi-cavity molds (assuming identical cavities) and ensuring identical duration of holding pressure for each cavity. The (gate-) sealing times should be identical, assuming identical configuration of the gating areas – such as identical gate diameters, for instance.
Figure 1.10 illustrates types of runner systems often used with multi-cavity molds. Thanks to its identical flow paths, the *star-shaped runner* is *naturally* balanced and, to that degree, preferable with respect to flow behavior. If slides have to be used, this configuration is often not possible. In such cases, *in-line runners* can be used which, however, are disadvantaged by unequal flow paths, i.e., varying degrees of pressure loss. Since the degree of process shrinkage depends largely on pressure, they cannot produce molded parts with uniform performance characteristics. This weakness can be compensated to some extent by *calculated* balancing, e.g., using mold flow analysis. This is done, for example, by varying the flow-channel diameter so as to fill each cavity at the same pressure level. In contrast to natural balancing, calculated balancing *depends* on the point in the cycle. Frequently required changes in processing conditions vis-à-vis the underlying calculated data call the reliability of such analyses into question.
Therefore, as much as possible, an at least *partial*, better yet: *entirely natural* balancing is to be preferred. However, it cannot be denied that such a configuration often leads to a relatively unfavorable ratio of molded part volume to flow channel.

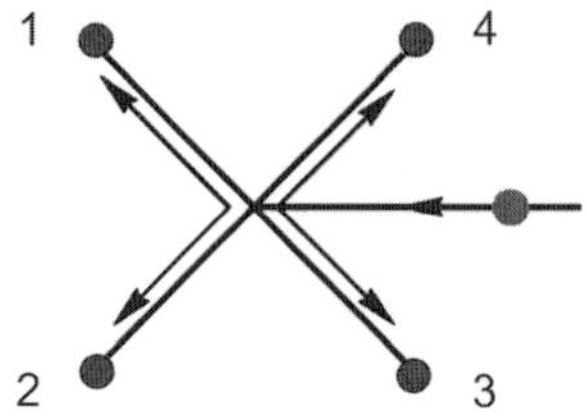

Figure 1.11 Relatively fast melt flow in directions 1 and 2 in a naturally balanced runner system

Problems of this kind can be solved by using appropriate hot runner systems, although not without additional technical complications. In spite of natural balancing, anomalies can occur in flow behavior, Fig. 1.11. It has been observed, for instance, that low viscosity melts tend to flow faster in flow directions 1 and 2 than in directions 3 and 4.

1.2.2 Hot Runner Systems

A hot runner system is the connection between the injection-molding unit and the gate of the cavities, functioning as a feed system for the hot melt. It is one component of an injection mold. In contrast to the frozen sprue in standard molds, the thermoplastic polymer "dwells" for the length of one injection cycle in the hot runner system and remains in a molten state. It is not removed together with the part. That is why this technology is commonly referred to as "sprueless injection molding", Figs. 1.12 and 1.13.
The active principle of the melt feed system corresponds to that of communicating pipes: no matter how large the cross-section of the feed lines or the length of the "pipes" in the hot runner system, the melt remains in direct contact with the gate. Thus it is innately capable of starting to fill all

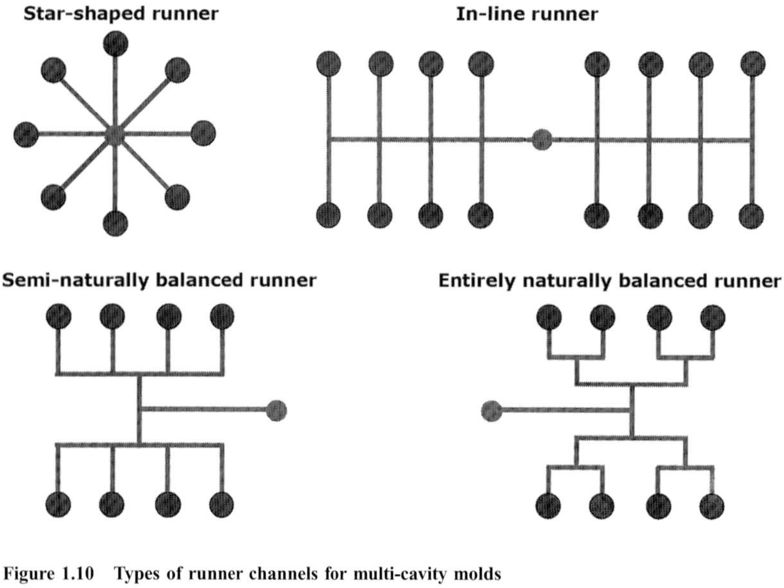

Figure 1.10 Types of runner channels for multi-cavity molds

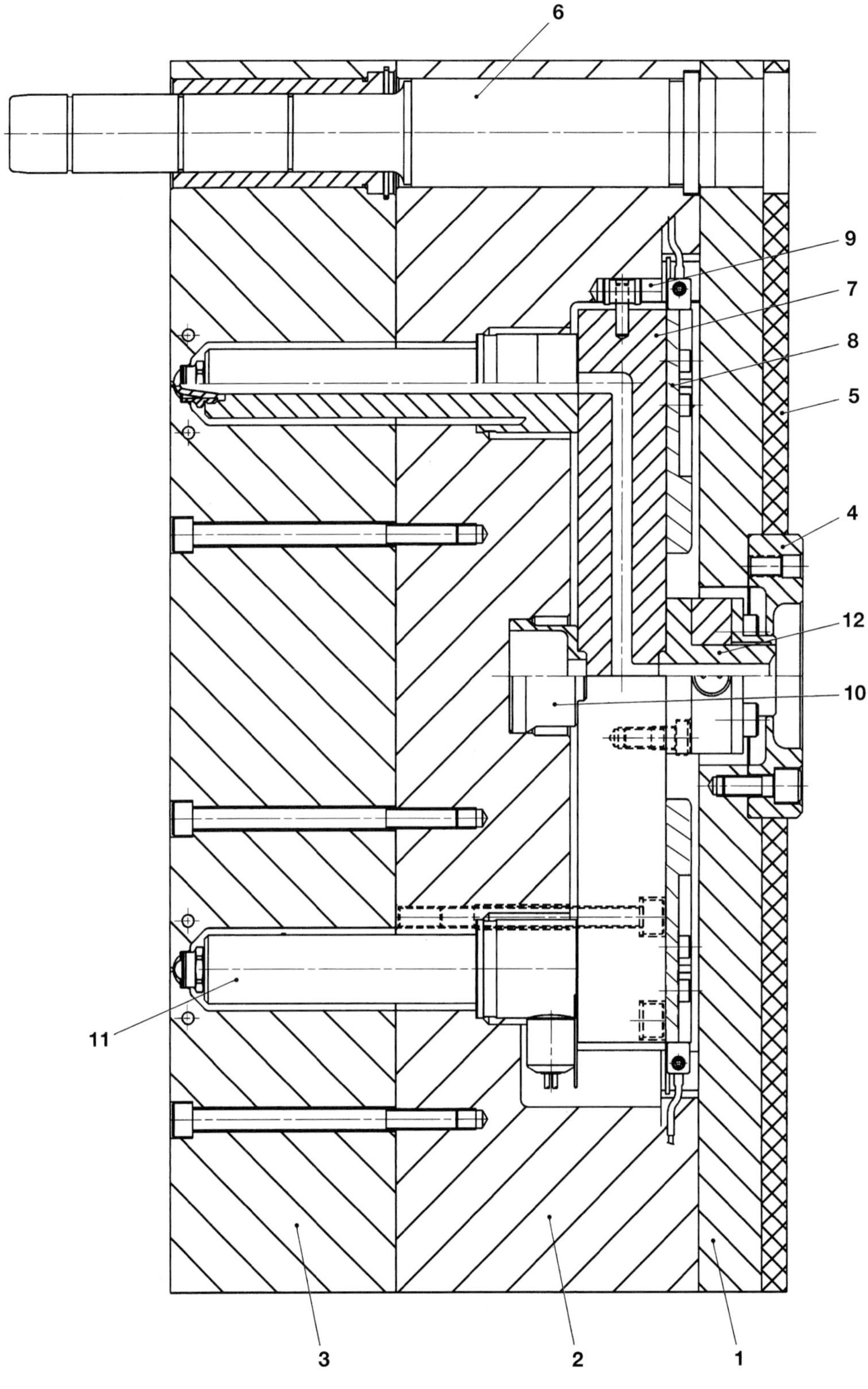

Figure 1.12 Hot side with open sprue nozzles
1: platen, 2: frame plate, 3: nozzle retainer plate, 4: centering flange, 5: insulation sheet, 6: guide pillar, 7:hot runner manifold, 8: heating plate, 9: twist lock: 10: supporting and centering disk, 11: heated, open sprue nozzle 12: heated distributor bushing
(Courtesy: Mold-Masters)

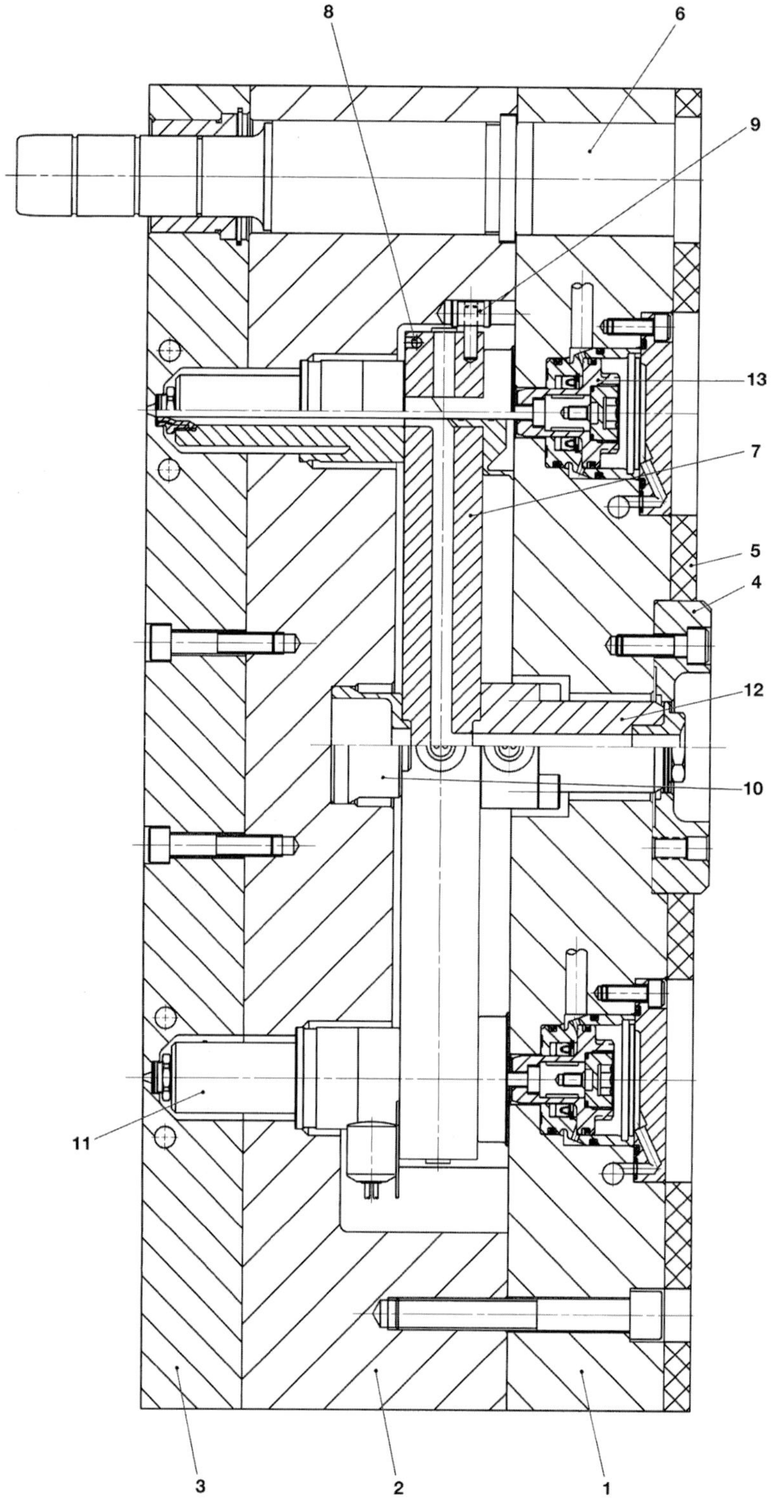

Figure 1.13 Hot side with needle valve-system
1: platen, 2: frame plate, 3: nozzle retainer plate, 4: centering flange, 5: insulation sheet, 6: guide pillar, 7:hot runner manifold, 8: tubular heater, 9: twist lock, 10: supporting and centering disk, 11: heated sprue nozzle with value gating, 12: heated distributor bushing 13: pneumatic/hydraulic-needle valve system
(Courtesy: Mold-Masters)

Table 1.1 1: Types of components in hot runner systems

Component	Type
Hot-runner manifold	Externally heated Internally heated Self-insulating
Manner of heating the hot-runner nozzles	Externally heated, indirect Externally heated, direct Internally heated, indirect Internally heated, direct Internally and externally heated Self-insulating
Centering for the sprue nozzle	Indirect via hot runner manifold Form-sit connection
Transition to cavity	Open nozzles Thermally conductive tip Needle shut-off Thermo seal

cavities in the system simultaneously. This also means that the designer has considerable freedom in creating and configuring the flow channels (e.g., arrangement of the channels in several levels within the hot runner manifold). It is both normal and sensible to equip the hot runner system with heat control. The design principles employed for various hot runner systems can differ considerably. This applies to both the hot runner manifold and the hot runner nozzles, the type and design of which can have considerable influence on the properties of a molded part (Table 1.1).

The various hot runner systems are not necessarily equally well suited for processing of all thermoplastics, even though this may be claimed occasionally. The system that processes the melt as gently as possible should be considered a particular criterion for selection. From a heat transfer standpoint, this requires very involved design principles. Accordingly, hot runner systems satisfying such requirements are more complex, more sensitive, and possibly more prone to malfunction than conventional injection molds. As for the rest, the guidelines of precision machining must be observed to a very high degree when manufacturing such molds. Further aspects for consideration include:

- Since there is no sprue to remove, its (longer) cooling time cannot influence the steps for removal, i.e., cycle times can be shortened.
- No costs are incurred for removing, transporting, regranulating, storing, drying, etc., the sprue. Another point is that regranulate may impair part characteristics. Nor should the contamination problem be underestimated.
- Reduced injection melt volume, due to the absence of sprues, often permits use of a smaller injection molding machine.
- The absence of sprues reduces the projected surface. Holding force, as well as the melting capacity of the injection molding unit can be reduced.
- Hot runner technology offers maximum freedom of gate configuration geometry.
- Since no cooling effects occur, as they do when the sprue solidifies, the pressure requirement can be kept low, even at extremely low flow rates.
- Considering the maximum permissible holding time of the melt in the hot runner system, the channel cross-sections in the hot runner system can be increased. This reduces shear load on the melt.
- Cascade injection molding (sequential injection molding, needle shut-off controlled so that the melt is forced to flow in one preferred direction), multiple-component injection molding, co-injection molding, back-injection molding, multi-daylight molds, as well as family molds would be unthinkable today without hot runner technology.
- The gate area of a hot runner nozzle can be controlled in such a way that the (holding) pressure time can be reduced. This applies not only to the design techniques (e.g., appropriate design of contact surfaces in separate temperature areas) used, but also for the selection of suitable materials (materials as required with high or low heat conductivity), as well as to separate gate heat control. This affects part quality and can lead to a reduction in processing shrinkage.
- Mold costs can be significantly higher when hot runner systems are used. This is especially the case for needle shut-off systems.
- If only a negligible gate vestige is allowed on the surface of the molded part, the cross-section of flow at the gate must be correspondingly small. The high level of shear together with the danger of thermal damage to the melt may necessitate a needle shut-off system in order to enable larger gate cross-sections without noticeable gate vestige on the part surface. Mold costs are thereby increased.
- The time and expense for servicing and maintaining a hot runner system are higher, demanding specially trained and qualified personnel. Trouble-free functioning hot runner systems require care and a high degree of precision, demanding appropriately qualified mold makers, for one.
- Hot runner systems, compared to standard molds, are much more difficult to create [1].

When processing abrasive and/or corrosive molding compounds, the hot runner system must be suitably protected. For instance, the incompatibility of the melt with copper and copper alloys may have to be taken into consideration, since it may lead to catalytically induced degradation (e.g., molding POM, homopolymer). Suitably protected systems are available from suppliers. For the sake of better temperature control, hot runner systems with closed-loop control should be given preference to those with open-loop control.

In medium-sized and, especially, large molds with correspondingly large hot runner manifolds, natural or artificial balancing of the runners is successfully employed with the objective of obtaining uniform pressures or pressure losses. With natural balancing, the flow lengths in the runner system are designed to be equally long. With artificial balancing, the same result is achieved by varying the diameter of the runner channels as necessary. Natural balancing has the advantage of being independent of processing parameters such as temperature and injection rate, for example, but it also means that the manifold becomes more complicated, since the melt must generally be distributed over several levels. This is done, for example, by diffusion welding of several hot runner block levels. An optimum hot runner system must permit complete displacement of the melt in the shortest possible period of time (color changes), since stagnant melt is prone to thermal degradation and thus results in reduced molded part properties.

Open hot runner nozzles may tend to drool. After the mold opens, melt can expand into the cavity through the gate and form a cold slug that is not necessarily remelted during the next shot. In addition to surface defects, molded part properties can also be reduced in this manner as well. In an extreme case, a cold slug can even close the gate.

With the aid of melt decompression (pulling back the screw before opening the mold), which is a standard feature on all modern machines, or with an expansion chamber in the sprue bushing of the hot runner manifold, this problem can be overcome. Care must always be taken, however, to keep decompression to a minimum in order to avoid sucking air into the sprue, runner system or region around the gate (i.e., to avoid the "diesel-effect").

1.2.3 Cold Runner Systems

In a manner analogous to the so-called runnerless processing of thermoplastic resins, thermosets and elastomers can be processed in cold runner molds. This is all the more important, because crosslinked, or cured, runners generally cannot be regranulated. The feed channel in a cold runner system has a relatively low, "colder" temperature in order to keep the thermoset or elastomer at a temperature level that precludes crosslinking of the resin. As a result, the requirements placed on a cold runner system are very stringent: the temperature gradient must be kept to an absolute minimum and the thermal separation of the mold and cold runner must be complete in order to reliably prevent such crosslinking. If, nevertheless, difficulties occur during operation, the mold must be so designed that it is easily accessible to correct problems without a great deal of work. For example, an additional parting plane can allow crosslinked runners to be removed easily.

1.2.3.1 Molds for Processing Elastomers

Elastomer processing is comparable in principle to thermosets processing. Both differ from thermoplastics processing primarily in that the material is brought into heated molds and undergoes crosslinking (it cures) and cannot be reprocessed. The statements made in Section 1.2.3.2 for thermoset molds thus also apply in general to molds for elastomer processing.

Nevertheless, the design details of elastomer molds differ according to whether rubber or silicone is to be processed [1]. For economic reasons, runnerless or near-runnerless automatic molding and largely flash-free parts with perfect surfaces are expected here as well. Gating techniques and mold design are critical and require a great deal of experience. To prevent flash from forming during the processing of elastomers, which become very fluid upon injection into the cavity, molds must be built extremely rigid and tight with clearances of less than 0.01 mm. To vent the cavities, connections for vacuum pumps or overflow channels need to be provided at all locations where material flows together. Computer-aided mold designing [2] offers significant advantages since everything required to optimize process management can be taken into consideration during the design stage [3]. Just as in molds for thermoplastics and thermosets, the runner system in multiple-cavity molds has to be balanced. The cold runner principle together with important details relating to the design of elastomer injection molds is described in [1]. Standardized cold runner systems (CRS) are preferred on account of risk distribution, better availability, far superior quality and fast return on investment (Fig. 1.14).

To change the complete part-forming section (PFS) (1), the mold is disassembled in the mold parting line (MPL) with the aid of quick-clamp elements (2) [8]. Thermal insulation between the part-shaping section and cold runner system is achieved with the insulation sheet (3). Pneumatic needle-valve nozzles (4) offer many economic, qualitative and production advantages over open nozzle systems. Large cross-sectional areas in gate regions (6) that can be sealed by needles place minimum stress on the melt and lead to parts of consistent quality. Closing the gate orifice prevents the material from crosslinking in the nozzle despite the high temperature in the part-shaping section. The throttles (5) for the feed channels ensure optimum balancing of the multiple cold runners by regulating the melt flow in each cavity.

This cold runner system is ideal for processing liquid silicone rubber (LSR). Under certain conditions, solid silicone rubber and natural rubber may also be processed with the aid of standardized cold runner systems [8]. While rubber materials, due to their high viscosity, generally require very high pressures in the cold runner and injection unit, silicone materials, especially the addition-crosslinking two-component liquid silicones, can be processed at relatively low pressures (100 to 300 bar). Low injection pressure is essential for minimizing flash formation. In addition, the molds must be built

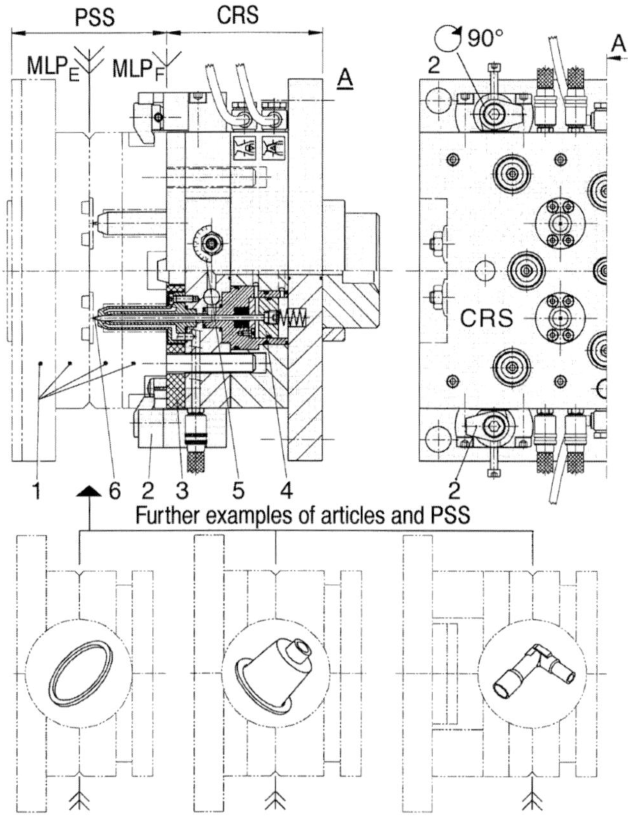

Figure 1.14 Cold runner system (CRS) with pneumatic needle-valve nozzles and throttles for balanced cavity filling, replaceable part-forming sections PSS; MLP_E, MLP_F: parting planes
1: part-forming section, 2: quick clamp elements, 90° turn, 3: thermal insulation sheet, 4: pneumatic needle valve, 5: throttle, 6: gate
(Courtesy: EOS (now DME))

extremely precise and leak-proof. Silicones cure very quickly, so that the cycle time is considerably shorter than for other types of rubber.

The part-forming sections (PFS) of the molds are best heated electrically, with the various mold sections divided into several heating circuits. Insulation sheets (3) should be provided between the mold and the machine platens as well as in the mold itself in order to keep the temperature within narrow limits. The mold steel must also be selected for the relatively high operating temperature of 170° to 220°C. Chrome-alloy steels are used for part-forming sections and often are given an additional hard and/or soft surface coating, such as chrome plating, nickel plating, TiN, CrN or Lamcoat® (WS_2) finish. The surface finish has an effect on the flow properties of the material processed as well as on the release of the molded parts, depending on the part geometry and specific elastomeric material. A slightly roughened part-forming surface is often advantageous. Demolding of elastomeric parts is not without its problems, since such parts are instable and often have undercuts. If positive demolding by means of ejector pins and air assist is not possible, the molded parts can also be removed from the cavity by an auxiliary device (e.g., brushes) or robotic part extractors. The special nature of elastomers requires specific measures with regard to flow properties, temperature control, and part demolding, so that elastomer processing still remains a case for specialists. With improved machine technology, optimization of material characteristics, availability of trial molds [9], substantial user support from system suppliers for filling elements (cold runners) and the increased use of computers, the designing of molds for and processing of elastomers into precision parts pose no difficulties today.

1.2.3.2 Molds for Processing Thermosets

Molds for processing thermosetting molding compounds are comparable in principle with those used for processing thermoplastics, bearing in mind, however, that peculiarities specific to these molding compounds must be taken into consideration.

Molds for processing of thermosetting molding compounds are generally heated electrically. The heat needed for the crosslinking reaction is drawn from the mold. Once in contact with the cavity surface, the viscosity of the melt passes through a minimum, i.e., the melt becomes so low in viscosity that it can penetrate into very narrow gaps and produce flash. The molds thus have to fit very tightly, while at the same time providing for adequate venting of the cavities. These largely opposing requirements are the reason why flash cannot be completely eliminated. Molds should be designed to be extremely stiff so that breathing and the resulting deformation that promotes the formation of flash are avoided. The use of pressure transducers is recommended to determine and monitor the injection pressures, on the basis of which the mechanical properties of the mold are calculated. The pressure actually required depends on the size and geometry of the molded parts. Material selection is of great importance with regard to the life expectancy of the molds. Through-hardening steels are to be preferred for the part-forming surfaces and must exhibit a resistance to tempering consistent with the relatively high operating temperatures of the molds. For molding compounds that tend to stick, e.g., unsaturated polyester resins, steels with $\geqslant 13\%$ chrome content have proven useful, e.g., no. 1.2083 tool steel. Since the thermosetting molding compounds are sometimes modified with abrasive fillers, special attention must be paid to the resulting wear. Fillers, such as stone flour, mica, glass fibers, and the like promote wear. In wear-prone regions of the mold such as the gate, for example, metal carbide inserts should be provided. Other wear-prone mold components should generally be designed as easily replaceable inserts.

Hard chrome plating has proven useful as a means of increasing the wear resistance of part-forming surfaces. At the same time, a certain corrosion protection is achieved. Titanium nitride coatings increase the service life of molds noticeably. Improvements by a factor of 5 have been reported. In addition to improved wear and corrosion resistance, the few microns thick layer facilitates part release and mold cleaning. Stainless steels with more than 18% chrome are also suitable for corrosion protection, but are limited in terms of achievable hardness.

Depending on the geometry of the molded part and the type of molding compound, different amounts of draft for part release must be provided, usually between 1 and 3°. At the time of ejection, thermoset parts exhibit very little shrinkage due to the relatively high temperature. As a result, parts are not necessarily retained on the mold cores, but rather may be held in the cavity by a vacuum. As a rule, thermoset parts are not yet fully cured at the time of ejection and are thus relatively brittle. Accordingly, an adequate number of ejector pins or a suitably large surface area for other ejection means should be provided to avoid damaging the parts during ejection. Undercuts should be released by means of movable cores or slides, which, in addition to being designed for long-term operation, should permit easy removal of any cured resin that might possibly collect. For complicated parts with internal undercuts, fusible core technology is employed. The vent channels should be approx. 0.01 to 0.03 mm wide and highly polished in order to simplify the removal of any flash occurring in them.

Combination heating utilizing heater cartridges for the mold plates in conjunction with frame heaters has proven useful in achieving satisfactory temperature homogeneity. The bulk of the heat is provided by the cartridge heaters, while the heated frame serves as a heat shield against the surroundings. Depending on mold size, 30 to 40 W/kg of mold weight have been found to be adequate for the required heating capacity. Each heating circuit should be provided with its own thermocouple, to be placed between the heating element and the part-forming surface. As a general principle, molds should be provided with insulation sheets to prevent heat losses and the resulting temperature differences. Besides being placed between the mold clamping plates and machine platens, such insulation sheets can also be positioned between mold plates and possibly even other areas of the mold. With the aid of computer programs it is further possible to simulate mold heating and thus to specify the location of heating elements. It has also proven helpful to determine the mold temperature during operation by means of thermophotography and then to use this information to make any necessary changes in new molds.

In general, thermoset molding compounds are not regranulated. Accordingly, an attempt should be made to keep the size of the runner system small relative to the size of the molded parts. (Note: Small qualities of regranulate can improve part surface, although they do impair melt flow behavior). The gate should be located such that it can be easily removed without damaging the part. In principle, all of the gate types commonly used in thermoplastics processing can be employed. As with thermoplastic molding compounds, the type and location of gates will affect the physical properties of the molded parts. In contrast to gates for thermoplastic injection molding, which have to be as large as possible in order to avoid material degradation as the result of shear and friction, gates for thermoset processing are intended to increase the melt temperature via friction. Thus, the appropriate gate size and number of gates must be established on the basis of the type of part to be molded and the type of molding compound to be processed. As a rule, the molding compounds are modified by the supplier to meet the set criteria, which means that joint discussions early in the planning phase for a mold are advisable.

Even though injection molding may be the most economical means of producing thermoset molded parts, there is still interest in process variations such as injection compression molding, for instance, which can be employed to produce very high-quality parts automatically. Injection compression molding combines the advantages of compression molding and those of injection molding. Figure 1.15 shows a multiple-cavity mold with runner pinch-off. Given the appropriate design of the runner system, runners are pinched off during compression.

Multiple-cavity molds with a common filling chamber (Bakelite/Common Pocket System) as

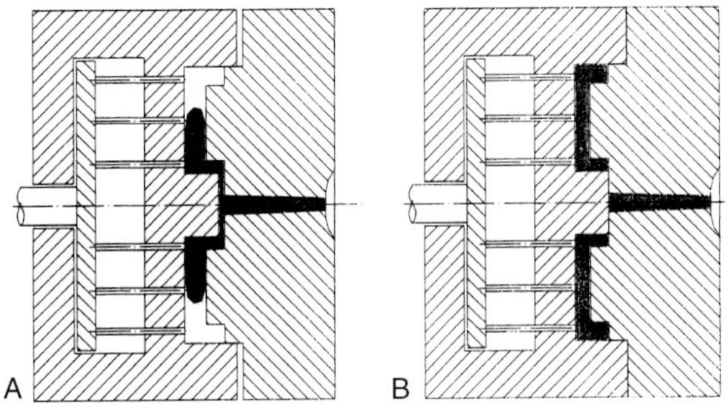

Figure 1.15 Multi-cavity injection-compression mold with immersion gating
A: during injection, B: mold closed

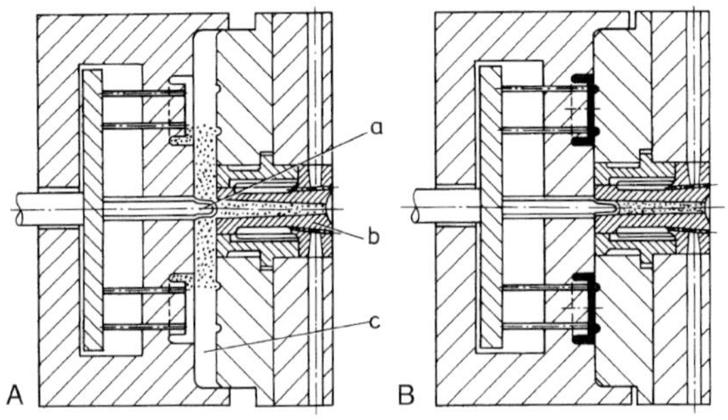

Figure 1.16 Injection-compression mold (Bakelite Common Pocket System)
A: during injection, B: mold closed, a: melt spider, b: sprue bush, c: common pocket

shown in Fig. 1.16 represent relatively simple-to-build, proven designs where the molded parts have minimal flash, which is ejected along with the parts. The sprue bushing is designed to function as a cold runner. The three-plate design (Bucher-Guyer System; Fig. 1.17) is a relatively involved design, but does permit center gating, which is particularly advantageous for round parts. The so-called HTM process (Bakelite **H**igh-**T**emperature **M**olding) represents a development in which cavity conditions similar to those for injection molding thermoplastic resins are achieved during the injection of

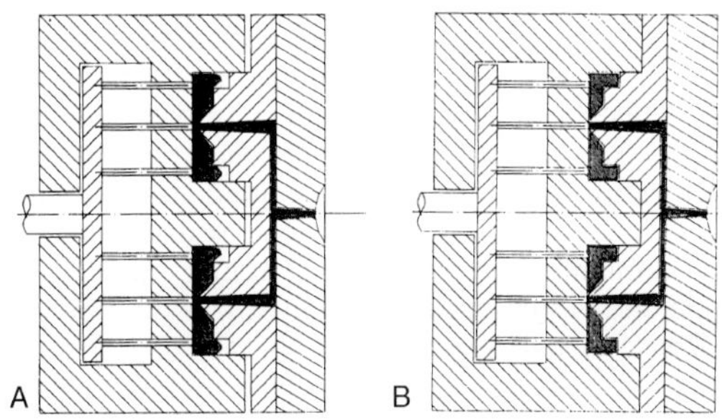

Figure 1.17 Two-cavity, 3-plate injection-compression mold (Bucher-Guyer System)
A: during injection, B: mold closed

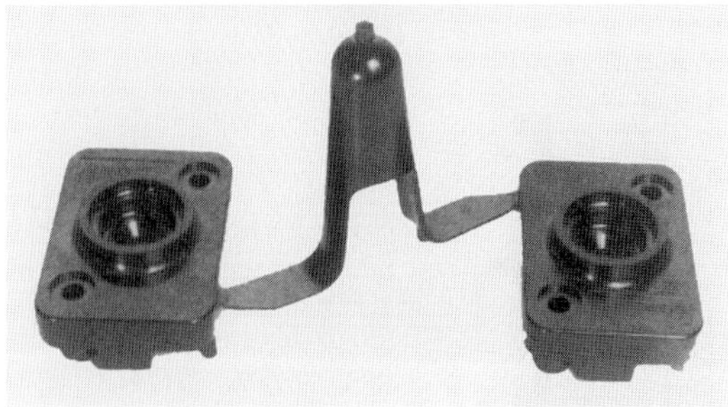

Figure 1.18 HTM-molded parts with runner (Bakelite High-Temperature Molding); shot weight 96b, runner portion 12 g, parts dimensions 45 × 70 mm

thermosetting resins. Whereas during conventional thermoset processing the resin in the cavity briefly becomes more fluid through contact with the heated mold (e.g., 170°C), thus flowing into the smallest gaps and generating flash, with the HTM process the resin is overheated in the runner system so that it cures immediately upon entering the cavity. With this type of runner system, also known as the hot-cone method, dimensionally very accurate, almost flash-free parts can be produced in multiple-cavity molds. A significant reduction in cycle time is one major advantage of this technique. Figure 1.18 shows finished parts molded by the HTM process along with their runner system. Figure 1.19 provides a diagram of the HTM runner system.

So-called cold runner systems are used to process thermoset resins in a manner analogous to the use of hot runner systems employed for processing thermoplastic resins. Whereas the mold plates and part-forming inserts of thermoset molds have an operating temperature of, for instance, 170°C which initiates curing of the material, the sprue bushing and runner channels in a cold runner system are kept at a lower temperature by means of circulating cooling fluid. The temperature is set such that the material does not cure, yet still has a viscosity suitable for processing. For example, the temperature

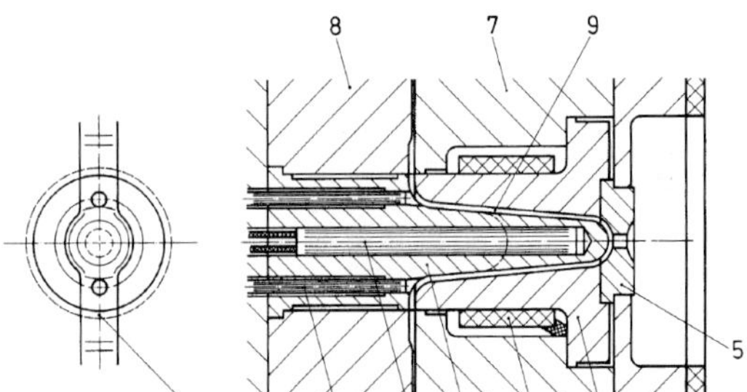

Figure 1.19 Diagram of HTM process
1: sprue bush, 2: spiral heating cartridge, 3: spreader, 4: heater cartridge with thermocouple, 5: sprue disk, 6: ejector, 7/8: cavity plates, 9: runner system

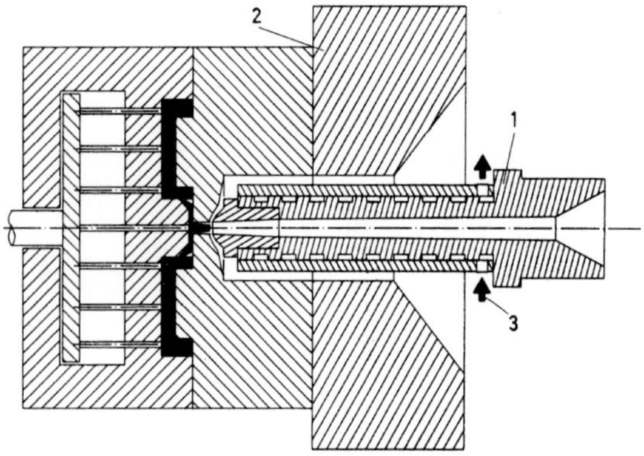

Figure 1.20 Thermoset injection mold with cold runner sprue bush belonging to 1: the machine injection unit that immerses into the nozzle side of 2: the cavity plate and fits non-positively during injection, 3: cooling circuit intake/outlet

set in a cold runner system may be 100°C. Figure 1.20 shows a sprue bushing designed to operate on the cold runner principle. The cold runner system does not necessarily have to be part of the mold. It is often practical to design the machine nozzle to function as the cold runner. In this way, excellent thermal separation of the mold from the cold runner is ensured. This relatively inexpensive solution provides a well-defined break-off point and in addition is easy to maintain. The diagram in Fig. 1.20 shows how the nozzle extends into the mold, which must be suitably enlarged. This technique can be used with single-cavity molds to injection mold quasi-runnerless thermoset parts. In multiple-cavity molds, the cold runner system is usually incorporated into the mold itself in a manner similar to that employed in hot runner systems for injection molding thermoplastics.

Figure 1.21 illustrates a multi-cavity mold with a cold runner system located at the secondary mold parting line (2), which can be released and opened for servicing. A short sprue is required on the molded parts to connect the cold runner system to the cavities. The gate must be dimensioned

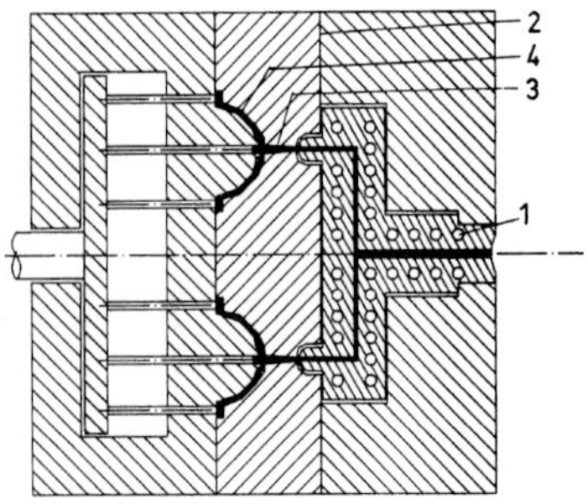

Figure 1.21 Multi-cavity cold runner injection mold with a cold runner system
1: (Bucher-Guyer system) arranged in the secondary parting plane 2, 3: sprue, 4: molded part

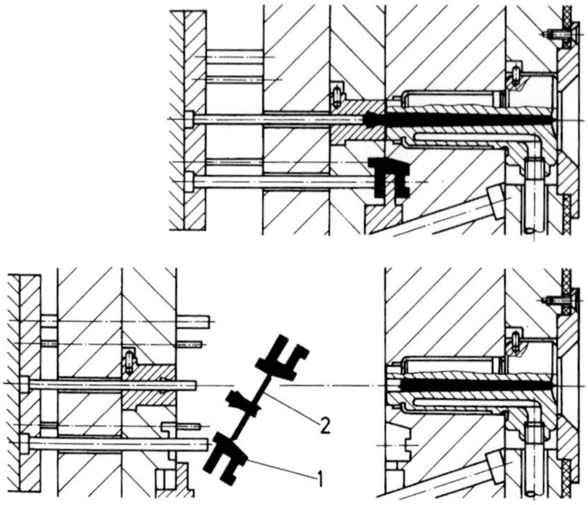

Figure 1.22 Example of a standardized cold runner bush applied to a two-cavity thermoset injection mold
1: molded part, 2: sprue

according to the material to be processed. In order to obtain smooth separation on the part surface, a hydraulic or pneumatic needle shut-off (valve gating) system has to be employed, the use of which, however, is not unproblematic. Figure 1.22 illustrates how the sprue can be eliminated in a multi-cavity thermoset injection mold by using a cold runner bushing. The contact surface between the cold runner bushing and the mold plate should be kept as small as possible in order to minimize heat transfer. In addition, the face of the sprue bushing should not come in contact with the movable mold half; an air gap of approx. 0.3 mm should be provided for thermal separation.

1.3 Temperature Control in Injection Molds

The wide range of different polymers that can be injection molded brings with it a correspondingly wide range of mold wall temperatures. In addition, whereas extremely low temperatures are required for mass production articles, parts demanding high operational reliability require higher, sometimes even very high temperatures. That means that mold temperature control may involve "cooling" in some instances and "heating" in others. The *temperature of the shaping surface* (mold wall temperature) of the injection mold is of major significance. It is achieved by suitable means for maintaining temperature. Actual mold wall temperatures influence

- Part dimensions
- Part weight
- Shrinkage
- Dimensional imperfection
- Warping
- Mechanical characteristics

- Surface quality
- Mold filling
- Pressure requirements
- Demolding stiffness, and especially
- Cycle time

The number of influenced variables indicates how important it is to maintain mold wall temperature to satisfy all requirements. To this end, a temperature control medium is required that should fulfill the demand for thermal equilibrium (thermal uniformity) and constancy of the temperature field. Such a system has the task of transporting heat. Given the same flow velocity and cross-section of the cooling channel, the best possible heat removal is achieved with water, by comparison with other fluid heat control media (e.g., water/ethylene glycol mixture or oil). However, we cannot overlook the fact that the steam pressure of water increases at higher temperatures. For instance, the steam pressure of water at 300°C is approx. 90 bar, whereas the corresponding value for thermal oil is approx. 0.115 bar. Temperature control with fluid media reaches its limits when, for example, there is no or insuffcient room for cooling channel bores between thin ribs. In this case, cooling may be possible with a – CO_2 – system. The increased effort and expense, however, require precise, prior cost-effectiveness analysis.

One essential influence on, for example, the degree of shrinkage, is the cooling rate of the melt in the cavity: the higher the cooling rate, the less the shrinkage, and vice versa. As a consequence of their lower–medium cooling rate, thick-walled areas of a molded part shrink more than do thin-walled areas. If the wall thickness of a molded part varies, shrinkage will vary correspondingly; this can, for instance, lead to warping. This is one reason to require uniform wall thickness in molded parts. This behavior can be seen in the pvt diagram (p = pressure, v = spec. volume, t = temperature). The progression of the pressure/temperature curve is characteristic. The amount of difference in specific volume between points 4 and 6 in Fig. 1.23 is a measure of the *volume* shrinkage of a molded part. The higher the cooling rate (°C/min), the smaller it is, Fig. 1.24.

For molded parts requiring high operating reliability, the goal is to obtain a *uniform* cooling rate at every point. This can be achieved, for example, by appropriate techniques when designing the part, as well as the mold, Fig. 1.25 (corner area, separate cooling circuit). In order to produce molded parts

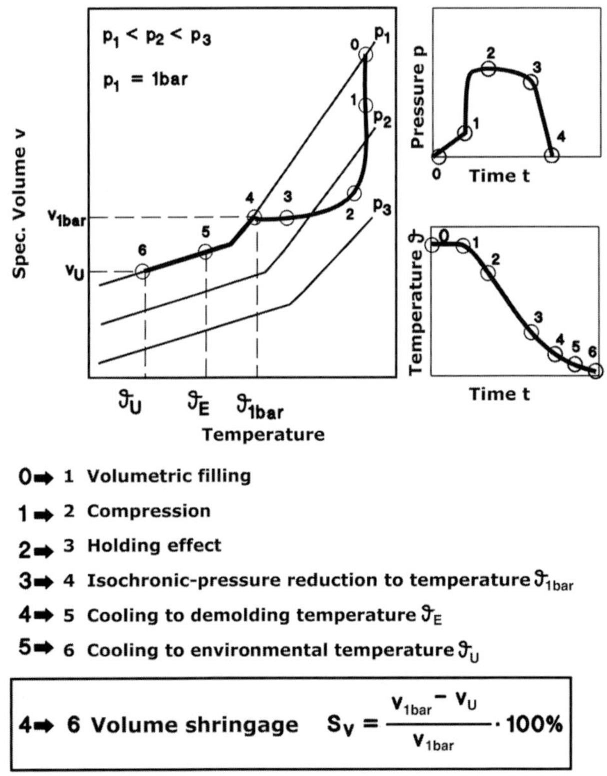

Figure 1.23 Condition curve in the p-v-ϑ diagram for a semi-crystalline thermoplastic material (Courtesy: Ticona)

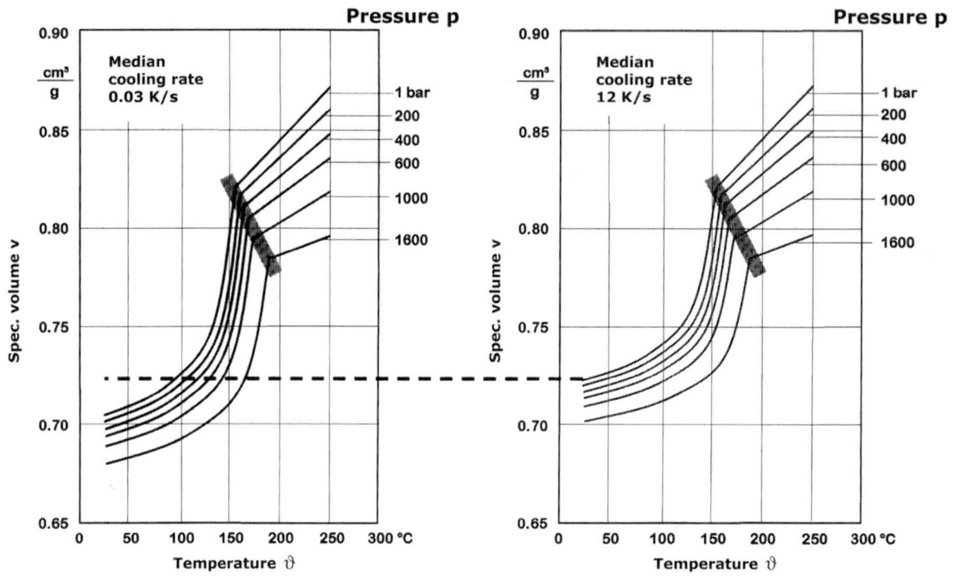

Figure 1.24 p-v-ϑ diagram for POM C 9021, varying cooling rate, resulting in different specific volumes for each case ie, the spec. volume $v = 1/\rho$, with ρ = density
(Courtesy: Ticona)

with *uniform and stable* features, the isochoric pressure drop to room pressure (1 bar), that is, from 3 to 4 in Fig. 1.23, has to take place at unvarying temperature δ_{1bar}, and also at a uniform cooling rate (Fig. 1.24). This task has to be performed mainly by controlling mold temperature. The temperature field in the mold is determined essentially by the heat penetrability of the cavity material employed – such as steel or aluminum. Temperature differences decrease with increasing heat penetrability. Environmental influences, such as room temperature, as well as (strong) air movement influence the thermal economy of an injection mold. Insulation for the external walls of the mold should, therefore, be a fundamental consideration, in order to ensure uniform production and to keep energy costs down. System cooling capacity can thus be reduced.

The data from Table 1.2 can be used to design a first version of cold runner geometry in an injection mold for processing thermoplastics, depending on more precise calculations. However, the cooling channel bores often cannot be optimally adapted to the part contours, thus not conforming to the requirement for thermal homogeneity. By contrast, joining technology can be applied to separate a mold core into sections and cut cooling channels to fit the profile, typically by milling. In order to maximize the fluid contact surface in the cooling channel, the channels usually take the form of rectangles, not circles. The sections thus created are bonded using vacuum welding techniques. With such a state-of-the-art cooling channel system, cycle times can be significantly reduced in many cases, and operational reliability will be improved when parts are molded by this method [9].

1.4 Types of Ejectors

As a consequence of processing shrinkage, molded parts tend to be retained on mold cores (this does not

Table 1.2 Cooling channel geometry, guidelines for preliminary design, distance from center of cooling channel to mold wall 1 to 5 D_{kk}

Part Wall Thickness mm	Cooling Channel Diameter (D_{kk}) mm	Cooling Channel Center Distances mm	Distance Cooling Channel – Part Center mm
≤1	4.5 to 6	10 to 13	11 to 15
1 to 2	6 to 9	13 to 19	15 to 21
2 to 4	8.5 to 11	19 to 23	21 to 27
4 to 7	11 to 14	23 to 31	27 to 35

Specified shape

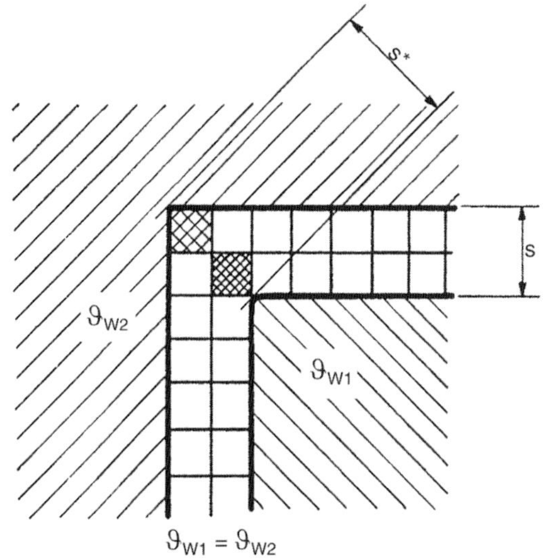

$s^* = \sqrt{2} \cdot s \approx 1.4 \cdot s$

Dimensional/shape deviation

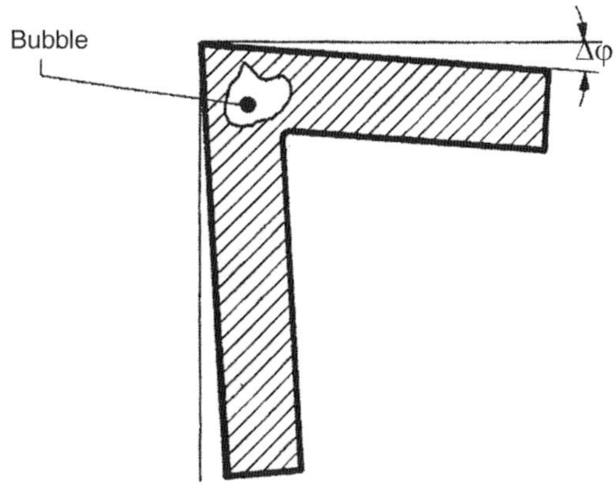

Explanations

Shape deviation (angle of warp $\Delta\varphi$) due to varying shrinkage, danger of bubble formation.

Cause:
Despite uniform mold wall temperatures $\vartheta_{W1} = \vartheta_{W2}$ local cooling of the molded part varies due to varying mold wall contact surfaces:

- single mold wall contact "normal cooling"
- double mold wall contact, "increased" cooling
- no mold wall contact, reduced cooling

Solutions:
- Intensify mold temperature control in corner area.
- Reduce material accumulation (s*) in corner area (Fig. 1.22c and d: $s_1 < s_2$)

Note:
Due to reduced wall thickness, an undercut "H" results in the corresponding demolding direction.

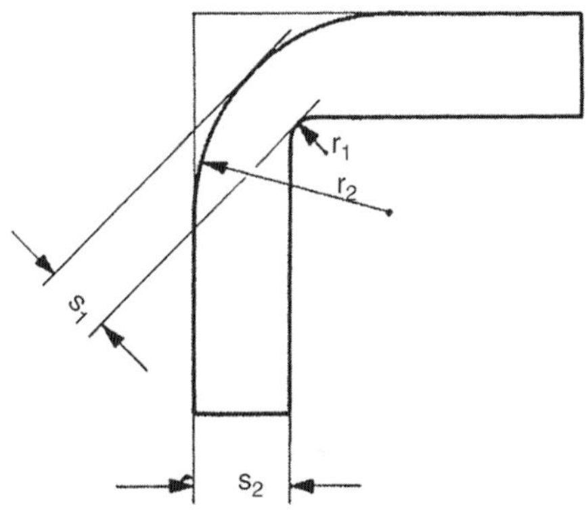

Relatively simple design version.

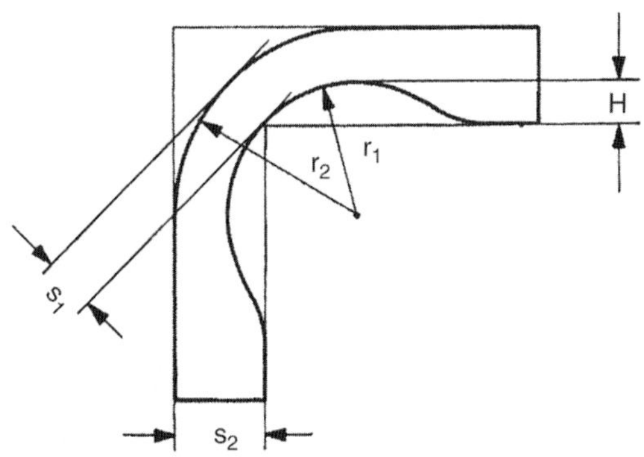

More sophisticated, more effective version.

Figure 1.25 Corner warping resulting from uneven cooling; problem-solving measures
(Courtesy: Ticona)

necessarily hold true for parts molded from thermosetting resins). Various types of ejectors are used to release molded parts:

- Ejector pins,
- Ejector sleeves,
- Stripper plates, stripper bars, stripper rings,
- Slides and lifters,
- Air ejectors,
- Disk or valve ejectors, etc.

The type of ejector depends on the shape of the molded part. The pressure on the surface of the section of the molded part to be ejected should be as low as possible in order to avoid deformation. Profiled ejector pins should be secured against twisting.

Usually, the mold cores and thus also the ejector mechanisms are located on the movable platen of the injection molding machine. In certain cases, it may be advantageous to attach the core to the stationary platen. In this case, special ejector mechanisms are required. To release undercuts, slides are generally needed. Internal undercuts can be released by collapsible cores or internal slides. Threads may be released by:

- Slides,
- Removable inserts,
- Collapsible cores,
- Unscrewing cores, etc.

Undercuts which are intended to act, for instance, as snap fits can also be (forcibly) released directly, i.e., without the use of slides, lifters etc. It must be ensured, however, that the ejection temperature is considerably above room temperature and that the material stiffness is correspondingly low. The ejection forces must not lead to stretching of the molded part, nor should ejectors be forced into the molded part. The permissible deformation during such forced ejection depends on the physical properties of the particular resin at the ejection temperature and on the design of the undercuts. A general statement with regard to the possibility of using (cost-reducing) forced ejection cannot be made. In principle, however, forced ejection should be taken into consideration when laying out a suitable mold.

Textured or grained surfaces generally act like undercuts. Unless a certain minimum draft is provided, they can result in visible damage to the surface. As a guideline to avoid such damage, approximately 1° of draft is required per 1/100 mm of texture depth. Ejectors serve not only to release the molded parts and transfer heat, but are also needed to vent the cavity. Inadequate venting can lead, for instance, to

- Incomplete cavity filling,
- Inadequate welding where flow fronts meet,
- The so-called diesel effect, i.e., thermal degradation (burning) of the molded part, etc.

Problems with venting occur far from the gate, especially in the vicinity of weld lines. The ejection of core holes can lead to vacuum formation and thereby to sink marks on the part surface. Adequate ejection draft, better yet: venting of the core holes during ejection, are required measures.

1.5 Types of Undercuts

Release of undercuts (see also Section 1.4) generally requires additional design features in the mold such as several opening planes. Additional release surfaces can be provided by slides and split cavities. Molds equipped with slides release external undercuts with the aid of

- Angle pins
- Cams
- Hydraulically or pneumatically actuated mechanisms.

Internal undercuts can be released through the use of

- Lifters
- Split cores, which are actuated by means of a wedge
- Collapsible cores, which have smaller outside dimensions in the collapsed state than in the expanded state.

If threads cannot be released by means of split cavities or slides, or if the witness line is undesirable, unscrewing molds are employed. These may utilize

- Replaceable cores that are unscrewed outside the mold
- Threaded cores or threaded sleeves that release the threads in the molded part as the result of rotation during ejection. They are actuated either by the opening motion of the mold (lead screws, gear racks, etc.) or by special unscrewing units.

For short production runs, undercuts can also be released through the use of so-called lost cores (see also Section 1.6.1). When threads intended for fastening are involved, it is often more economical to mold through-holes instead of threads and then use commercially available self-tapping screws.

1.6 Special Designs

1.6.1 Molds with Fusible Cores

Fusible core technology is employed to produce molded parts with cavities or undercuts that could not otherwise be released. Low melting point, reusable alloys on the basis of tin, lead, bismuth, cadmium, indium, or antimony are employed. Depending on the composition, very different melting points result (so-called Wood's alloy, lowest melting point approx. 50°C). By introducing heat, e.g., inductive heating, the metallic core can be melted out of the molded part, leaving almost no residue.

1.6.2 Prototype Molds of Aluminum

Heat-treatable aluminum-zinc-magnesium-copper alloys (material no. 3.4365) have proven useful as a material for injection molds used to produce prototypes or small to medium run molds, see also

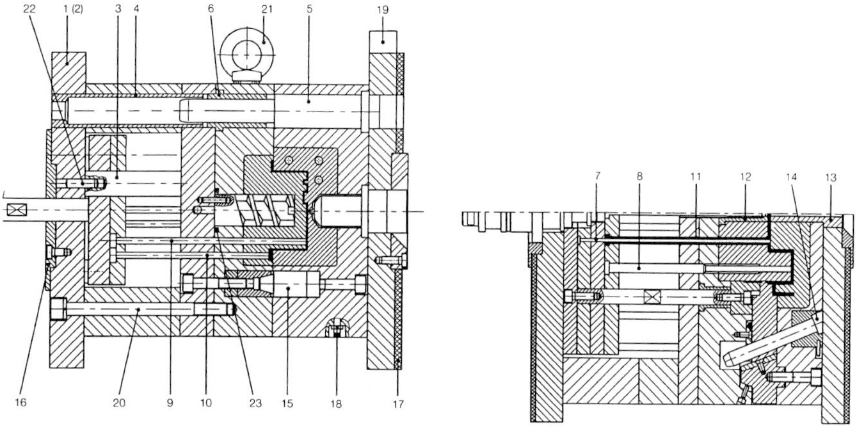

	Parts list		
Item no.	Description	DIN	ISO
1	Plate, no bore	DIN 16760-1	ISO 6753-2
2	Plate bored	DIN V 16760-2	./.
3	Support roll	DIN ISO 10073	ISO 10073
4	Centering sleeve	DIN 16759	ISO 9449
5	Guide pillar	DIN 16761	ISO 8017
6	Guide bush	DIN 16716	ISO 8018
7	Ejector, straight, cyl. head	DIN ISO 6751	ISO 6751
8	Ejector, offset	DIN ISO 8694	ISO 8694
9	Ejector, tapered head	DIN 1530-3	./.
10	Flat ejector	DIN ISO 8693	ISO 8693
11	Ejector sleeve	DIN ISO 8405	ISO 8405
12	Sprue retainer bush	DIN ISO 16915	ISO 16915
13	Sprue bush	DIN ISO 10072	ISO 10072
14	Angle puller pillar	DIN ISO 8404	ISO 8404
15	Centering unit,round	DIN ISO 8406	ISO 8406
16	Centering flange	DIN ISO 10907-1	ISO 10907-1
17	Thermal insulation sheet	DIN 16713	ISO 15600
18	Coolant connection	DIN 16766-1	./.
19	Electrical connection	DIN 16765	./.
20	SHC screw	DIN EN ISO 4762	ISO 4762
21	Ring bolt	DIN 580	ISO 3266
22	Dowel pin	DIN EN ISO 8734	ISO 8734
23	O-ring	DIN 3771-1	ISO 3601

Figure 1.26 Standardized mold components, drawing, and parts list

Section 1.10.3.3. The advantages of this material, such as weight reduction, ease of machining, good thermal conductivity compared to tool steel, must be weighed against its lower strength, reduced wear resistance, low stiffness resulting from its low modulus of elasticity and relatively high coefficient of thermal expansion. In some cases, the properties of aluminum can be used to advantage in combination with steel. A surface coating (e.g., electroless nickel-plating) can substantially increase wear resistance.

1.6.3 Prototype Molds Made of Plastics

To save on the cost-intensive machining needed to produce the part-forming surfaces in molds, curable casting resins can be employed. When strengthened by metal inserts or reinforced with glass fibers, etc., such casting resins can also meet more stringent requirements, within certain limits. The low wear resistance of casting resins must always be taken into consideration. Generally, such molds are used only to produce prototypes or small numbers of parts by means of injection molding.

Molds and/or mold inserts can also be made using stereolithography (STL). The polymer materials used in this process are UV curable (laser beam). With this method, high dimensional accuracy can be achieved.

1.7 Status of Standardization for Injection Molds[1]

ISO standards valid worldwide for the area of mold and die making are being developed by the ISO/TC29/SC8 Technology Committee. Thanks to the active cooperation of many experts on this committee, the goals of the highly developed Central

[1]Revised by H. Lange

European mold making industry are largely being realized.

1.7.1 Standardized Mold Components (as of Mid-2005)

Figure 1.26 shows the standardized components of an injection mold, as well as the corresponding parts list with their standard designations.

1.7.2 Standardized Electrical Connections for Hot Runner Molds

This standard provides an optimum degree of safety in the market, since users and suppliers can follow a standardized terminal configuration for control circuits. DIN standard 16765 (see Fig. 1.26, parts list item no. 19) defines the electrical connections for hot runner molds and temperature control facilities. It distinguishes two types of connection.

- *Connection A:*

Both for the control equipment within the injection molding machine, as well as for external control equipment on molds with their load and signal lines in *separate plug sockets*

- *Connection B:*

For external control equipment of the injection molding machine when used with molds having load and signal lines *in one plug socket.*
Figure 1.27 reproduces an illustration from DIN 16765 (type B: load and signal lines in one plug/socket for max. six control points).

1.7.3 Terminology Standards for Injection Molds

1.7.3.1 DIN ISO 12165 "Tools for Molding-Components of Compression and Injection Molds and Die-Casting Dies"

The assignment of mold types is defined as follows:
- Conventional mold (two-plate mold)
- Split cavity mold (sliding split mold)
- Stripper mold
- Three-plate mold
- Multi-cavity mold
- Hot runner mold

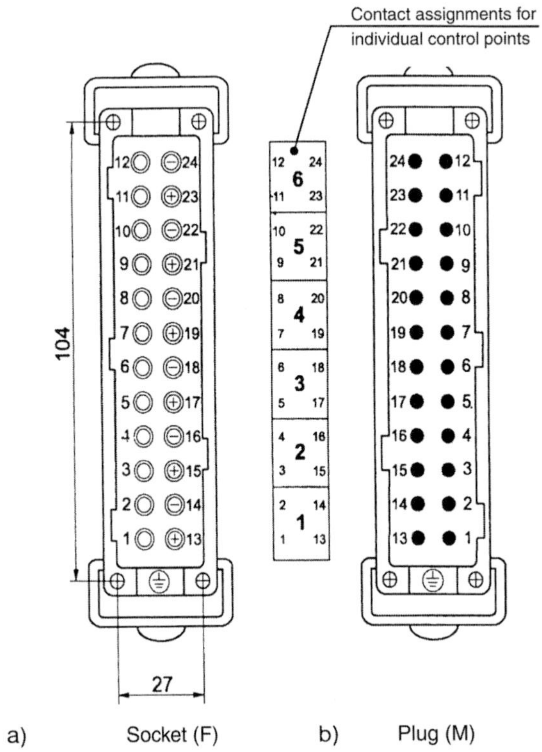

No. of control points	Contact no.	Assignment	Empty pole
6	1 to 12 13 to 14	Load lines, AC 250V Signal lines, positive pole at 13,15,17,19,21,23	None

Figure 1.27 Example of electrical connections for hot runner molds (excerpt from DIN 16765, type B)
A: control equipment connector, B: mold connector

For better orientation, all designations of mold components are subdivided according to the following product groups:

- Platen
- General accessories
- Feeding
- Cooling, heating
- Ejection, demolding
- Other mold-relevant parts
- Mold parts pertaining to die-casting dies

Examples are provided of designs in the area "Types of Molds for Injection Molding and Die-Casting Dies". The item number corresponds to the component listed.

DIN ISO 12165 provides users in the area of mold and die making with standards that authoritatively define the designations for their most commonly used components in English, German, French, and Swedish.

1.7.3.2 DIN 16769 "Components for Gating Systems – Terms"

The various gating systems are subdivided as follows:

- Gating systems for frozen sprues
- Hot sides

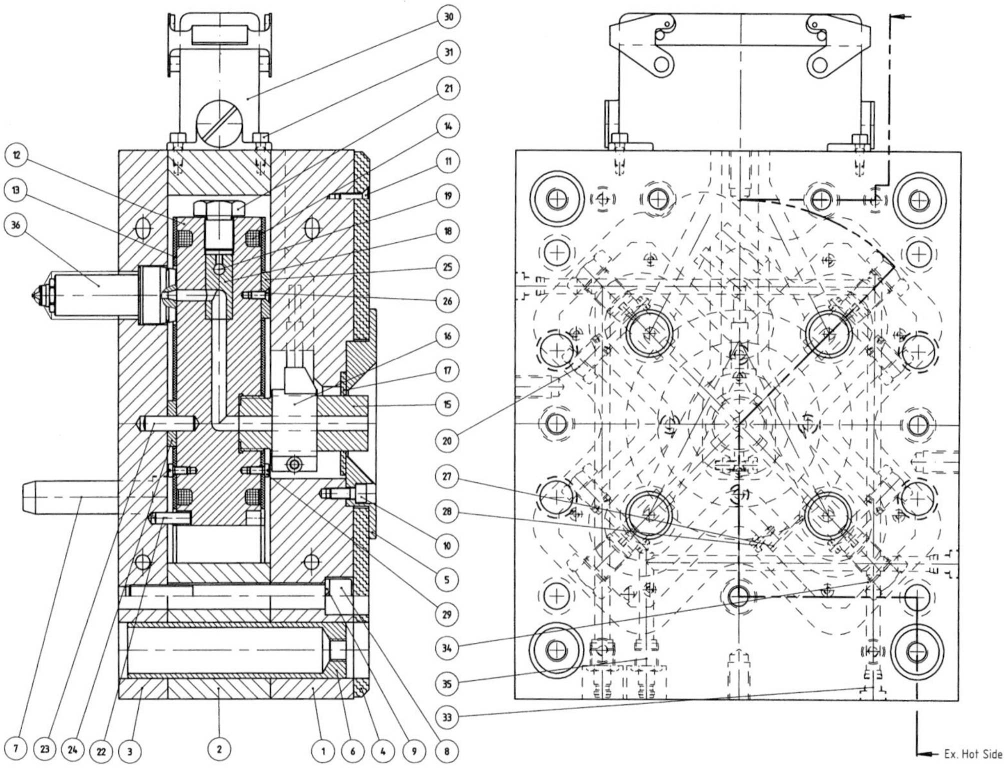

Item no.	Terms	Standard
1	Clamping plate	DIN 16760-1
2	Bolster plate	
3	Nozzle mounting plate	
4	Centering flange	DIN ISO 10907-1
5	Thermal insulation plate	DIN V 16713
6	Centering sleeve	DIN 16759
7	Guide pillar	DIN 16761
8	SHC screw	DIN EN ISO 4762
9	Locking edge washer	
10	SHC screw	DIN EN ISO 4762
11	Countersunk screw, hexagonal head	DIN EN ISO 10642
12	Connector housing	

Figure 1.28 Components for the hot side on a four-fold hot runner mold with a list of the standard designations (excerpt from DIN 16769)

4.5 Demolding					
Slide	☐	Position horizontal	☐	Position turning wedge	☐
Slide drive					
Angular pillar	☐	Hydraulic	☐	Puller	☐
Other	☐				
Ejector system	☐	Fixed side	☐	Moving side	☐
Ejector plates, guided					
Slideway	☐	Ball traveler	☐		
Other	☐				

	Manufacturer	National norm or ISO
Two-way ejector	☐	☐ ....
Latch lock	☐	☐ ....
Angular ejector	☐	☐ ....
Thread demolding via:		
— Drive spindle	☐	☐ ....
— Rack & pinion / hydraulik	☐	☐ ....
— Collapsible core	☐	☐ ....
Drive		
— Hydromotor	☐	☐ ....
— Hydralic	☐	☐ ....

Figure 1.29 Tool specification sheet for injection molds, excerpt from DIN ISO 16916

- Externally heated gating systems
- Internally heated gating systems
- Cold runner gating systems.

Figure 1.28 provides an example of the hot side on a four-fold hot runner mold (excerpt from DIN 16769). The hot side forms a functional element that contains all hot-channel components for the gating system and is supplemented by a mold platen (1), a frame platen (2), a fixed platen (3), as well as guiding and centering elements.

1.7.4 DIN ISO 16916 "Tools for Molding – Tool Specification Sheet for Injection Molds"

The basis for ISO 16916 valid currently worldwide was provided by the DIN standard 16764 of 1998. During the offer and ordering phases is has been rather diffcult to obtain all requirements involving the design and making of injection molds. Therefore, the publication of this standard in English and German satisfies the market demand for uniform definition of specifications. At last, offers from different suppliers can be compared objectively (Fig. 1.29).

1.8 Standard Mold Components

In order to produce injection molds effciently and economically, a very wide range of standard components are available that have, in many cases, been pre-machined to near-finished dimensions. By the term *standard components* we mean elements that have been standardized both by the supplier and standards committees for the basic design of an injection mold. They can be classified into various functional categories:

- Mold rack
- Gating systems
- Guiding and centering elements
- Cooling systems
- Ejector systems
- Accessories
- Clamping systems
- Sliding mechanisms
- Measuring and control devices
- Mold inserts, etc.

Depending on specific requirements, some of these components are available in a range of materials. Using computer programs can *expedite* the work of designing the mold and optimizing the part to be molded. For machining molds by electrical discharge, standard blanks of graphite or electrolytic copper are available.

1.9 Injection Mold for Producing Test Specimens from Thermoplastic Resins

In order to directly compare the physical properties of thermoplastic resins – as determined from test specimens – originating from different materials suppliers, the plastics database CAMPUS was developed in 1988. To supplement it, DIN EN ISO 2.94–1...4 standard of 1998 "Injection molding of test specimens of thermoplastic materials; General principles, and molding of multipurpose and bar test specimens" was worked out [10] (see also Example 6, "Standard Mold Base with replaceable inserts for the production of standard test specimens"). For the production of test specimens from thermoplastics, whose melting and mold-wall temperatures are relatively high, it is advisable to use tool steels with high tempering properties.

1.10 Materials Selection

1.10.1 General Requirements for Materials

In order to maximize functionality, the requirements placed on materials of mold components differ; they include:

- *Easy machineability*
  Cutting tools should be subject to minimum wear, and cutting forces (the cutting task) should be minimal.
- *High wear resistance*
  Polymers are often modified with fillers and reinforcing materials, depending on their intended application. These, as well as some coloring pigments, aggravate wear. Thus the selection of a suitable mold material and, if required, surface treatment or coating, is of considerable importance.
- *High corrosion resistance*
  Corrosion is the destruction of metal materials beginning on their surface, caused by chemical (or electrochemical) processes. The chemical agents required for it may be cleavage products, special additives such as flame retardants, but also the melts themselves. For example, hydrogen chloride (HCl) can be generated during PVC processing and, in a humid atmosphere, produce hydrochloric acid. When polyacetal (POM) is processed, formic acid can develop if the melt has contact with atmospheric oxygen, such as in vented moulding machine when the vents are open. This chemical reaction can cause pitting corrosion both in the injection molding unit and in the injection mold, including the hot runner system. The most common cause of corrosion damage is thermal damage to the melt due to, for example:
  - shear related, undue temperature increase,
  - considerable pressure loss in the melt-free system leading to temperature increase
  - unduly long holding time under processing conditions,
  - excessively damp granulate (regranulate), e.g., when piles of polymer material are stored in the open and are subject to atmospheric condensation,
  - when using chemical gas-developing agents (e.g., to obtain finely porous structure).

A "complex load" acts on a part when, in addition to a chemical attack *at the same time* the metal surface is being worn down, for example, mechanically. Thus the damage to the mold can be cumulative. It is advisable to use corrosion resistant steels and/or, if possible, *gas-tight* surface coatings.

- *Good dimensional stability*
  For example, the processing of high-temperature resistant polymers requires melting temperatures up to approx. 420°C (e.g., PEEK). This presumes tool steels with correspondingly high hardness retention. *Heat resistant* steels are suitable when they are capable of tolerating constantly high temperatures without undergoing structural transformation and associated changes in their mechanical characteristics and/or dimensional alterations. Dimensional variations during heat treatment, such as case hardening, must be kept small, but usually cannot be entirely avoided. If *pre-tempered* tool steels are used, heat treatment subsequent to metal-removing machining can be dispensed with. Thus, problems such as dimensional changes due to warping can be avoided. However, their relatively low Rockwell hardness (approx. 40 HRC) must be considered. By contrast, when *through-hardened* tool steels are used, hardness values up to 62 HRC can be achieved. Pre-hardened steels, due to their high return on investment, remain one of the most important mold steels. If necessary, wear protection can be improved by surface treatment, such as with a PVD coating.
- *Good weldability*
  It is not uncommon that, *subsequent* to completion of a mold, corrections have to be made to it which can only be accomplished by build-up welding. Also, in production, repair welding often becomes necessary. The tool steel used needs to have a low carbon content and be as low-alloy as possible. Surface coatings impair welding work.
- *High-gloss polishability*
  To achieve mirror-bright, glossy part surfaces (e.g., for optical lenses), the tool steel used should have a hard, homogenous surface with a high percentage of purity. The sulfur content in particular should be extremely low.
- *High texturing capability*
  These demands on the tool steel resemble those of polishability. For instance, a surface textured by photo-etching presupposes additional materials with low carbide content.
- *Good thermal conductivity*
  Intensive and uniform mold temperature regulation is extremely important in order to meet quality requirements with regard to performance capability of molded parts, and also for economical reasons. Thermal conductivity as a measure of the rate of temperature change directly affects cooling time, and thereby cycle time, as well. Thermal conductivity is especially decisive for achieving thermal uniformity in a mold. In order to influence heat transfer in a particular manner, various alloyed steels can be

employed. The effect of this measure on thermal conductivity, however, is relatively modest. The noticeably higher thermal conductivity of copper and aluminum and their alloys stands in contrast to their relatively low modulus of elasticity, low strength, relatively low hardness and low wear resistance. Depending on the type and quantity of alloying constituents, higher thermal properties can be balanced against higher strength. Wear resistance can be significantly improved by various surface coatings. However, it must be realized that, in the presence of surface or Hertzian pressure, a relatively hard surface layer can become indented if they lack suffcient support from softer substrates (much like a layer of thin ice on a fluid). This problem, among others, can at least be minimized by composite structures, such as aluminum/steel. Care must be taken with regard to the considerable differences in thermal expansion between steel and the non-ferrous materials mentioned.

1.10.2 Tool Steels

The stiffness of a mold is independent of the steel selected, since the modulus of elasticity is practically identical for all common tool steels. Nevertheless, depending on the importance given to the various requirements, different materials may meet particular requirements better than others:

- Case-hardened steels
- Prehardened steels
- Through-hardening steels
- Corrosion-resistant steels
- Special materials

The following describes a selection of common and proven tool steels.

1.10.2.1 Case-Hardening Steels

Low-carbon steels ($C \leqslant 0.3\%$) are used that are given a hard, wear-resistant surface through case hardening (Table 1.3). During case hardening or carburizing (treatment temperature approx. 900 to 1000°C), carbon diffuses into the near-surface regions of the material. The hardening depth is a function of temperature and time. After case hardening for up to several days, a carburizing depth of approx. 2 mm can be achieved. A hard, wear-resistant surface is achieved by quenching the carburized workpiece, while the core in general remains tough, assuming adequate workpiece thickness.

Case-hardening steels are highly polishable and well suited for texturing. Hardening of the carburized surfaces can achieve up to 62 HRC. Changes in dimensions and shape are unavoidable due to the differing heat treatments (carburizing, hardening), i.e., the heat treatment has to be followed by finishing. Metal removal from the extremely hard boundary layer can only be done by polishing. For further details see also DIN 17022 and DIN 17210.

Table 1.3 Case-hardened steels

Abbreviation	Material No.	Surface Hardness Rockwell C	Remarks
CK 15	1.1141	to approx. 62	For parts subject to low loads
21 MnCr5	1.2162	58 to 62	Standard case-hardening steel, good polishability
X6CrMo4	1.2341	58 to 62	Preferred for hobbing
C19NiCrMo4	1.2764	60 to 62	Very good polishability, high standard of surface quality

1.10.2.2 Prehardened Steels

Prehardened steels are hardened by heat treatment, generally martensite tempering, or raised to the desired degree of strength by austempering. In this way, properties such as yield point, tensile strength, and toughness can be precisely determined. As the tempering temperature increases, strength decreases, for example, but toughness rises, on the other hand. When prehardened steels are used, it must not be overlooked that the carbon content and alloying constituents are largely responsible for the progression of the hardening process through the cross-section of a part. Thus some prehardened steels leave much to be desired, while others are almost uniformly through-hardenable. Component alloys capable of improving through-hardenability include chrome, manganese, molybdenum, nickel, and vanadium. Manganese and silicon increase yield point and tensile strength. Nickel improves toughness characteristics.

The form and dimensions of a component influence its cooling rate. When cooling takes place very

Table 1.4 Prehardened steels

Abbreviation	Material No.	Notes
40CrMnMo7	1.2311	Good cut- and polishability
40CrMnMoS8-6	1.2312	Very good cutability, not suitable for polishing, among other things
X36CrMo17	1.2316	Good corrosion resistance
54NiCrMoV6	1.2711	Creep-resistant and tough, polishable to high gloss
40CrMnNiMo8-6-4	1.2738	Rather like 1.2311, but more through-hardenable

quickly, martensitic structure is obtaind which is characterized by high strength and hardness, but noticeably reduced toughness. If cooling is very slow, martensite formation can be totally suppressed. The material exhibits toughness. Depending on the cooling rate required, water, oil, or air are used for quenching (thus, e.g., the term "oil hardeners"). When a workpiece is hardened, internal stress arises due to non-uniform cooling that can lead to warping and, in extreme cases, to heat-treatment cracking. Heat-treatment cracking is usually promoted by the specific mold design, e.g., when junctures are not rounded off, or by sharp-edged thread run-outs, etc. This is caused by increased stress due to *notching*, see also Section 1.11.

Prehardened steels with hardness up to approx. 40 HRC are machined as manufactured without having to be subjected to any further hardening treatment. Warping is thereby largely eliminated. Table 1.4 lists the common available prehardend steels.

Through-hardening steels (see Tables 1.5 and 1.6) are hardened up to 62 HRC, but not until after being largely finished. These materials exhibit fewer tendencies to warp than do case-hardening steels.

In order to achieve a uniform microstructure throughout even larger cross-sections, through-hardening (alloyed) steels are used whose hardness strength and toughness can be matched to the particular requirements through heat treating (quenching and tempering). By selecting the temperature at which tempering takes place, these properties can be optimized. The through-hardening steels have proved to be very well suited for processing abrasive molding compounds, e.g., with glass fibers as filler.

Due to their high achievable compression strength, through-hardened steels are suitable even at high edge-pressure loads. These tool steels can be divided into two groups:

- *Cold*-work steels and
- *Hot*-work steels.

Cold-work steels are those that can be used at room temperatures, or somewhat warmer, for instance, in mold building. Maximum application temperature is approx. 200°C. At temperatures above 200°C, hot-work steels have to be used. The demands placed on this material group include high heat resistance, high hardness retention and high heat/wear resistance. Injection molds for processing engineering polymers should be manufactured from hot-work steels due to the specification of high mold wall temperatures. Figure 1.30 illustrates the progression of hardening as a function of tempering temperature for cold- and hot-work steels, among others [11]. Cold-work steels exhibit high original hardness which, however, being a function of the

Table 1.5 Cold-work steels

Abbreviation	Material No.	Hardness HRC	Remarks
X45NiCrMo4	1.2767	50–54	Very good polishability, high toughness
90MnCrV8	1.2842	56–62	Normal wear resistance
X155CrVMo121	1.2379	58	Good wear resistance and toughness, not easily polishable
X210CrW12	1.2080	60–62	High wear resistance
X165CrMoV12	1.2601	63	Highly wear-resistant steel

Note: For components with low requirements, the non-alloy tool steel C45W3, material no. 1.1730 can also be used in non-hardened condition

Table 1.6 Hot-work steels

Abbreviation	Material No.	Hardness HRC	Remarks
X38CrMoV5-1	1.2343	48–50	Standard hot-work steels
X40CrMoV5-1	1.2344	48–52	Slightly higher hardness than 1.2343
X40CrMoV5-1	1.2344 ESU*	48–52	Like 1.2344, but almost entirely isotropic characteristics
56NiCrMoV7	1.2714	50–56	Good toughness

*The steel is smelted by the so-called "electroslagremelting" process to obtain the highest possible purity and homogeneity. With this process technology, steels are obtained with largely isotropic characteristics (uniform materials behavior in all three dimensions). Such materials are also characterized by improved dimensional stability subsequent to heat treatment.

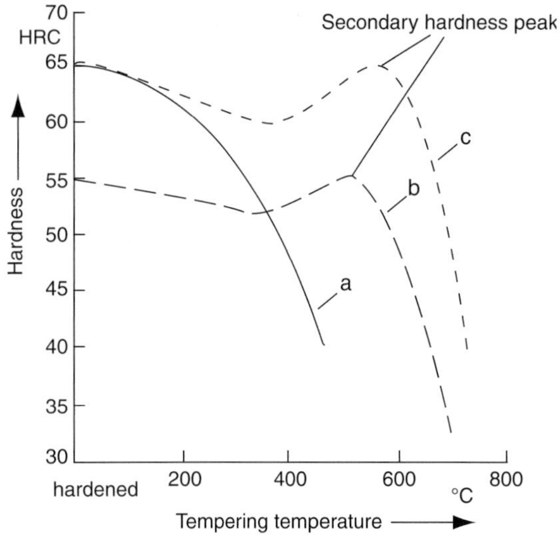

Figure 1.30 Progression of hardness as a function of the tempering temperature of different tool steels (according to [11])
a: Cold-work steel b: Hot-work steel c: High-speed steel

tempering temperature, falls continuously. In hot-work steels, the original hardness is significantly lower, but progresses almost uniformly to approx. 500°C. Thanks to their high obtainable compression strength, through-hardeners are suitable even at high edge-compression loads. Tables 1.5 and 1.6 provides an overview of the commonly available cold- and hot-work steels (see also DIN 17350).

1.10.2.3 High-Speed Steels

Based on their alloying constituents (Co, Cr, Mo, V, W), high-speed steels have very high tempering resistance and heat hardness. They can be used at temperatures up to 600°C. Although, such high temperatures are not achieved in injection molds, but high hardness and wear resistance provide an excellent basis for producing, for example, nozzle tips subject to heavy wear, guide pin bushings for valve gates, or replaceable gate bushings. Figure 1.30 illustrates the progression of hardness as a function of tempering temperature for high-speed steel. The secondary peak is the result of special carbides being precipitated ("precipitation hardening"). Table 1.7 lists several high-speed steels. High-speed steels (**HSS**) have proven especially good in cutting technology as lathe chisels, reamers, etc., due to their high wear resistance.

Table 1.7 High-speed steels

Abbreviation	Material No.	Rockwell Hardness C
S 6-5-2	1.3343	approx. 64
S 6-5-2-5	1.3243	
S 10-4-3-10	1.3207	

1.10.2.4 Corrosion Resistant Steels

Corrosion is the destruction of metal materials beginning on the surface of a component, caused by chemical or electrochemical processes. As far as there is any corrosion activity when polymer materials are being processed, mainly *chemical* attack on melt feeding components has to be considered, since their surfaces are always the weakest area. To that extent, corrosion has considerable significance for the functionality of an injection mold.
The occurring forms of corrosion are essentially:

- Surface corrosion (largely uniform)
- Pitting (penetrates the surface at random),
- Crevice corrosion (occurs in crevices between pressure joints or between interlocking and/or friction locked (screwed together) components), and
- Vibrational crevice corrosion (so-called corrosion fatigue resulting from interaction between corrosion and alternating mechanical stress), see also DIN 50900.

Corrosion is always accompanied by material destruction that can be aggravated (so-called cumulative damage) by erosion (mechanical wear, i.e., surface erosion, e.g., by glass fibers).
The most common cause of corrosion damage is thermal damage by polymer melts, see Section 1.10.5. To protect against corrosive plastics or additives, there is always the possibility of electroplating the molds. One possible disadvantage, however, is that the deposited layer may delaminate at shut-off edges, for example, as the result of high surface pressure. The use of corrosion-resistant steels is thus recommended in such cases.
Non- and low-alloy tool steels are not resistant to corrosion load. A minimal chrome content of >12% leads to passivation of the steel surface, making it resistant to oxidizing media. The condition of the surface to be protected is very important. Even colors generated by welding have a negative effect on corrosion resistance. The surface should be as uninjured, smooth, and clean as possible. For example, pickling can be used to "clean" and simultaneously improve passivation. Commercial corrosion resistant steels are supplied in a soft-annealed or hardened and tempered state. By appropriate heat treatment subsequent to extensive metal-removing machining, the required in-service hardness can be achieved. It needs to be kept in mind that the type of heat treatment influences corrosion resistance. By means of surface treatments, the wear properties of corrosion resistant steels can be improved. Nitriding, however, reduces corrosion resistance, for example. It should be clear that the "corrosion resistant" steels are in no way immune to corrosion. The influencing factors mentioned, such as surface and heat treatments, etc.,

Table 1.8 Corrosion-resistant steels

Abbreviation	Material No.	Surface Hardness Rockwell C	Remarks
X42Cr13	1.2083	54–56	Corrosion-resistant only when polished, hot-work steel
C36CrMo17	1.2316	50	Machining after heat treatment, high corrosion resistance
X105CrMo17	1.4125	57–60	Rust- and acid-resistant steel, wear-resistant

can actually have a negative effect on corrosion stability. Table 1.8 lists commonly available corrosion-resistant steels.

1.10.2.5 Powder-Metallurgical (PM) Steels

The molten phase can be eliminated when powder-metallurgical finished parts or semi-finished products, such as plates, blocks, etc., are produced by powder metallurgy (e.g., by the metal injection method, MIM). Production involves three steps:

- Metal powder is produced,
- The powder is compressed, e.g., to produce blanks and
- Subsequently heat treated, i.e., *sintered.*

Two application areas for sintering technology have been known for a long time: permanent magnets and friction bearings. Some polyphase (pseudo-) alloys can also be produced by sintering, whose constituents would otherwise become segregated and/or are insoluble in the melt. Others have melting points or vapor pressures so different that one constituent would evaporate before the other had melted. Thereby, melt-technologically impossible material properties can be combined, as for instance the property combination "hard" and "tough", "(highly) thermoconductive" and "(highly) wear-resistant", etc. In the respective company publications, the range of properties achievable with powder-metallurgic steels can be found such as:

- High to very high wear resistance
- Good toughness
- High hardness combined with good toughness
- Good polishability
- High thermal conductivity, etc

In the following, a short description of three exemplary PM steels is given.

- *Ferro-Titanit* (Thyssen Edelstahlwerke), Titanium carbide steels, hardenable to max. 71 HRC, depending on chemical composition, high tempering retention, high wear resistance, good corrosion resistance
- *Vanadis-Superclean* (Uddeholm), Cold-work steels with good wear resistance, good toughness, high dimensional stability following heat treatment, polishable, isotropic properties
- *TZM* (Metallwerke Plansee), Molybdenum alloy with good thermal conductivity, low thermal expansion, high corrosion resistance, low heat warping, and good wear resistance

Sintering makes use of a very modern technology: the Rapid Tooling Process. Using direct metal-laser sintering (DMLS), sintered mold inserts can be produced from 3D CAD data which are apparently suitable for use even in standard molds for mass production. The metals thus utilized in addition to steel include, for example, aluminum (alloys), copper, and nickel. For one thing, the relatively quick availability of laser-sintered mold-inserts can help cut the cost of mold making compared to those produced by conventional metal-removing methods.

1.10.2.6 Cast Ferrous Materials

The ferrous materials that can be shaped directly into building components are divided into two groups:

- Cast steel and
- Cast iron

Iron-carbon alloys with and without alloying constituents qualify as *cast steel.* The carbon content of these materials can reach approx. 2%. *Cast iron* has a carbon content of $\geqslant 2\%$ (to approx. 4.5%). The carbon is often precipitated as free graphite when the melt solidifies. We can speak of cast iron with

- Lamellar graphite (see also DIN 1691)
- Ball graphite (see also DIN 1693), and as
- Chilled cast iron
- Malleable cast iron (see also DIN 1692) and as
- Special cast iron

Thereby quite different material cost properties are sought. A cast design component (a mold cavity, etc.) can be advantageous compared with one shaped in large part by metal-removing methods. However, other than for a few exceptions, casting technology has not found wide-spread application for polymer mold making. Considering that almost every injection mold is not a standard item, but one of a kind, the scope for design offered by casting technology ought to be given close consideration. This is especially the case for large dimension molds.

1.10.3 Non-Ferrous Metals

1.10.3.1 Aluminum Alloys

Molds or mold components made from aluminum alloys have found a niche of their own in recent years. It is to be expected that the specific properties of this material group long established in the airplane industry will continue to gain acceptance. Special interest focuses on the following properties:

- Low density, approx. 2.8 g/cm^3 (approx. 7.9 g/cm^3 for steel)
- Reduced modulus of elasticity compared with steel, approx. 70000 $N/2 \times mm^2$ (approx. 210000 N/mm^2 for steel)
- High thermal conductivity, approx. 165 W/mK (approx. 14 to 40 W/mK for steel)
- High temperature conductivity, approx. 0.3 m/h (approx. 0.02 to 0.06 m/h for steel)
- Thermal expansion coefficient is twice that of steel
- High quantity of metal removal possible
- Wire- and spark erosion possible with higher removal rates than with steel (Note: no so-called "white (crevice-prone) layer" develops)
- Low corrosion resistance
- Surface treatments, e.g., with chemical nickel can be done easily
- Depending on the type of material, good to very good properties for polishing, etching and welding
- Aluminum with a bright surface emits the lowest radiant energy compared to almost all other technical surfaces, see also Section 1.1.2.

Due to its high thermal conductivity and also its high temperature conductivity (the degree of temperature conductivity is a measure of the rate of temperature change) shorter cycle times can be achieved than with comparable steel molds. However, if considerable wall thicknesses are dictated to obtain required strength, this advantage may be minimized. The lower strength and rigidity values of aluminum materials cannot easily be counteracted by design measures. Another disadvantage is its low tolerance of surface or Hertzian stress. Composite designs offer practical solutions, e.g., using a steel frame in order to relieve the stressed mold components made from aluminum alloy. Composite aluminum-iron die casting, for example, has been state-of-the-art in engine design for some time. Finally, comprehensive calculation and comparison of the pros and cons are needed to decide whether it makes technical sense to use aluminum alloys in injection molds as well as in large tool construction. Complete standard molds made from aluminum alloys are available on stock for finishing [12]. There is a large selection of standardized and special aluminum alloys that cover a wide range of specific properties, see also DIN 1712, DIN 1725, DIN 1745, DIN 17007. Table 1.9 lists several commercially available aluminum materials. The low hardness values refer to the "soft" state, the top value to "heat hardened".

Table 1.9 Aluminum alloys

Abbreviation	Material No.	Brinell Hardness HB 2.5/62.5
AlCuMg1	3.1325	50 to 100
AlZn4,5Mg1	3.4335	45 to 105
AlZn4,5MgCu1,5	3.4365	140

1.10.3.2 Titanium Alloys

Titanium and its alloys, e.g., with aluminum or vanadium, are characterized by high strength, low density, high corrosion resistance and – what makes them important for certain components in hot runner systems – very low thermal conductivity. In order to achieve the strictest thermal separation possible, e.g., of a hot runner manifold from the mold platen surrounding it, where temperature differences can run over 100°C, support discs made from titanium alloy TiA16V4, material no. 3.7165, are finding ever wider use. This material has a very low thermal conductivity coefficient of $\lambda = 6.5$W/mK. By contrast, tool steel, depending on its chemical composition, has a thermal conductivity coefficient of $\lambda = 14$ to 40 W/mK. It is worth considering, however, the relative difficulty of metal removal and the high price of titanium alloys. When combined with steel, the lower thermal elongation coefficient of titanium alloy ($\alpha = 8.6 \cdot 10^{-6}$1/K) is worth considering.

1.10.3.3 Copper Alloys

The technically most important property of (unalloyed) copper is its high electrical conductivity and thus its very good thermal conductivity (Wiedemann-Franz's Law). Pure copper is a very soft material that can be strain-hardened. Strain-hardening, however, reverts to zero under heat treatment (recrystallization annealing), e.g., annealing time 1h/100°C (!). Thus, in an injection mold, the recrystallized structure is always present with reduced strength. To achieve higher strength values at elevated temperatures, it can be alloyed with various additives. Besides pure copper, the following alloys are technically interesting for mold making:

- Copper alloyed with cadmium, zirconium, beryllium, cobalt, nickel, chrome, silicon,
- Brass (copper and zinc),
- Bronze (copper and tin) and its alloys.

In contrast to pure copper, all alloys exhibit enhanced strength properties and reduced thermal conductivity. Beryllium particles are classified as carcinogenic. This problem field can be avoided without health consequences in almost all cases by suitable machining methods (wet grinding, etc.) Moreover, a beryllium-free CuNiSi alloy is available under the trade name Albromet W 164 (Albromet Handelsgesellschaft, Geretsried) that, compared to half-hard $CuBe_2$, is characterized by nearly comparable mechanical strength properties and considerably improved thermal conductivity. Copper and its alloys are used in mold making especially in order to provide for rapid (selective) heat transfer. In addition to the reduction of sliding friction, one of the main aims is especially to reduce cycle time and avoid warping in the molded part. Composite designs, e.g., those with a supporting framework of steel and a composite with copper combine two required characteristics: high strength and high thermal conductivity. Examples:

- Hot runner manifold from steel with copper-cast melt-feed and heating system (Unitemp system),
- Copper cast core and ejector pins (Hasco system).

The potential for shortening cycle times originates particularly from two outstanding properties of copper alloys: their high thermal and temperature conductivity (this statement also holds for aluminum and its alloys). Table 1.10 shows the differences between a hardened steel and copper, aluminum, and a copper alloy (Albromet W 164).
Table 1.11 lists several common copper alloys utilized for mold making.

Table 1.10 Thermal and temperature conductivities of different materials by comparison

Material	Thermal Conductivity λ W/mK	Temperature Conductivity a m^2/h
X45NiCrMo4 (1.2767)	31	0.03
Copper, pure	395	0.42
Aluminum, pure	229	0.34
Albromet W 164	164	0.15

Table 1.11 Common copper alloys used for mold making

Material	Thermal Conductivity λ W/mK	Tensile Strength Rm N/mm^2
Electrolytic copper 2.0060	395	250
Elmedur X CuCrZr* 2.1293	320	590
CuCoBe 2.1285	197	1000
Albromet W 164** CuNiSi alloy	164	900
CuBe2, half-hard 2.1247	130 105	1170 1300

* Thyssen Edelstahlwerke
** Albromet Handelsgesellschaft, Geretsrield, Germany

1.10.4 Anorganic Nonmetallic Materials

1.10.4.1 Ceramic Materials

Today, ceramic materials find application for injection molds exclusively as support discs in hot runner systems. Technical ceramics exhibit great hardness and strength even at high temperatures, as well as low thermal expansion and very low thermal conductivity. In contrast to these properties which are desirable in certain applications, ceramics also exhibit undesirable properties:

- Increased brittleness
- Notch sensitivity, as well as
- Inability to release stress peaks by local plastic deformation.

In order to eliminate shear stress on ceramic support discs, mainly due to heat expansion differences, composite designs with an outer steel frame have proven useful. Even the slightest deflection by they platen they are supporting can – under unfavorable conditions – cause the ceramic support discs to break. Material removal by machining of *ceramic support discs* is diffcult at the least, e.g., to make heights match. On composite designs, however, the *metallic frame* can be easily machined.
Their coefficient of thermal conductivity of e.g. $\lambda = 3$ W/mK is very low. Thus ceramic support discs can play an important role in minimizing heat loss in a hot runner system.

1.10.5 Surface Treatment Methods

The surface of any material or component is usually its weakest area. At the same time, the surface is often the region of greatest load. The type of load can vary considerably and be extremely complex. In the broadest sense, the surface has decisive influence on the functionality of a component. Some types of load are:

- mechanical
- corrosive and
- abrasive/erosive

They often interact with one another and thus are *cumulative*. For example, glass fiber reinforced PA 66 can cause corrosive damage to a metallic surface. This is especially the case when, among other things, regranulate is used that has not been predried. A more or less protective corrosion layer has virtually no opportunity to form permanently, since the glass fibers erode the layer as the melt flows through.
The most common cause of *corrosion damage* in a mold is process-related *thermal damage* by the melt, e.g., due to:

- Shear-related temperature increase
- Thermal inhomogeneity in the hot runner system
- High pressure losses in the melt feed system
- Excessively long holding time in the injection molding unit (vented machines!) and/or in the hot runner system

- Moisture in the granulate, e.g., when piles of plastics are stored out of doors (condensation)
- Insufficient treatment and/or wrong selection of material for the mold surface in contact with the melt
- Chemical foaming agents (e.g., to obtain highly porous structures)
- Flame retarding substances
- Processing of certain polymers on *vented machines* in which the melt has at least occasional contact with oxygen in the air at the vent opening, and
- Catalytic degradation of unprotected metal surfaces by the melt (such as chemical attack due to incompatibility of PP and POM homopolymer melts with copper)

Damage to metallic surfaces due to *wear* (abrasion, erosion) are brought about essentially by fillers and reinforcing materials. Such wear is a loss of material due to surface attrition, mainly caused by sliding friction. The wear resistance of material is, among other things, proportional to its hardness [13].

It is generally the case that the condition and/or the type of surface treatment of a mold have essential influence on its functionality. In mold making, surface treatments should be used with the aim of improving such properties as:

- Surface hardness
- Compressive strength
- Wear resistance
- Corrosion resistance
- Sliding properties
- De-molding behavior.

The following surface treatment methods have proven useful in mold making in particular:

- Nitriding
- Carburizing
- Hard chrome plating
- Hard nickel plating
- Hardcoating.

1.10.5.1 Nitriding

In the nitriding process, atomic nitrogen penetrates materials surfaces at temperature varying from 350 to 580°C, depending on the method. This way, and *not* by martensite formation, a significant increase in the hardness of the materials surface is achieved and with it a clear improvement in wear and fatigue resistance. Nitriding can be performed in gas flow, in saline melts, in powder or in plasma (highly ionized, electrically conductive glass at high temperatures). The nitriding depth generally amounts to just a few tenths of a millimeter. Nitriding is used, for instance, on molds with thin spider legs, but also on ejector pins in order to improve their dry-sliding properties [14]. Practically all steels commonly used in mold making can be nitrided. Nitride forming materials in particular include such alloying elements as chrome, molybdenum, vanadium and aluminum. The nitriding of corrosion resistant steels is not recommended since it reduces their stability.

1.10.5.2 Carburizing

Carburizing is employed on steels with a low carbon content ($C \leqslant 0.3\%$), whereby carbon diffuses into the near-surface regions. Steel “case-hardened” in this manner can be hardened in the usual processes (i.e., hardening and tempering) and exhibits increased surface hardness. At the same time, the core generally remains ductile if its material has a suffciently large cross-section. The result is a significant improvement in component properties under wear as well as interactive loading (see also Section 1.9.2.1).

1.10.5.3 Hard Chrome Plating

The electrolytic deposition of hard chrome layers is largely used to achieve hard and wear-resistant surfaces that have proven effective for mold components used for processing abrasive plastics. Moreover, the hard chrome coating serves to reduce the tendency to gall and considerably improves corrosion resistance (multi-layer chrome plating). Hard chrome plating also finds application for the repair of worn surfaces. In the event of repeated plating and deplating, hydrogen embrittlement of the near-surface regions should be considered. Along edges and similar surfaces, the formation of raised bead and the delamination of the chrome coating are to be expected (Fig. 1.31).

Hard chrome coatings generally exhibit microfissures due to high internal tensile stress (note: chrome plated cylinder sleeves are utilized in engine design so that the microfissures can serve as oil pockets). Hard chrome coatings are sensitive to reducing substances, such as hydrochloric acid, and thus are not suited for molds in which PVC is to be processed.

1.10.5.4 Hard Nickel Plating

Nickel can be deposited both galvanically and chemically (without an external current source). Depositing the nickel without external current eliminates the irritation caused by the formation of varying coat thickness, especially on the edges (raised bead). This results in trouble-free nickel plating through openings, in holes, on profiled surfaces, etc., as well as on the inner walls of pipes of any length. The internal stress on the nickel layer is significantly lower than that on galvanically deposited hard chrome layers. Nickel plated

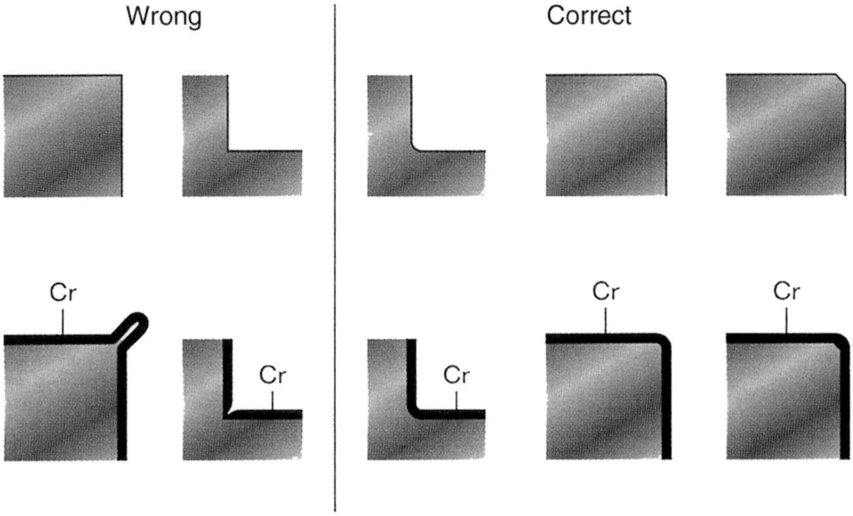

Figure 1.31 Structure of hard chrome layers Cr at various junctures
(Courtesy: Buderus)

surfaces are generally fissure-free, i.e., also gastight. Depending on the process, phosphorus, for example, can be embedded in the nickel layer. This enables a heat treatment to increase hardness; as a result of the depositing of nickel phosphide, the progression of hardening reaches a peak at a heat treatment temperature of 400°C (approx. 1100 HV 0.1). Thus, approximately the same hardness values can be achieved as in hard chrome layers.
The film thickness usually employed is 20 to 50 µm. Nickel-phosphorus-silicon carbide dispersion has also proven useful for depositing electroless coatings on surfaces for protective measures. The above-mentioned methods are characterized especially by their good performance in providing protection against corrosion as well as wear and can also be employed with nonferrous metals such as copper. It must not be forgotten, however, that the nickel plate, which is much harder than the substrate material, can be damaged under a compressive load and tends to delaminate.

1.10.5.5 Hard Materials Coating

The very positive results obtained by hard-coating metal-removing tools such as drills using PVD (golden tools), i.e., improved service-life, have had notable influence on injection mold making, as well. In particular, wear resistance and thus the service life of injection molds can be improved. Hard coatings provide very good wear resistance due to their very high hardness. Further advantages include:

- (Conditional) corrosion protection
- Very good contour fidelity

Table 1.12 Properties of various PVD hard coatings [15]

	Balanit A	Balanit B	Balanit C	Balanit D	Balanit Futura	Steel
Chem. designation	TIN	TICN	WC/C	CrN	TiAIN	1.2343
Microhardness [HV0.05]	2300	3000	1100	1750	3000–3500	600
Coeffi. of friction vs. steel [dry]	0.4	0.4	0.2	0.5	0.4	0.8
Thickness [µm]	1 to 4	1 to 4	1 to 4	1 to 10	1 to 4	–
Max. temperature [°C]	600	400	300	700	800	200
Color of coating	Bright yellow	Blue gray	Black	Silver gray	Violet	–

- Good protection against dragging (pick-up)
- Good sliding properties and
- Reduced mold deposit formation (loss adhesiveness)

Hard coatings with materials by the PVD method (*P*hysical-*V*apor-*D*eposition) are state-of-the-art, of course, but require a high level of experience. The coater should know as much as possible about the entire process of manufacture, steel selection, treatment methods, etc. of the mold to be coated. The properties of various hard materials coating are listed in Table 1.12. For mold making, TiN, TiCN, and WC/C coatings are especially interesting. In order to achieve a good coating, the mold is often heat treated at a temperature of approx. 450 °C. This presupposes a suitable choice of steel. PVD coatings are not gastight, i.e., they provide only limited protection against corrosion. Chemical nickel plating with a layer thickness of approx. 20 µm in combination with a PVD coating increases corrosion resistance significantly. It should be noted that bore holes cannot generally be PVD coated to any depth. Layer adhesion and thickness decrease with increasing bore depth. The bore diameter/depth ratio should be approx. 1:1.

1.11 Material Properties under Mechanical Stress

Injection molds are mechanically stressed both statically and dynamically (interactive) by various forces, such as holding force, injection pressure, etc. Additional forces can arise, e.g., due to inhibited thermal expansion, and also internal stress due to non-uniform cooling in a hardening procedure with subsequent dimensional changes (warping) and, in extreme cases, heat-treatment cracking [16]. As soon as the internal stress is superimposed by loads, component failure may result due to mechanical overstressing. Also, overstressing can take place as a result of design related notching: heat-treatment cracking often develops from sharp edges (*notches*). Not only on molded parts, but just as commonly on molds we find notched areas, e.g., sharp-edged junctures, drill holes, etc., that can noticeably reduce shape strength. It is amazing that this state of affairs gets such little attention and is scarcely treated even in the latest technical publications. Damage to a mold, even at uniform load, may not arise until after a certain number of molding cycles – i.e., time independent – without any defects having come to light when the mold was inspected. The so-called fatigue endurance limit has been termed such because the sustainable amplitude of stress on a mold subject to alternating stress from injection and holding pressure is a function of endurance. The functionality of such a mold, therefore, has a variable time limit [17].

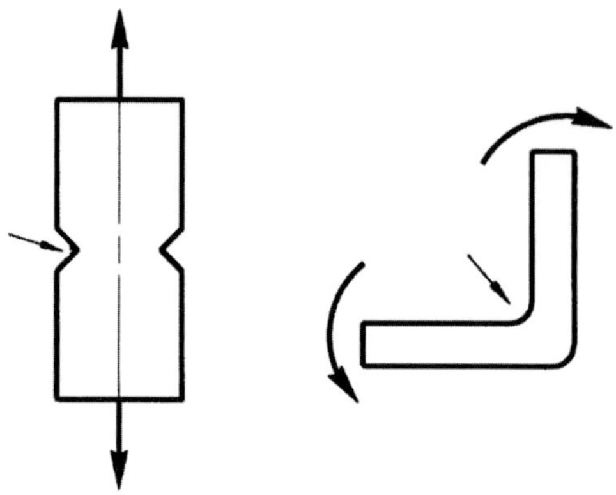

Figure 1.32 Structural parts with various notches

1.11.1 Notch Effect under Static Stress

Due to notching, stress and stress distribution (uni- or multiaxial) are altered in any structural part. Notches are regions of sudden change in cross-section, or places where the component shape changes direction at a relatively sharp angle or edge [18]. Figure 1.32 shows examples of various notches.

At the notch root of a stressed part, an overstressing arises that is defined by the shape factor (stress concentration factor) α_S:

$$\alpha_S = \frac{\text{stress in notch root}}{\text{nominal stress}}$$

or

$$\alpha_S = \frac{\sigma_{\text{max. clast}}}{\sigma_N}$$

The shape factor α_K is valid for ideal elastic material behavior (validity of Hooke's Law $\sigma = \varepsilon \times E$). If the yield point of the part is exceeded (partial plastification), the overstressing is relaxed.

It is valid:

$$\sigma_{\text{max., plastic}} < \sigma_{\text{max., elastic}}$$

Thus soft steel, for example, can better relax overstressing $\alpha_{max.}$ by local yield than a hardened steel which is (significantly) more notch-sensitive.

The shape factor α_S and, thereby, its size depends on

- The design shape of the part (e.g., flat or round).
- Part and notch dimensions (e.g., notch radius r, notch depth t) and
- The type of stress.

Overstressing can amount to twice that of the calculated stress σ_N (so-called nominal stress):

$$\sigma_{max} = \sigma_S \cdot \sigma_N$$

with $\sigma_S > 1$,

The shape fidelity of a structural part can, therefore, be decisively (negatively) influenced by notching.

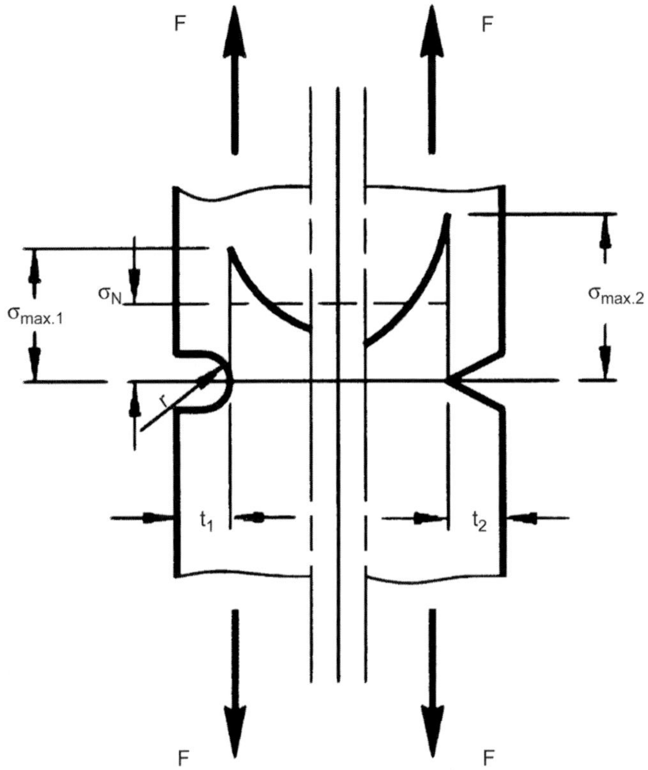

Figure 1.33 Progression of tension on tensile stressed and notched sheet material
$t_1 = t_2, r_1 \gg r_2 \rightarrow \sigma_{max2} > \sigma_{max1}, \sigma_N = const.$

The goal should be, for example, to always design junctures in cross section with an edge radius, and never square-edged.

If the gating area is produced by spark erosion, it should be considered that, at high current intensity and long pulse duration, there can be surface alteration (structural change) as well as microfissures (micronotching with very small notch radius, i.e., large α_S). Thereby, the shape fidelity of the part is noticeably reduced. The summary results in Table 1.13 were determined on variously treated test bars made from hot-work steel 1.2343 (X38 CrMoV51) [19].

In order to obtain a fissure-free surface with low unevenness, reworking and/or subsequent treatment of the eroded surface is required. This can be done, e.g., by spark-erosive polishing following the final fine-machining [20].

Table 1.13 Relative impact energy depending on the type of treatment of flat test bars made from 1.2343 [19]

Type of Treatment	Relative Impact Energy %
Universal machining	100
Smooth-eroded on the tensile face in the annealed state, subsequently hardened	83
Coarse-eroded on the tensile face in the annealed state, subsequently hardened	77
Smooth-eroded on the tensile face after hardening	69
Coarse-eroded on the tensile face after hardening	36

1.11.2 Notch Effect under Dynamic Stress

Components such as sprue nozzles, hot runner manifolds, or cavities are dynamically (alternately) stressed (here: so-called fluctuation stress σ_F) according to the the injection molding cycle, see Fig. 1.34. The most important material property for the shape fidelity of, for example, a sprue nozzle is the pulsating fatigue strength σ_P of the material applied at *processing temperature*. This material parameter is determined in Wöhler tests under dynamic stress, see Fig. 1.35. The damage curve indicates from what number of cycles onward the corresponding stress σ_1 reduces the fatigue limit σ_P. Due to fluctuating stress, breaches can occur at

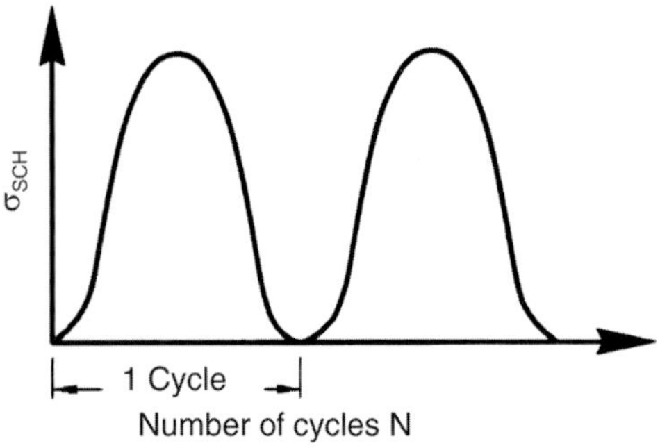

Figure 1.34 Fluctuating stress on a component in sync with the injection-molding cycle (schematic)

stress levels that, given certain conditions, are far below the yield point of the material involved [21]. From the progression of the Wöhler curve it can be qualitatively realized that the pulsating fatigue strength σ_F is significantly lower at a high number of cycles N than under one-time or short-term stress. If, for example, only the bursting pressure of a gate nozzle at room temperature is determined, all the result shows is its short-term behavior and in no way its *long-term behavior* in actual use *at processing temperature*.

The surface condition affects component strength negatively in the following order:

- Ground
- Polished
- Scrubbed
- Notched
- Corroded

[21]. Higher temperature also lowers (sometimes considerably) the alternating stressability of a component [22]. The type of material, its composition, and technical treatment have considerable influence on its fatigue limit. The fatigue limit σ_P is deter-

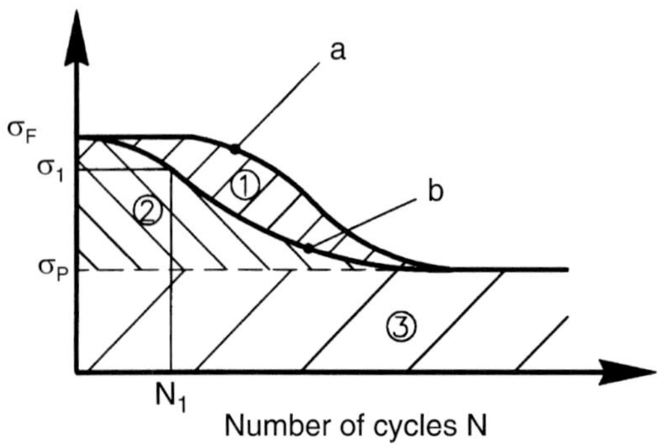

Figure 1.35 Wöhler curve, relationship between maximum tolerable stress and number of cycles
1: damage area, 2: overloadable area, 3: area of fatigue limit, a: Wöhler curve, b: damage curve

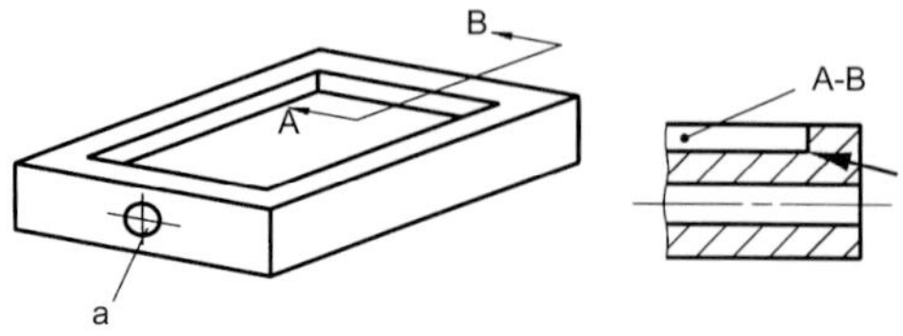

Figure 1.36 Hot runner manifold block with area recessed to hold a thermal insulation plate lying perfectly flat (diagram), notch ←
a: Melt channel

mined on smooth, polished and notched σ_{Pnot} test bars. The quotient of these factors is defined as the fatigue strength reduction factor β_n:

$$\beta_n = \frac{\sigma_P}{\sigma_{Pnot}}$$

σ_{Pnot} is always smaller than α_S. In general, the following is valid: $\beta_n > 1$ [23].

Figure 1.36 shows the example of a design for a hot runner manifold. In the area indicated by the arrow (circulating), the recessed area has been cut to shape (high fatigue strength reduction factor β_n). Pulsating stress load was in the damage range of the Wöhler curve. The component failed after a relatively low number of cycles N due to fissuring in the sharp corner area.

1.12 Thermal Insulation and Reflector Plates

Thermal protection in injection mold making usually is limited to thermal separation between the mold mounting platen and the injection molding machine, as well as occasional insulation of the hot runner manifold block, see Fig. 1.37. Insulation of the external mold portions remains the exception.

Some advantages of thermal insulation are:

- Reduced heat loss, e.g., to the surroundings, thus reduced energy costs
- Improved temperature distribution (thermal homogeneity) in the mold, minimized temperature fluctuation in the face of changing environmental influences (e.g., temperature fluctuations at the production site), and
- Reduced heat-up times.

Depending on demands made especially on insulating and strength properties, various thermal protection systems are available. The thermal conductivity factor of the various insulation materials lies between 0.05 und 0.3 W/mK. Compression resistance values up to 650 N/mm^2 are achieved [24]. The bases of the insulation plates are high-temperature resistant, inorganically (e.g., glass fiber) reinforced polymers. Insulation by means of

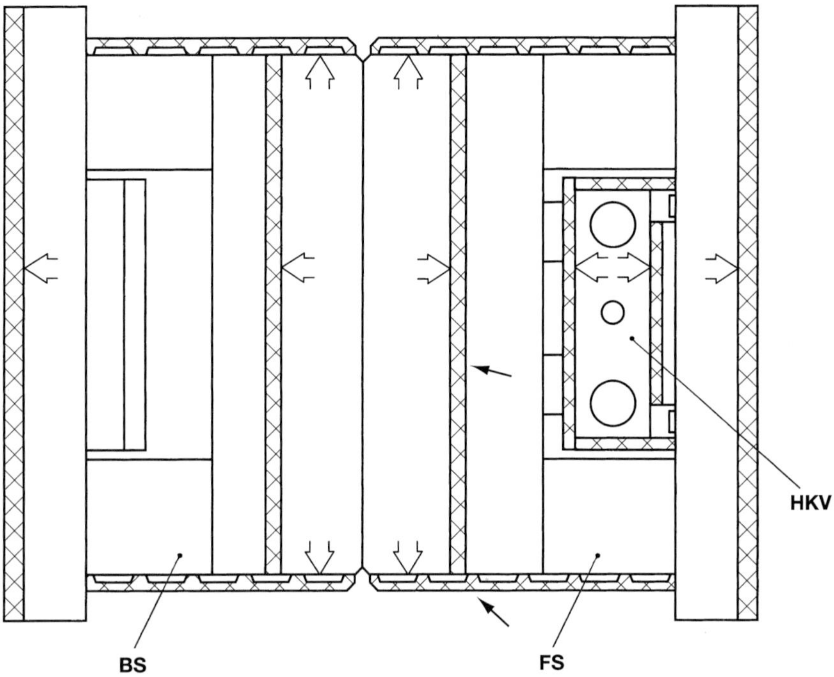

Figure 1.37 Insulation and/or reflector sheets for heat protection of an injection mold, heat flow
BS: movable side, FS: fixed side, HKV: hot runner manifold block
(Courtesy: Strack)

compartmented profiling and aluminum covering is remarkably effective at reducing heat loss by convection and radiation, e.g., on mold surfaces. In order to insulate hot runner manifold blocks, for instance, heat resistant thermal protection plates with aluminum surfaces are utilized. In order to minimize the radiation energy exchanged between the hot runner manifold block and, for example, the cavity plate, bright aluminum plates are used, material, e.g., AlMg3.

References in Chapter 1

1. Kunststoffe 93 (2003) 2, p. 68
2. Gotzmann, G.: Zerschellt Deutschland an zweiter Garde? Kunststoffe 94 (2004) 7, p.3
3. Company publication: Exaflow, Groß-Umstadt, Germany
4. Hofmann, W.: Werkzeuge für das Kautschuk-Spritzgießen, Kunststoffe 77 (1987) 12. p. 1211-1226
5. Benfer, W.: Rechnergestützte Auslegung von Spritzgießwerkzeugen für Elastomere. Dissertation RWTH/Aachen, 1985
6. Janke, W.: Rechnergestütztes Spritzgießen von Elastomeren. Dissertation RWTH/Aachen, 1985
7. Emmerichs, H.; Giesler, D.: Kappen aus LSR, 16fach-Spritzgießwerkzeug mit Kaltkanal-Angußsystem, Kunststoffe 87 (1997) 9, p. 1150 and 1191
8. Emmerichs, H.: Spritzgießwerkzeuge für die Silikonverarbeitung, Fachtagung 7. Würzburger Werkzeugtage SKZ Würzburg 1996
9. Company Publication: Literaturzitat fehlt((S. 36 MS)) des Fa:. Contura-MTC, Menden, Germany
10. Kunststoffe 79 (1989) 8, p. 713
11. Bargel, H.-J., Schulze, G.: Werkstoffkunde. 8. Ed., Springer, 2004
12. Company publication: USK Normteile, Kierspe
13. Ondracek, G.: Werkstoffkunde. 4. Ed., Expert-Verlag, 1994, p. 187
14. Auswerferstifte aus nitriertem Warmarbeitsstahl mit (schwarzer) Oxidationsbeschichtung. Company publication:, Drei-S-Werk, Schwabach
15. Mumme, F.: PVD-Beschichtungen für die Heißkanaltechnik. Vortrag, SKZ-Seminar Heißkanaltechnik beim Spritzgießen, Würzburg, 1998
16. Handbuch der Kunststoffe-Formenstähle. Edelstahlwerke Buderus (Hrsg.), Wetzlar, 2002
17. Unger, P.: Hot Runner Technology Hanser 2006. Hanser, München, 2004
18. Thum, A. et al.: Verformung, Spannung und Kerbwirkung, VDI.-Verlag, Düsseldorf, 1960
19. Meilgen, R.: Beeinflussung der Werkzeugoberfläche von Warmarbeitsstählen durch Funkenerodieren. Mitt. Fa. Saarstahl
20. König, W.: Jörres, L.: Funkenerosives Polieren. Industrie-Anzeiger 65/66 (1989)
21. Tauscher, H.: Berechnung der Dauerfestigkeit, Einfluss von Werkstoff und Gestalt. VEB Fachbuchverlag, Leipzig (1961)
22. Tauscher, H. et al.: Dauerschwingversuche an warmfesten Stählen bei erhöhter Temperatur. Materialprüfung 8 (1969) 12, p. 458-464
23. Welling, K. et al.: Festigkeitsberechnung, Alfred Kröner Verlag, Stuttgart, 1976
24. Brandenburger Folientechnik Company Publication, Landau, Germany

1.13 Further Reading on Injection Mold Construction

Menges, G., Mohren, P.: How to Make Injection Molds, 3rd Ed. Hanser Publishers, Munich, 2001

Rees, H.: Mold Engineering, 2nd ed., Hanser Publishers, Munich, 2002

Kennedy, P.: Flow Analysis of Injection Molds, Hanser Publishers, Munich, 1995

Pye, R.G.W.: Injection Mold Design, Longman Scientific and Technical, New York, 1989
DIN-Taschenbuch 262, Press-, Spritzgieß- und Druckgießwerkzeuge
Stoeckert K. / Mennig, G.: Mold-Making Handbook, 2nd Ed., Hanser Publishers Munich 1999
Knappe, W., Lampl, A., Heuel, O.: Kunststoffverarbeitung und Werkzeugbau, Carl Hanser Verlag, München, Wien, 1992
Beaumont, J.F.: Runner and Gating Design Handbook, Hanser Publishers, Munich, 2004
Unger, P.: Hot Runner Technolgy, Hanser Publishers, Munich, 2006

2 Special Design Features of the Example Molds

Stripper Plate Mold

A stripper plate surrounding the mold core presses on a large portion of the part rim during demolding. Thus the part is not subject to high local stress during demolding, such as individual ejector pins would generate.

Examples: 1, 14, 22, 31 to 33, 41, 43, 49, 51, 57, 59, 74, 77, 79, 84, 86, 102, 112, 115 to 117, 120.

In special cases, the mold core lies on the fixed side of the mold. The stripper plate is then driven from the moving mold side, or by special activating elements.

Examples: 4, 29, 46, 50, 75.

Unscrewing Mold

Threaded portions of molded parts that cannot and/or must neither (because of markings) be demolded by external or internal splits or slides, nor by collapsible cores, are demolded by unscrewing threaded cores and/or sleeves. In the simplest case, the thread-forming mold sections are inserted into the mold, demolded with the part and, once outside the mold, unscrewed.

Example: Thread core inserted 27.

(See also "Screw core")

Aluminum Mold

For prototypes, pilot runs, but also for mass production molds, the part-forming mold platens are manufactured from high-strength, easily workable aluminum alloy. The high thermal and temperature conductivity of aluminum alloy can also be of economical value.

Examples: 99, 124.

Sprue Punching in the Mold

The sprue is located where a bore is to be made in the molded part. During part demolding, this area is punched out, the bore formed and the sprue separated.

Examples: 23, 61.

Gating through the Core

The mold core lies on the fixed side of the mold. The sprue passes through the core to the inside of the molded part.

Examples: 46, 50, 93.

Ejection, Multi-stage

Removal of the molded part from the core takes place in several sequential steps.

Examples: 1, 16, 31, 45, 50, 51, 54, 56, 60, 66, 70, 82, 98, 101, 103, 104, 110, 112, 122.

Ejector Safety Mechanism

If the ejectors in the mold are actuated via electronic or hydraulic control systems, instead of mechanical machine stops and retraction devices, there is a danger that, in the event of any disruptions, they will retract too soon or too late, and the mold will be damaged. An additional mechanical safety mechanism built into the mold will eliminating this danger.

Examples: 42, 118, 122.

Split Cavity/Slide Mold

In addition to its parting lines, the mold has demolding elements (splits, slides) that move laterally to the direction of opening. Splits serve to demold larger surfaces, while slides are used for smaller areas of the mold contour.

Examples:

External splits and/or slides: 3, 5, 11 to 13, 15 to 27, 29, 31 to 33, 37, 43 to 45, 47, 48, 62, 66, 70, 83, 84, 89, 100, 103, 108 to 111, 113, 115, 118, 122, 123, 125, 128.

Internal splits and/or slides: 11, 15, 17, 37, 45, 73, 82, 97, 100, 104, 108, 110, 113, 118, 119.

Three-Plate Mold

The mold consists of several mold plate assemblies that open at two or three parting lines. As a rule, the molded part is demolded from one parting line and the sprue from another.

Examples: 12, 32 to 35, 56, 65, 66, 78, 98.

Thermoset and Elastomer Molds

The molding compound consists of two or more reactive components that do not react with one another and crosslink or cure until the mold cavity has been filled. The cavity wall temperature is higher than the molding compound temperature.

Examples: 62 to 67, 114, 121, 125.

Inserts

Components fabricated outside the mold are inserted into it to be encapsulated with molding compound.

Examples: 16, 69, 88, 98, 99, 105.

Venting Components

These serve to release air from areas in the mold cavity where it can become trapped by inflowing compound.

Examples: 4, 7, 12, 44, 67, 69, 70, 116, 117, 120.

Stack Mold

The mold has two parting lines (planes) at wide angles to the clamping direction, each of which contain cavities. The parts are generally gated via a hot runner located between the two parting lines. The holding force required for the mold is determined by the greatest buoyancy generated by either of the parting lines.

Examples: 15, 36 to 38, 40, 41, 43, 44, 113.

Collapsible Core

A mold core (with folds) that enables demolding of inner undercuts by collapsing (at its folds).

Examples: 9, 60, 72, 80, 107.

Assembly in the Mold

The various parts of a multi-sectional molded part are injected separately in the mold and assembled prior to demolding by a special movement sequence.

Example: 89.

Gas-Assist Injection Molding

Gas is forced into the cavity already partly filled with melt. The result is a molded part with compact outer skin and gas-filled interior.

Example: 127.

Hot-Runner Mold

The melt conduction channels in the mold are heated between the sprue bushing and the cavity gates so that the molding compound in them remains flowable.

– With Cold Secondary Runners

Where direct hot-runner gating is problematical, e.g., lack of space, the final portion of the runner is bridged with a cold secondary runner (with e.g., a tunnel or film gate).

Examples: 8, 11, 17, 23, 26, 28, 33, 37, 45, 47, 55, 56, 59, 69, 75, 89, 109, 120, 123.

– With Decompression

Prior to mold opening, the pressure on the molding compound in the hot runner is released in order to prevent drooling from the nozzles.

Examples: 36, 40, 41, 43, 44, 54, 69, 83, 92, 99.

– With Needle Shut-off

The gates are sealed by needle shut-off mechanisms and, if required, squeezed flat.

Examples: 42, 51, 55, 77, 83, 92, 122, 129.

– With Naturally Balanced Runners

The runners are designed such that all paths and flow channels leading to the gates are of equal length.

Examples: 15, 36, 42, 44, 46, 49 to 55, 76, 83, 88, 104, 120, 129.

Additional Examples: 33, 38, 42, 47, 48, 57, 60, 70, 75, 80, 86, 93, 108, 110, 111, 112, 123.

Hot-Runner, Self-Insulating

The melt feed channels are so thick that during continuous, uninterrupted injection, a free-flowing "core" remains inside, although the channels are unheated. A more or less frozen layer of melt forms on the channel walls, protecting the core from freezing for a time. Subsequent to downtime and prior to start-up, the frozen channel content must be removed and replaced by free-flowing melt. The same procedure is followed for a color change.

Example: 76.

Cold-Runner Technology (Thermoset and Elastomer Molds)

In contrast to the hot-runner principle, the melt-feed channels in the mold are at a temperature lower than in the mold cavities. The molding compound in the channels does not cure between molding cycles.

Examples: 62, 63, 67, 114, 125.

Core Cooling by Air Blast

Thin mold cores in which none of the standard cooling systems can be accommodated are cooled by blasts of compressed air between molding cycles.

Examples: 54, 101.

Core Centering Prior to Injection

Long, thin mold cores are supported mechanically before molding compound is injected; this prevents bending under the force of the inflowing melt. Shortly before the mold cavity is completely filed, the support elements are removed and the support areas are also filled with molding compound.

Examples: 4, 91, 101.

Core Pullers

Core pullers serve to demold bores in molded parts which do not lie in the opening direction of the mold.

Examples: 7, 12, 24 to 26, 29, 30, 60, 84, 90, 97, 99, 101, 106, 109, 113, 117, 119, 122, 128.

Core Puller, Curved

The mold core is curved and must be demolded in a curve.

Example: 7.

Cooling Pins (Heat Pipes, Thermal Pins)

They serve to remove heat from mold regions that are, for example, diffcult to supply with coolant.

Examples: 26, 54, 96.

Cooling System, Face-Joined

The part-forming mold sections are split where cooling channels are required. After the channels have been machined mirror-image on both faces, the parts are bonded together by a special joining technique (e.g., high-temperature vacuum welding). Advantages: The channels can be configured uniformly with the part-forming mold surfaces, and the distances to slides, ejectors and venting components can be kept quite small.

Examples: 61, 77, 95.

Copper/Bronze Inserts for Cooling

Heat dissipation from the mold cavity is increased by inserts made from metal with good thermal conductivity.

Examples: 26, 32, 36, 44, 45, 48, 60, 89.

Pneumatic Ejectors

Demolding is effected completely or in part using compressed air acting directly on the molded part.

Examples: 14, 22, 32, 36, 44, 54, 108, 114.

Multi-Component Mold

A molded part consisting of two or more different types of or different-colored molding compounds is usually produced in consecutive injection sequences. Either the part is formed gradually in different stations of the mold, or certain cavity regions are initially covered by mechanical devices and filled later. The various sections of the molded part can be firmly attached or move independently of one another.

Examples: 55, 59, 82, 87, 102, 129.

Metal Injection Molding (MIM)

Metal powder is mixed with a thermoplastic polymer and thereby, under the influence of heat, rendered flowable for injection molding. The polymer fraction is precipitated from the molded part (green part), for example, by heating (thermal degrading). This causes dense sintering of the "brown part".

Examples: 128, 130.

Microstructures

Dimensions of the mold and molded part structures lie in the micron range.

Example: 101.

Outsert Technology

Functional parts (bearing sockets, spacers, retaining clips, etc.) made from thermoplastic polymer are injected form- and/or force-fit into pre-punched holes in a metal blank.

Examples: 97, 98, 99.

Pneumatic Nozzle

Prior to demolding, the sprue is separated at the nozzle from the molded part and ejected by a pneumatically actuated device.

Examples: 13, 73, 97.

Injection-Compression Mold

The molding compound flows with relatively low resistance when injected, since the mold cavity is enlarged. When the mold is finally closed, the melt is distributed tightly over the entire cavity. The result is a part true to dimension and with low molecular and/or filler orientation.

Examples: 62 to 64, 95, 125.

Unscrewing Core

If the screw-components are a composite part of the mold, they have to rotate for the part to be demolded. Rotation either results from the opening and closing of the mold, or is actuated by special drive elements (electric, hydraulic).

Examples: Rotation by mold motion: 74, 78, 112, 113.

Rotation by motor drive: 18, 28, 46, 53, 65, 70, 71, 96, 116, 120.

Standard Mold Base with Variable Inserts (Cassette Mold)

Mold bases are especially suited for producing prototypes as well as test and standard specimens. The mold base remains on the machine while the (pre-heated) inserts can be quickly interchanged.

Examples: 6, 124.

Submarine Gate (Tunnel Gate)

This type is mainly used for smaller molded parts. The sprue is separated from the molded part by the opening motion of the mold.

Examples: 4, 7, 11, 17, 25, 28, 31, 32, 37, 47, 55, 56, 59, 74, 75, 82, 84, 89, 96, 101, 103, 105, 107, 109, 115, 116, 117, 120, 123, 124, 130.

Displacement Slides

Accumulated melt and wall thickening in the mold can lead to sink marks that are unacceptable for either optical or functional reasons. Such melt accumulations are displaced by slides traversing into them.

Example: 42.

Rank-and-Pinion Slides

The sliding motion usually results from the opening motion of the mold. Power is transmitted either via toothed wheels or by two gear racks engaging their helical gears that mesh at a certain angle.

Examples: 16, 42, 104.

Forcible Demolding of Undercuts

Depending upon the elasticity of the molding compound and the size of the undercut, it is sometimes possible to demold an undercut in the molded part by stripping or with compressed air.

Examples: 1, 3, 11, 14, 49 to 51, 70, 85, 104, 110, 114, 120.

3 Examples

Example 1, Single-Cavity Injection Mold for a Polyethylene Cover

The cover with dimensions 141 mm × 87 mm × 12 mm high (Fig. 1) has an approximately oval shape. On the upper side, it has an inwardly projecting lip that forms an undercut around the entire part. The elasticity of polyethylene is used to release this undercut, thereby permitting release from the core without the use of complicated part release mechanisms.

Mold

The cavity half of the single-cavity (Figs. 2 to 5) consists essentially of the mold plates (1, 2), the heated sprue bushing (41) and the cavity insert (46). The mold is based on the use of standard mold components, except for the core backup plate (47), core plate (48), core ring (50) and stripper ring (49). Final and accurate alignment of the two mold halves is ensured by four locating pins (37).

Part Release/Ejection

The mold opens at I; the molded part is retained on the core as it is withdrawn from the cavity. As the knockout bar (14) is pushed forward, the ejector rods (33) attached to the ejector plate (7) actuate plate (3) with the attached stripper ring (49; parting line II). At the same time, plate (8) with the attached core (47, 48) moves forward through the action of the compressed springs (39).
Plate (4) with the attached core ring (50) remains stationary, because it is attached to the clamping plate (5) via the bars (6) (Fig. 5). Both the molded part and the core are now free of the core ring (50). After a distance W, plate (8) comes up against plate (4); the core (47, 48) comes to a stop and the spring (39) is compressed further. The stripper ring, however, continues to move and can now strip the molded part off the core. During this stripping action, the rim of the molded part, along which the stripper ring (49) acts, is expanded. Accordingly, the stripper ring must not hold the molded part too tightly in order not to hinder its expansion.

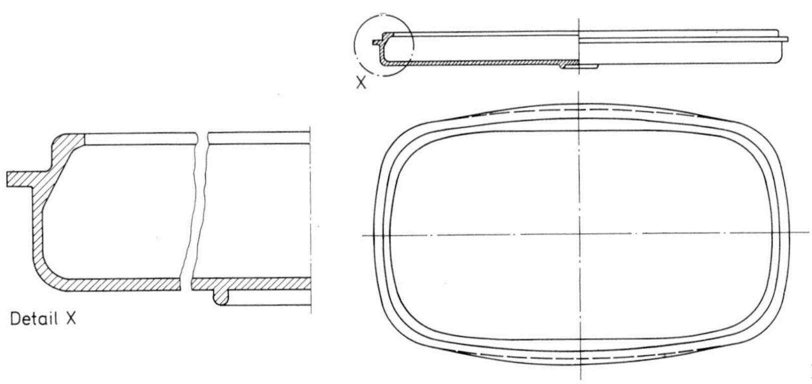

Figure 1 Polyethylene (PE) cover

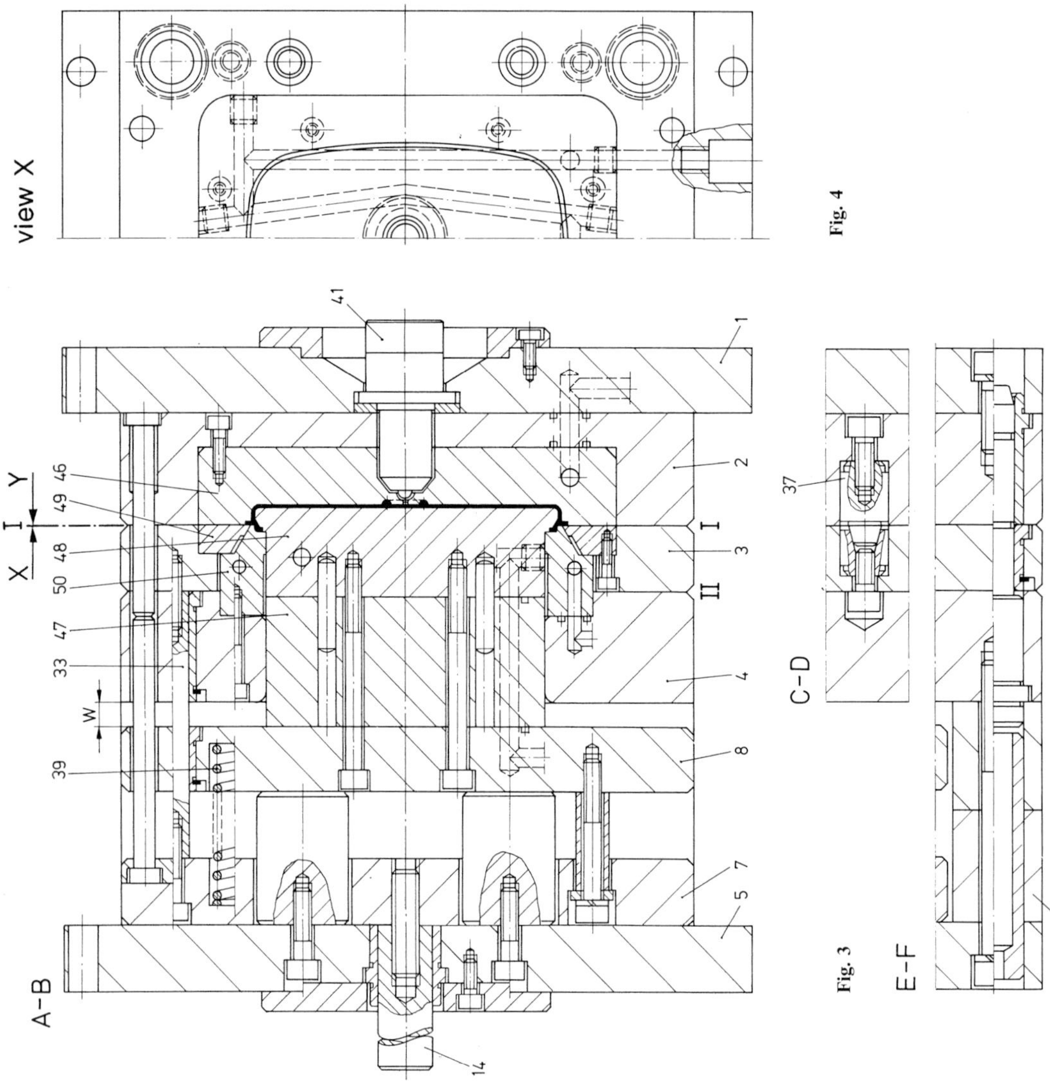

Fig. 4

Fig. 3

Fig. 5

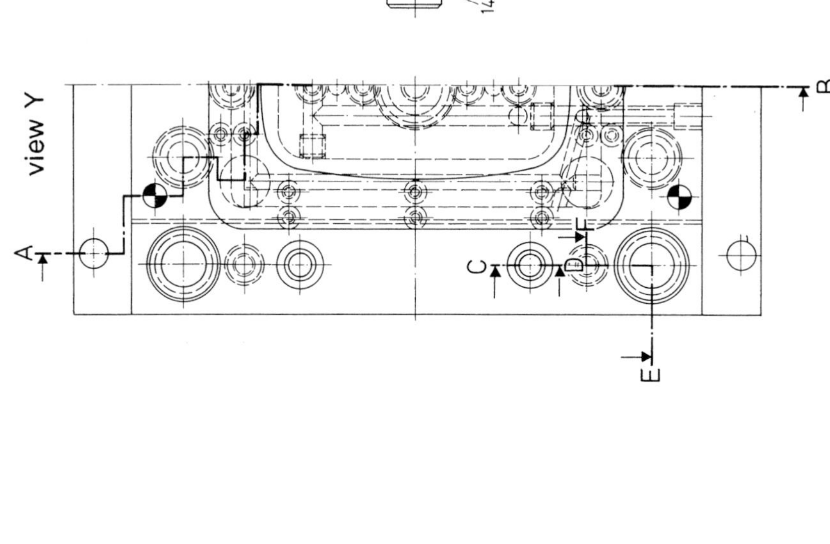

Fig. 2

Figures 2 to 5 Single-cavity injection mold for polyethylene cover
1: clamping plate; 2, 3, 4: mold plates; 5: clamping plate; 6: bars; 7: ejector plate; 8: ejector plate; 14: knockout bar; 33: ejector rods; 37: locating pin; 39: spring; 41: heated sprue bushing (Hasco); 46: cavity insert; 47: core backup plate; 48: core plate; 49: stripper ring; 50: core ring
(Courtesy: Hasco)

Example 2, Two-Cavity Injection Mold for Elbow Connector Made from PA 66

The article consists of two half-shells (Fig. 1) that are fitted and bonded together outside the mold. Average wall thickness is approx. 2.5mm. Process shrinkage was calculated at 1% of cavity-dimensional layout. In order to fasten cable clamps for strain relief, suitably shaped universal slots are provided. Surface quality is that of technical polishing.

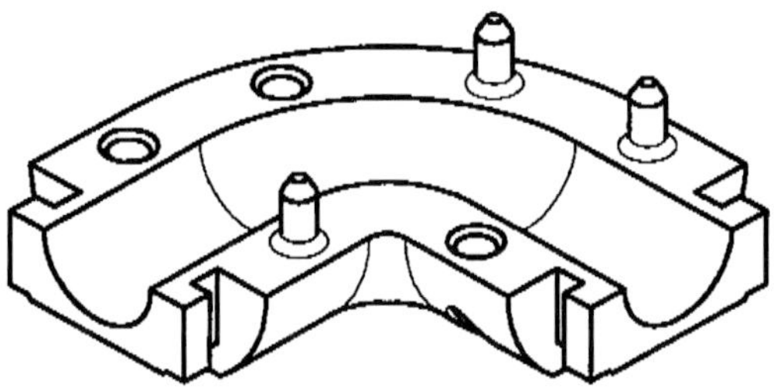

Figure 1 Half-shell of an elbow connector, diagram

Mold

The design corresponds to a standard DIN ISO 12165:2002-06 mold with a *single* parting line, Fig. 2. Changeable two-piece mold inserts (4a, b) and (5a, b) made from 1.2767 throughhardened steel are screwed to both cavity plates made from prehardened steel. The outer contour of the half-shells is shaped in mold inserts on the fixed side (4a, b), the inner contour in those on the moveable side (5a, b). Mold dimensions are 156 × 156 × 257mm. The relatively large installation height results, for one, from the dimensions of the two-stage ejector. The clamping plates (1) and (10) are equipped with thermal insulation sheets (6) in order to improve thermal efficiency of the mold. The ejector assemblies (7a, b) and (8a, b) are moved by a centrally mounted, standardized two-stage ejector (11). The ejector rod (12) engages the ejector system via an automatic ejector coupling. The ejector assemblies are guided by four pillars Ball cages are used for the ejector assemblies (7a, b).

Gating

The externally heated sprue bush with tip (14) is equipped with a screwed-on screwed on gate bush (Fig. 3). A spacer ring (16) serves to attach the gate nozzle to the centering flange (15). Via a short sprue carrot and a sub runner, which is also incorporated parabolically into the gate bush, the cavities are each filled via submarine gates (see also detail BB). The gating nozzle is secured against twisting by a dowel pin (17). The three holes on each half-shell are formed by core pins (18). To form the pegs, contoured ejector sleeves (19) with core pins (20) are used (detail D). The insert (21) recognizable on the moveable side is used as a core retainer plate for another variant of the molded part (not illustrated). To eliminate the possibility of a cold slug being injected through the gate into the cavity when filling begins, there is a catch-hole in the subrunner.

Demolding

Spring-loaded ejector pins (22) pre-loaded by return pins (23) during mold closing assist demolding on the fixed side (section B-B and detail E). Due to the undercut in the ejector (24), the gating system remains at first on the moveable side. When the mold opens, the frozen sprue is pulled from the nozzle and the gate is sheared off. The ejector assemblies perform two strokes per cycle according to the sequence: stroke 1 of the two-stage ejector causes the sprue to demold, and stroke 2 enables the molded part to demold.

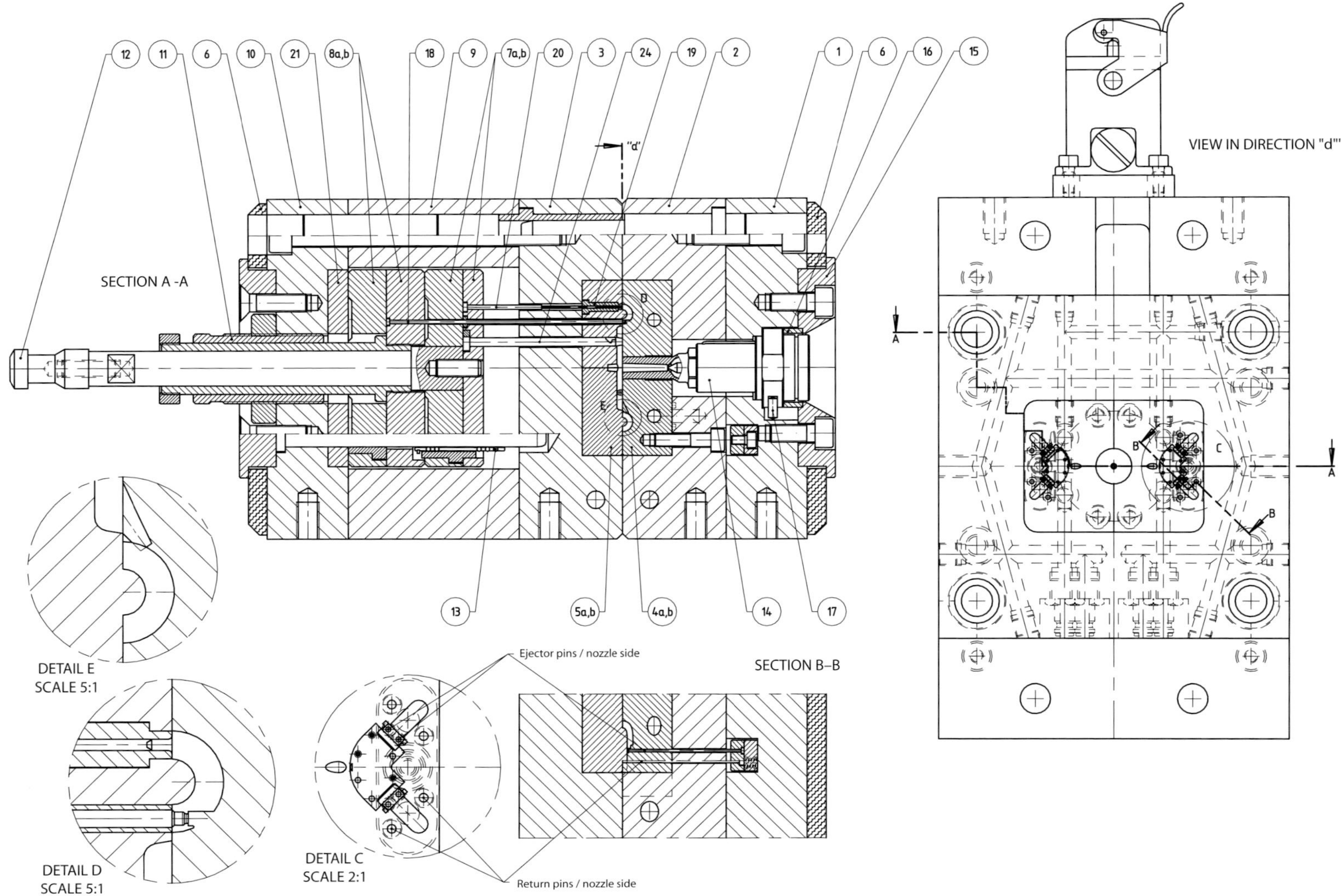

Figure 2 Two-cavity injection mold for elbow connector
1: clamping plate FS, 2: cavity plate FS, 3: cavity plate BS, 4a, b: mold inserts FS, 5a, b: mold inserts BS, 6: thermal insulation sheet, 7a, b: front ejector assembly, 8a, b: rear ejector assembly, 9: spacer strip, 10: clamping plate BS, 11: two-stage ejector, 12: ejector rod, 13: ball-bearing traveler, 14: gating nozzle with antechamber, 15: centering flange, 16: spacer ring, 17: dowel pin, 18: core pin (drills hole), 19: ejector sleeve, 20: core pin (forms peg), 21: insert, 22: spring-loaded ejector pin FS, 23: return pin FS, 24: ejector with undercut
(Courtesy: Hasco, Lüdenscheid; Möller, Bad Ems)

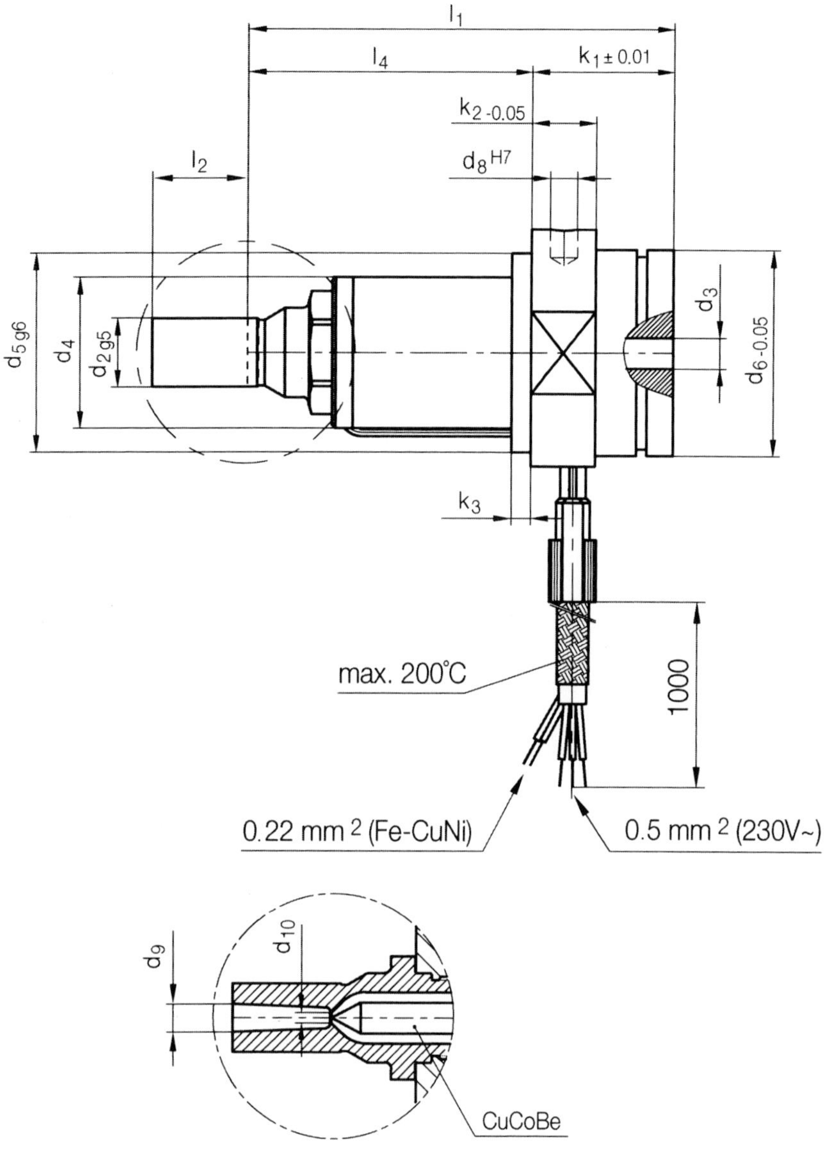

Figure 3 Heated sprue nozzle with antechamber and tip

Example 3, Injection Mold for the Body of a Tape-Cassette Holder Made from High-Impact Polystyrene

Molded Part: Design and Function

A cubic molded part of impact-resistant polystyrene (Fig. 1) forms the main body of a tape-cassette holder (Fig. 2) consisting of a number of injection-

Figure 1 Main body for a cassette holder, Z: Spring latch

Figure 2 Finished, assembled cassette holder with the main body from Fig. 1 and several cassettes inserted

molded parts. Several cassette holders can be stacked on top of each other by snap fits to yield a tower that can accommodate more cassettes. The molded part, which has a base measuring 162 mm × 162 mm and is 110 mm tall, consists of a central square-section rod whose two ends are bounded by two square plates. Between these plates, and parallel to the central rod, are the walls, forming four bays for holding the cassettes.

Single-Cavity Mold with Four Splits

The mold, with mold fixing dimensions of 525 mm × 530 mm and 500 mm mold height, is designed as a single-cavity mold with four splits (Fig. 3). The movable splits (9) are mounted on the ejector side of the mold with guide plates (21) and on guide bars (20). The splits form the external side

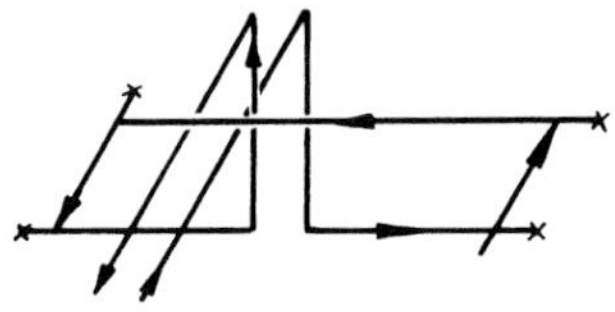

Figure 4 Cooling of the punch (7)

walls of the molded part while the internal contours of the bay's comprising ribs, spring latches and apertures are made by punches (34) that are fitted into the splits and bolted to them. Core (6), which is mounted along with punch (7) on platen (23), forms the bore for the square-section rod. The punch (7) and the runner plate (14) form the top and bottom sides of the molded part.
When the mold is closed, the four splits are supported by the punch (7) and each other via clamping surfaces that are inclined at less than 45°. Furthermore, the apertures in the molded part ensure good support between punches (34) on the splits, core (6) and runner plate (14).
The closed splits brace themselves outwardly against four wedge plates (12) which are mounted on the insert plate (18) with the aid of wear plates (13). Adjusting plates (11) ensure accurate fitting of the splits. Each slide is driven by two angle pins (8), located in insert plate (18) on the feed side. Pillars (39) and bushings (37) serve to guide the mold halves. The plates of each mold half are fixed to each other with locating pins (27).
The molded part is released from the core by ejector pins (25), which are mounted in the ejector plates (3, 4). Plate (23) is supported on the ejector side against the clamping plate via two rails (40) and, in the region of the ejector plates beneath the cavity, by rolls (2).

Feeding via Runners

The molding compound reaches the feed points in the corners of the square-section rod via sprue bushing (16) and four runners. The rod's corners

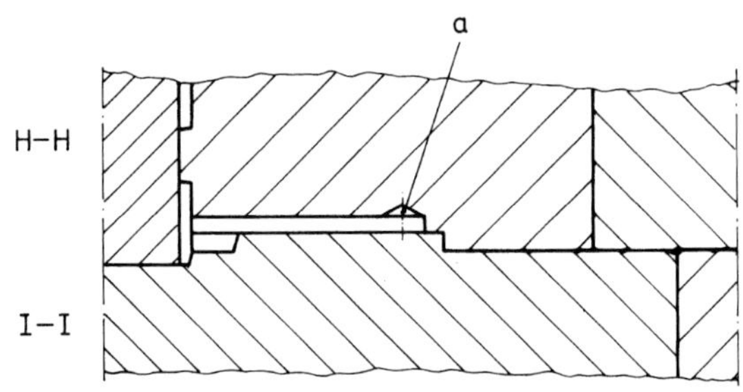

Figure 5 Detail of latch Z in the main body along H-H and I-I in Fig. 3

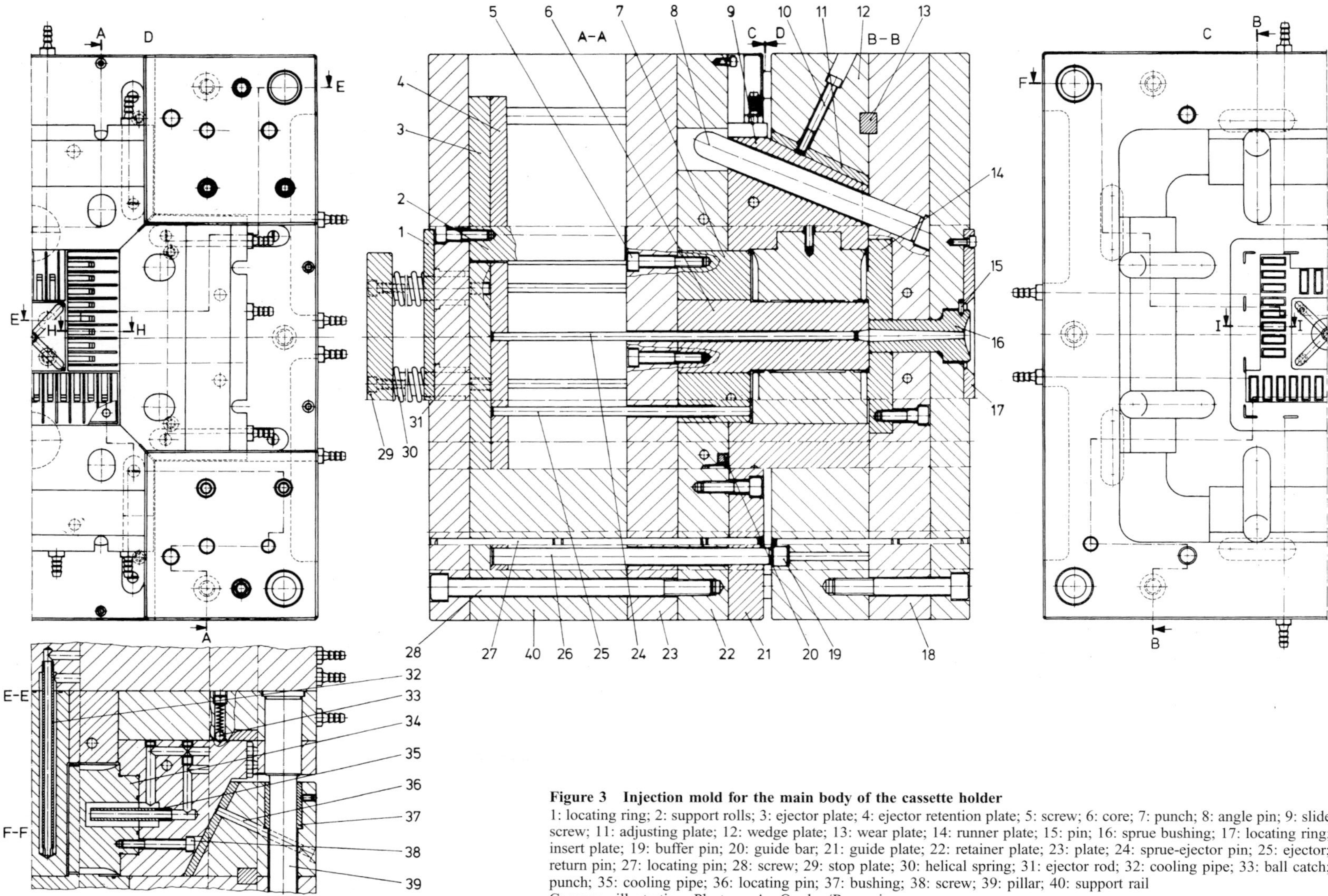

Figure 3 Injection mold for the main body of the cassette holder
1: locating ring; 2: support rolls; 3: ejector plate; 4: ejector retention plate; 5: screw; 6: core; 7: punch; 8: angle pin; 9: slide; 10 screw; 11: adjusting plate; 12: wedge plate; 13: wear plate; 14: runner plate; 15: pin; 16: sprue bushing; 17: locating ring; 18: insert plate; 19: buffer pin; 20: guide bar; 21: guide plate; 22: retainer plate; 23: plate; 24: sprue-ejector pin; 25: ejector; 26: return pin; 27: locating pin; 28: screw; 29: stop plate; 30: helical spring; 31: ejector rod; 32: cooling pipe; 33: ball catch; 34: punch; 35: cooling pipe; 36: locating pin; 37: bushing; 38: screw; 39: pillar; 40: support rail
Company illustration: Plastor p.A., Oradea/Romania

have a slightly larger flow channel than the other walls of the molded part. The sprue bushing is secured against turning by pin (15).

Mold Temperature Control

Cooling channels are located in the core retainer plate (22) and the insert plate (18). Punch (7) is cooled as shown in Fig. 4. Core (6) is fitted with two cooling pipes, while punch (34) is fitted with cooling pipe (35). Furthermore, the slide (9) are cooled.

Demolding

As the mold opens, the slides (9) are moved by the angle pins (8) to the outside until the punches (34) are retracted from the side bays of the molded part. As Fig. 5 shows, the cavities of the spring latches Z are located on the one hand between the faces of the four punches (34) and runner plate (14) and, on the other, between the two adjacent side faces of the punches (34).

On opening of the mold, the ratio of the distance moved by the slides to the opening stroke between runner plate (14) and slides is the tangent of the angle formed by the angle pins and the longitudinal axis of the mold. Thus, when the mold opens, enough space is created behind the latches Z to enable them to spring back when the punches (34) slide over the wedge-shaped elevations (a) of the latches (Fig. 5). The situation is similar for ejecting latches between adjacent punch faces. As the mold opens further, the angle pins and the guide bores in the slides can no longer come into play. The open position of the slides is secured by the ball catches (33). The molded part remains on core (6) until stop plate (29) comes into contact with the ejector stop of the machine and displaces ejector plates (3, 4) with ejector pins (24, 25). The molded part is ejected from the core, and the sprue from the runners. When the stop plates are actuated, helical springs are compressed (30) that, as the mold is closing, retract the ejector pins before the slides close. Return pins (26) and buffer pins (19) ensure that the ejector system is pushed back when the mold closes completely.

Example 4, Five-Cavity Injection Mold for Tablet Tubes Made from Polystyrene

It has been found that especially with tubes which are relatively long in relation to their diameter, it is extremely difficult to prevent displacement of the core and avoid the resulting variation of wall thickness with all the detrimental consequences. As the result of uneven melt flow, the core may become displaced toward one side even when a centrally positioned pinpoint gate is used on the bottom.

In the following, an injection mold is described, in which displacement of the core is reliably prevented. It has been determined that gating from two opposite points on the open end of the tube already leads to considerably less displacement of the core than occurs when gating on the bottom. It is useful to design these two points as tunnel gates so that they are automatically sheared on opening of the mold which eliminates the need for any secondary operations.

With long tubes, however, even this type of gating is not enough to ensure completely uniform wall thickness. The core must be held in position until the melt reaches the bottom.

This is accomplished in the mold shown in Figs. 1 to 4 as follows:

To avoid an unnecessarily long sprue, the water-cooled cores (*a*) are fastened on the stationary mold half. The face of the core has a conical recess about 0.5 mm deep into which a conical protrusion on the movable core (*b*) is pressed by means of spring washers (*c*) when the cavity is not filled. As soon as the plastics melt fills the cavity to the bottom and flows into the annular space around the protrusion, the injection pressure overcomes the force exerted by spring washers and displaces the movable core (*b*) by an amount corresponding to the thickness of the bottom. The entire bottom now fills with melt. A vent pin (*d*) with running fit in the movable core (*b*) to permit the compressed air to escape is provided to ensure that the melt will flow together properly at the center of the bottom.

As the mold opens, the spring washers assist in ejecting the tablet tubes from the cavities as well as in shearing off the two tunnel gates. The tubes are supposed to be retained on the cores, from which they are stripped by the stripper plate (*e*) during the final portion of the opening stroke. The runner system is initially retained by undercuts on the sucker pins (*f*). However, as soon as the stripper plate (*e*) is actuated, the runner system is pulled off the sucker pins (*f*) and drops out of the mold separated from the molded parts.

Example 4

Fig. 1

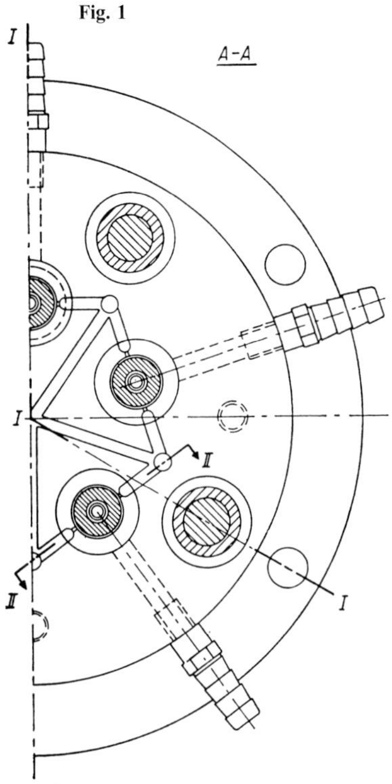

Fig. 2

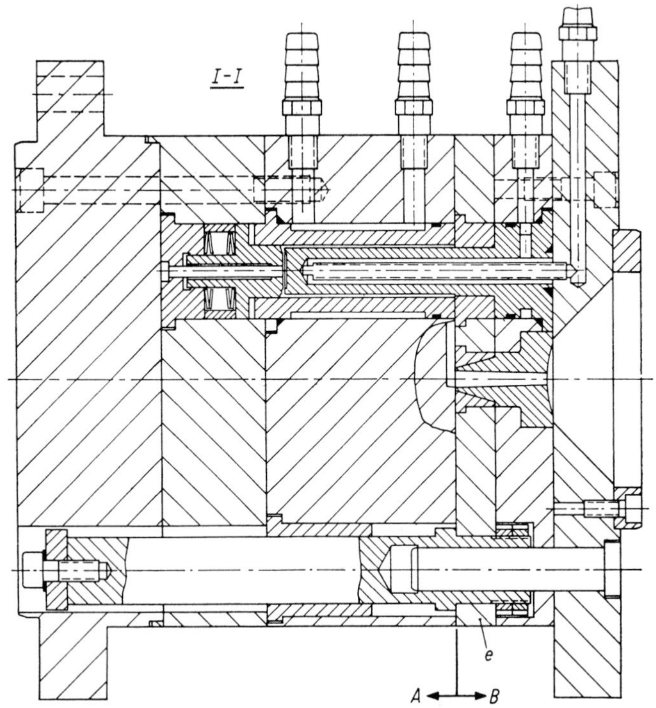

Fig. 3

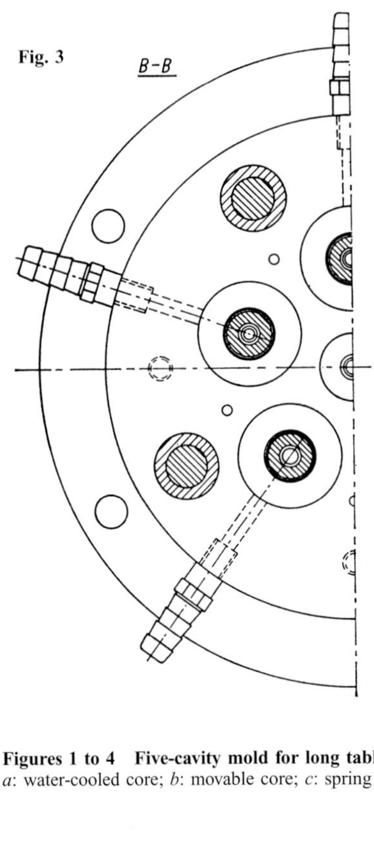

Fig. 4

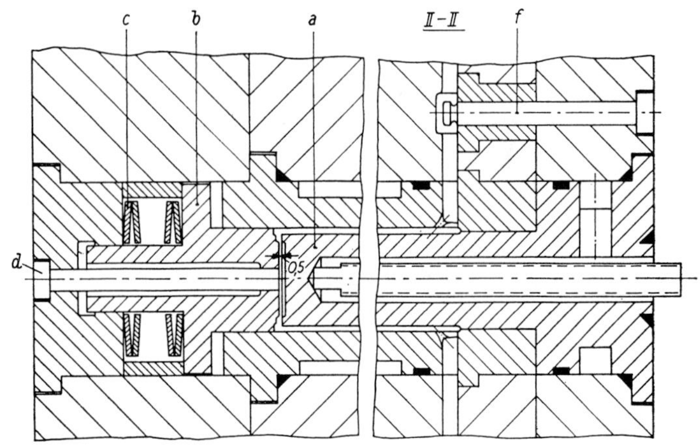

Figures 1 to 4 Five-cavity mold for long tablet tubes
a: water-cooled core; *b*: movable core; *c*: spring washers; *d*: vent pin; *e*: stripper plate; *f*: sucker pin

Example 5, Four-Cavity Injection Mold for a Polyamide Joint Element

The element (Fig. 1) is similar to a pipe fitting. It has four socket openings, two of which form a through-hole. The other two openings are located in the plane perpendicular to this hole such that their axes enclose an angle of 84°. The 84° branch contains a rib with a hole.

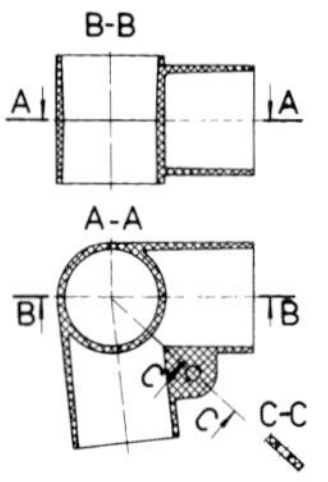

Figure 1 84° joint element

Mold

The mold with a size of 560 cm × 560 cm × 345 cm high (Figs. 2–13) is designed with four cavities such that the cavities enclosing the 84° angle lie within the parting plane, whereas the through-hole extends in the opening direction of the mold.
The four mold cavities formed in the mold insert plates (12, 13) are arranged in the parting plane in such a way that each of two mutually parallel cores of a pair of cavities can be actuated by a common core puller. Six slide bars are thus available for pulling the eight cores.
The core slide bars (24, 28) run on the mold plate (6) in guides (35, 38) and on slide rails (32, 36). The closed slide bars are locked by locking wedges (21, 30). Angular columns (22, 29), which are fixed to

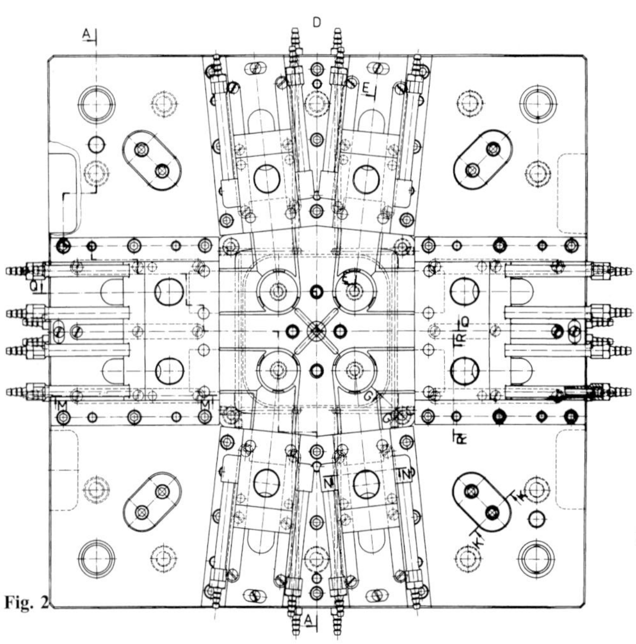

Fig. 2

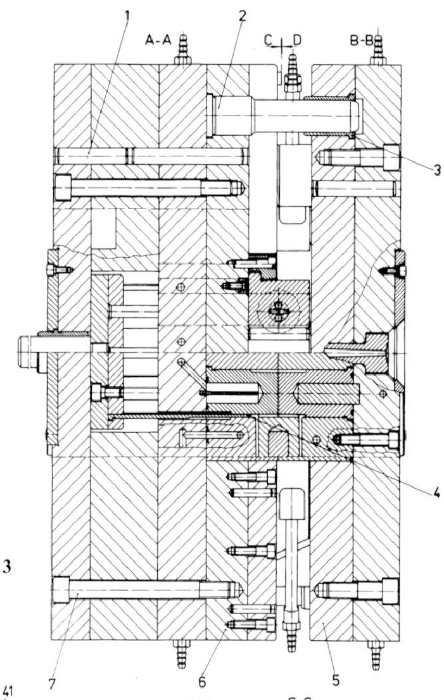

Fig. 3

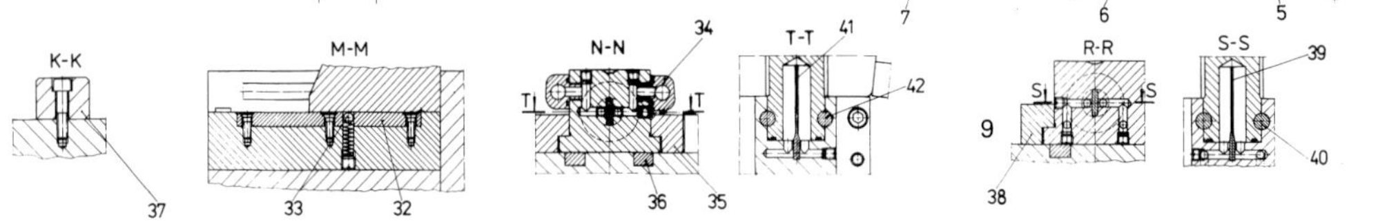

Fig. 4 Fig. 5 Fig. 6 Fig. 7 Fig. 8 Fig. 9

Figure 2 View of the movable parting plate of the mold at the ejector side (cf. Fig. 3, view D)

Figure 3 Longitudinal section A-A (cf. Fig. 2) and B-B (cf. Fig. 10)
1: locating pin; 2: guide column; 3: guide bush; 4: cavity ejector; 5: fixed mold plate; 6: movable mold plate

Figure 4 Section K-K (cf. Fig. 3)
37: check buffer

Figure 5 Section M-M (cf. Fig. 3)
32: slide rail; 33: ball detent

Figure 6 Section N-N through the individual slide bar (cf. Fig. 2)
34: cooling water connector; 35: slide-bar guide; 36: side rail

Figure 7 Section T-T through the slide core (cf. Fig. 6)
41: partition wall (for cooling water diversion); 42: cylindrical pin

Figure 8 Section R-R through the double slide bar (cf. Fig. 2)
38: slide bar guide

Figure 9 Section S-S through the slide-bar core (cf. Fig. 8)
39: partition wall; 40: cylindrical pin

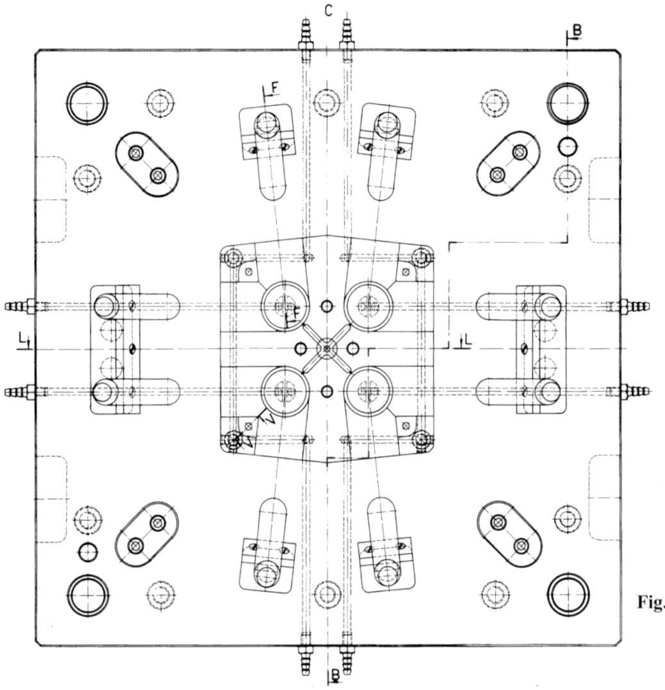

Fig. 10

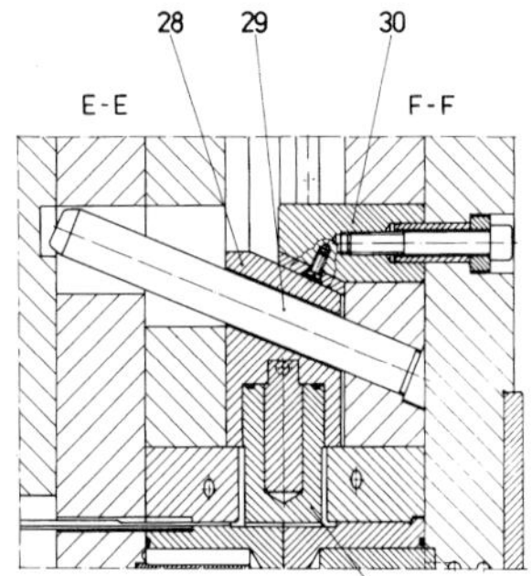

Fig. 11

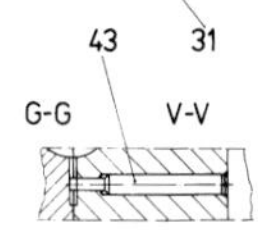

Fig. 12

Figure 10 View of the feed-side (fixed) parting plane (cf. Fig. 3, view C)

Figure 11 Longitudinal section E-E (cf. Fig. 2) and F-F (cf. Fig. 10)
28: individual slide bar; 29: angular column; 30: locking wedge; 31: core insert

Figure 12 Sections G-G (cf. Fig. 2) and V-V (cf. Fig. 10) through the pin for the "string hole"
43: pin

the mold plate (5) at the feed side, engage in the slide bars and actuate them as the mold is opened and closed. Ball detents (33) secure the position of the opened slide bars when the angular columns are moved out of the slide bars. The core inserts (18, 31) are fixed in the slide bars by means of cylindrical pins (40, 42).

Because of the large number of slide bars, the clamping area of the mold is comparatively large compared with the closing area between the mold insert plates (12, 13), which is determined by the mold cavities. To ensure uniform loading of the parting plane during closing, check buffers (37) are mounted on both mold plates (5, 6).

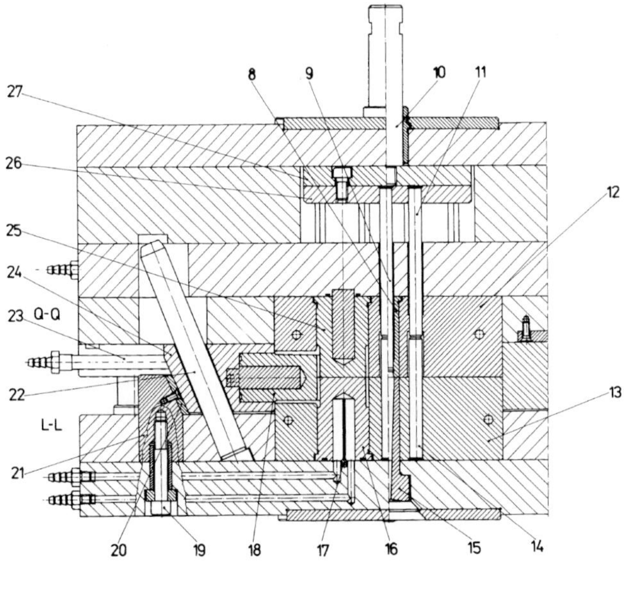

Figure 13 Longitudinal sections L-L (cf. Fig. 10) and Q-Q (cf. Fig. 2)
8: sprue puller bush; 9: sprue ejector; 10: ejector bolt; 11: ejector-plate return pin; 12, 13: mold insert plates; 14: buffer pin; 15: sprue bush; 16: core insert; 17: partition wall; 18: core insert; 19: screw; 20: locating strip; 21: locking wedge; 22: angular column; 23: connection tube; 24: double slide bar; 25: core insert; 26, 27: ejector plates

Gating

The melt is fed, via conical sprue in the sprue bushing (15) and via cruciform runners located in the parting plane, to the pinhole gates at the side walls of the four cavities.

Cooling

To cool the cavities, cooling bores are incorporated into the mold insert plates (12, 13). All four cores of each mold cavity are efficiently cooled by means of a central bore containing an inserted partition wall (17, 39, 41). The seat surfaces of the core inserts are sealed by O-rings. The cores inserted into the slide bars are supplied with water via connection pipes (23) and flexible hoses. The water for the fixed cores is fed and discharged via bores in the respective mold plates lying below the cores.

Demolding

When the mold opens in the parting plane C-D (Fig. 3), the moldings remain on the ejector side, where they are first held by the core slide bars. This pulls the moldings off the cores (16) at the feed side and out of the cavity parts. The sprue cone is demolded by the sprue puller.
During the opening action, the slide bars are pushed outward by the effect of the angular columns and pull the cores located in the parting plane. During this process, the moldings are still held firmly by the cores (25) at the ejector side. Finally, cavity ejectors (4) (three per mold cavity) and the sprue ejector (9) eject the molding completely.
As the mold closes, ejector-plate return pins (11), which strike against buffer pins (14), push back the ejector plates, and thus the cavity ejectors and sprue ejector. The core pullers are brought back into the ejection position by the angular columns. The mold is operated semi-automatically.
The prime objective in describing this injection mold was to demonstrate the arrangement and operation of the core pullers. To save sprue material, it would of course be possible to use a hot-runner nozzle instead of the conical sprue bush. It would then be possible to separate the moldings from the cruciform sprue automatically by means of submarine gates.

Example 6, Mold Base with Replaceable Inserts to Produce Standard Test Specimens

Component testing, product development and short production runs often require injection molded parts that have been produced under defined and reproducible conditions. Conventional molds have long mold change times, with the disadvantages of lengthly idle times and excessive residence time of the melt in the barrel. Purging of the melt would mean a material loss that could not be justified with the often small quantities of expensive experimental materials.
In order to avoid these disadvantages, a mold base was developed that meets all of the requirements with regard to processing, economy and reliability of operation. This mold base with interchangeable plug-in inserts is also suitable for production of flat molded parts, e.g. gears, small plaques etc., and is characterized by the following features.
The mold cavity is located in the interchangeable mold plate (1) on the ejector side (plug-in insert). The cavity is machined only into this plate, which seals against a flat mating plate (2) bolted to the stationary-side clamping plate. The plug-in inserts can be removed and stored without any aids within approximately one minute. The weight of each plug-in insert is approx. 6 kg.

Mold Temperature Control

The cooling lines for mold temperature control are located in the plug-in insert and mating plate. Self-closing quick disconnects (3) in the supply lines facilitate replacement of the inserts. With a suitably sized mold temperature control unit, the insert reaches operating temperature after only 8 shots thanks to optimum positioning of the cooling lines.

Cavity Pressure and Cavity Wall Temperature

Cavity pressure and cavity wall temperature are measured and recorded along with additional important process variables. Only similar test specimens are produced in a given insert. The mold design permits simultaneous filling of all cavities and is based largely on the use of commercially available standard mold components. The materials used, heat treatment (core 64 RC) and surface treatment (CVD for the mating plate, surface 72 RC) ensure high wear and corrosion resistance.

Fig. 1

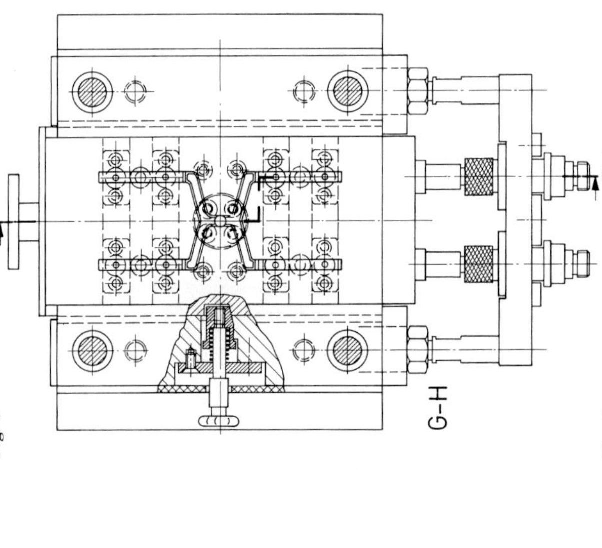

Fig. 2

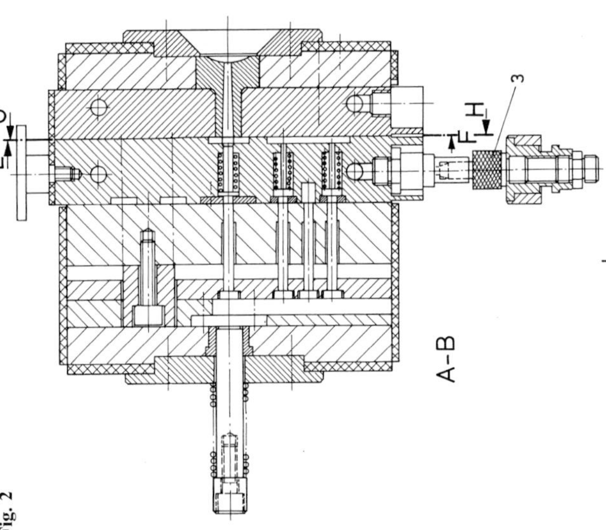

Fig. 3

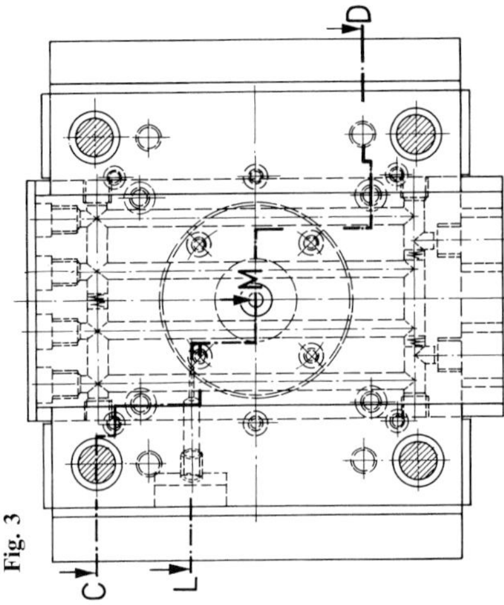

Fig. 4

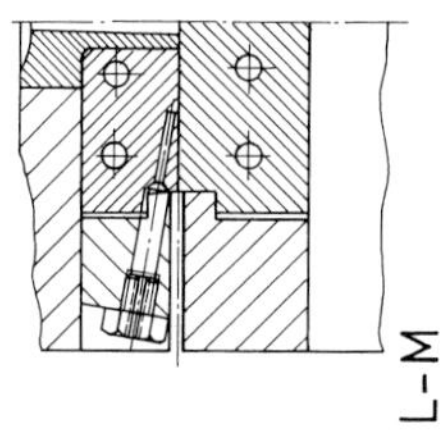

Fig. 5

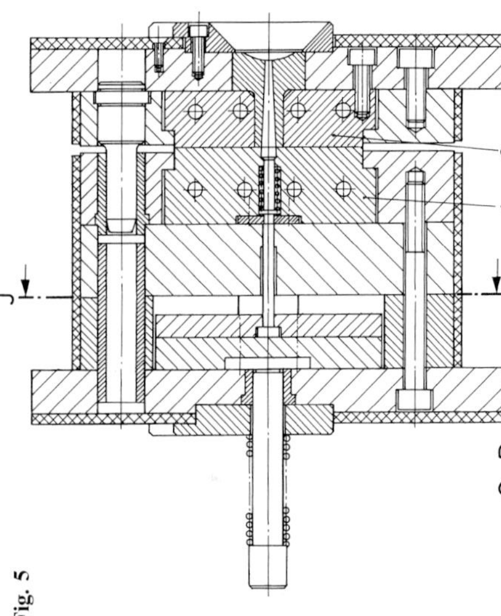

Fig. 4

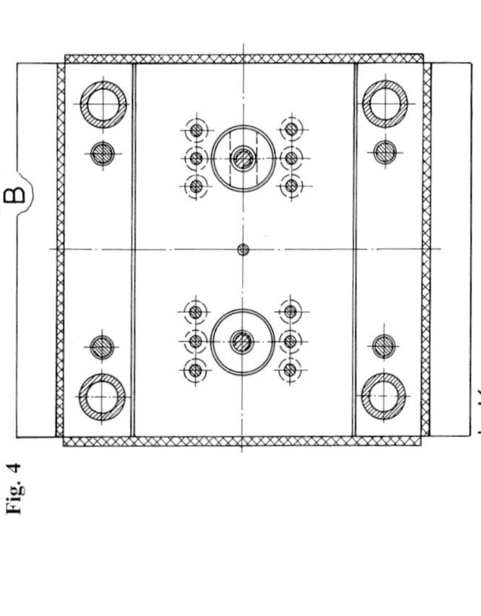

Figures 1 to 6 Mold base with interchangeable inserts for the production of standard test specimens
1: plug-in insert; 2: mating plate; 3: quick disconnect

Example 7, Two-Cavity Rotary Core Mold for a Polyacetal Pipe Elbow

The pipe elbow in Fig. 1 is about 70 mm in length and has a bore 5.2 mm in diameter. The user has stipulated that the bore may have hardly any draft. It consists of a straight section 22 mm long and a curved section with a radius of 102 mm describing an angle of 25°. The bore has a rounded inlet on the flange side. A snap-fit hook (H) for attaching the pipe elbow during assembly is located on the outside of the flange. At the opposite end, the pipe elbow has a connecting nipple, whose first ribbed section must not have any axial parting marks. Along the outside of the curved section of the pipe elbow are four ribs.

Mold Requirements

Two core pullers are required for ejecting the pipe bore – a straight one and one for the curved section. Core pullers that are actuated hydraulically or by gear drives are usually employed for ejecting curved internal contours. However, they are sometimes so large that they determine the required machine size. Moreover, they involve high production and repair outlay. The task was therefore to find a reasonably priced, rugged actuator for the curved core puller that would fit the size of the machine appropriate to the molded part. A straight angle pin which engages a specially designed bore on the slide provides the movement of the curved slide. When designing the bore, consideration had to be given to the fact that, on mold opening and closing, an orbitally guided slide additionally executes a transverse movement to the direction of movement dictated by the angle pin. The two active surfaces of this bore are the surfaces of the lines of contact between angle pin and slide during mold opening and closing. Figure 2A shows the design of the slide and the bore. For greater clarity, all invisible lines have been omitted from Fig. 2B, except for that of the bore. In the injection position (a), the column axis projects along lines (a_o and a_u). In the retracted position (b), the lines are displaced to b_o and b_u (o: top of slide, u: bottom of slide).

Mold Design and Construction

The mold (Fig. 3) is a two-cavity design with external dimensions of 296 mm × 296 mm and a

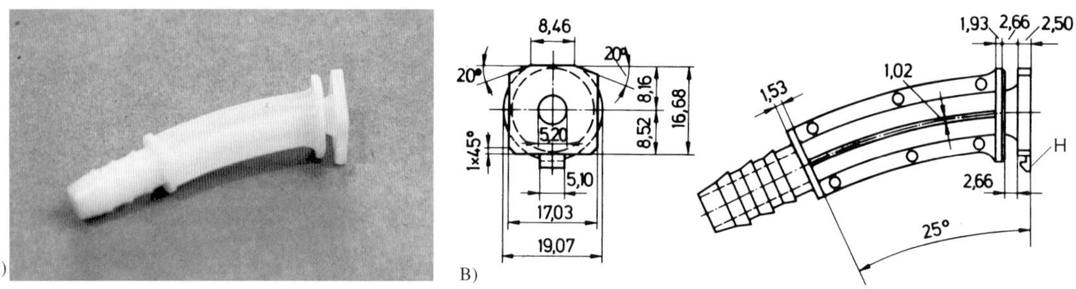

Figure 1 The pipe elbow (A) has a curved bore as shown in drawing (B)

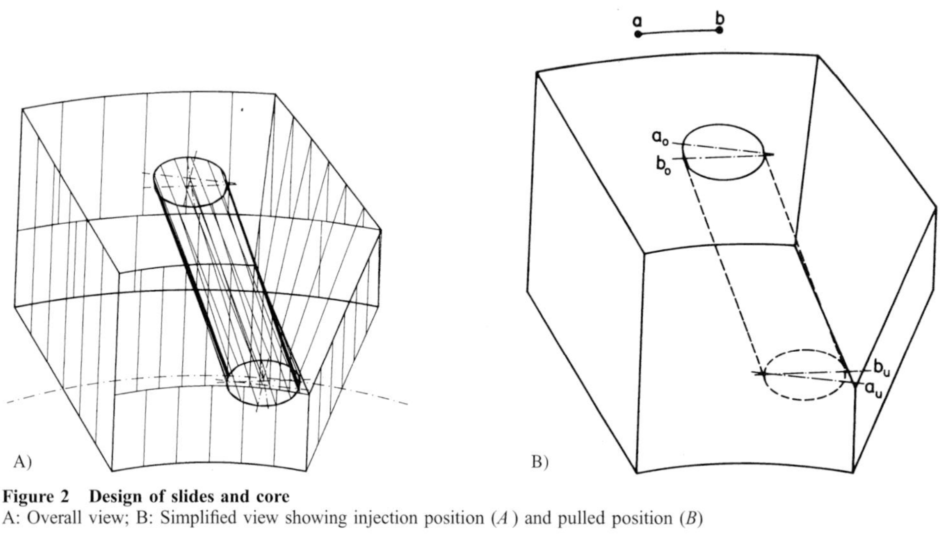

Figure 2 Design of slides and core
A: Overall view; B: Simplified view showing injection position (*A*) and pulled position (*B*)

Figure 3 View of the opened mold
Left: Fixed mold half with main runner; right: movable mold half with distributor

height of 240 mm. The cavities are incorporated symmetrically on the same axis into the hardened plates (1, 2) around the longitudinal mold axis as shown in Fig. 4. Further details of the mold may be seen in the cross-sectional drawing in Fig. 5, which essentially shows the mold design with the guide pillars, and in Fig. 6, which shows the slides with the angled guides and the ejectors. Ejector-side

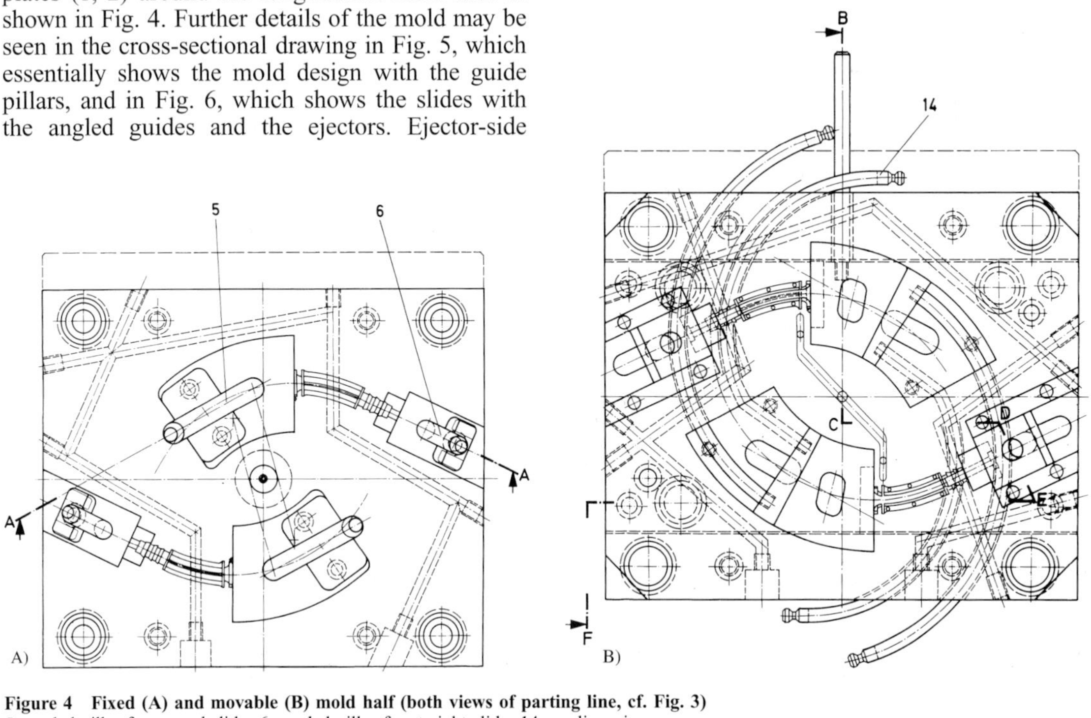

Figure 4 Fixed (A) and movable (B) mold half (both views of parting line, cf. Fig. 3)
5: angled pillar for curved slide; 6: angled pillar for straight slide; 14: cooling pipe

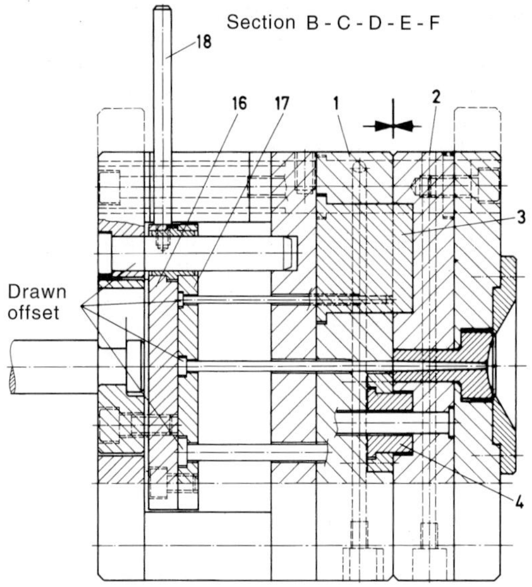

Figure 5 Section B-C-D-E-F (cf. Fig. 4) through the closed mold
1, 2: mold plates; 3: curved slide; 4: straight slide; 16, 17: ejector plates; 18: control pin

mold plate (1) contains the two curved slides (3) and straight slide (4), which are actuated by angle pins (5) and (6). Locking of the slides when the mold is closed is effected by compressors (7) and (8). The mold is so designed that the maximum possible number of parts, such as slides, guides and compressor fits, may be produced by wire EDM. The aperture in the curved slides is also made by wire EDM. The core inserts (9, 10) for forming the bore in the molded part are bolted to the slides. The mold insert (11) in the straight slide partly surrounds the core (10). It ensures that the section of the molded-part spout formed inside it remains flash-free. It is made of a special, hardenable "vented steel" (powder metallurgical alloy) which has a special microstructure that rules out the risk of entrapped air.

Runner System/Gating

The two pipe elbows are gated via a sprue, a runner and individually by a submarine gate.

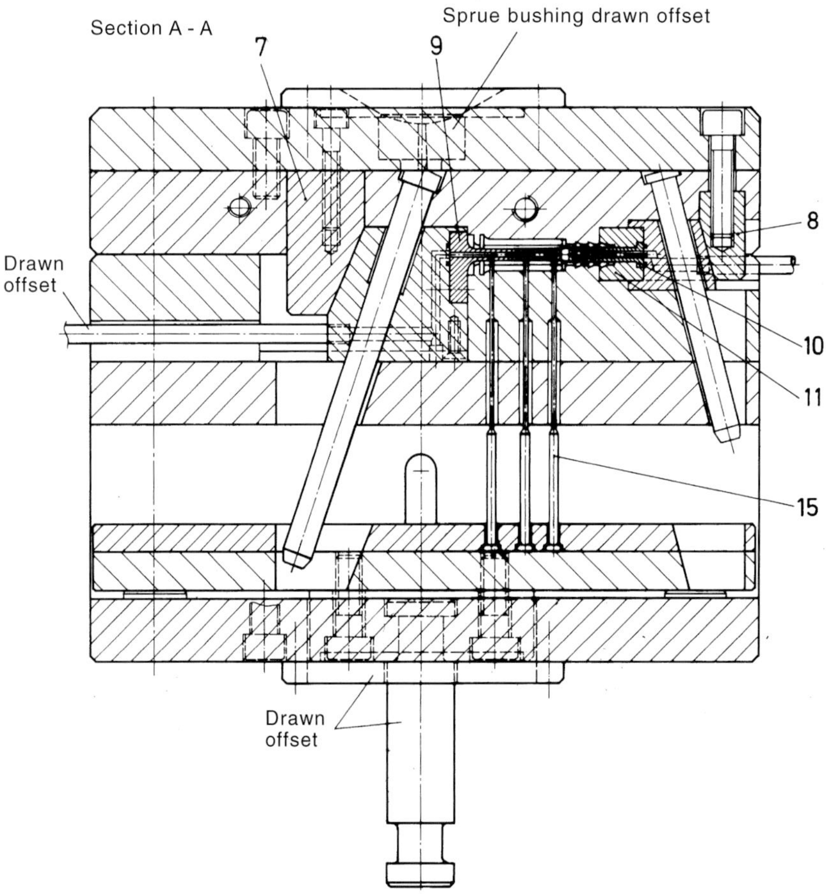

Figure 6 Section A-A (cf. Fig. 4) through slides, angled bolts and ejector in closed mold
7, 8: compressor; 9, 10: core insert; 11: mold insert; 15: ejector pin

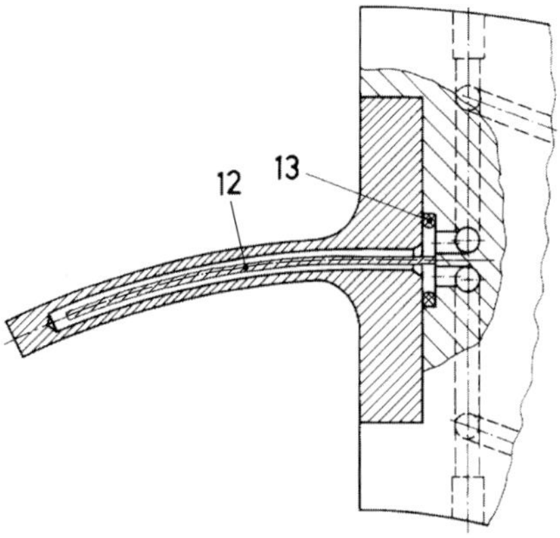

Figure 7 Cooling of the slide cores is a special feature (cf. Fig. 4)
12: baffle; 13: O-ring
Company illustration: Geiger Technik, Garmisch-Partenkirchen, Germany

Temperature Control

Cooling channels are incorporated into the two mold plates (1, 2) in the usual fashion.

Worth noting is the cooling of the cores in the two slide pairs (Fig. 7). The cores have bores each measuring 2.4 mm in diameter into which baffles (12) of 0.3 mm thickness are inserted for guiding the cooling water. O-rings (13) seal the connecting points between the slides and the core inserts. The cooling medium is fed to the cores via bores in the slides and via attached curved pipes (14) or tubes that copy the movements of the slides on mold opening and closing. The inserts of the curved cores were bored straight, then bent warm, stress-annealed and hardened. Final machining was done with spark erosion. Despite the low flow cross-section, this core cooling works exceptionally well.

Part Release/Ejection

When, after mold opening, the slides or cores are retracted, the parts are ejected by groups of six ejector pins (15). The pins press against the two ribs of the molded part in the mold parting line. The position of the ejector plates (16, 17) can be checked with a control pin (18) and a proximity switch not shown in the diagram.

Example 8, Hot Runner Injection Mold for Car Front Fender

Mold

This trial mold was used to injection mold prototype car fenders in different thermoplastics, such as PC/ABS blend, modified PA, elastomer-modified PC, PC/PBTP blend and modified PBTB for a number of manufacturing companies.

The single-daylight mold (Figs. 1 to 4) for a left front fender consists essentially of sprue half (1) with the hot runner system and the ejector mold half (2, 3, 4). Pillars (9) and bushings (10) serve to guide the two mold halves. When the mold is closed, the mold halves are fixed relative to each other via angled surfaces and wear plates (110, 113). The external dimensions of the mold are 1750 mm × 990 mm, the mold height is 752 mm, and its total weight is 9520 kg. The mold plates (1, 2) are made of steel (material number 1.2311) and annealed to a strength of 1000 N/mm^2.

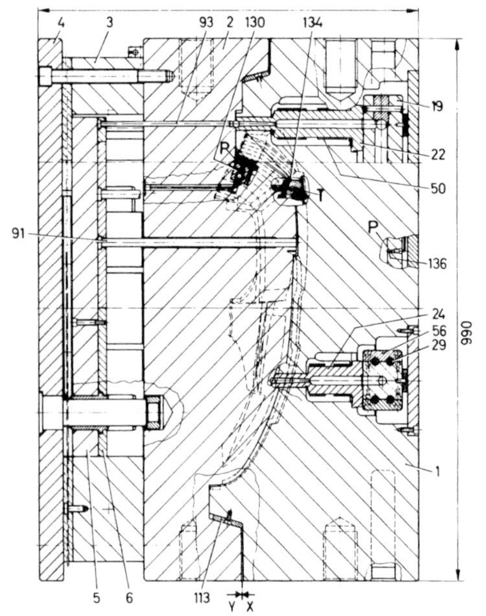

Figure 1 Side view of mold, section
1: machine-side mold plate; 2: ejector-side mold plate; 3: spacer bars; 4: clamp plate; 5, 6: ejector plates; 9: guide pillar; 10: bushing; 18: support roll; 19: hot runner distributor; 20: sprue bushing; 21, 22, 24, 25: heated nozzles; 26, 27, 28: distributor manifold centering; 29: cartridge heaters; 50: heating collars; 56: cartridge cover strips; 76: hot runner covering plate; 81: spacer rings; 85: guide component; 130: pressure sensor; 134: temperature sensor; 136: pressure sensor

Gating and Temperature Control

The melt passes through a sprue bushing (20) into the hot runner (19) made of steel (material number 1.2311). From four heated nozzles (21, 22, 24, 25), also of steel (material number 1.2162), it passes through feed channels (Fig. 4) to the various gates of the mold cavity. The gates measure 15 mm × 1.2 mm.

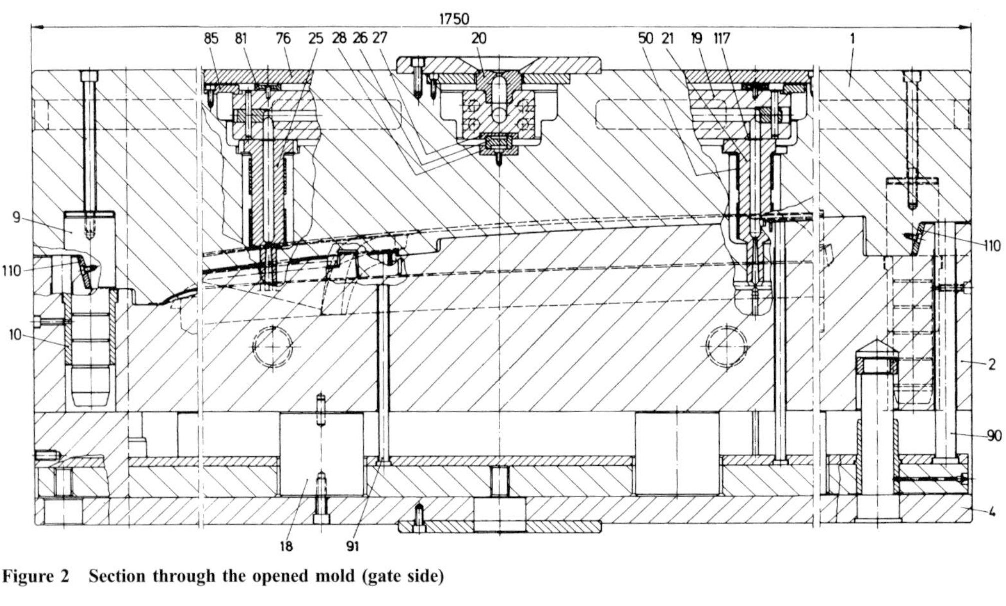

Figure 2 Section through the opened mold (gate side)

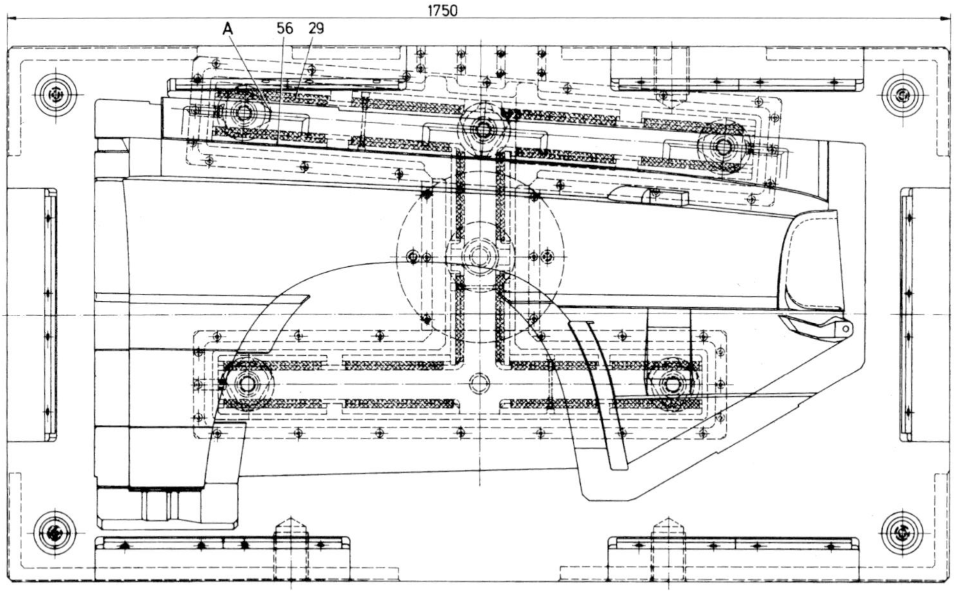

Figure 3 Plan view of the mold, section

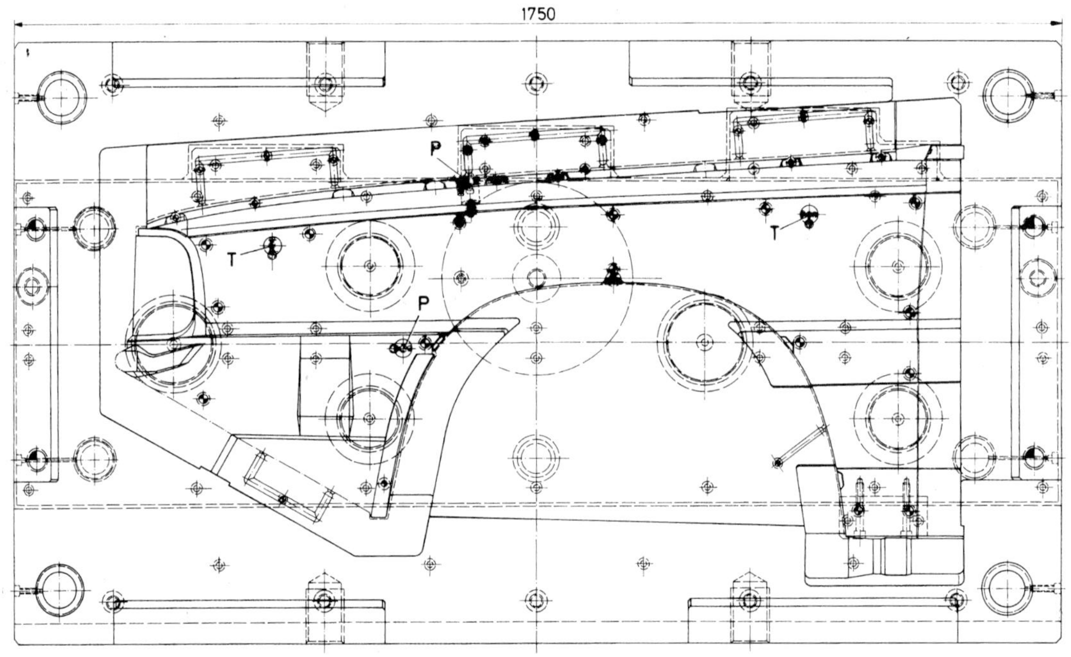

Figure 4 Section through the opened hot runner injection mold for a car front fender (clamping side)

Figure 5 shows the location of the feed system on the molded part and Fig. 6, the dimensions of the feed channels.

The H-shaped hot runner manifold block (19) is heated by cartridge heaters (29) which are pressed against it by laterally bolted cover strips (56). The distributor lies in a recess of the feed-side mold plate (1) and is pressed by the hot runner covering plate (76) against the nozzles via spacer rings (81). Guide components (85) prevent the manifold from twisting about the locating device (26, 27, 28). They also allow the manifold to expand and contract during heating and cooling. The hot runner nozzles are each centered in their locator bores in mold plate (1) by means of three contact surfaces (Fig. 2) in such a way that they do not experience the expansion and contraction undergone by the manifold. The contact surfaces facing the manifold are convex. Transfer areas for the molding compound are sealed with hot runner O-rings (117). The ends of the hot runner

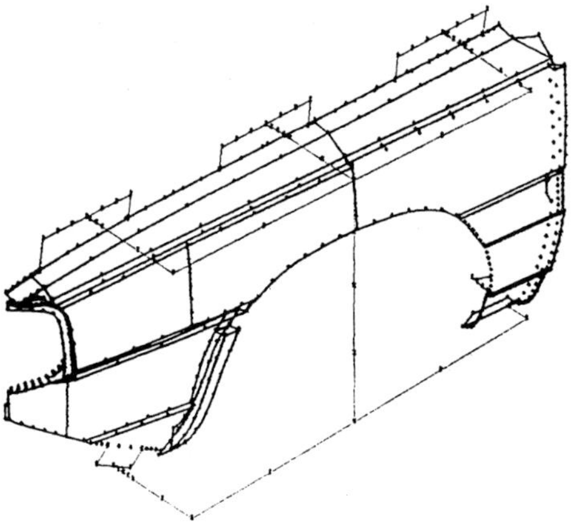

Figure 5 Fender plus gating system

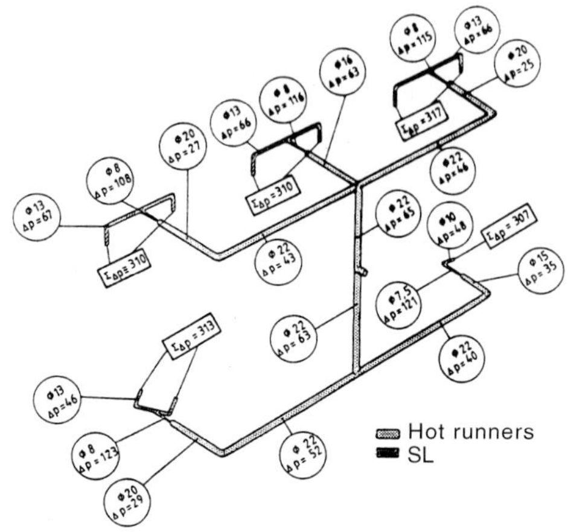

Figure 6 Balanced gating system (diameters in mm, pressures in bar)
Company illustrations: SMV Formenbau, Ellerau, Germany

nozzles on the mold side have conical bores. In this region, they fit snugly into the mold platen to ensure good thermal conduction and thus rapid cooling of the sprue formed therein.

On mold opening, the frozen sprue separates from the hot melt in the manifold at the narrowest point of the conical bore. The nozzles are heated with heating collars (50).

Two temperature probes are attached to each nozzle to monitor and regulate the temperature. The cartridge heaters in the manifold are arranged in seven groups, each of which is individually regulated. The cartridges have an installed rating of about 40 kW.

The melt pressure in the manifold is registered by pressure sensor P (136), and that in the cavity, by sensor P (130), and can therefore be used for control purposes. The temperature-control system consists of a network of bores in the two mold halves. These bores are not shown in Figs. 1 to 4. Heat sensors T (134) measure the temperature in the mold.

Ejection of the Car Front Fender

When the mold opens, the molded part remains in the moving mold half. The sprues are also held in place by sprue pullers above the ejector pins (93). As soon as the stripper plates (5, 6) are pushed forward by the ejector of the machine, the ejector pins (91, 93) push the molded part along with the sprues from the mold core. On mold closing, the ejector plates and thus also the ejector pins are pushed back by ejector-plate return pins (90). The mold plate (2) is supported by rolls projecting right through the ejector plates (18) to the clamping plate (4).

Literature

1. Bangert, H.; Dung, T.; Staeblein, P.: Kunststoffe 77 (1987) 12, p. 1227–1231

Example 9, Injection Mold for Magnifying Glass Frame with Handle

Examples of injection moldings with inner peripheral grooves are frames for spectacles and magnifying glasses (Fig. 4). Grooves may also be required for instrument housings that have to be glazed. In standard practice, two solutions exist for designing articles of this nature.

- The part of the mold in which the groove is formed (actually a plate) is divided into four segments that are withdrawn in pairs with the aid of toggles.
- While it is still capable of plastic deformation, the injection molded part is forced out of the segment of the mold in which the groove is formed.

The first case demands a mold of complicated design and therefore incurs high costs. Owing to the toggles and pivots required, it is susceptible to breakdown. In the second case, the rough treatment of the molded part may cause easily deformation of the groove. This entails high reject rates, or, if correction is possible, considerable expense.

The design shown in Figs. 1 to 3 overcomes both disadvantages. It represents a two-cavity mold for the production of magnifying glass frames. At four points on the outside of the molded part, there are small lenticular protrusions: three on the frame itself and one on the handle. They allow a grip for the ejector pins (11) and are subsequently removed.

There is a spiral slot on the plate (19) in which the peripheral groove is formed. The slotted plate is secured in the center by two bolts with hemispherical heads. A helical spring is thus formed. When the molded part is ejected, this spring is withdrawn from the peripheral groove without causing any deformation.

A mold of this comparatively simple design has given good results in practice.

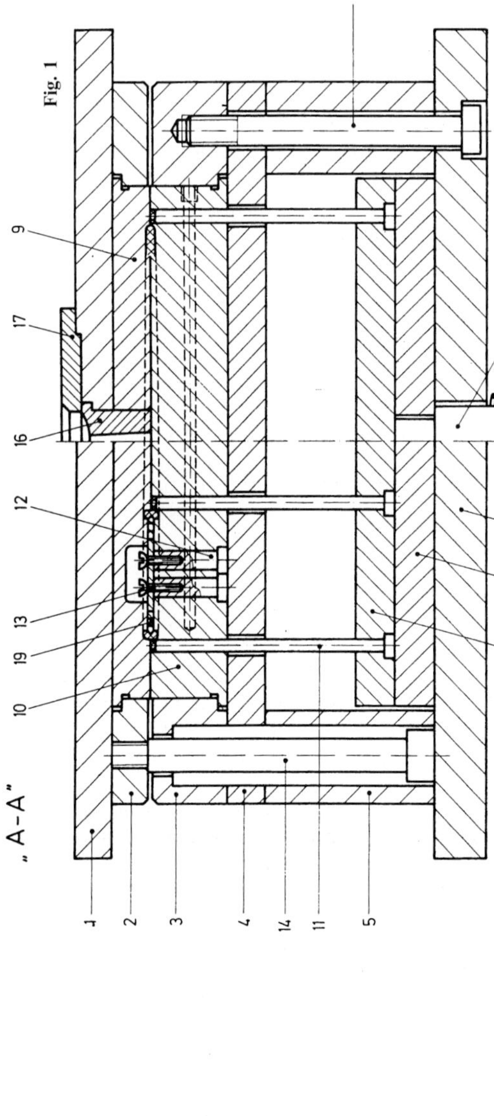

Fig. 1

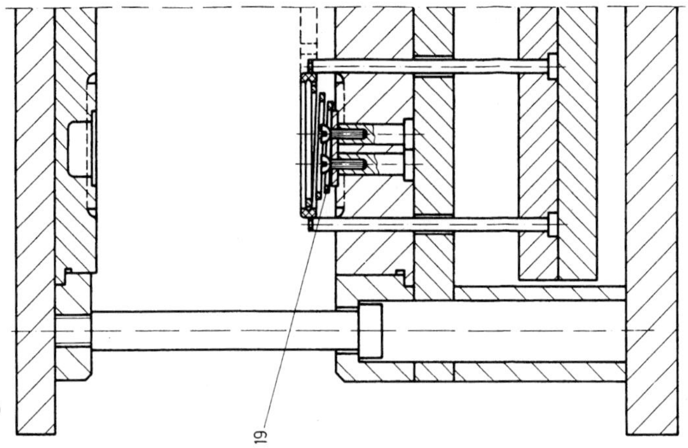

Fig. 2

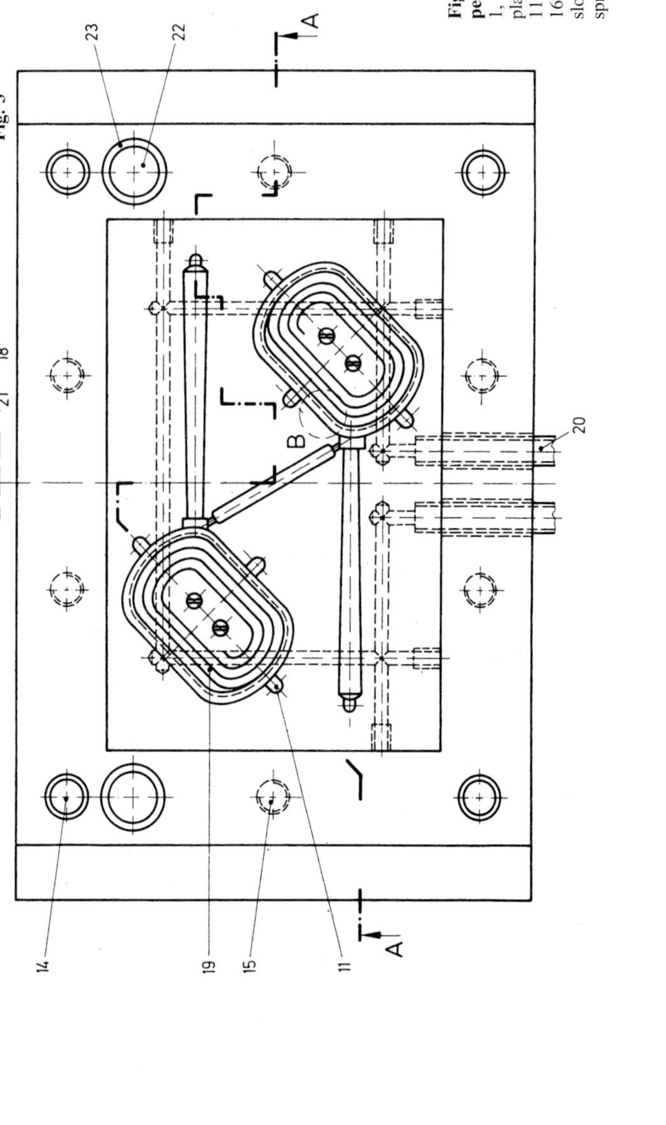

Fig. 3

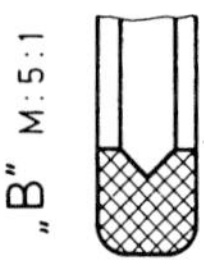

Fig. 4

Figures 1 to 4 Tooling for the injection molding of articles with inner peripheral grooves; Figure 2 shows the mold in open and ejection position
1, 2: upper clamping plate; 3: body of mold; 4: base plate; 5: spacer; 6, 7: ejector plate; 8: lower clamping plate; 9: fixed mold insert; 10: lower moving mold insert; 11: ejector pin; 12: bolts; 13: bolts with hemispherical heads; 14: stop; 15: bolt; 16: sprue bushing; 17: washer; 18: ejector rod; 19: spring steel plate with spiral slot for forming the peripheral groove; 20: connection for cooling water; 21: spring; 22: guide pins; 23: bushings for guide pins

Example 10, 16-Cavity Hot-Runner Mold for Cover Caps with Segmented Internal Contours Made from Polypropylene (PP) or Polyethylene (PE)

Cover caps are staple goods in packaging technology, mass-produced in technically sophisticated multi cavity molds. In general, they are articles with high standards of surface quality and design. Top-quality, decoratively textured surfaces and internal undercuts are specified for the cover cap presented here.
The article has a diameter of approx. 22mm, a height of 11mm and a wall thickness of 0.7/0.8mm and a shot weight of 1g. Cycle times of 10 s are achieved (Fig. 1).

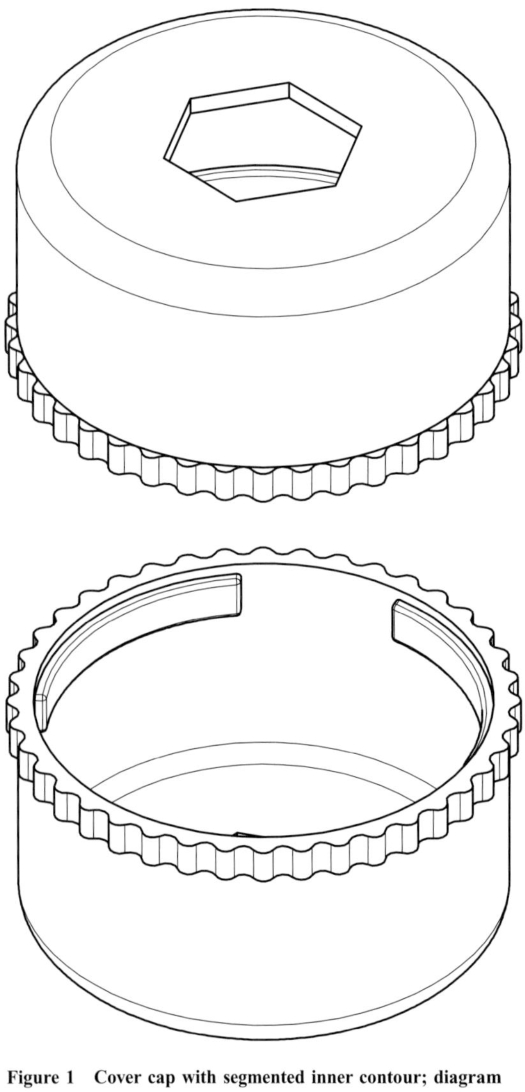

Figure 1 Cover cap with segmented inner contour; diagram

Mold

The concept of the mold is basically a 3-plate system, Fig. 2. To expedite mold designing and machining, standardized mold plates are used that fit the mold dimensions 446 × 496mm. The standardized platens and components are available for this CAD design as 3D volume elements in modular software programs from the standard units' manufacturers. The cavity plate on the nozzle side supporting the inserts (1) is screwed together from the parting line in order to make the hot-runner nozzles easy accessible.

Gating

The insertion geometries for the 16 self-closing sprue nozzles (2) are incorporated into the hardened mold inserts (3) according to manufacturer specifications. Because of the hexagonal opening in the middle of the article, the gate position is laterally offset.
The externally heated hot-runner manifold (4) is naturally balanced. The four heating circuits are individually controlled. The distributor bush (5) is heated by a band heater (6). The heat losses due to convection into the mold are minimized by titanium support disks (7). The entire hot nozzle side is protected on the side to the machine clamping plate by an insulation sheet (8). The high surface quality requirements on the mold part made a needle shut-off system necessary. The shut-off cylinders (9) are mounted in the clamping plate (10) made from prehardened 1.2312 steel. The sixteen cavities are closed by pneumatically operated needles with cylindrical gate bores. The result is a perfectly smooth gate.

Demolding

The cover cap is laid out with three segments forming an undercut in the inner diameter region. Standardized collapsible cores are used for molding and demolding these contours (11).
The undercut contour is formed in the front region of the collapsible core. The segments of this assembly are screwed into the core retainer plate (13) with the mounting elements (12). The conical inner cores (in 11) are held in the ejector-side clamping plate (14) and positioned by a dowel pin (16). The first demolding sequence is initiated by the stripping plate (17) moving forward with the core retainer plate on two diagonally mounted latch

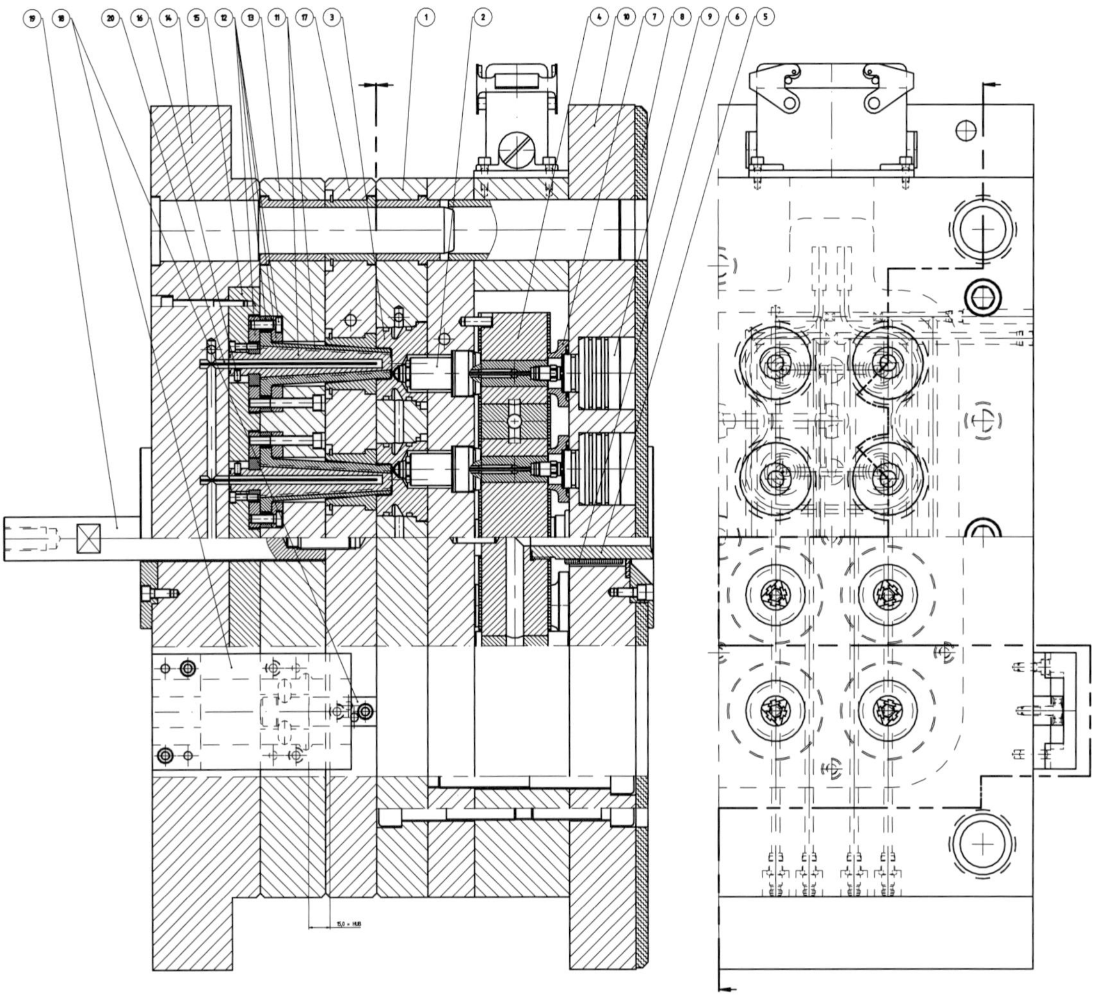

Figure 2 16-cavity hot-runner mold with needle-shutoff nozzles
1: cavity plate FS, 2: needle-shutoff nozzles, 3: mold insert FS, 4: hot-runner manifold, 5: distributor bush, heated, 6: nozzle heater band, 7: support disk, 8: thermal insulation sheet, 9: pneumatic cylinder, 10: clamping plate DS, 11: collapsible core, 12: mold insert BS, 13: mounting elements, 14: core retainer plate, 15: clamping plate BS, 16: internal core retainer plate, 17: stripper plate, 18: latch lock, 19: ejector rod, 20: core head cooling.
(Courtesy: Hasco, Lüdenscheid)

locks (18). This stroke motion is performed via the ejector rod (19) that is connected to the ejector system. During this sequence, the V-guided profile cores and V-ledges (in 11) leave their injection position and collapse over the inside core taper. After approx. 15mm stroke, the undercuts have been demolded. In the second demolding sequence, the latch locks release the block on both platens so that the stripper plate can remove the article.

Cooling

The small tolerances permitted for the molded part and the 10 s cycle time are made possible by very intense circulation cooling in the mould platens and the separation of insert cooling into several circuits. This has been contributed to by dividing the serial core head cooling of the collapsible cores into four groups (20).

Example 11, Four-Cavity Injection Mold for a Housing Made from Acrylonitrile-Butadiene-Styrene (ABS)

The four-cavity mold is used to produce two each of the upper and lower halves of a cosmetic device housing. The two halves of the housing are joined together by means of snap fits. These hook-shaped connections form internal and external undercuts on both parts that are released by lifters.

Mold

The four cavities are arranged in a rectangle at the mold parting line (Fig. 3). The mold inserts (11, 14) are made of hardened steel; the lifters (7, 8, 20) are case-hardened.
The lifters move sideways in the ejector plates (9, 10) and are attached to slide blocks (12) that run in corresponding guide grooves. The four leader pins for the two mold halves as well as the guide pins for the ejector plates are not shown in the drawing.
Mold dimensions are 296 mm × 547 mm, with a mold height of 290 mm. The mold weighs approximately 280 kg.

Runner System/Gating

The melt flows from a heated sprue bushing (13) through an H-shaped runner system to the four submarine gates feeding into the cavities (Figs. 1, 3).

Mold Temperature Control

The cavity inserts are provided with cooling channels for temperature control, while the cores contain bubblers in which baffles (19) ensure that the cooling water is directed to the tip of the bubbler well. The temperature in the cavity inserts is monitored with the aid of thermocouples (16).

Part Release/Ejection

Upon opening of the mold, the molded parts and runner system are retained on the core half. As soon as the machine ejector actuates the ejector plates (9, 10), the runner system and molded parts are released and pushed off the cores by the lifters (7, 8, 20) and ejectors (17). The submarine gates shear off, while the lifters release the undercuts formed by them.
Inside the housing halves (H) there is an undercut boss formed by the end of a core pin (18) and the surrounding hole in the core insert. The core pin (18) is fitted into the core insert with a certain play (V). As the parts are ejected, this pin is carried along by the molded part until the boss is clear of the hole in the core insert. The core pin (18) now stops and the boss is stripped off the end of the core pin by the remaining ejector motion. Before the mold closes, the ejectors must be pulled back.

Fig. 1

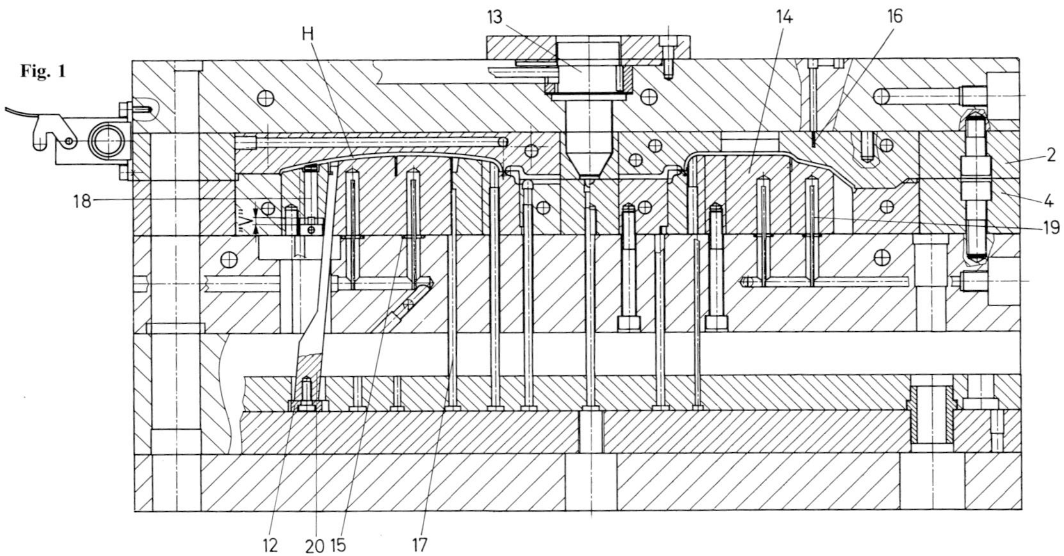

Fig. 2

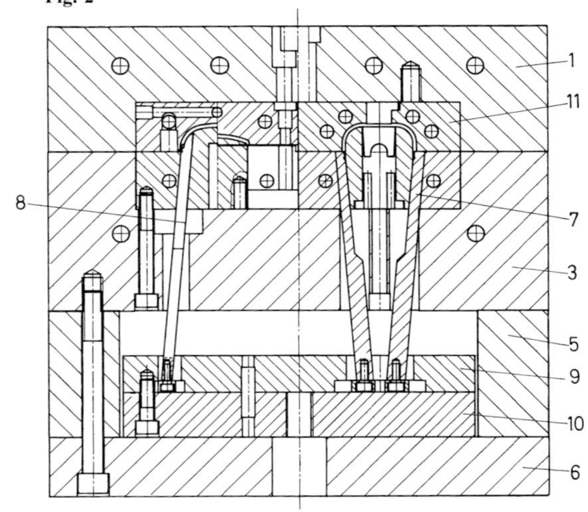

Fig. 3

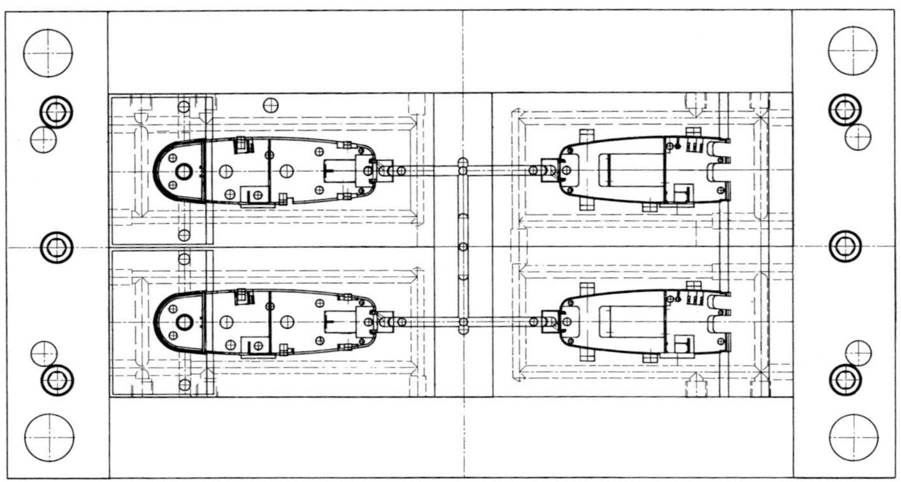

Figures 1 to 3 Four-cavity injection mold to produce a housing with integrated snap fits
1: mold plate; 2: bar; 3: mold plate; 4, 5: bars; 6: clamping plate; 7, 8: lifters; 9, 10: ejector plates; 11: mold insert; 12: sliding block; 13: heated sprue bushing; 14: mold insert; 15: O-ring; 16: thermocouple; 17: ejector; 18: core pin; 19: baffle; 20: lifter

Example 12, Four-Cavity Injection Mold for a Nozzle Housing Made from Polyamide

The nozzle housing for a windshield washer system has three interconnected channels and two external mounting clips. Core pulls and slides are required to release the molded parts.
The nozzle housing (Fig. 1) is used to distribute the water coming from the water pump to the two spray nozzles of an automobile windshield washer system. The hose from the pump is slipped over the conical tip on the end of a tubular extension A. The channel in this extension divides inside the housing into two channels D that lead to the spray nozzles. A hook H and a snap spring F are used to attach the housing. The ribs R serve to stiffen the tubular extension.

Mold

The mold (Figs. 2 to 4) contains four cavities. To release the hook and snap springs, four pairs of slides (12, 13) actuated by cam pins (28) are used. The cavity surface is formed by the slide inserts (32). The slides are held in place in the closed mold by the heel blocks (9) and wear plates (10).
The channels D are formed by core pins that are attached to a total of eight slides (22). The cam pins (18) are employed to actuate the slides, while the pins (26) hold the slide (22) in place when the mold is closed.
The channel in the tubular extension A is formed by the core pin (19), the conical tip of the extension by the bushing (23). The adjacent core pins (19) and bushings (23) on one side of the mold are attached to a bridge (15, 16). The base of the cylinder (24) is also bolted to the center of the bridge between the two core pins. The rod of cylinder (24) is attached to the backup plate (4). To pull the core pins (19) the piston in the cylinder is pressurized and the cylinder, along with the bridge and core pins, moves away from the mold. The bushing (23) are held in place by the pins (25).

The cavity insert (21) is made of through-hardened steel, while the slide inserts (32) and slides (22) utilize case-hardened steel.
The drawings were produced on a CAD system; Figs. 2 to 4 two dimensionally (2D), Fig. 1 three dimensionally (3D).

Runner System/Gating

The melt flows from the sprue bushing (11) into the mold via an H-shaped runner system through the cavity insert (21) to a pinpoint gate at the parting line associated with each pair of slides and enters the cavity at the hook H of the molded part (Fig. 3).

Mold Temperature Control

Cooling channels in the plates (1, 3 and 4) are provided for mold temperature control.

Venting

The ejector pin (31) is located in the cavity where air entrapment may occur. It thus also aids in venting.

Part Release/Ejection

Part release and ejection are accomplished by means of a latch mechanism (17) and stripper bolt (27) not described in further detail.
The mold first opens at parting line I, at which point the gates shear off the molded parts and are pulled out of the cavity insert (21). The runner system is held on the sucker pins (30). The cam pins (18)

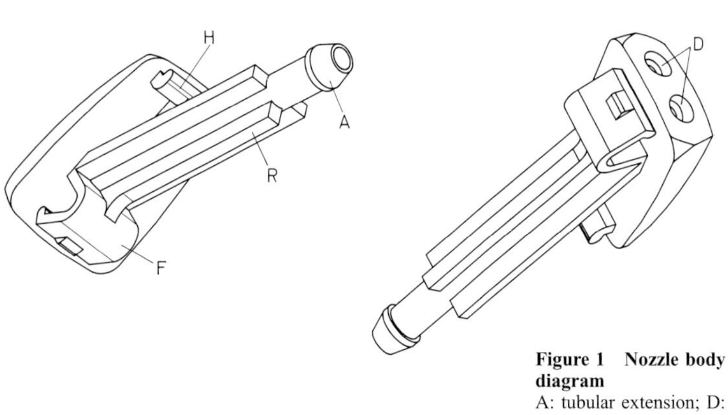

Figure 1 Nozzle body for an automobile windshield washer system, diagram
A: tubular extension; D: channels; F: snap spring; H: hook; R: rib

actuate the slides (22), which withdraw the short core pins from the molded parts.

As the mold opens at parting line II, the slides (12, 13) on mold plate (4) are spread apart by the cam pins (28). At the same time, the undercuts on the hooks and the snap springs are released along with the pins (25) blocking the hydraulic side cores, so that the core pins (19) can now be withdrawn. The parts are now ejected by ejectors (29) and (31). Finally, the runner stripper plate (2) is actuated by the stripper bolts (27) (parting line III) and the runner system is stripped off the sucker pins (30). Before the mold closes, the ejectors must first be retracted and then the hydraulic cores set.

Fig. 2

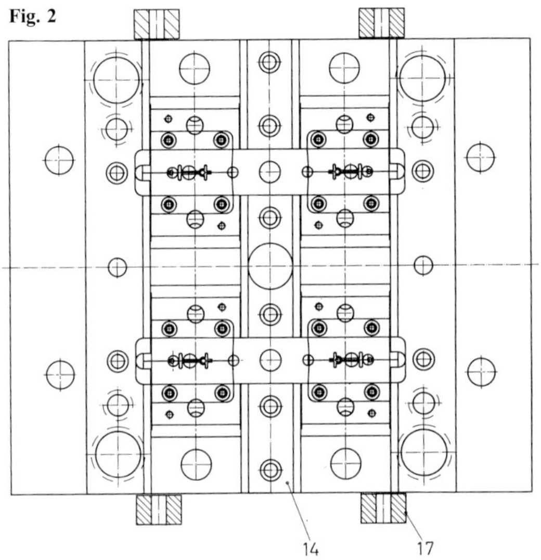

Fig. 4

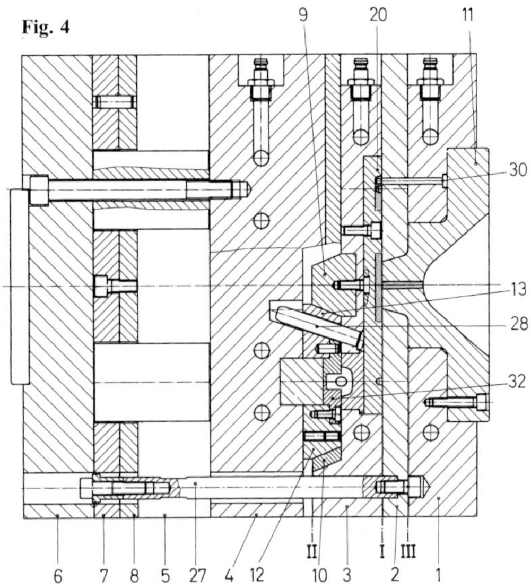

Fig. 3

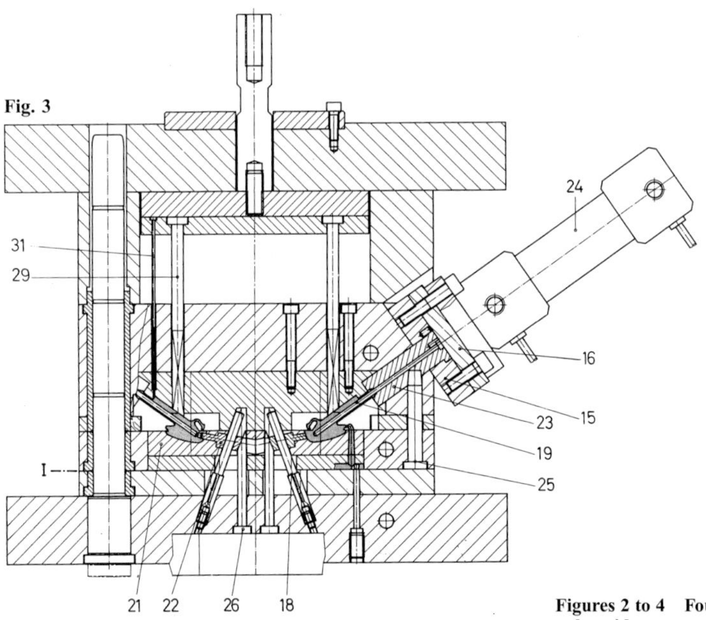

Figures 2 to 4 Four-cavity injection mold for a nozzle housing of polyamide

1: clamping plate; 2: runner stripper plate; 3: mold plate; 4: backup plate; 5: bar; 6: clamping plate; 7, 8: ejector plates; 9: heel block; 10: wear plate; 11: sprue bushing; 12, 13: slides; 14: slide guides; 15, 16: bridge; 17: latch mechanism; 18: cam pin; 19: core pin; 20: runner plate; 21: cavity insert; 22: slide; 23: bushing; 24: hydraulic cylinder; 25, 26: locking pins; 27: stripper bolt; 28: cam pin; 29: ejector; 30: sucker pin; 31: ejector pin; 32: slide insert

Example 13, Single Split Cavity Mold for a Threaded Plug Made from Polyacetal (POM)

The threaded plug is a cylindrical body 65 mm in diameter and 23 mm high with a trapezoidal thread having a pitch of 3.5 mm. A split cavity is used to form the threads. The necessary number of split cavity segments depends on the thread pitch and its profile as well as on the material used to mold the part.
Figure 1 shows the plan view of a thread with a trapezoidal profile. If one attempted to form this thread in a split cavity with two halves, i.e. the parting line lay in the plane of the figure, the mold would damage the thread upon opening, because of the undercuts at the positions H.

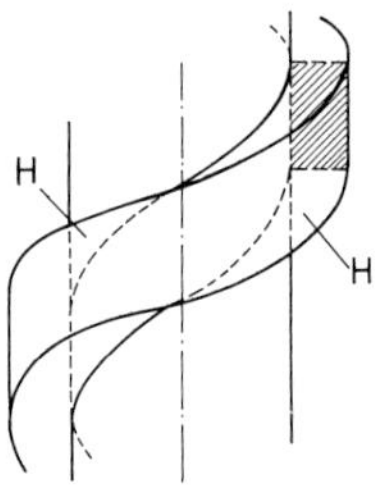

Figure 1 Thread with rectangular profile (plan view)

The more pronounced the flanks of the thread profile are inclined (trapezoidal/triangular thread) and the smaller the thread pitch and depth, the smaller are the undercuts. With injection molding, the size of the undercut is decreased by the shrinkage of the resin up to the moment of ejection. In addition, many resins are still elastic enough to withstand minor deformation without damage. If in spite of all these factors the undercut is still too large, the number of segments forming the split cavity must be increased.
The investigation of the situation with regard to undercuts and the determination of the necessary number of cavity segments is best carried out with the aid of a computer which can be used to search out the regions endangered by the undercuts on the basis of the thread geometry.

Mold

The present mold (Figs. 2, 3) has four cavity segments (1) each of which is actuated by two cam pins (2). The cavity segments are guided on the mold plate (3) and when closed are held in position by means of wear plates (4) attached to the mold plate (5). The mold is constructed of standard mold components.

Runner System/Gating

The part is gated centrally via a pneumatic nozzle (6).
Operation of the nozzle is described in Example 97 The sprue (8) in the nozzle tip (7) is attached to the molded parts via 2 pinpoint gates next to the central hole.
Before the mold opens, the sprue (8) is separated from the molded part and ejected by actuating the pneumatic nozzle (6).

Part Release/Ejection

As the mold opens, the four cavity segments (1) are spread apart by the eight cam pins (2) and the threads are released. Now the ejectors (9) can strip the molded part off the core.
Ejection takes place hydraulically via the molding machine. The molded part is blown off the ejectors (9) by a blast of air.

Mold Temperature Control

The mold plates (3) and (5) as well as the cavity segments (1) are provided with cooling channels. The hollow core (10) contains a cooling insert (11) with grooves to guide the cooling water.

Fig. 2

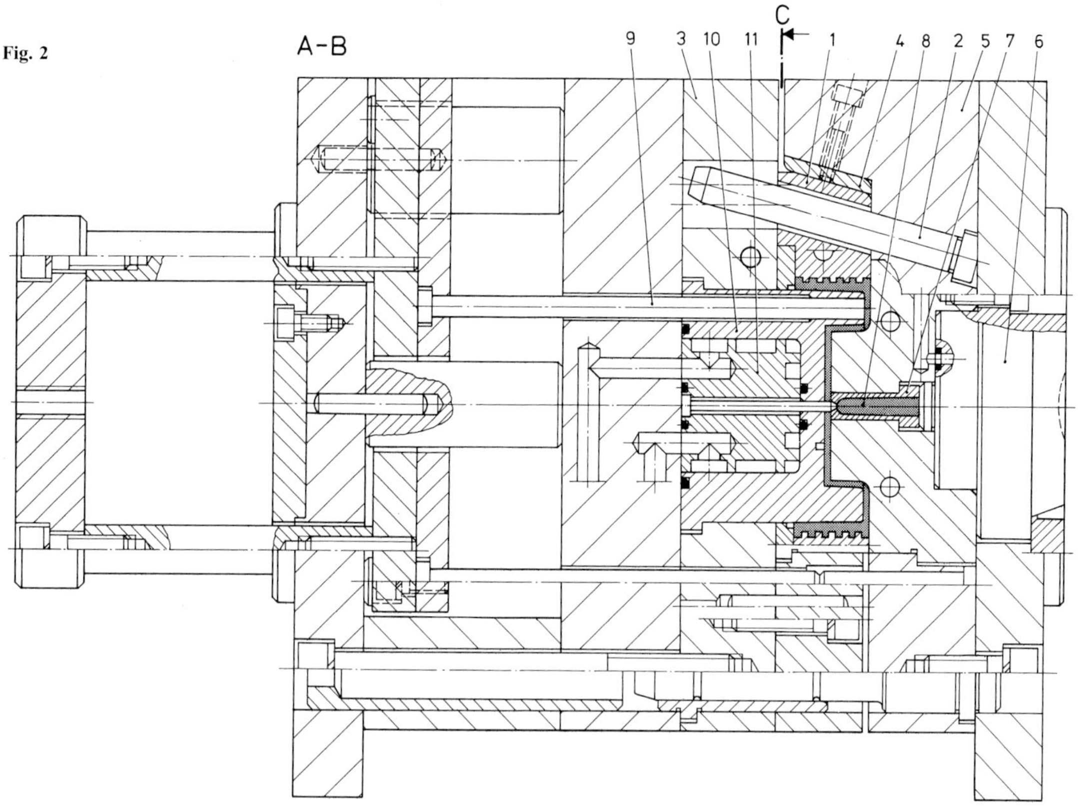

Fig. 3

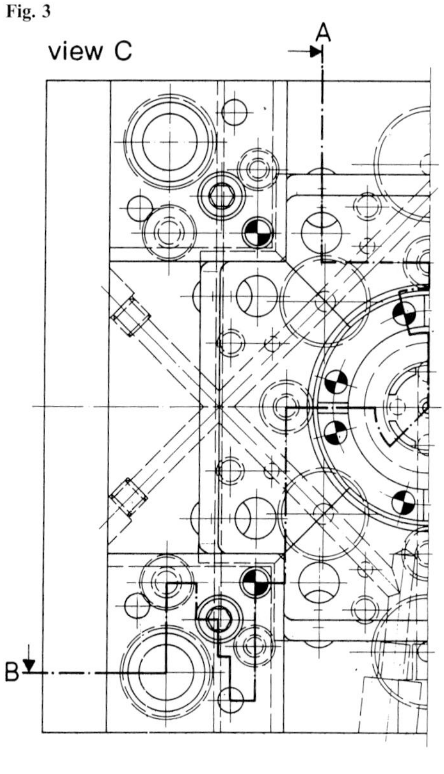

Figure 4 Threaded plug with sprue

Figures 2 and 3 Single split cavity mold for a threaded plug of POM

1: cavity segments; 2: cam pin; 3: mold plate; 4: wear plate; 5: mold plate; 6: pneumatic nozzle; 7: nozzle tip; 8: sprue; 9: ejector; 10: hollow core; 11: cooling insert

(Courtesy: Hasco)

Example 14, Demolding a Polyethylene Container with External Undercuts

The twenty-liter container (US Patent 4648834) shown in the mold drawing (Fig. 1) has several external rims that normally require side action in the mold to be released.
Such side action significantly increases the cost of a mold. This example shows that with clever use of the shrinkage of the molded part and for moderate undercut depth mold costs and manufacturing time can be saved while simplifying mold handling (weight, volume, mechanics).

Mold

The mold consists of a cavity half (1) and a core half (2) which are guided by means of leader pins (3) and aligned with respect to one another by means of a taper lock (4).
A ring (5) that forms the underside of the rim on the outside of the container is attached to the cavity half (1).
Stripper rings (7, 8) that give the shape of the external undercuts move on guide pins (6) attached to the core (2) and passing through the tapered alignment section (4). Stripper ring (7) is actuated by ejector rods (9), while stripper ring (8) is attached to stop bolts (10) that limit its motion.

Part Release/Ejection

Part release and ejection are described in Fig. 2 in four steps.

Step I

The mold has opened. The molded part and core have separated from the cavity.

Step II

The stripper ring (7) is pushed forward by the ejector rod (9). The molded part is pushed off the core, while the undercuts continue to pull stripper ring (8) along. The taper of the core now releases the inner surface of the molded part, the diameter of which decreases as the result of shrinkage, so that the rim formed in the ring (8) can now be snapped out of the recess. The ring (8) now stops.

Step III

Stripper ring (7) now completely pulls the rim on the molded part out of the ring (8).

Step IV

Ring (7) comes to a stop. The molded part is ejected by air.
The air used to eject the part is directed to the inside through the valve insert (11). Before the mold opens, compressed air is forced under the bottom edge of the container through the cavity half in order to facilitate removal from the cavity.

Example 15, Injection Mold with Reduced Opening Stroke for Milk Crates from Polyethylene

Beverage crates (US Patent 4731014) usually have a grid-like structure on their exterior surfaces as a result of the stacking rim, reinforcing ribs and handles, the release of which requires the injection mold to have external slides (side action). If the slides are located in the stationary cavity half of the mold, the opening stroke required equals twice the crate height plus the axial stroke of these slides in order to be able to eject the molded part.
The ejection principle described here needs a shorter opening stroke. It is thus well suited for especially deep parts or for stack molds.
The milk crate shown in Fig. 1 has dimensions of 300 mm × 300 mm and a height of 280 mm. Its grid-like structure forms external undercuts.

Figures 2 to 4 illustrate the ejection principle along with the additional possibility of releasing internal undercuts (on the core).
The mold (Fig. 2) consists of the core (1) with core lifters (2), cavity bottom plate (3) with sprue bushing (4) and the cavity frame (5) with movable external slides (6). The cavity frame (5) can be moved in the direction of mold opening by means of hydraulic cylinders (7).
During opening (Fig. 3), the cylinders (7) hold the bottom plate of the cavity (3) and cavity frame (5) together. The molded part (8) is held in the cavity by virtue of its external undercuts; the core (1) is withdrawn. Any undercuts on the inside of the molded part are released by the displacement of the

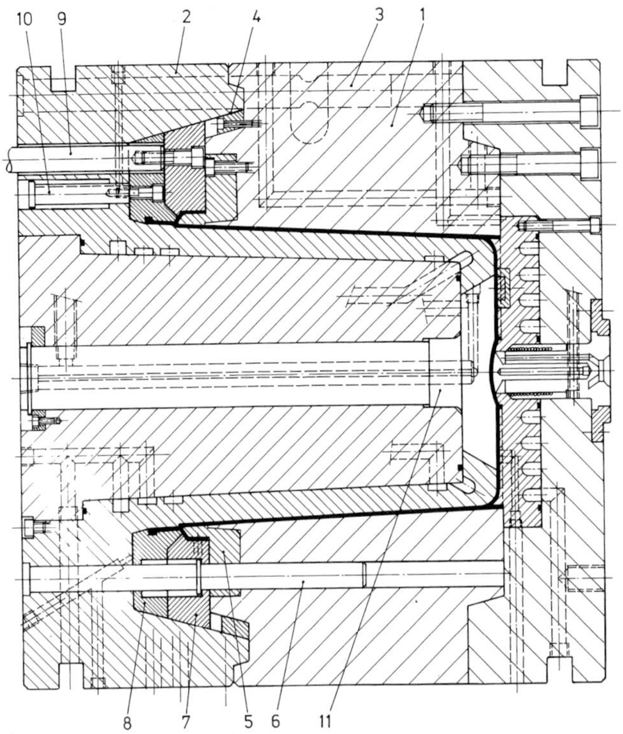

Figure 1 **Mold for a 20-liter container with external undercuts**
1: cavity half; 2: core half; 3: leader pin; 4: taper lock; 5: ring; 6: leader pin; 7: stripper ring; 8: stripper ring; 9: ejector rod; 10: stripper bolt; 11: valve insert

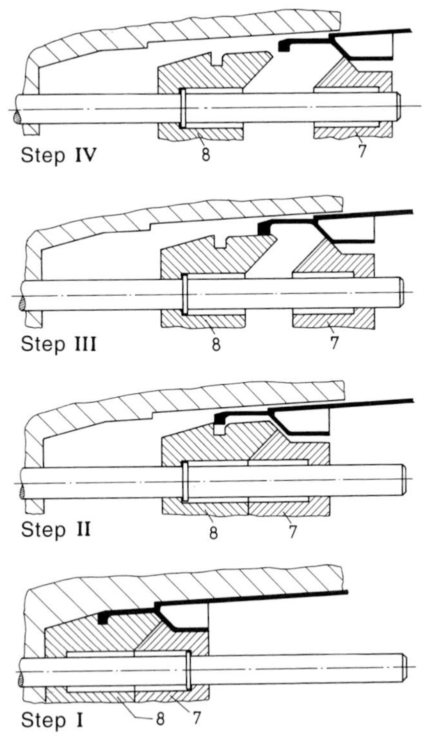

Figure 2 Ejection sequence

Example 14

Example 15

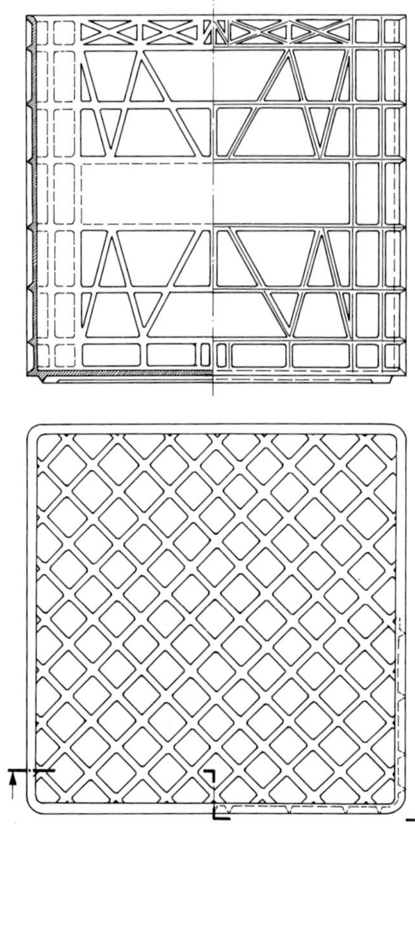

Figure 1 Milk crate

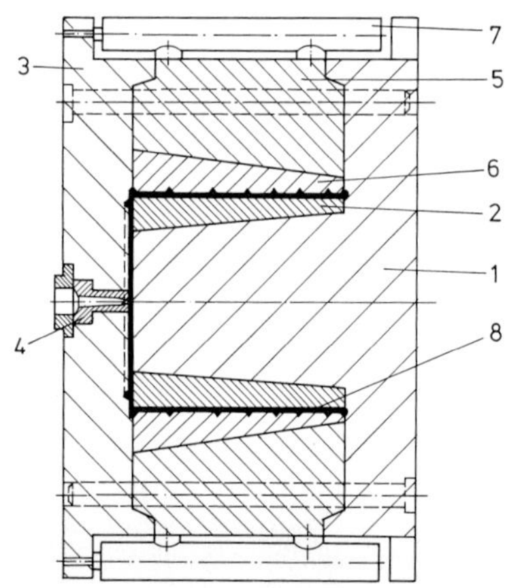

Figure 2 Single-cavity mold for a milk crate
1: core; 2: core lifters; 3: bottom plate of cavity; 4: sprue bushing; 5: cavity frame; 6: external slides; 7: hydraulic cylinders; 8: molded part

Fig. 3

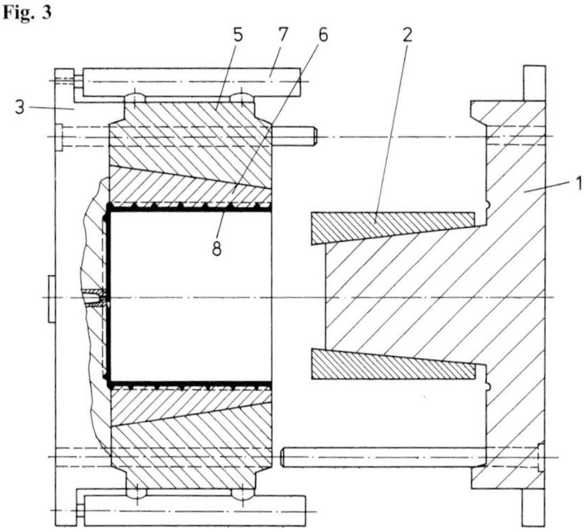

Fig. 4

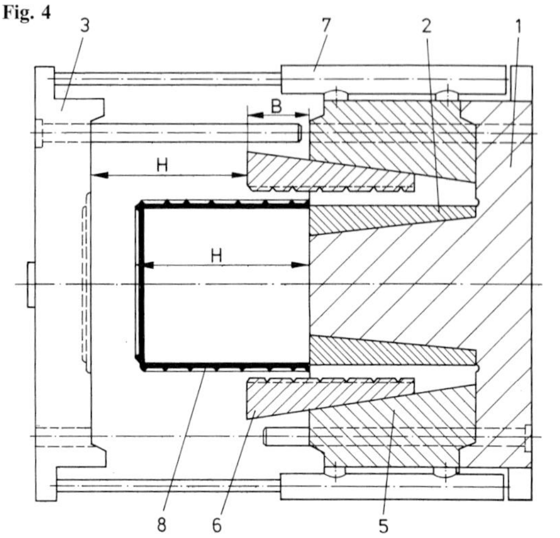

Figures 3 and 4 Single-cavity mold for a milk crate
1: core; 2: core lifters; 3: bottom plate of cavity; 4: sprue bushing; 5: cavity frame; 6: external slides; 7: hydraulic cylinders; 8: molded part

core lifters (2) on the core (1). The molded part can now shrink freely and becomes smaller than the cross-section of the core. The cylinders (7) then push the cavity frame (5) toward the core (Fig. 4). The rim of the molded part – which is now smaller – pushes against the core lifters (2) or core (1) in which case the core lifters (2) – if present – are pushed back.

The external slides (6) located in the cavity frame (5) do not follow the axial movement of the frame until they are far enough apart to release the external contour of the molded part (8). The molded part can now drop free.
The opening stroke of the mold is thus only somewhat larger than the crate height H plus the distance B required for the side action.

Example 16, Two-Cavity Injection Mold for Recessed Refrigerator Handles Made from Polyamide

A two-cavity injection mold had to be made for injection molding recessed handles for refrigerators of polyamide reinforced with 35 wt.% glass fibres. The recessed handles (Fig. 1) have a grooved internal structure, three flat channels from the outside to the inside, two metal inserts to be encapsulated, and recesses, into which the case of the refrigerator door engages when the handle is mounted.

Construction of the Mold (Figs. 2 to 5)

Because of the function of the molded part the main axis of the handle indentation is set at an angle of 45° to the recesses which engage with the case walls. Since the recesses and the attached and encapsulated metal inserts are to demold on opening the mold (Fig. 2), an ejection motion with an angle of less than 45° to the handle must release the molding

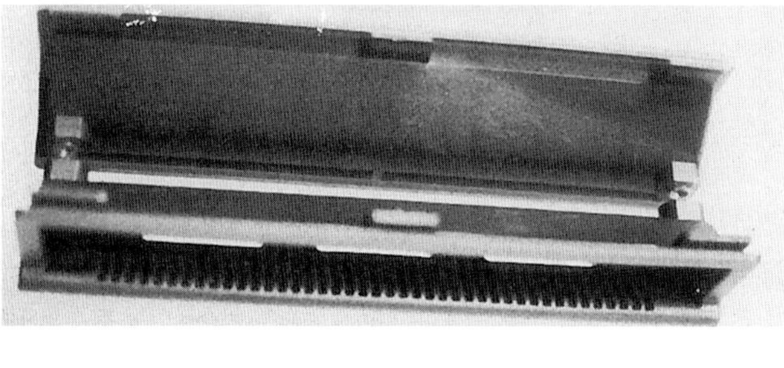

Figure 1 Recessed refrigerators handles of polyamide reinforced with 35 wt.% glass fibers and two metal inserts
top: front; bottom: rear

from the core (11). Further, a mechanical slide (13) is required for releasing the flat channels and the beaded edge of the mold.

Ejector Mechanism

The handles must be pushed away from cores (11) without any tilting; thus, hydraulically operated ejectors are not acceptible since, because of possible differences in forward motion, they do not guarantee exactly parallel guidance. It was decided to operate the ejector by means of rack and pinion mechanisms (23), which are driven through pinions (19), shaft (25), external gear wheel (21), and racks (24), by the opening movement of the mold. In order to ensure the necessary delay in the ejector motion until release of the molding by the mechanical slide (13), the block (39) in the top half of the mold, which encloses the outer racks (24), runs freely along a distance of 24 mm in the recess in the mold plate on the nozzle side, until meeting the stop. The loosely inserted spring (42), which is tensioned by mounting the mold on the machine, acts as support. Only when the block (39) is stopped by the spring (42) on the opposite side does the relative movement of the outer rack (24) begin, rotating the outer gear wheels (21), which again operate the internal rack drive.

The block (39), outer racks (24), outer gear wheels (21), and spring (42) were economically mounted in milled grooves on the top side of the mold, partially enclosed by the cover plate (41). The shafts (25) were mounted in bearings (36) under the outer gear wheels (21) to maintain a low bending moment in the spindles. Their exact position is achieved by bearings, fitting the inner racks (23) to the actual ejector (32), as well as by mutual displacement of the outer racks (24) made possible by means of slotted holes in these. The racks are finally connected to the block (39) by pins (40). Subsequently, the outer racks are finally calibrated along their length in order to ensure a precisely defined ejector position in the case of a closed mold.

The slides (13) are made of steel with material no. 1.2541, while the mold components (10, 11, 12) utilize steel no. 1.2343.

Runner

The sprue opens into an S-shaped runner formed in the cavity block (12). The S-shape provides a central sprue for both cavities, which are displaced because of the rack and pinion ejector mechanism. An overlapping gate connects with a central lug of the respective molded part, which is concealed when the handle is mounted, so that the mark caused by this is unobtrusive.

Mold Operation

As the opening motion begins, the mechanical slides (13) are moved outward by the cam pins (15) and release the three flat channels. Simultaneously, the sprue begins to be released from the sprue bushing (31). After an opening distance of 24 mm the open recesses of the molded part and the metal inserts are withdrawn from their cores. Then the motion of the outer racks (24) begins relative to the ejector-side mold half. The ejectors (32) are advanced, effecting a movement of the molded part at an angle of 45° to the mold axis. The resulting movement vertical to the mold axis pulls off the overlapping gate. The axial component of the ejection movement carries with it the strip (10), such that after a distance of 14 mm the recesses formed by the strip (10) are also released. The moldings are now pushed further until they fall from the core (11). Finally, the sprue, which in the meantime has also been fully released, is also ejected by the machine ejector through the sprue ejector (27).

On closing the mold, the spring (42) ensures that the ejectors (32) have returned before the mold finally closes. The return pins (28) for the sprue ejector have the same effect, but in this case synchronously with the closing action.

Fig. 2

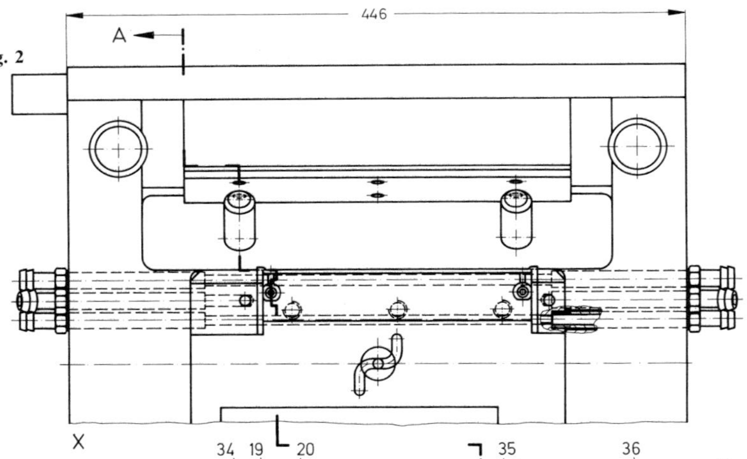

Fig. 3

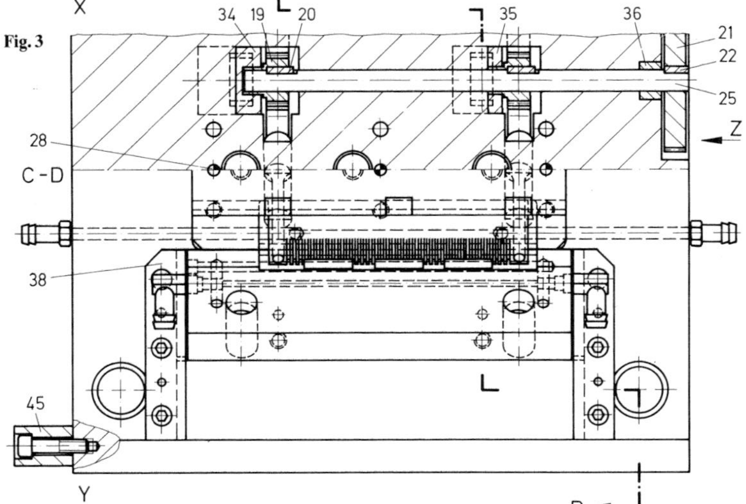

Fig. 4

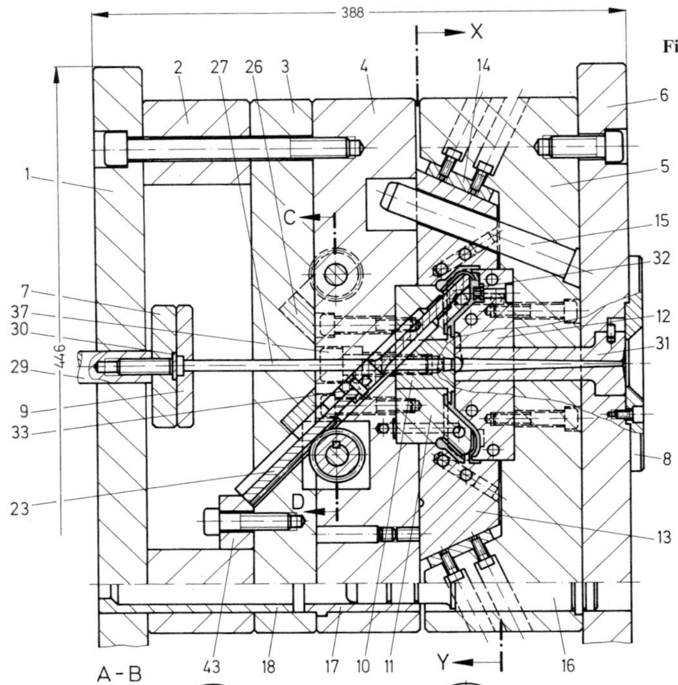

Fig. 5

Fig. 4

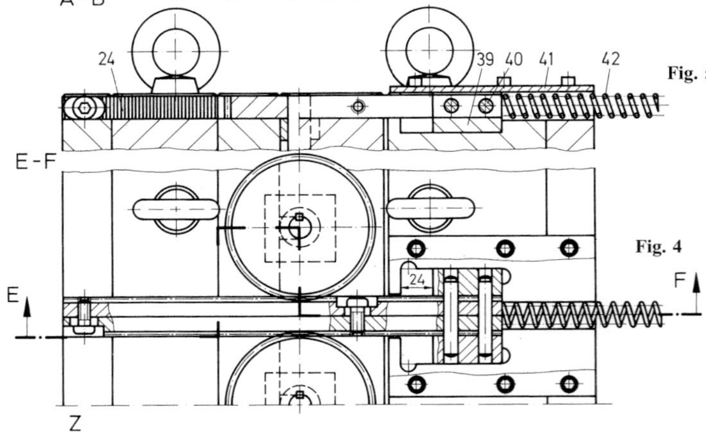

Figures 2 to 6 Two-cavity injection mold for recessed handles for refrigerators
1: moving-half base plate; 2: ejector frame; 3: moving-half mold plate; 4: core retainer plate; 5: fixed-half mold plate; 6: fixed-half base plate; 7: ejector plate; 8: fixed-half locating ring; 9: ejector retaining plate; 10: strip; 11: core; 12: cavity block; 13: slide; 14: ejector plate; 15: cam pin; 16: guide pin; 17, 18: guide bushing; 19: pinion; 20: feather key; 21: gear wheel; 22: feather key; 23, 24: rack; 25: shaft; 26: sliding block; 27: sprue ejector; 28: return pins; 29: ejector rod; 30: retainer; 31: sprue bushing; 32: ejector; 33: bolt; 34, 35, 36: bearing; 37: stop screw; 38: locating slide; 39: block; 40: pin; 41: cover plate; 42: spring; 43: stop; 45: spacer block

Example 17, Injection Mold for a Grass Catcher Made from Polypropylene

The grass catcher consists of two halves that are produced in a common mold and joined to one another by means of snap fits. A metal rod that functions as a hinge for the grass catcher cover is inserted when assembling the two part halves. The plug-in connection at the handle is secured by means of a self-tapping screw. The weight of the molded grass catcher (without steel rod and screw) is 1525 g, with wall thicknesses ranging between 2.5 and 4.5 mm. The outside dimensions are 440 mm × 370 mm × 315 mm. An injection molding machine with a clamping force of 10,000 kN is required to produce the grass catcher halves.

Figure 1 Grass catcher of polypropylene for a lawn mower

The injection mold shown in Fig. 2 is 1100 mm high, 790 mm wide and 884 mm long. The opening stroke is approximately 800 mm. With about 300 individual components, the total weight is 4.3 t.

Both grass catcher halves have been oriented in the mold in such a manner that the interiors are formed by cores (5, 6). The ejectors are located on the core half. The screen-like sections at the back of the grass catcher are formed by two mechanically actuated slides (39). In addition, two small slides (41) that are also mechanically actuated serve to release the undercuts associated with the snap hooks in this area. Two additional snap hooks for the rear screen-like section are formed by the slide in this mold half. Undercuts in the direction of draw in the interior of the grass catcher are released by a total of six lifters (26, 27) which simultaneously push the parts off the cores (5, 6) during ejection.

Runner System/Gating

Each grass catcher half is filled via two submarine gates located on the outside lower surface and connected to a four-arm runner system. This runner system is machined into the core half and connected to the machine nozzle via a heated sprue bushing (35, 36). The heated sprue bushing has five heater bands (37) with a total heating capacity of 500 W.

Mold Temperature Control

Water lines 15 mm in diameter are provided wherever possible for mold temperature control. Connections to this system of water lines are made via quick-disconnect fittings. Sections in which it was not possible to place water lines are cooled via bubblers.

An insulating plate is attached to the mold clamping plate (1) on the stationary half in order to avoid the undesirable heating of the machine platen by the heat lost from the heated sprue hushing (35, 36).

Part Release/Ejection

In addition to the lifters (26, 27), 34 ejector pins have been provided. The ejector plates (9) and (10) are located and guided by means of guide pins (19) and bushings (20). To increase the rigidity of the mold, the ejector housing contains seven support pillars (21) in addition to the spacer bars (7).

Figure 2 Injection mold for producing both halves of a grass catcher simultaneously

1: mold clamping plate; 2, 3: cavity insert; 4: core retainer plate; 5, 6: core insert; 7: spacer bar; 8: base plate; 9: ejector plate; 10: ejector retainer plate; 11: stationary-side locating ring; 12: movable-side locating ring; 13: ejector rod; 14: leader pin; 15: guide bushing; 16: insert; 17: return pin; 18: spring washers; 19: guide pin; 20: guide bushing; 21: support pillar; 22, 23, 24, 25: sliding plate; 26, 27: lifters; 28: guide block; 29: sliding stone; 30: insert; 31: ejector; 32: sleeve; 33: extension; 34: ejector retainer bushing; 35, 36: heated sprue bushing; 37: heater band; 38: insert; 39: slide; 40: cam pin; 41: slide

a: shown without insert ►

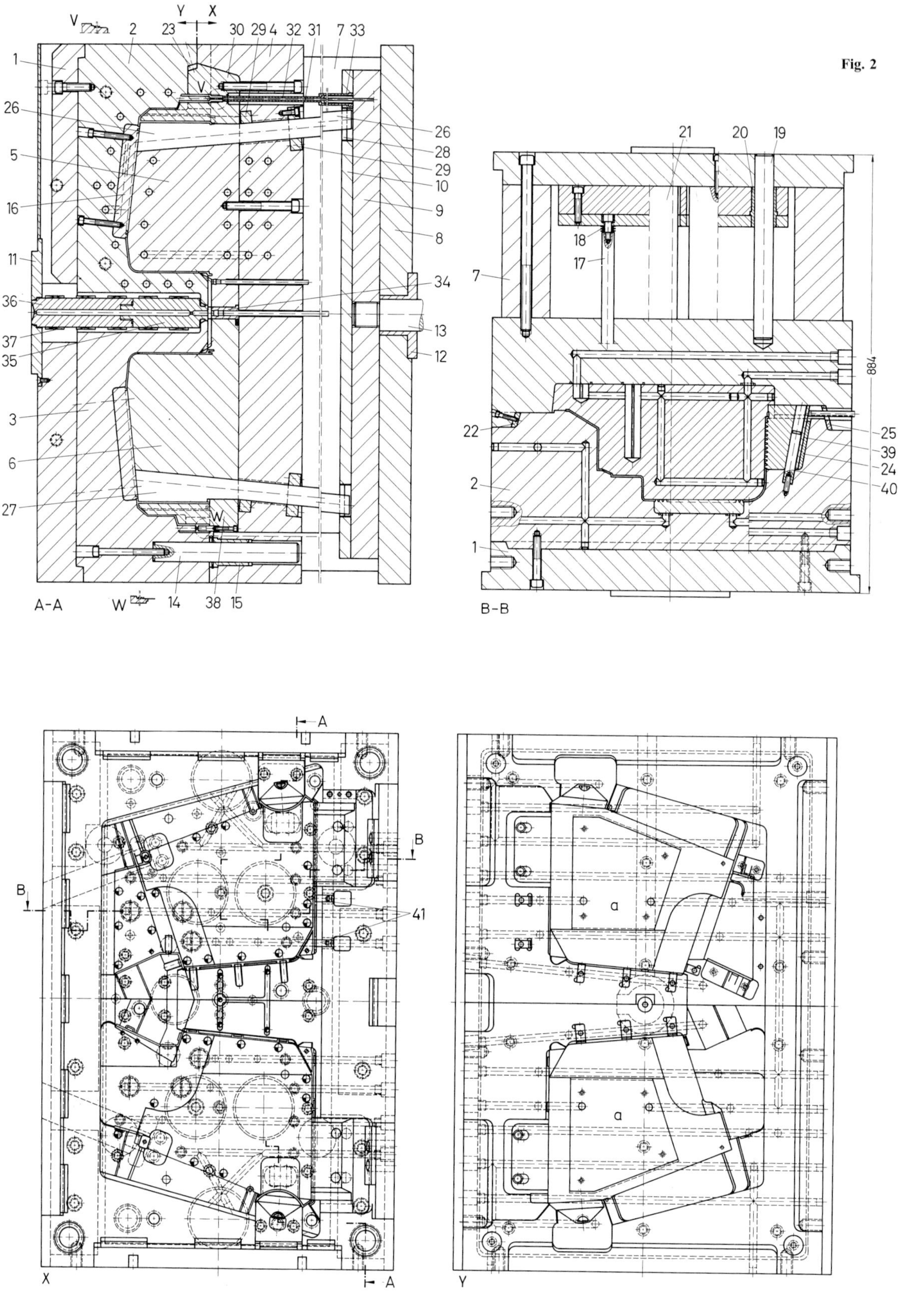
V
W
Y
X
1
2
3
4
5
6
7
8
9
10
11
12
13
14
15
16
17
18
19
20
21
22
23
24
25
26
27
28
29
30
31
32
33
34
35
36
37
38
39
40
41
884
a
A-A
B-B
A
B

Fig. 2

Example 18, Injection Mold for Hose Connectors Made from Polyamide 6.6

The object of the hose connector illustrated in Fig. 1 is to connect extensions to garden or household hoses which are too short or to repair broken ones. It consists of a center section and two compression nuts. The center section is designed as an outer cylinder with a concentric inner segment attached to it by means of a ring-shaped rib midway along the length of the part. Starting at this rib, the inner segment tapers conically to each end. The outer cylinder is provided with internal threads at either end (see Section C-D, Fig. 4). The ends of the hose are pushed over the conical section, to be compressed against them and clamped by the compression nuts. A single-cavity mold is used to produce this center section on an injection molding machine with a vertical injection unit.

Figure 1 Three-piece hose connection

Figure 2 Single-cavity injection mold for the center section of a hose connector

Mold

Threaded cores (13, 14) are installed in the fixed as well as in the moving half of the mold (Figs. 3 to 6). These cores can be synchronously operated by means of the gears (35, 36, 37) and the splined shaft (38), which extends through both mold halves. They are driven by an unscrewing unit attached to the shaft of the threaded core (13), which protrudes centrally from the fixed mold half. Slides (11, 12) are moved by cam pins (41) to release the outer surface of the molded part. The cavity is filled from above through a sprue machined into the mold parting line and the faces of the slides.

Operation of the Mold

Once the molding compound has cooled sufficiently, the core (13) is driven directly by the unscrewing unit via the shaft (19) and is unscrewed from the internal thread of the molded part with the aid of the threaded bushing (17) and lead threads on the threaded core (13). The turning motion of the core (13) is simultaneously transmitted to the core (14) by gears (35 to 37) and the splined shaft (38). Through the action of core (14), whose direction of rotation is opposite to that of core (13) due to the idler gear (36), and with the aid of threaded bushing (23) and the lead threads on the core (14), the threads in the moving half of the mold are released. Upon completion of unscrewing, the mold opens, the slides (11) and (12) releasing the outer surface of the molded part and the sprue. Finally, the molded part – still sitting on the cooled inner core (22) – is stripped off the latter by the stripper plate (15) and ejected. Once the stripper plate (15) has been returned by the hydraulic ejector of the molding machine and the cores (13) and (14) have been reset, the mold closes and another cycle begins.

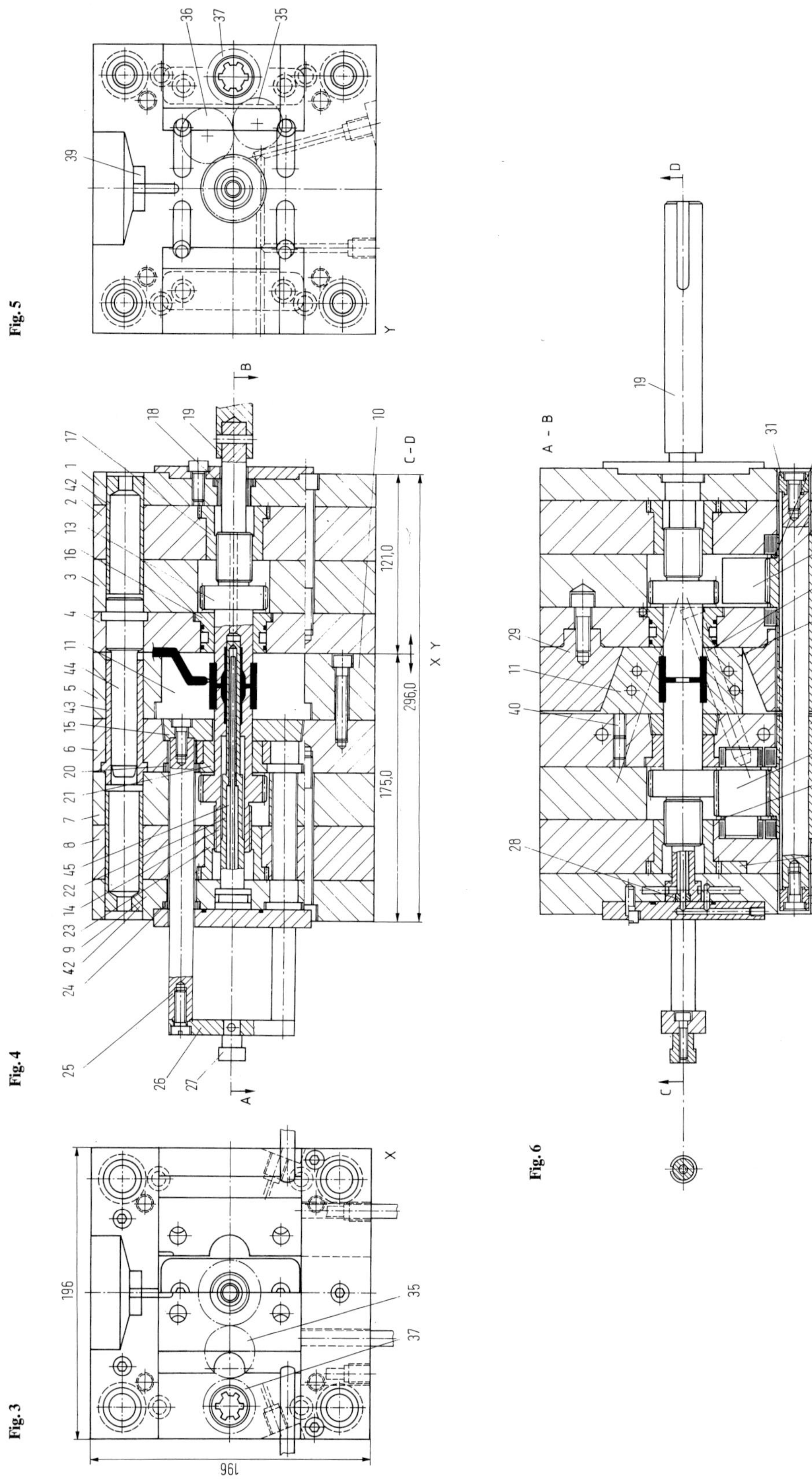

Figures 3 to 6 Single-cavity injection mold for the center section of a hose connector
11, 12: slides; 13, 14: threaded cores; 15: stripper plate; 17: threaded bushing; 19: shaft; 22: inner core; 23: threaded bushing; 35, 36, 37: gears; 38: splined shaft; 41: cam pin

Example 19, Two-Cavity Injection Mold for the Coil Form of an Auxiliary Relay

On the coil form for an auxiliary relay, sharp-edged guidance of wires and feed-through openings as well as parting line flash on the coil form surface had to be avoided. Since the flange edges of the coil form must not be distorted, large square ejectors were employed. Flash formation in the coil area can only be prevented from occurring at the parting line between the slides (16) and (17) by precision guidance and completeness of shutoff. Wear plates (18) and (19) have been provided to compensate for any possible wear during the time the mold is in operation; if required, these plates can be replaced.

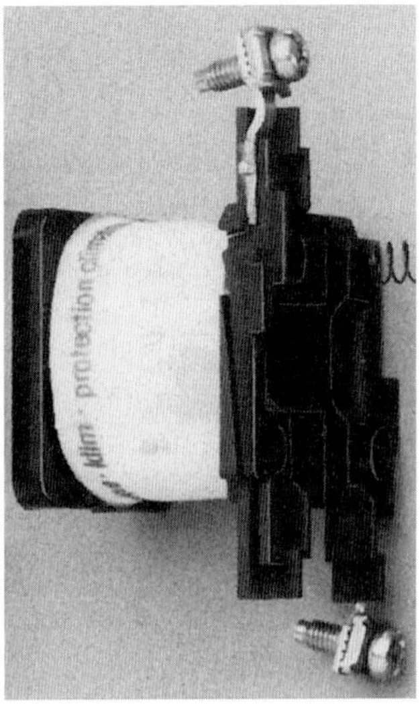

Figure 1 Coil form for an auxiliary relay

Mold Operation

Operation of the slides is essential for forming the area taken up by the coil as well as for releasing the undercuts on the connecting flange of the coil. The ejector system of the molding machine is actuated only after the mold has opened completely at the parting line D/E, when the slides (16) and (17) are retained in their final position by the detents (15). The coil forms, now resting freely on the mold cores (13), can be ejected by the large square ejectors (24). These ensure that the coil form is not distorted during ejection. The sprue, which still adheres to the coil form and also lies in the parting line between the slides, is simultaneously ejected by the ejector pins (25). By being interlocked with the machine controls, the microswitch (32) mounted on the ejector housing (5) prevents the injection molding machine from closing the mold before the ejectors (24) and (25) have been retracted completely. Only then is the microswitch actuated by the stop (33) connected with the ejector plates (7) and (8). This precaution prevents ejectors still protruding from being bent over by the slides and subsequent destruction of the mold cavity surface in the slide. In addition, check scales verify that the sprue and molded parts have been completely ejected from the mold.

Fig. 2

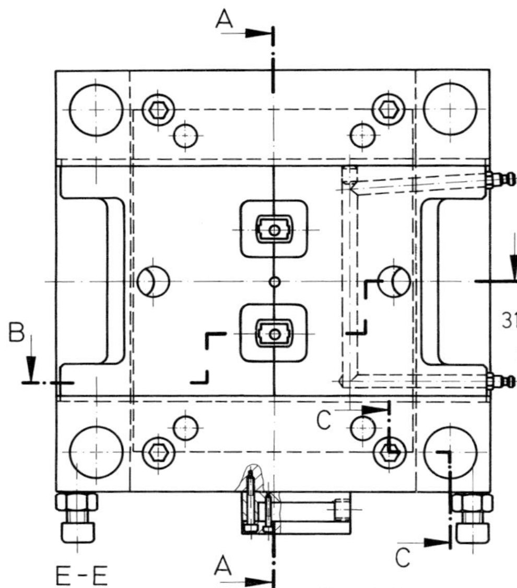

Fig. 3

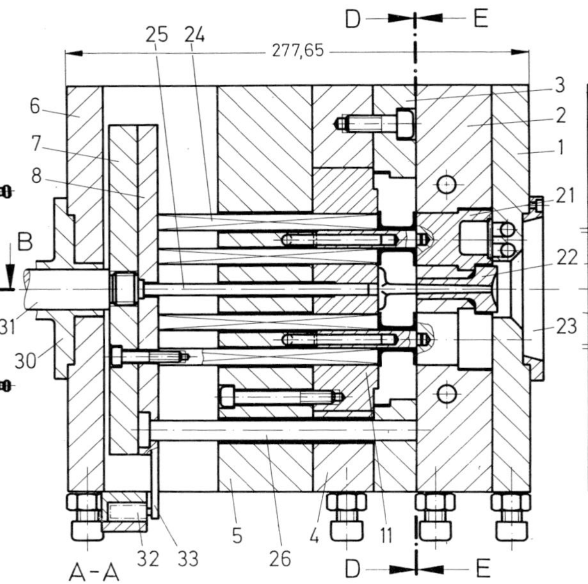

Fig. 4

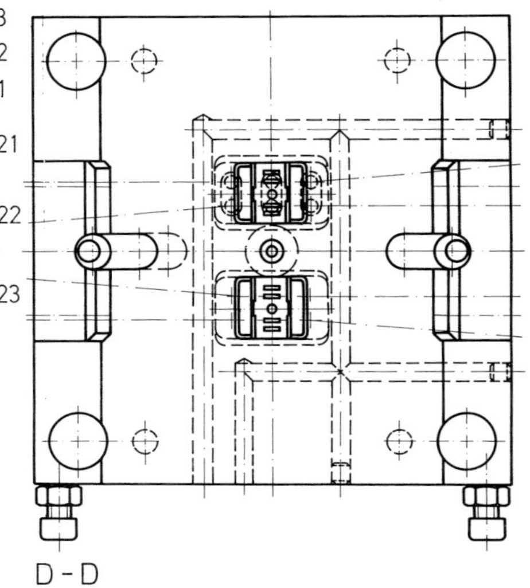

Fig. 5

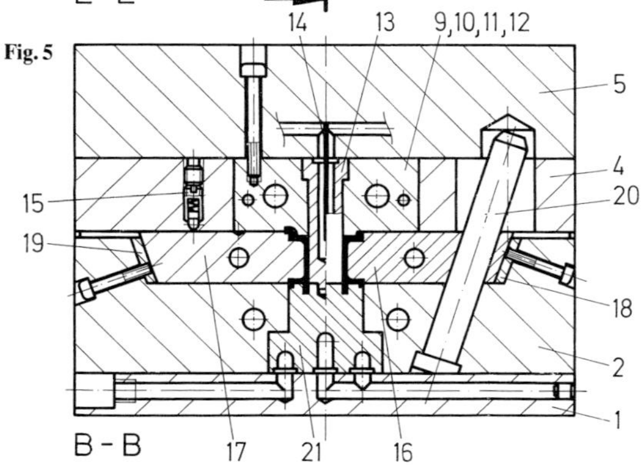

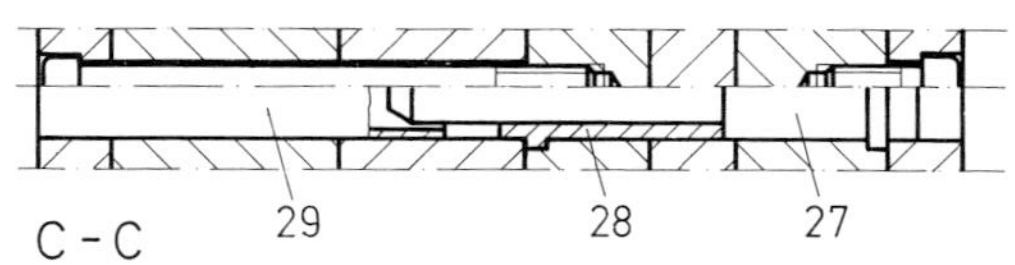

Fig. 6

Figures 2 to 6 Two-cavity injection mold for coil forms
1: stationary-side clamping plate; 2: stationary-side core retainer plate; 3: guide strip; 4: moving-side core retainer plate; 5: ejector housing; 6: moving-side clamping plate; 7, 8: ejector plates; 9, 10, 11, 12: core insert plates; 13: core; 14: baffle; 15: detent; 16, 17: slides; 18, 19: wear plates; 20: cam pins; 21: stationary-side core; 22: sprue bushing; 23: stationary-side locating ring; 24: square ejector; 25: sprue ejector; 26: pushback pins; 27: leader pin; 28: guide bushing; 29: locating sleeve; 30: moving-side locating ring; 31: ejector rod; 32: microswitch; 33: stop

Example 20, Single-Cavity Hot-Runner Mold for Business Card Boxes Made from Polypropylene

The business card box has dimensions of 9mm × 93mm × 125mm, a wall thickness of 1.5mm, and a weight of 18g, as well as a film hinge (thickness 0.3mm) for opening and closing the lid (Fig. 1). Its surface is textured (lasered cavity inserts) and inscribed. Internal slides mounted on angular pillars serve to demold the undercuts. Mold size is 246mm × 296mm. An injection molding machine with a clamping force of 600kN is utilized for production. The molded part is removed by a removal device. The actual cycle time is 12 s.

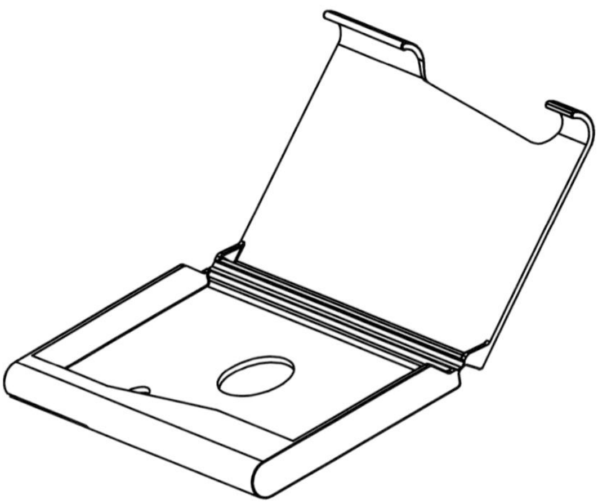

Figure 1 Business card box with film hinge made from polypropylene, diagram

Mold

The mold is constructed mainly from standard components. The externally heated and controlled hot-runner manifold is equipped with aluminum reflector plates to minimize heat loss due to radiation. Tubular heaters pressed in on both sides provide heat. In order to avoid convective heat loss, a seal disk has been mounted on the fixed side between the distributor bush and the centering flange. It also protects the mounting space of the hot-runner system from possible flashing from the feeding machine nozzle. The Z3210 front-mounted open hot-runner nozzle with a shaft diameter of 11mm is equipped with replaceable nozzle tip, heating element and thermo couple. The nozzle head is connected to the hot-runner manifold by a sliding seal face. The melt flow element is friction-locked to the hot-runner manifold block (shrink joint) and secured against twisting (Fig. 2). The melt-chamber insert can be cooled separately. The molded part is gated at one point on the bottom of the box. Since the core for the "finger-hole" is on the nozzle side, an unavoidable flow line stretches through the film hinge. To ensure that this does not create a shear point, the flow line has to lie precisely at a *right* angle to the hinge. During the planning phase, a mold flow simulation was done in order to estimate pressure requirements in addition to flow behavior. The prime objective was to obtain as simultaneous and uniform a through-flow at the "bottleneck" film hinge as possible. In order to eliminate part warpage, for one thing, core cooling was provided near the contour. Special attention should be paid to the cooling system for the internal slides mounted on inclined pillars that are bored-out and face-bonded by a method such as high-temperature vacuum welding. The cooling connections are moveable, corresponding to mold function. The variable inscription insert on the front has, for example, an engraved company logo. When doing the engraving, attention must be paid that all junctures are rounded off and not sharp-edged, in order to eliminate for instance so-called air streaks (air pockets dragged in by the melt and washed to the part surface where they spread out) that cause surface defects. The hot-runner system selected is well flushed, an important requirement for quick color changes.

Demolding

After the mold opens, controlled ejector assemblies are actuated in two stages by the ejector rod via a push-pull return mechanism consisting of rods, return sleeves and bushes in an eject/release sequence. The molded part has to be removed by a removal device; "dropping it", although quite possible, is, considering the high demands on quality, out of the question. When the mold closes, the ejector assemblies are returned to their starting position by return pins.

Literature

Der Stahlformenbauer (2005) 3, p. 132–134

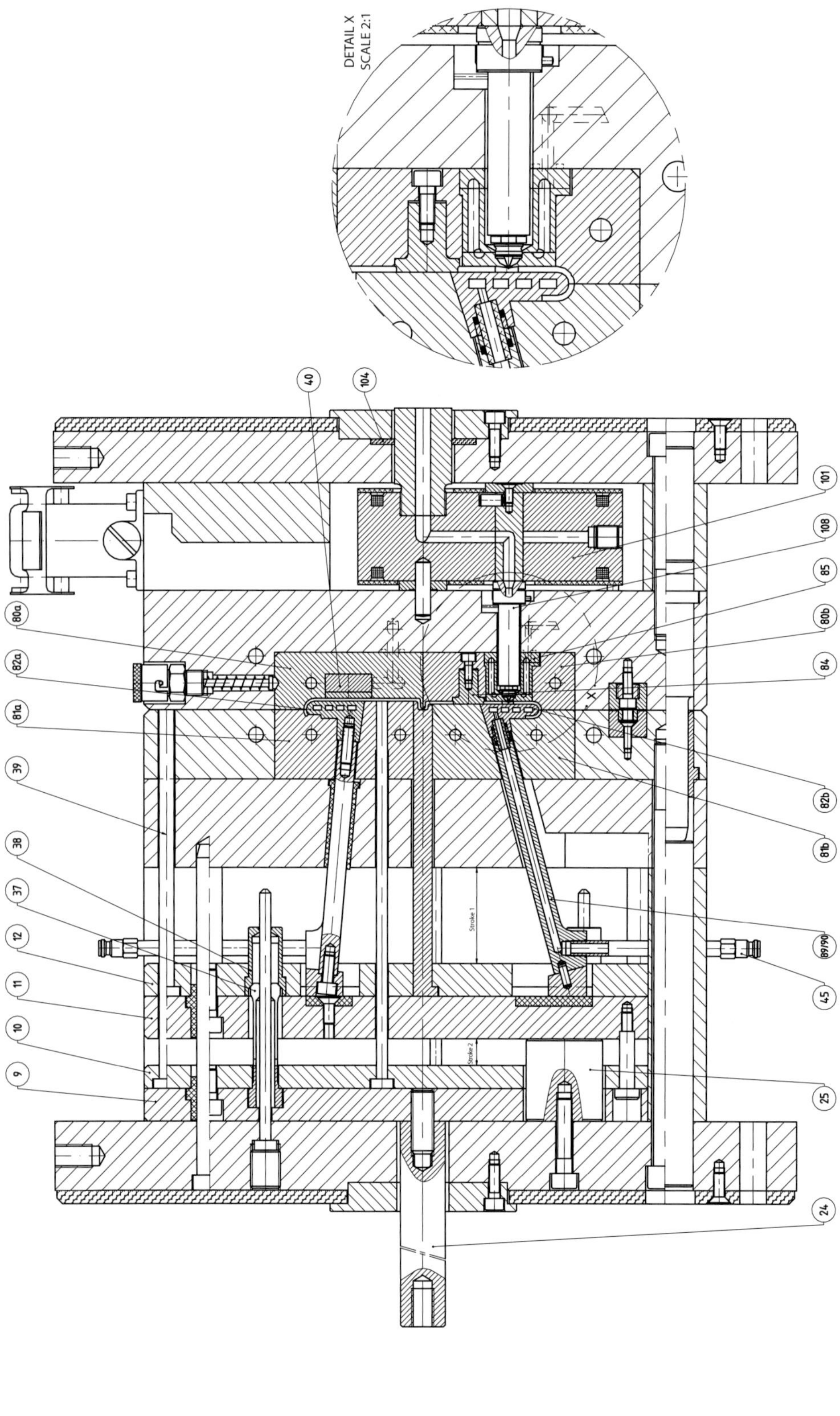

Figures 2 to 5 Single-cavity hot-runner mold for calling card box
9, 10, 11, 12: ejector plate assemblies, 24: ejector rods, 25: dowel pins, 36: shoulder screw, 37: return-pin sleeve, 38: return-pin bush, 39: return pins, 40: inscription insert, 45: extension nipple, 80a, b: mold inserts, 82a, b: moveable mold inserts, 84: core punch, 85: antechamber bush, 89, 90: angular pullers with hidden slide, 101: hot-runner single diverter, 104: central bushing, 108: heated sprue bush
(Courtesy: Hasco, Lüdenscheid)

Example 21, 8-Cavity Injection Mold for Threaded Rings Made from Polyacetal (POM)

Rings with completely finished, M22 internal threading with a threaded length of 4mm are produced in this mold for the electrical industry. Average wall thickness is 2.5mm. The outer diameter is stepped and bears lateral ridges with a pitch of $16 \times 22.5°$ for optimal grip. Molded part weight is approx. 2g; maximum shot size of the heated sprue nozzle is 200g. The threaded ring must exactly circular. An easy flowing POM is a suitable choice for this job. Tolerances should be considered critical. There are no special surface quality requirements.

Mold

The design shows a three-plate, standard size 218mm × 296mm mold with two parting lines (Fig. 2). The third plate consists of two separately guided cavity plates (19, 21) that are required by sequential demolding strokes and an additional stripping function. The mold inserts for the outer contour (3) are arranged symmetrically and equidistant to each other and the center. Corresponding to them on the ejector side, there are eight threaded cores (4), each carried in two axial-needle ball bearings (5), for forming the internal contour. The threaded cores are centered with snug-fitting centering bushes (6, 7). The mold cores (8) on the fixed side in the inserts (3) and on the threaded cores (9) center each other in injection position. To ensure convergence of the cavity plates, pre-centering devices (10) are mounted in the first parting plane. They prevent possible wear or damage during mold closing. The gearing of the threaded cores engages the central spur wheel (11) directly. The spur wheel rests on the ball-bearing drive shaft (12) and is driven by a chain from a hydromotor (13). The drive is controlled by the program of the injection molding machine.

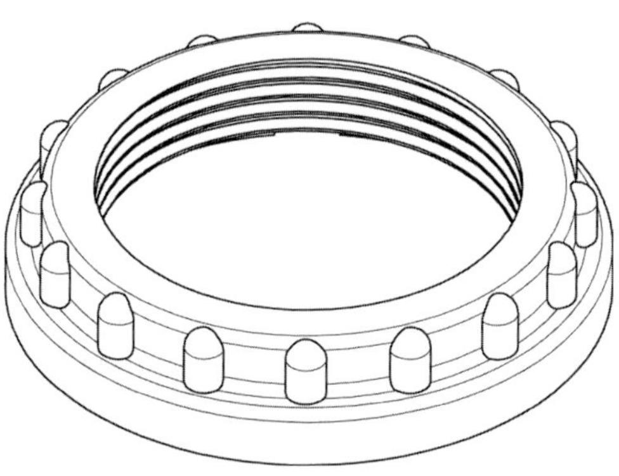

Figure 1 Threaded ring made from polyacetal, diagram

Gating

The requirement that the parts be uniformly filled using a minimum of control devices is met by the star-shaped secondary distributor with equally long flow paths and a central, heated sprue nozzle. The nozzle (14) is equipped with an elongated screwed on gate bush insert and injects to the runners through a short, tapered bar gate. The molded parts are gated directly from the side.

Cooling

The mold inserts require intensive cooling in a very narrow space. Several cooling circuits are required to achieve uniform article quality. The usual axial shift cannot be used to despindle the thread cores in this example due to the core cooling. Sealing elements (15) in the core retaining plate (16) surround the rotating thread cores, protecting them from water under pressure, thus making core cooling possible. The core head cooling lines pass through the inside of the tubes (17), the entire length of the core is cooled by the returning water.

Demolding

High demolding forces and friction are reduced by Diocronite™ coating on the cavities and moving components. Demolding takes places in three sequences:

- The mold is opened at the first parting plane. Opening path is 70mm. The molded part and the distributor sprue are demolded together and remain at first on the ejector side.
- Subsequent opening of the second parting plane is limited to 6mm by four shoulder screws (18). The internal threading is demolded by the driven thread cores. The solidity required for demolding is provided by the frozen gate and runner. The spring-mounted cavity plate (19) pushes the article forward on the threading.
- The articles and the runner, which up to this point on retaining pins (20), are demolded by the stripper plate (21) which is moved by the machine ejector via ejector rods (22).

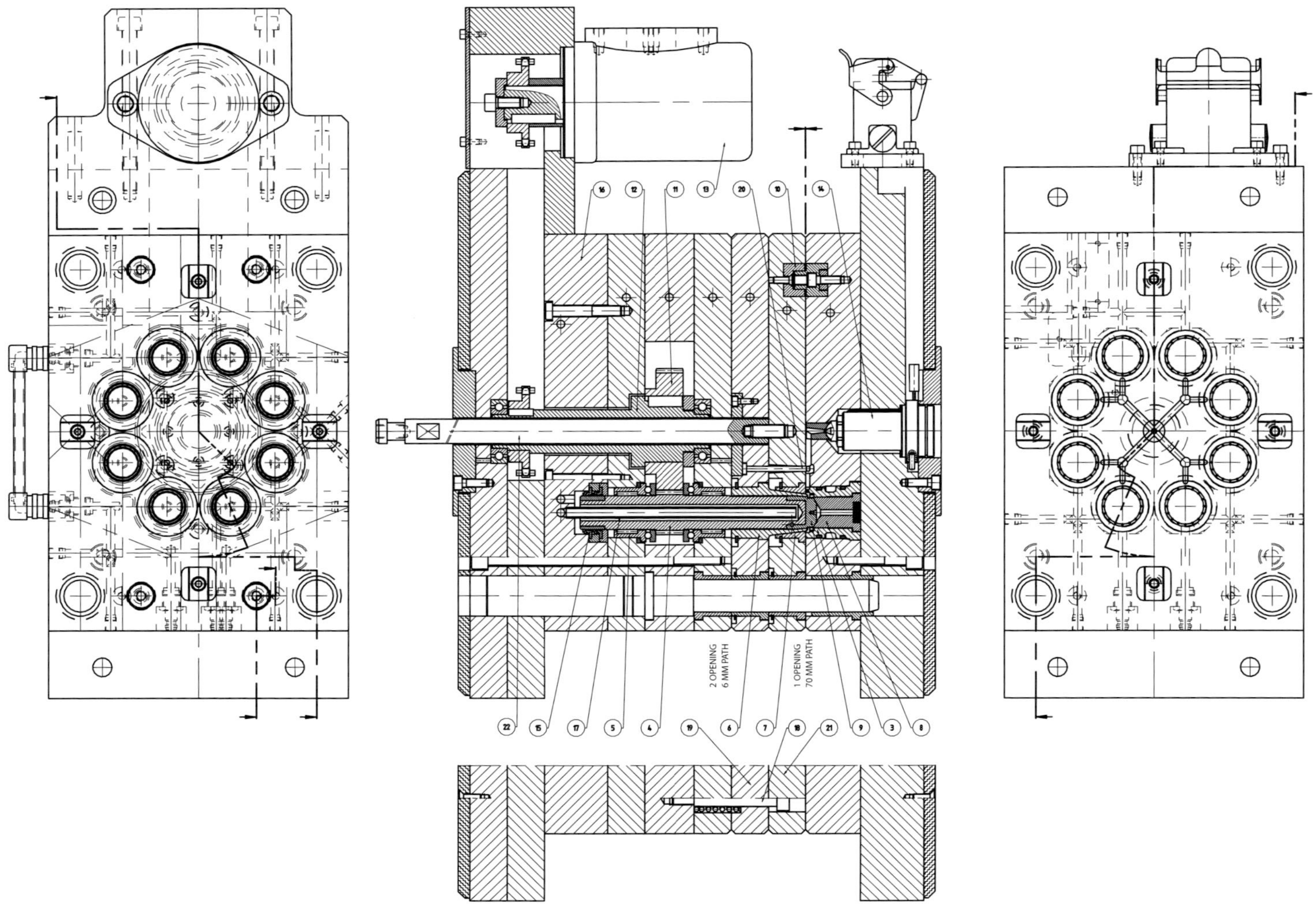

Figure 2 8-cavity injection mold for threaded rings made from polyacetal
3: mold inserts DS, 4: thread core, 5: axial needle ball bearing, 6, 7: centering bush, 8: mold cores DS, 9: mold cores AS, 10: pre-centering devices, 11: spur wheel, 12: drive shaft, 13: hydramotor, 14: heated sprue nozzle with tip, 15: seals, 16: core retainer plate, 17: tube, 18: snug screw, 19: spring-mounted cavity plate, 20: retainer pin, 21: stripper plate, 22: ejector rod.
(Courtesy: Hasco, Lüdenscheid)

Example 22, Mold for a Pump Housing and Pump Piston Made from Polyacetal

The pump housing (*a*) and the pump piston (*b*) are rotationally symmetrical parts with various external undercuts and lateral perforations. The molding material is polyacetal. Due to the similarity of both parts, a mold was designed (Figs. 1 to 9) in which the piston (*b*) and the pump housing (*a*) can be produced simultaneously (so-called family mold). The outer surfaces of the molded parts are formed exclusively by a split cavity arrangement consisting of the cavity insert retainers (*c*) and the cavity inserts (*d*). When damaged, only the inserts have to be replaced. Various laterial perforations are molded by core inserts (*e*) housed in the cavity inserts.

Due to the height of the split cavity arrangement (146 mm, danger of tilting) and to achieve a large supporting cross-section for the taper lock (*f*), the cam pins (*g*) have been mounted outside the taper lock area. The split cavity halves are guided by the stripper plate (*h*). A cylinder secured on one side (*i*) of the split-cavity parting line additionally serves to locate the split cavity half. The runner (*k*) and the center gate (*l*) for the molded parts are also located in the split-cavity parting line. The inner surfaces of the molded parts are formed by the cores (*n*, *o*) held in the core retainer plate (*m*). They are secured against rotating by pins. Because of their liability to damage, the core tips (*p*) are screwed in and are thus replaceable. They, too, are secured against rotating by fixing.

Bushings (*q*) that are wear parts are fitted into the stripper plate (*h*). To ensure the reliability of the stripping operation, the bushings (*q*) and the fixed cores (*n*, *o*) have conical seating surfaces.

The temperature of the mold is controlled by five independent systems:

System A + B: temperature control of the fixed cores (*n*, *o*) by means of bubblers.

System C + D: temperature control of the split-cavity inserts (*d*) takes place via several channels linked together into a large-area circuit.

System E: temperature control of the mold plate (*r*) is by means of two parallel channels connected to each other by a cross channel.

The mold opens at the parting line I–I. The split-cavity arrangement is simultaneously opened by the cam pins (*g*), thereby releasing the external undercuts and lateral perforations. A relative movement of the cavity-insert-guiding stripper plate (*h*) with respect to the core retainer plate (*m*) is prevented by coupling the mold ejector bar (*v*) to the hydraulic ejector on the molding machine (not shown). Ejection takes place through the operation of the hydraulic ejector on the machine and the stripping action of the two stripper bushings (*q*) in the stripper plate (*h*). Discharge of the molded part is additionally ensured by an air blast (*s*).

The bubblers have different diameters as a result of the different core diameters. The supply line for the temperature control fluid should be equal to or greater in cross-sectional area than the remaining cross-sectional area for the return flow.

This is a function of the heat transmission coefficient a, which reaches its highest values at high flow rates (turbulence). The bubbler in core (*n*) illustrates that the discharge opening of small diameter supply lines should be cut at an angle to enlarge the cross-sectional area. With the pump piston, the stripper bushing contacts the piston skirt (*t*), the outside diameter of which is considerably larger than that of the piston itself. This could have lead to severe deformation when ejecting the piston. To create a more favorable flow of forces, nonfunctional ribs were placed at (*u*).

Figures 1 to 9 Mold for a pump housing and pump piston

a: pump housing; *b*: pump piston; *c*: split-cavity retainers; *d*: split-cavity inserts; *e*: core inserts; *f*: supporting cross-section for the taper lock; *g*: cam pins; *h*: stripper plate; *i*: cylinder; *k*: runner; *l*: gate; *m*: core retainer plate; *n*, *o*: pinned cores; *p*: core tip; *q*: stripper bushing; *r*: mold plate; *s*: connection for air blast; *t*: piston skirt; *u*: rib; *v*: ejector bar on mold; *x*: illustrated without split cavity arrangement; *y*: shown offset by 45° ▶

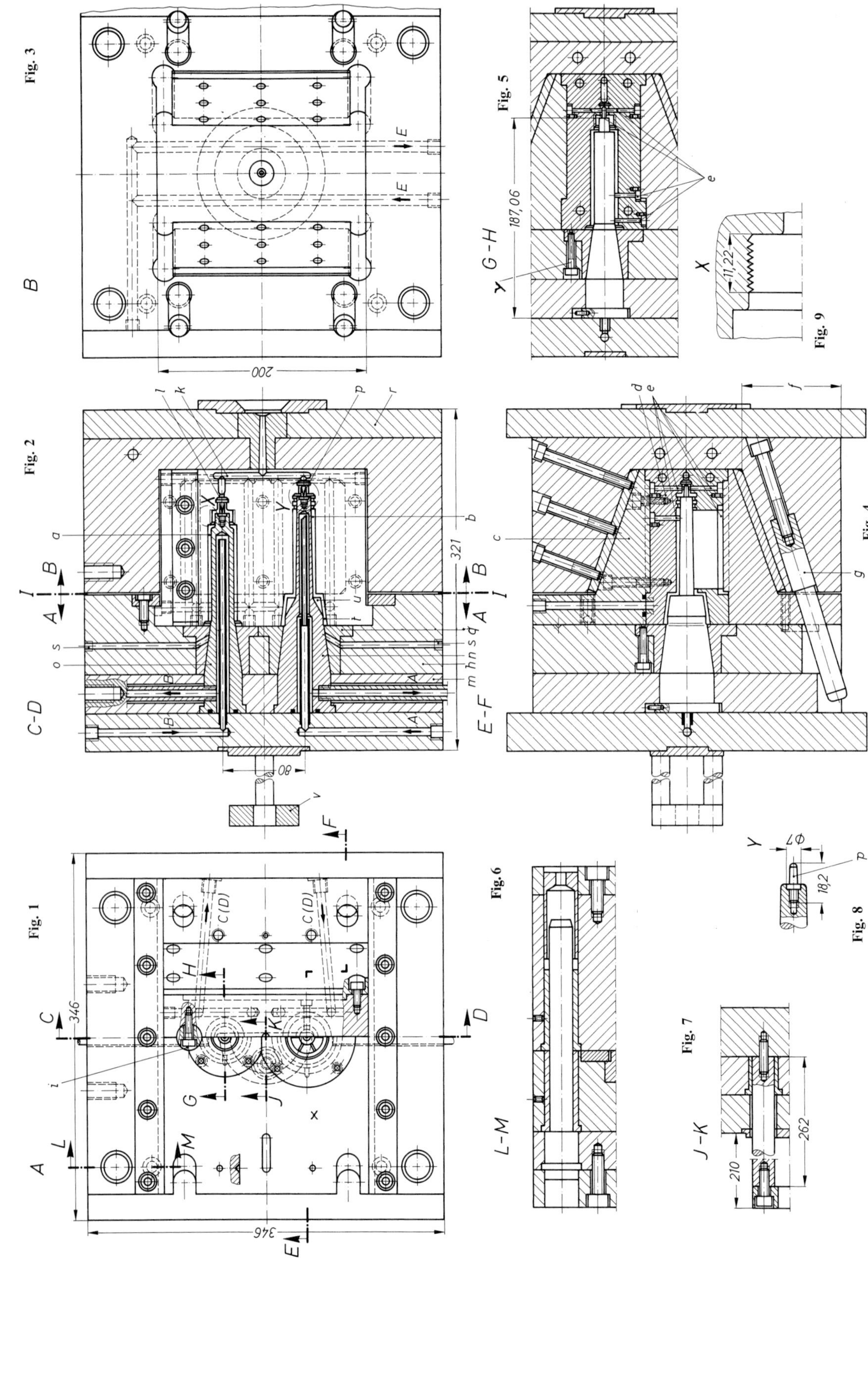
Fig. 1
Fig. 2
Fig. 3
Fig. 4
Fig. 5
Fig. 6
Fig. 7
Fig. 8
Fig. 9
C-D
E-F
B
G-H
L-M
J-K
X
Y
346
321
200
80
187,06
11,22
262
210
18,2

Example 23, Hot-Runner Injection Mold for Two Film Spools Made from High-Impact Polystyrene

Because of their geometrical shape, film spools require injection molds (Figs. 1 to 4) with slides (9, 10) which, in their closed state in the mold cavity, form the inner surface of the spool on which the film will later be wound. Suitable cores (2, 12) project inward into the double-walled spool from both sides to form the spokes and drive rosettes. The center holes in the spools are formed in the ejector side of the mold by movable cores that later function as sprue punches (16). A hot-runner system (15) with indirectly heated Cu-Be nozzles (14) is used to feed melt to the sprues (13) with three pinpoint gates.

Ejection of the spools takes place in such a way that first the sprues are separated from the nozzles (Fig. 5). Next the sprue is punched out of the hole in the hub (Fig. 6) before the spool itself is finally released (Fig. 7). This ensures that any further work on the spool is eliminated. Because of the need to punch out the sprues in the mold and thereby reduce finishing costs, it is only possible to use a hot-runner system with this two-cavity mold. A conventional three-plate mold with multiple cavities and a normal runner system would increase costs to an unacceptable level. At the same time, the hot-runner system reduces the material losses arising from a normal runner system.

On opening, the mold components and plates (4 to 8) that are bolted together and attached to the movable platen as well as plate (2), which is held by latch (23), move away from the mold half attached to the stationary platen of the molding machine. In this way, the sprues (13) are withdrawn from the sprue bushings. During this opening movement, the slides (9) and (10), which are guided on plate (5) by strips (3) and (4), are forced apart by the cam pins (11) so that each film spool is released around its circumference. At the same time, the hook at the end of latch (19) reaches the end of the groove cut in plate (1) and pulls the sprue punches (16) forward against the force of the springs (27) by means of punch plate (18). This pushes the sprues and runners out of the holes in the spool hubs until the hook at the end of latch (19) is lifted out of the groove by pin (20) moving onto the wedge (21). Further opening of the mold causes wedge (25) to engage pin (24), which then lifts latch (23), thus freeing plate (2), which has served to hold the molded parts. Stripper bolts (26) finally open the mold completely by drawing plate (2) away from the mold cavities. Lastly, actuation of the ejector (28) against the force of spring (29) pushes the film spools off the cores (12) through the action of ejector plate (30) and the attached ejector sleeves (32).

Fig. 5 Fig. 6 Fig. 7

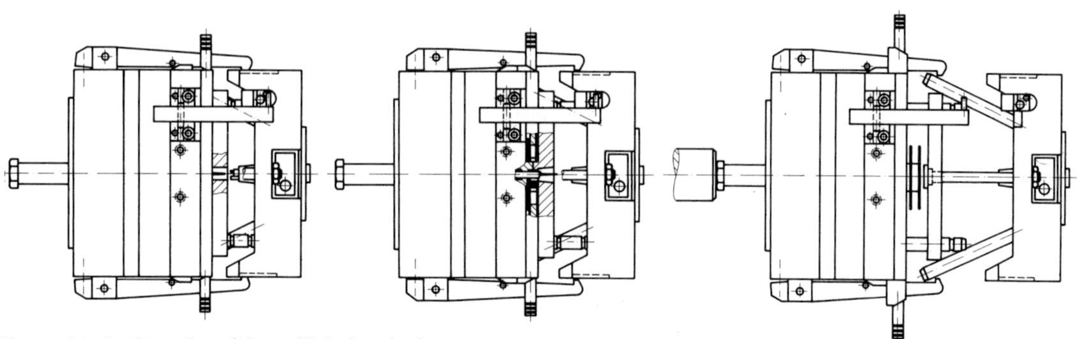

Figures 5 to 7 Operation of the mold during ejection
Fig. 5: sprue tears away
Fig. 6: sprue is punched out of the spool
Fig. 7: spool is ejected

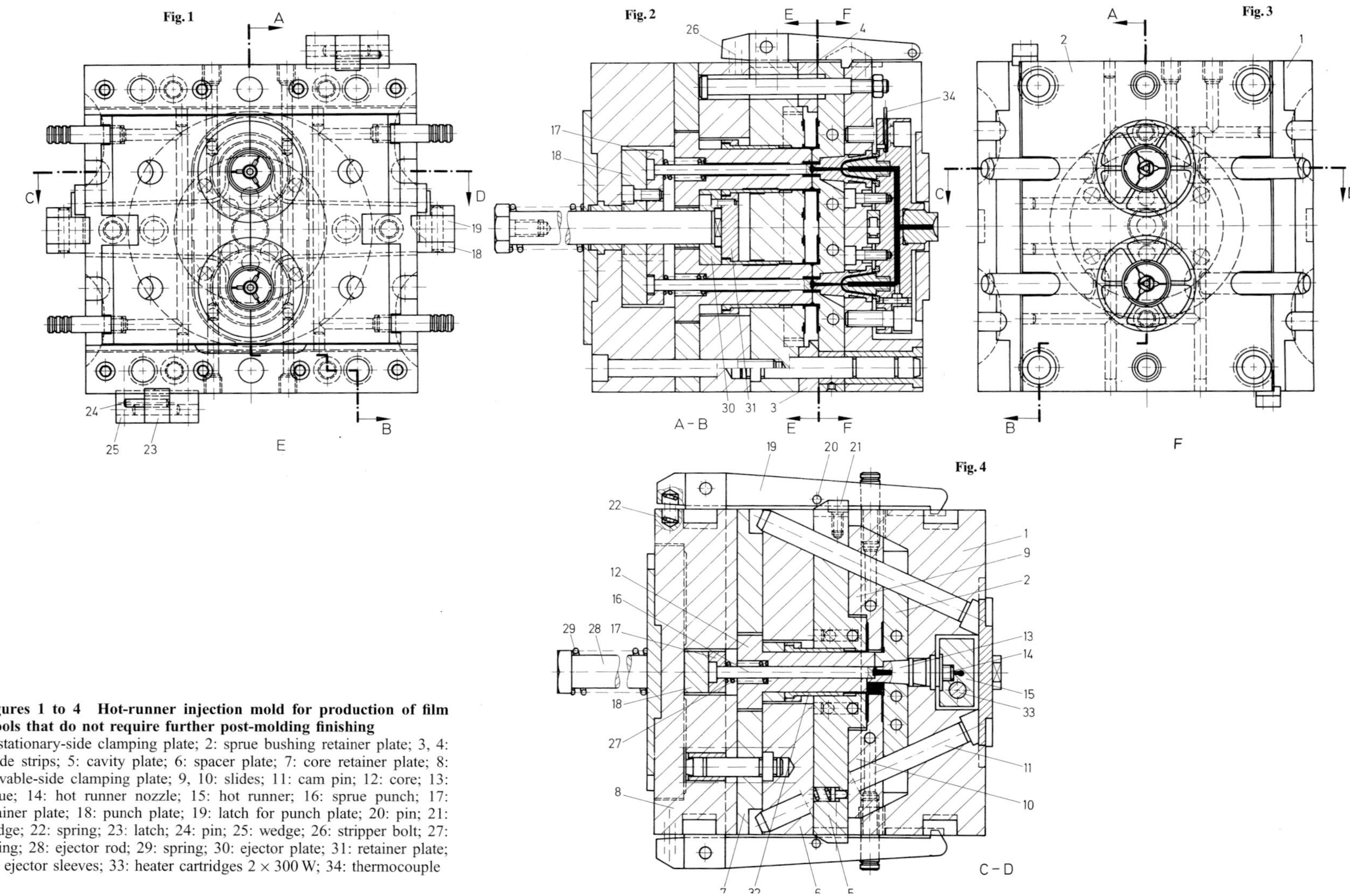

Figures 1 to 4 Hot-runner injection mold for production of film spools that do not require further post-molding finishing
1: stationary-side clamping plate; 2: sprue bushing retainer plate; 3, 4: guide strips; 5: cavity plate; 6: spacer plate; 7: core retainer plate; 8: movable-side clamping plate; 9, 10: slides; 11: cam pin; 12: core; 13: sprue; 14: hot runner nozzle; 15: hot runner; 16: sprue punch; 17: retainer plate; 18: punch plate; 19: latch for punch plate; 20: pin; 21: wedge; 22: spring; 23: latch; 24: pin; 25: wedge; 26: stripper bolt; 27: spring; 28: ejector rod; 29: spring; 30: ejector plate; 31: retainer plate; 32: ejector sleeves; 33: heater cartridges 2 × 300 W; 34: thermocouple

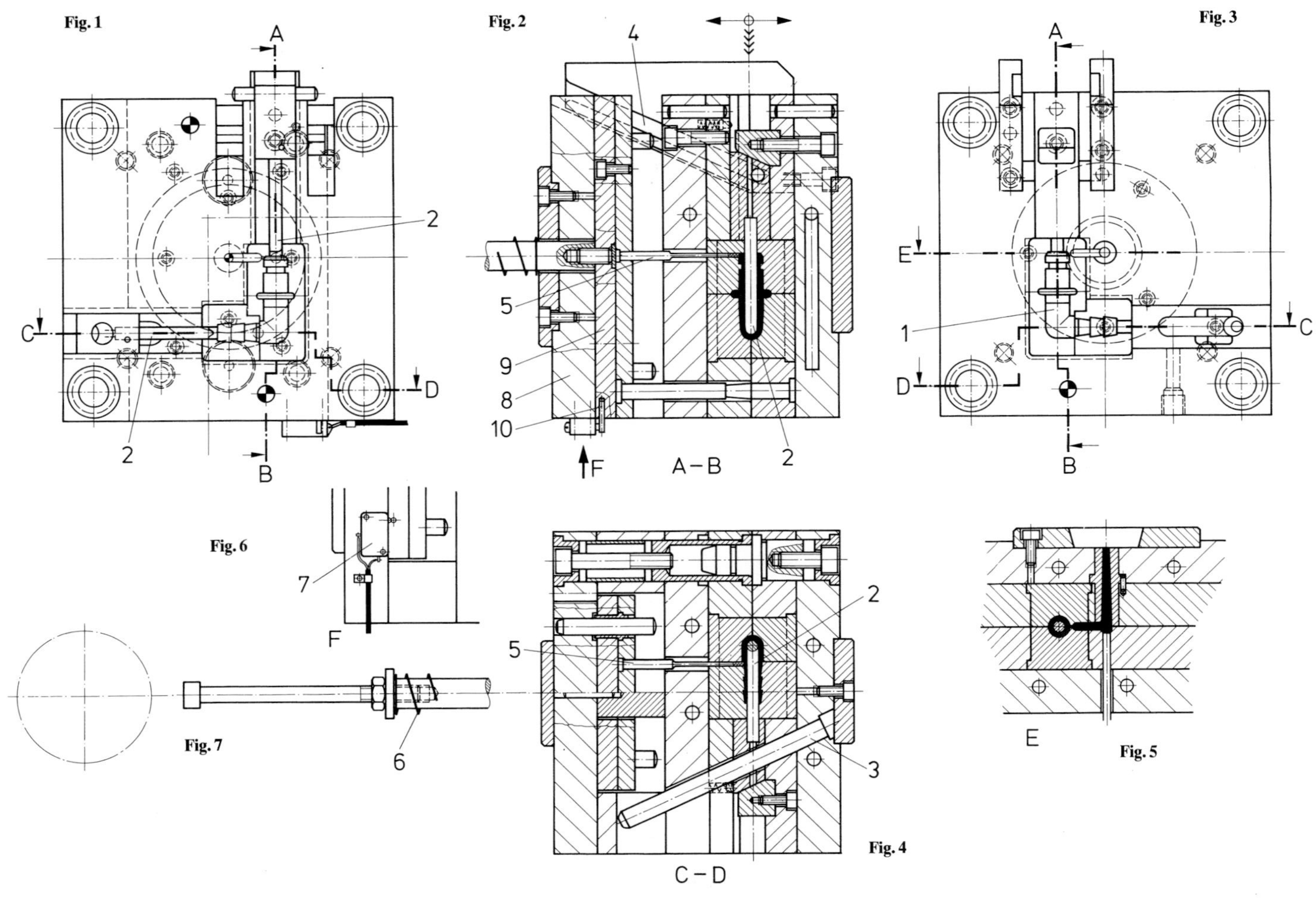

Example 24

Figures 1 to 7 Injection mold for angle fitting
1: angle fitting; 2: side cores; 3: cam pin; 4: slide; 5: blade ejector; 6: helical spring; 7: microswitch; 8: clamping plate; 9: ejector plate; 10: pin

Example 24, Injection Mold for an Angle Fitting from Polypropylene

If ejectors are located behind movable side cores or slides, the ejector plate return safety checks whether the ejectors have been returned to the molding position. If this is not the case, the molding cycle is interrupted.

This safety requires a switch on the mold that is actuated when the ejector plate is in the retracted position. The ejector plate return safety thus functions only if the molding cycle utilizes platen preposition, i.e., after the molded parts have been ejected, the clamping unit closes to the point at which the ejector plate is returned to the molding position by spring force. Only then does the control system issue the "close mold" command. In molds requiring a long ejector stroke, spring return of the ejector plate is often not sure enough. For such cases, there is an ejector return mechanism that fulfills this function. Attachment of the ejector plate return safety is shown in Figs. 1 to 7.

This single-cavity mold is used to produce an angle fitting (1). Two long side cores (2) meet at an angle of 90°. The somewhat shorter side core is pulled by a cam pin (3), while the longer core is pulled by a slide (4). The difficulty is that blade ejectors (5) are located under the two cores and must be returned to the molding position after having ejected the finished part before the two cores are set as the mold closes and possibly damage the blade ejectors. Possible consequences include not only broken blade ejectors but also a damaged cavity. Either of these could result in a lengthy interruption of production. For this reason, a helical spring (6) that permits operation with platen preposition is placed on the ejector rod. This spring then returns the ejector plate.

To ensure proper operation, a microswitch (7) is mounted to the clamping plate (8), while a pin (10) that actuates the switch is mounted in the ejector plate (9). After connecting the cable with the switch housing of the movable clamping plate, the ejector plate return safety is complete.

Example 25, Mold for Bushings from Polyamide with Concealed Gating

A flanged bushing is to be injection molded in such a way that any remnants of the gate are concealed or as inconspicuous as possible. The bushing would normally require a two-plate mold with a single parting line. The molded part would then be released and ejected along its axis, which coincides with the opening direction of the mold. The gate would be located on the outer surface of the flange since it is in contact with the mold parting line.

In order to satisfy the requirement for an "invisible" gate, the cavities (two rows of four) are placed between slides carrying the cores (Fig. 1) even though there are no undercuts. From a central sprue the melt flows through conical runners in the cores to pinpoint gates located on the inner surface of the bushings. As the slides move during opening of the mold the gates are cleanly sheared off flush with the adjacent part surface. The flexibility of the plastic selected is sufficient to permit release of the end of the runner from the angled runner channel. The parts are now free and can drop out of the mold.

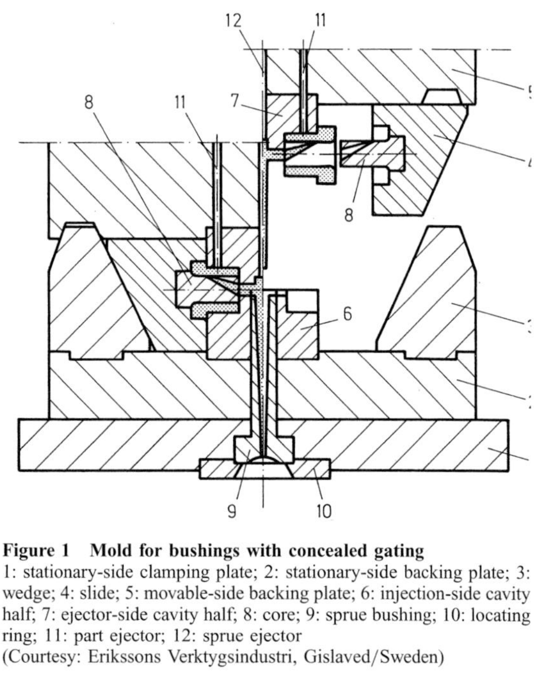

Figure 1 Mold for bushings with concealed gating
1: stationary-side clamping plate; 2: stationary-side backing plate; 3: wedge; 4: slide; 5: movable-side backing plate; 6: injection-side cavity half; 7: ejector-side cavity half; 8: core; 9: sprue bushing; 10: locating ring; 11: part ejector; 12: sprue ejector
(Courtesy: Erikssons Verktygsindustri, Gislaved/Sweden)

Example 26, Injection Mold for the Valve Housing of a Water-Mixing Tap Made from Polyacetal

A valve housing (Figs. 1 and 2) had to be designed and produced for a water-mixing tap. The problem when designing the tool (Figs. 3 to 7) resulted from the undercuts in four directions. Originally occurring considerable differences in wall thicknesses have been eliminated during optimization. Demands for high precision of the cylindrical valve seat in particular were negatively influenced by various recesses in the wall and adjoining partitions, which favored sink marks and ovalness.

Figure 1 View of the interior of the valve housing, showing the places of core penetration
Company photo: ARCU, Altermo/Sweden

Figure 2 View of the exterior of the valve housing
Company photo: ARCU, Altermo/Sweden

Polyacetal (POM) had been chosen as molding material. The complete molded part had to have homogeneous walls, and be free from flow lines if at all possible, as it would be subjected to ever-changing contact with hot and cold water during an estimated long life span. Inadequately fused weld lines would be capable of developing into weak spots and were therefore to be avoided at all cost.

Provision has been made for an electrically heated sprue bushing (30) (Fig. 6) in order to avoid a long sprue. The resultant very short runner leads to the gate on the edge of the pipelike housing, to be hidden by a part that is subsequently fitted to cover it.

Two cores each cross in the pipe-shaped housing, i.e. one core (16) each penetrates another core (19). This obviously presents a danger spot should the minutest deviation occur from the specified time- and movement-based coordination as well as from the accuracy in the mold.

The hollow cores (19) are kept in position by mechanical delay during the first phase of mold opening, while the crossing cores (16) are each withdrawn by an angle pin (31, 32). Mechanical actuation has been preferred over a hydraulic or pneumatic one in this case in order to exclude the danger of a sequencing error (the so-called human factor) during set-up and operation.

The cores (16, 33) consist of a copper-beryllium alloy. They are cooled by heat conducting pins (27, 28).

Figures 3 to 7 Injection mold for the valve housing of a water-mixing tap
1, 2, 3, 4: O-rings; 5, 6, 7: core clamping rings; 8, 9: core retainer with angle pin hole; 10: core retainer with angle guide; 11, 12: wedge; 13: guide rail; 14, 15: guide plate for core retainer; 16: internal core; 17, 18: external core; 19: core; 20: upper mold cavity half; 21: lower mold cavity half; 23: insert; 25: angle guide; 26: core baffle; 27, 28, 29: heat conducting pins; 30: heated sprue bushing; 31, 32: angle pins; 33: support core; 34, 35: ejector; 36: sprue ejector; 37: return pin; 38: locating ring; 38, 40, 41: stop; 43: screw; 44: lock nut; 45: fixed mold plate; 46: retainer plate for the upper mold cavity; 47: temperature control medium connection; 48: lower mold cavity retainer plate; 49: moving mold plate
(Courtesy: Seveko Fristedt & Sundberg, Karlskrona/Sweden, and Gustavsson Gravyr, Stockholm)

Fig. 3

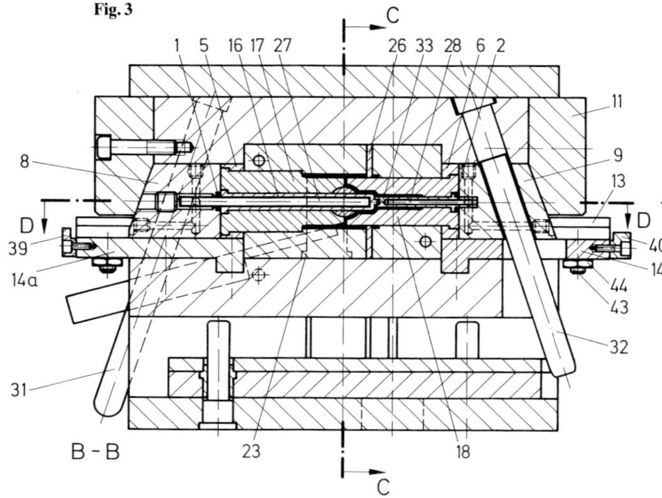

Fig. 4

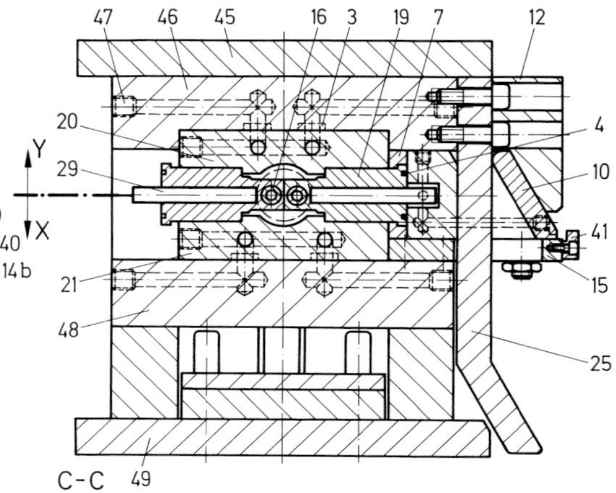

Fig. 6

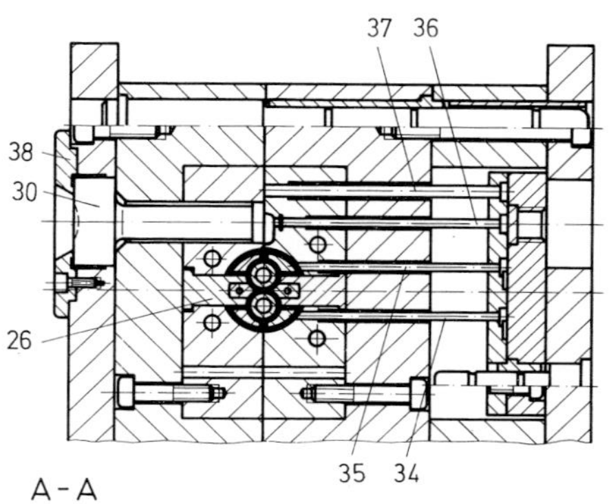

Fig. 5

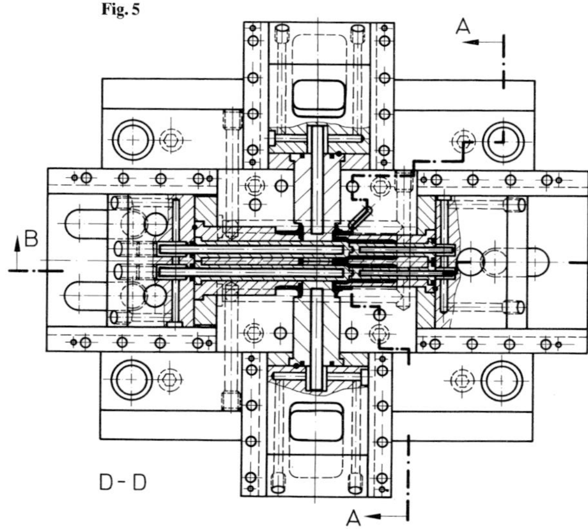

Fig. 7

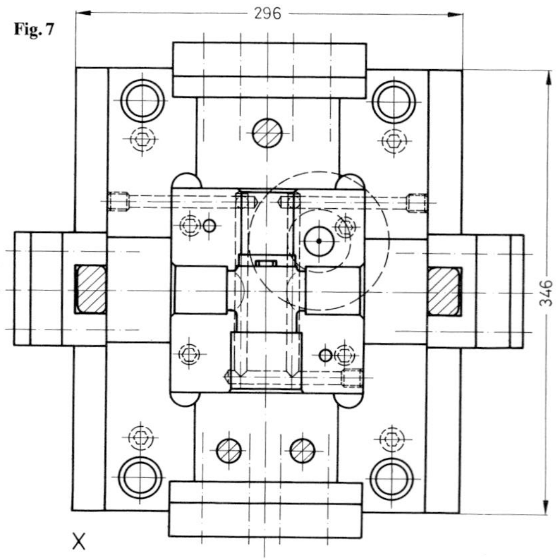

Example 27, Mold for a Lid with Three Threads Made from Polyacetal

The lid is a rotationally symmetrical part with three threads. Threads I and II are of the same pitch and can be formed by a single threaded core. The material employed is polyacetal. The total number of units to be produced is small. The mold (Figs. 1 to 5) is of simple design. The external shape of the molded part is formed by an insert (*c*), which is housed in mold plate (*b*) and secured against rotating. The temperature of this insert is controlled via a ring channel (heating/cooling system *A*). Thread III is formed by two slides (*d*). The part is injected through a diaphragm gate (*e*). The internal shape of the lid is obtained from a main core (*f*), which is housed in the mold plate (*p*) and is secured against rotating. The temperature of this core is controlled via an internal tube (heating/cooling system *B*). Its effectiveness is increased by the soldered-on spiral (*g*). The threads I and II are formed by a single threaded core (*h*). Because of the low number of moldings required, the mold has been designed for the threaded core (*h*) to be unscrewed outside the tool. The threaded core is inserted into an ejector ring (*i*) and is retained by three springloaded detents (*k*). It is located by the cone (*l*) of the fixed core (*f*). The mold opens at parting plane I–I positively assisted by two latches (*m*). The threadforming slides (*d*) are moved outward by this action. After a distance of 18 mm the latches are released by the control strips (*n*) and the mold opens at the main parting plane II–II. By actuation of the machine ejector the threaded core (*h*) is pushed in the direction of the fixed half by three ejector pins (*o*) and the ejector ring (*i*) for a distance of 90 mm (height of the molding plus 10 mm). During the movement the threaded core strips the molding off the fixed core (*f*). Then the molded part, with the threaded core (*h*), is pulled manually out of the stripping ring (*i*) without any danger of damaging the fixed main core (*f*). Unscrewing takes place outside the mold with the aid of an unscrewing device. To shorten the cycle time, several temperature controlled threaded cores are employed. While one part is being unscrewed, the next molded part is being produced.

Fig. 1

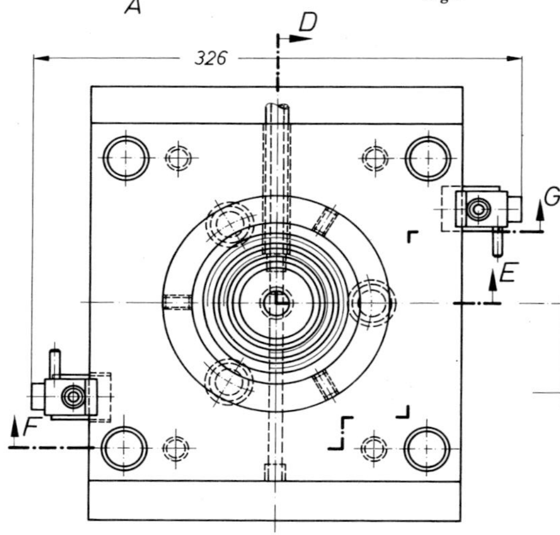

Fig. 2

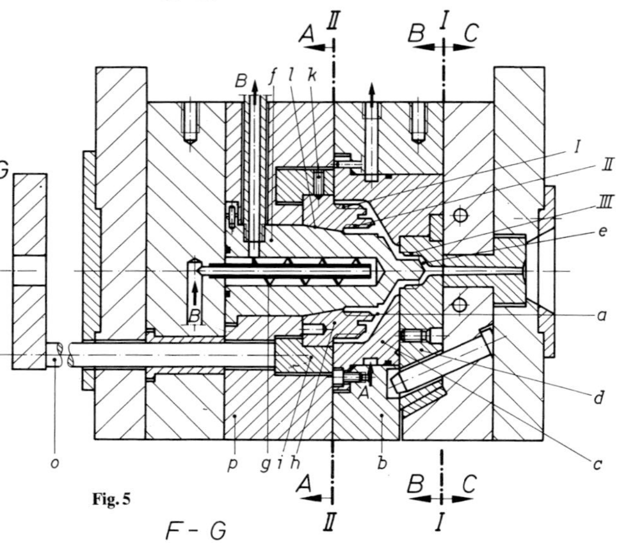

Fig. 3

Fig. 4

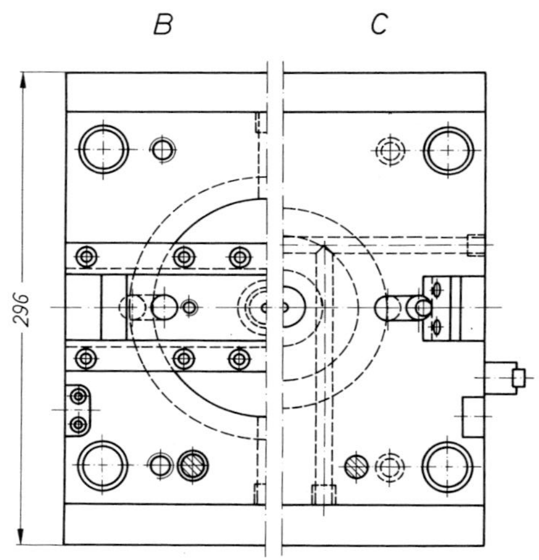

Fig. 5

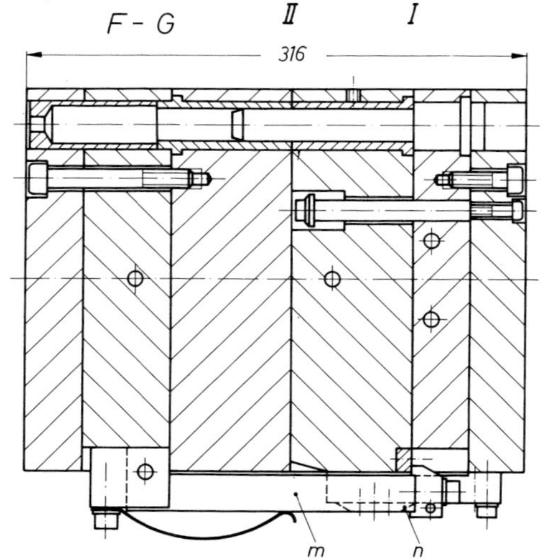

Figures 1 to 5 Mold for lid with three threads
a: molded part; *b*: cavity plate; *c*: insert; *d*: slides; *e*: diaphragm gate; *f*: main core; *g*: spiral; *h*: threaded core; *i*: ejector ring; *k*: spring-loaded detent; *l*: core; *m*: latch; *n*: control strip; *o*: ejector pin; *p*: mold cavity plate
I, II, III: threads

Example 28, Two-Cavity Injection Mold for Coupling Sleeves Made from Polyamide

The coupling sleeve in Figs. 1 to 5 had to be produced in a PA 66 with 30% by weight glass fiber reinforcement. The injection molded part has a center hole, entered by tapped M 10 holes that, starting from the peripheral surface, are opposite each other. As set screws are screwed into each tapped hole to push against a centrally fitted shaft, it is not necessary to have a continuous thread in both holes, which would have called for a bridging threaded core that would have had to cross the center core. Apart from problems with sealing, the unscrewing device also would have caused difficulty, as it would have had to perform a larger stroke. Use of the molded part allows for two separate threaded cores to be operated independently of each other, so that they can be driven by one rack each. To avoid further core pulls for the remaining shape of the molded part, it is put perpendicularly into the parting line of the mold by its axis of symmetry.

Concerning direct operation of the threaded cores by racks, a check must be made to ascertain that adequate transmission can be achieved or if intermediate stages are required to avoid an excessively long rack stroke.

The pitch of the metric thread M 10 is h = 1.5 mm. Allowing for a certain safety, an unscrewing distance of 11 mm must be taken up, which results in 11/1.5 = 7.33 rotations of the threaded core. For a

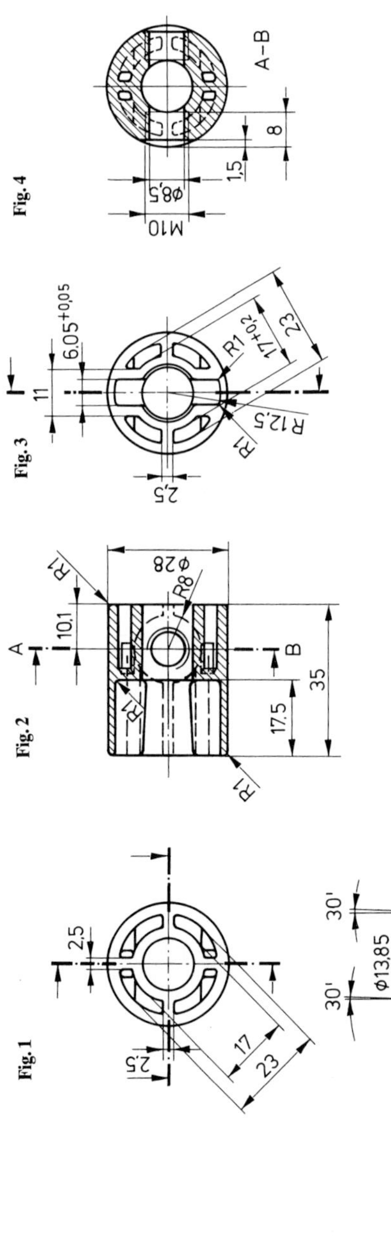

Figures 1 to 5 PA 66 coupling sleeve containing 30% by weight of glass fiber reinforcement

Figures 6 to 12 Two-cavity unscrewing mold for coupling sleeve
1: arrangement of mold cavities; 2: coupling sleeve; 3: fixed mold half; 4: fixed core for central hole in the molding; 5: moving core for the central hole in the molding; 6: locator for the screw cores; 7: M 10 thread on the threaded core; 8: threaded core guide; 9: threaded core pinion t = 15; 10: lead thread on the threaded core; 11: guide bushing; 12, 13: rack; 14: hydraulic cylinder; 15, 16: electric switches; 17: tunnel gate; 18: runner; 19: nozzle on the machine; 20: beryllium-copper nozzle tip; 21: ejector; 22: ejector plates; 23: ejector pins; 24: push-back pins; 25: support pillars; 26: insulating plate ▶

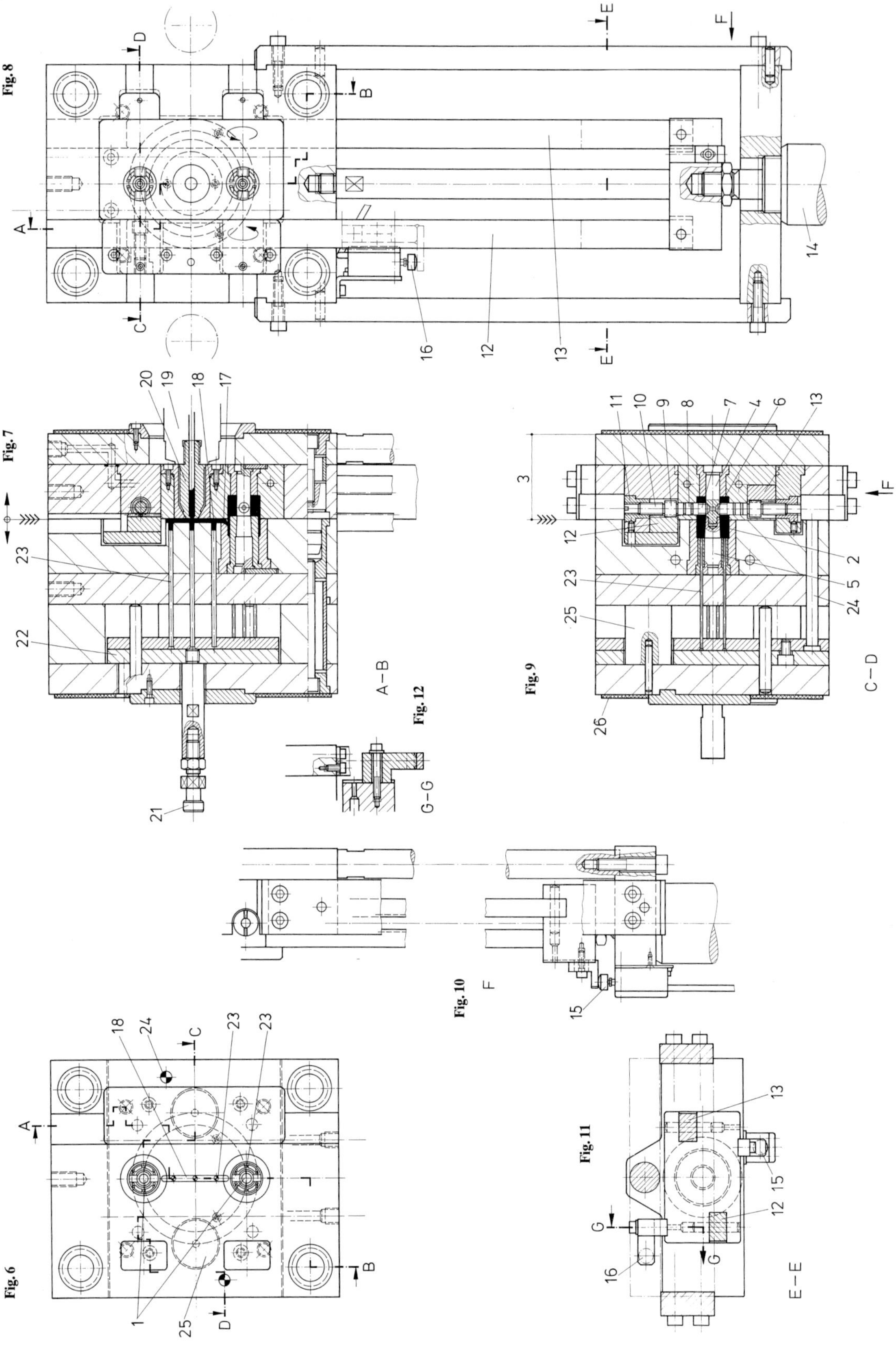

Fig. 6 Fig. 7 Fig. 8 Fig. 9 Fig. 10 Fig. 11 Fig. 12

pitch circle diameter of $d_0 = 12$ mm and a modulus of $m = 0.8$ mm, the pinion of the threaded core works out at $t = 12/0.8 = 15$ teeth and a pitch circle circumference of $12 \times \pi = 37.68$ mm. For 7.33 turns this results in a required rack stroke of $7.33 \times 37.68 = 276.19$ mm.

Standard hydraulic cylinders of 280 mm stroke are available. Divided by the pitch $d = \pi \times m = 2.5$ mm of the gear tooth system, this corresponds to 112 teeth on the rack, which with 15 teeth on the pinion turns the latter 7.46 times during one stroke. From this results an unscrewing distance of 11.19 mm, which is sufficient. It must be checked whether the space available on the injection molding machine allows installation of the mounting hardware and the hydraulic cylinder under the mold.

The mold design (Figs. 6 to 12) is such that two sleeves (1) can be produced at the same time. The unscrewing equipment has been installed in the fixed mold half (3) so that the hydraulic cylinder (14) does not have to participate in the opening and closing movement but can remain in position.

The center bore of the coupling, which tapers toward the moving mold half, is formed by two cores (4) and (5) which are self-centering.

The locators (6) for the threaded cores enter the core (4), which is held in the stationary side of the clamping plate, from both sides. The threaded cores are made up of the locators (6), the M 10 thread (7), a guide (8), the 15 gear teeth (9), and the guide thread (10) at the other end, which runs in the fixed guide bushing (11). The two racks (12) and (13) have been arranged offset to each other so that opposite directions of rotation can be transmitted to the opposing threaded cores. The hydraulic cylinder pushes the racks (12) and (13) up to unscrew the threaded cores. The upper racks protrude from the mold and need to be guarded by a screen. For interlocking with the machine's control circuit for the cycle sequence, the racks contact switch (15) in the lower and switch (16) in the upper position. By employing lateral submarine gating (17), the coupling sleeves are automatically degated from the runner (18). This is fed directly through a beryllium-copper nozzle tip (20), which is screwed into the female thread of the nozzle on the machine (19) to avoid the conventional tapered sprue penetrating the fixed mold half (3).

The operating sequence of the unscrewing mold takes place as follows:

The racks are moved in by the hydraulic cylinders, unscrewing the threaded cores from the molded parts. Then the opening movement of the mold starts. When finished, the hydraulic ejector of the machine, to which the ejector bar is coupled (21), pushes forward the ejector plates (22) and through them the ejector pins (23) for the coupling sleeves and runner. For safety, the push-back pins (24) also move out simultaneously. They have to return the ejector plate to the starting position in any case when the mold closes. Once the mold is closed, the racks are pulled up again and the new cycle can start with injection.

Example 29, Injection Mold for the Housing of a Polypropylene Vegetable Dicer

Molded Part

The housing accommodates a cutting disc that is driven by a hand crank (Fig. 1). The shaft of the crank drive is located in a bore in the housing. The underneath of the housing has a recess for accommodating a suction cap to attach the device to a table. The top of the housing has a filling shaft which supplies the cutting disc with the vegetables to be diced. A feed hopper will be attached to this filling shaft. The molded part weighs 386 g.

Figure 1 Housing for a polypropylene vegetable dicer

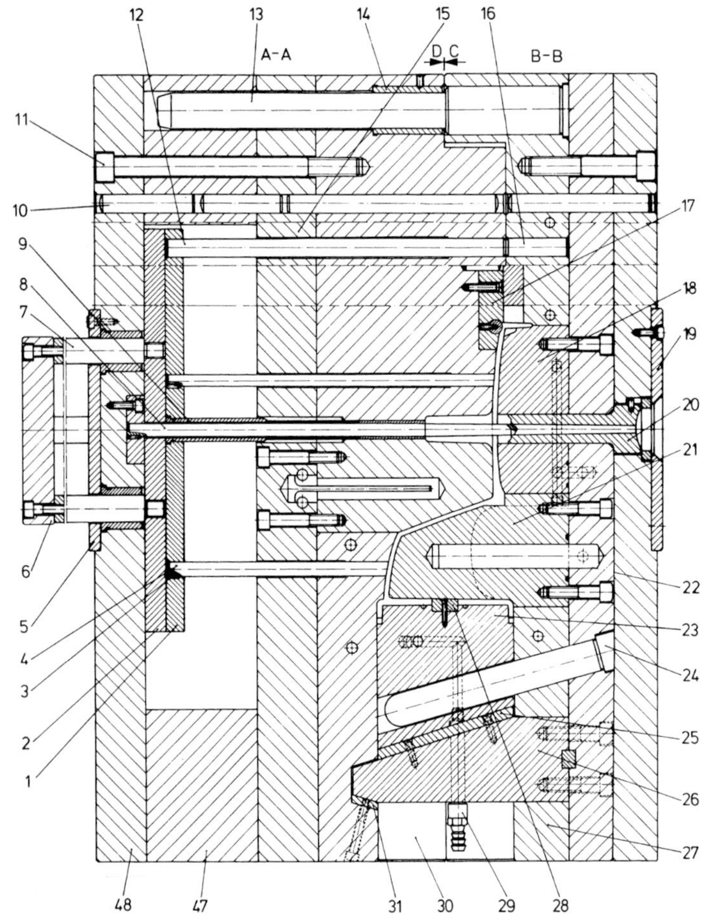

Figure 2 Longitudinal section through the injection mold for the housing
1: ejector retaining plate; 2: ejector base plate; 3: cylinder pin; 4: ejector pin; 5: locating ring; 6: stop plate; 7: ejector rod; 8: core pin; 9: ejector sleeve; 10: locating pin; 11: screw; 12: return pin; 13: guide pillar; 14: guide bushing; 15: support plate; 16: buffer pin; 17: mold insert; 18: punch; 19: locating pin; 20: sprue bushing; 21: punch; 22: punch retaining plate; 23: slides; 24: angle pin; 25: adjusting plate; 26: wedge: 27: mold plate (nozzle side); 28: split shoulder; 29: cooling pipe; 30: mold plate (clamping side); 31: adjusting plate; 47: bar; 48: mounting plate

Mold

The mold was designed so that the dicing chamber lies in the mold-opening direction. The housing base, the filling shaft and two other apertures are ejected with the aid of splits, a core puller and slides (Figs. 4 and 5).
The slide (23), moved by the angle pin (24), forms the inside contour of the housing base (Fig. 2). In the closed position, the split shoulder (28) lies against punch (21) and so forms the bore for attaching the suction cap to the housing base. The cylindrical slide lies in the mold parting line and each half is enclosed by the mold plates (27) and (30). Guide strips (50) (Fig. 3) lead the slide on the mold plate (30). The slide supports itself against the effect of the cavity pressure via the adjusting plate (25) and the wedge (26). Bending of the wedge is prevented by the adjusting plate (31) and the mold plate (30). The vegetable filling shaft and the passage to the dicing chamber are formed by the mobile core (33) (Fig. 6). Its movement is provided by the angle pin (32). Figure 7 shows the core guide in the guide strip (45). The inserted core is locked via the wedge (35) and adjusting plate (34). The guide strip (37) (Fig. 8) forms a rectangular opening in the side wall of the housing which lies half over and half under the mold parting line. It is moved by two angle pins (38) and is locked in the closed state by two bolts (39). A guide strip (49) which is bolted and doweled to the mold plate (30) is guided in a T-slot (Fig. 9).
Finally, a slit has to be formed in the housing wall that penetrates a reinforcement there. Rectangular aperture and reinforcement are formed by the slide (40) (Fig. 10) which is actuated by the angle pin (41) and locked by the wedge (42). Two bars (51) (Fig. 11) serve to guide the slide on the mold plate (30).

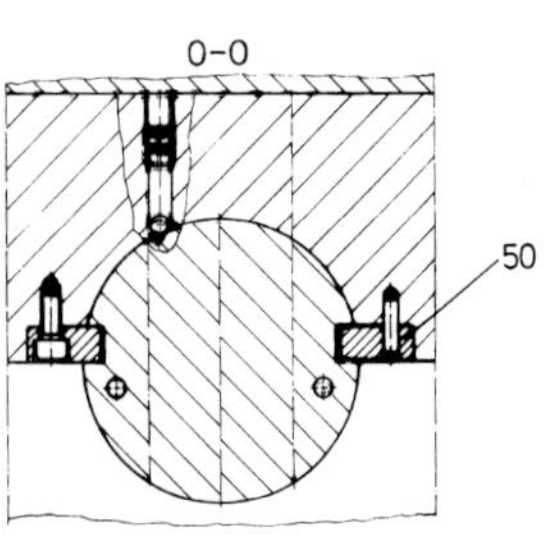

Figure 3 Guiding of mold slides in Fig. 2
50: guide bar

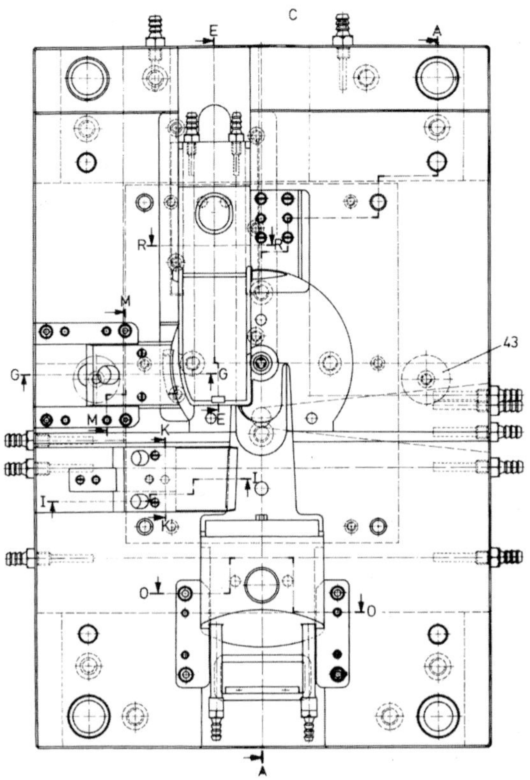

Figure 4 View of the moving side of the injection mold for a PP housing

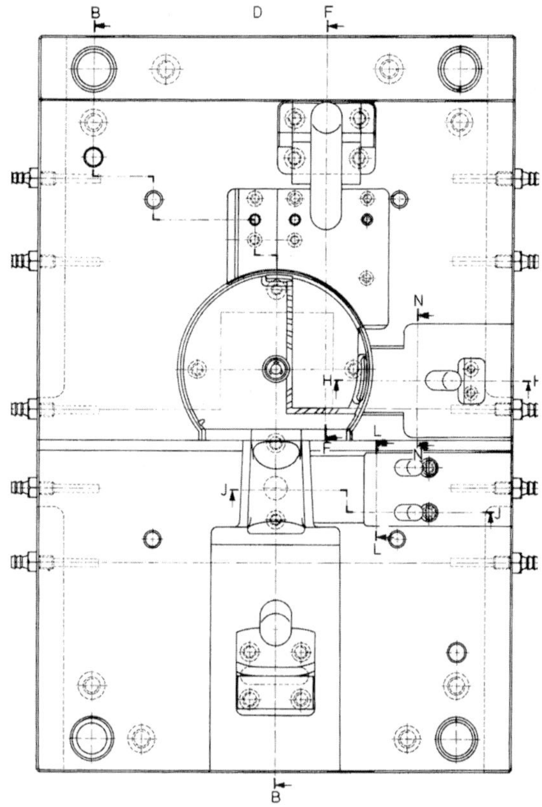

Figure 5 View of the stationary side of the injection mold for a PP housing

Since the angle pins traverse out from the slide, the core and the guide bars on mold opening, each is provided with ball catches that keep these guide elements in the "open" position. Bars (47) and rolls (43) support the plate (15) on the clamp plate (48).

Runner System/Gating

The sprue bushing (20) lies on the axis of the housing bore, which accommodates the blade drive shaft. The end of the sprue bushing forms the face of an eye inside the dicing chamber that is a part of the crankshaft mount. A core pin (8) protrudes into the bore of the sprue bushing and divides the sprue into three pinpoint gates.

Mold Temperature Control

The coolant is guided in bores and cooling channels in the mold plates, inserts and punches. The splits (23) and (33) offer sufficient space for accommodating cooling channels (33).

Part Release/Ejection

On mold opening, the angle pins on the fixed mold side push the splits, cores and slides on the

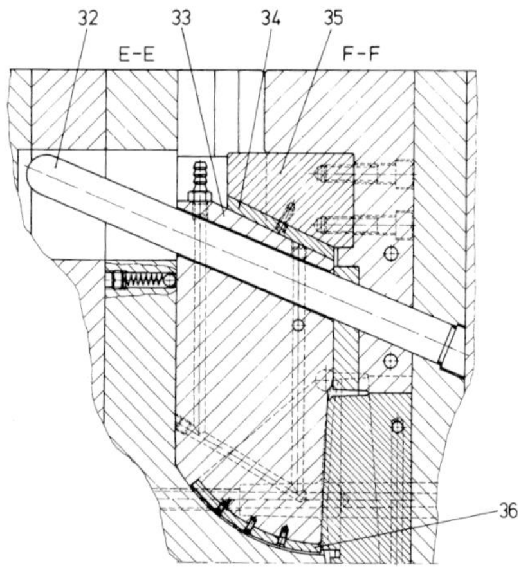

Figure 6 Demolding of the feed in a vegetable dicer
32: angle pin; 33: core; 34: adjusting plate; 35: wedge; 36: insert

Fig. 7

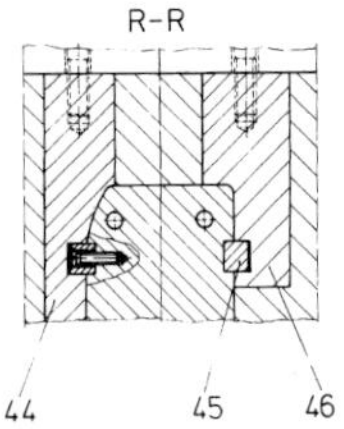

Fig. 8

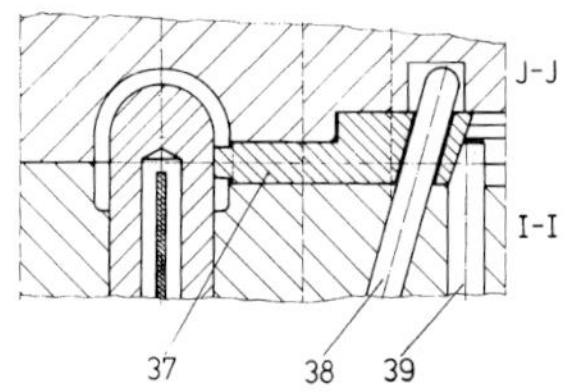

Fig. 9

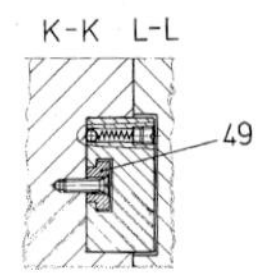

Fig. 10

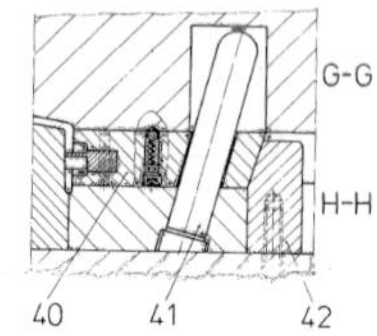

Fig. 11

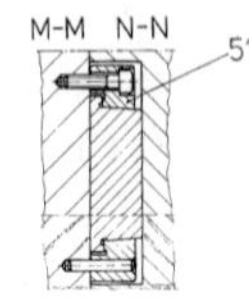

Figure 7 Core guiding for Fig. 6
44: punch; 45: guide bar; 46: punch

Figure 8 Demolding the rectangular aperture
37: slide; 38: angle pin; 39: locking bolt

Figure 9 Slide guide for Fig. 8
49: guide bar

Figure 10 Demolding a slot
30: slide; 41 angle pin; 42: wedge

Figure 11 Slide guide for Fig. 10
51: guide bar
(Courtesy: Plastor p.A., Oradea/Romania)

moving side so far outward that they release the undercuts of the housing. The molded part remains on the moving mold side. Ejector pins (4) and ejector sleeve (9) push the molded part out of the ejector-side mold cavities and off core pin (8). Since the ejector pins are contour-forming, they must be secured against twisting (pin 3). On mold closing, the ejector system is brought into the injection molding position by ejector-plate return pins (12) and buffer pins (16), and so too are the splits, cores and slides by their respective angle pins.

Example 30, Two-Cavity Injection Mold for a Polypropylene Toy Tennis Racket

The toy tennis racket shown in Fig. 1 measures 220 mm × 520 mm and is made of polypropylene. Handle and frame are extensively ribbed to provide stability, and the playing face also consists of a network of ribs. The handle is partly hollow. The bore in the handle is sealed with a plug.

Simple Die Design

The mold with the mold-fixing dimensions 700 mm × 480 mm is a two-part design (Fig. 2). This means that, under the influence of the mold buoyancy, stress is applied uniformly on the clamping unit of the injection molding machine, and that the gate is positioned in a favorable, material-saving location. The maximum mold height is 280 mm.

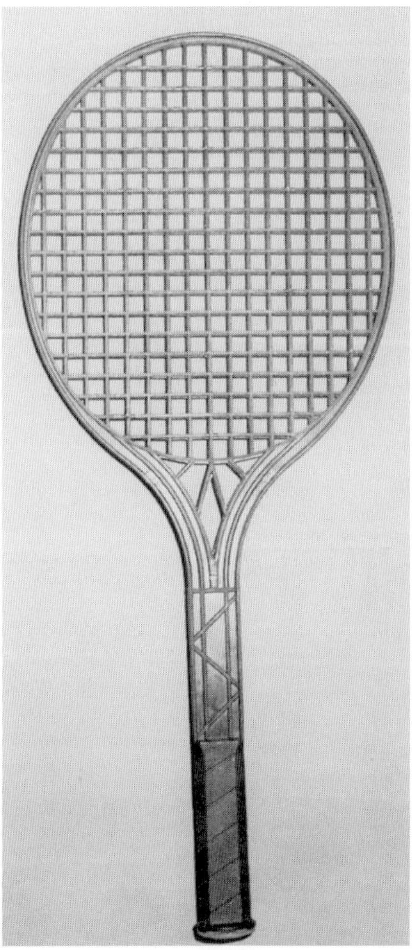

Figure 1 Injection molded, polypropylene toy tennis racket

Cavity plates (18, 19) with attached handle inserts (36) lie on both sides of the mold parting line. The tips of the core slides (3) project into the hollows formed by the handle inserts. The core slides can be moved on T-shaped guide plates (6) and guide bars (7) parallel to the mold parting line.
The ejector bars (37) are bolted to the guide plates (6). These are guided in bushings (38) and they bear two ejector plates (23, 24). The angle pins (4) actuate the two core slides (3) on mold opening and closing. They are locked in the injection position by wedge plates (1). In the open position, the ball catches (40) secure the core slides against unintentional movement. Rolls (26) brace the plate (20) against the ejector-side clamp plate (22).
The part-forming surface in the cavity plate and in the handle inserts are produced by spark erosion.

Operation of the Mold

The cavity wall temperature is 50 to 90°C (122 to 194°F). For ejection, the core slides (3) are partly pulled out of the handle bores of the rackets by the angle pins on mold opening (4). The force of the springs (33) is greater than the force exerted on the slides by the angle pin in the mold-opening direction, with the result that the guide plates (6) with their core slides (3) do not lift away from the guide bars (7) in the cavity plate (19).
The core slides therefore initially keep the two molded parts on the ejector side and pull them from the cavity plate on the feed side (18). Sprues and runners also remain in the moving mold half because of the undercut in the guide bushing for the sprue ejector pin (25).
After 60 mm of the mold opening stroke, the core slides (3) have moved outward by around 25 mm and stopped. The mold opens further till the ejector rod (30) meets the machine end stop. The two ejector plates (23, 24) now eject the molded part via ejector pins (12, 25) out of the ejector-side mold cavities and eject the sprue.
At the same time, the guide plates (6) are lifted out (3) from the cavity plate (19) with the core slides. The molded parts now sitting loosely on the slide lobes can be removed by hand. This process compresses the springs (33) further.
The springs (33) ensure during mold closing that core slide and ejectors get into the injection position before the mold closes completely. The ejector-plate return pins (31) secure the stop position of the ejector mechanism.
With this mold, toy rackets can be produced with relatively little mold outlay. However, an operator is required because the molded parts are removed manually from the injection molding machine.

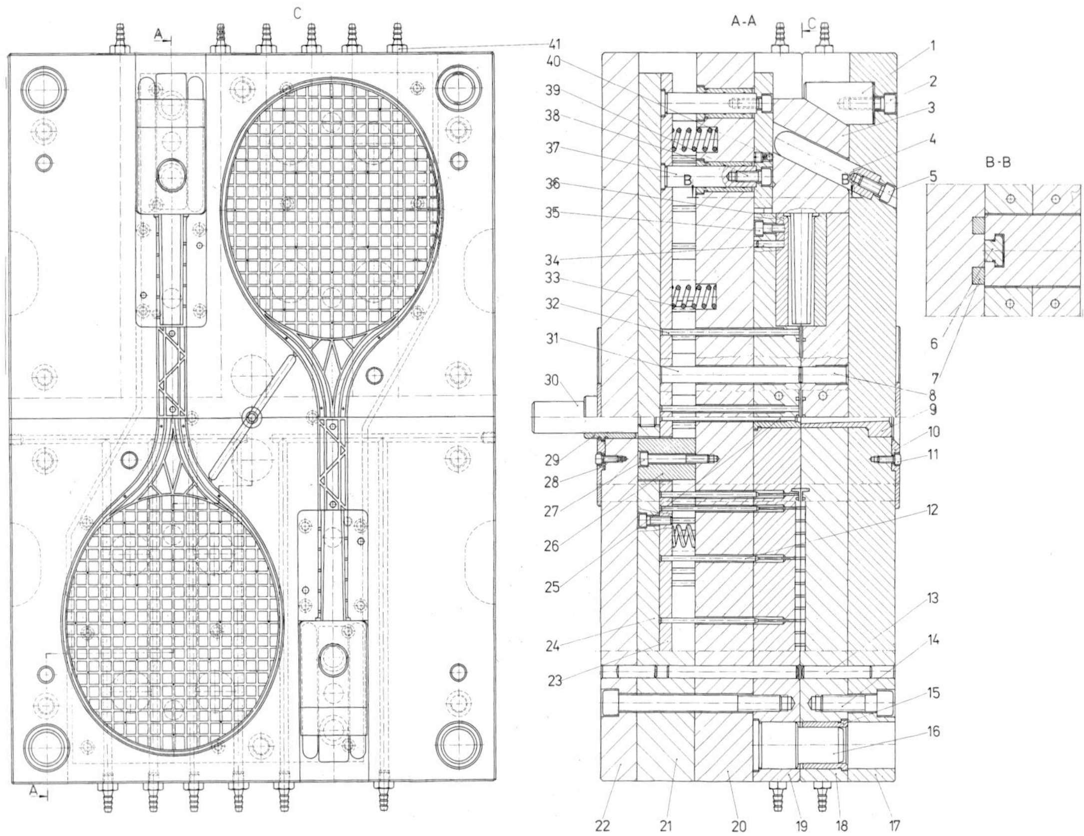

Figure 2 Two-cavity injection mold for making the racket shown in Fig. 1
1: wedge plate; 2: screw; 3: core slide; 4: angle pin; 5: screw; 6: guide plate; 7: guide bars; 8: buffer pin; 9: sprue bushing; 10: locating pin; 11: screw; 12: ejector pin; 13: locating pin; 14: screw; 15: guide bushing; 16: guide pillar; 17: mounting plate; 18, 19: mold cavity plate; 20: plate; 21: spacer plate; 22: mounting plate; 23: ejector retaining plate; 24: ejector plate; 25: sprue-ejector pin; 26: support roll; 27: screw; 28: locating ring; 29: guide bar; 30: ejector rod; 31: return pin; 32: ejector pin; 33: helical spring; 34: locating pin; 35: screw; 36: handle inserts; 37: ejector bar; 38: guide bushing; 39: screw; 40: ball catch; 41: cooling connection
(Courtesy: Plastor p.A., Oradea/Romania)

Example 31, Two Injection Molds with Two-Step Ejection Process for Housing Components from Polycarbonate

Injection molds and pressure die-casting molds, depending on the type of gate and the shape of the molded article, sometimes require a multi-step ejection process in which individual ejection elements have a different stroke in order to safely release the molded article (molded parts plus gate) from the mold. This is commonly achieved by employing mechanically controlled devices which, in conjunction with additional ejection plates, facilitate this releasing process.

The hydromechanical ejector accelerator described in this context allows a particularly simple mold design while requiring little effort in terms of assembly. Figure 1 shows its structure as well as its function. The ejector accelerator (1) as well as the ejector pin (4) are inserted into the two ejection plates (2, 3). Part (1) also receives the ejector pin (5). The ejection plates (2) and (3) are actuated by the ejector bolt (6).

A stroke H1 is available for the ejection plates. Activating the ejector bolt (6) initially gives rise to an equal forward motion of ejector pins (4) and (5), until – after a stroke H2 – the annular piston (7) of the ejector accelerator (1) strikes against the stopping face (8) (Fig. 1B). Once the ejection plates (2, 3) are moving further, the annular piston (7) is being pressed into the oil chamber (9), where it replaces the existing hydraulic oil. As a result, the central piston (10) dissociates from its piston stop (1). By the end of stroke H1 of the ejection plates (2, 3), the ejector pin (5) in the central piston (10) has passed a distance exceeding that of the ejector pin (4) by a distance H (Fig. 1C). When the ejection plates are retracted, the restoring springs (12, 13) are pushing the central piston (10) and the annular piston (7) and therefore also the ejector pin (5) back into their original position.

The ejector accelerator is available in two sizes (Table 1). The size 1 version can receive ejector pins with a tip diameter of up to 6 mm and a tip height of up to 4 mm; the corresponding tip sizes for the size 2 version are 12 mm in diameter and 10 mm in height, respectively.

Application in Injection Molding of Caps...

Figure 2 shows the application example of a double cap mold. The two molded parts are injected sideways via a sprue gate with a runner and tunnel gates (1).

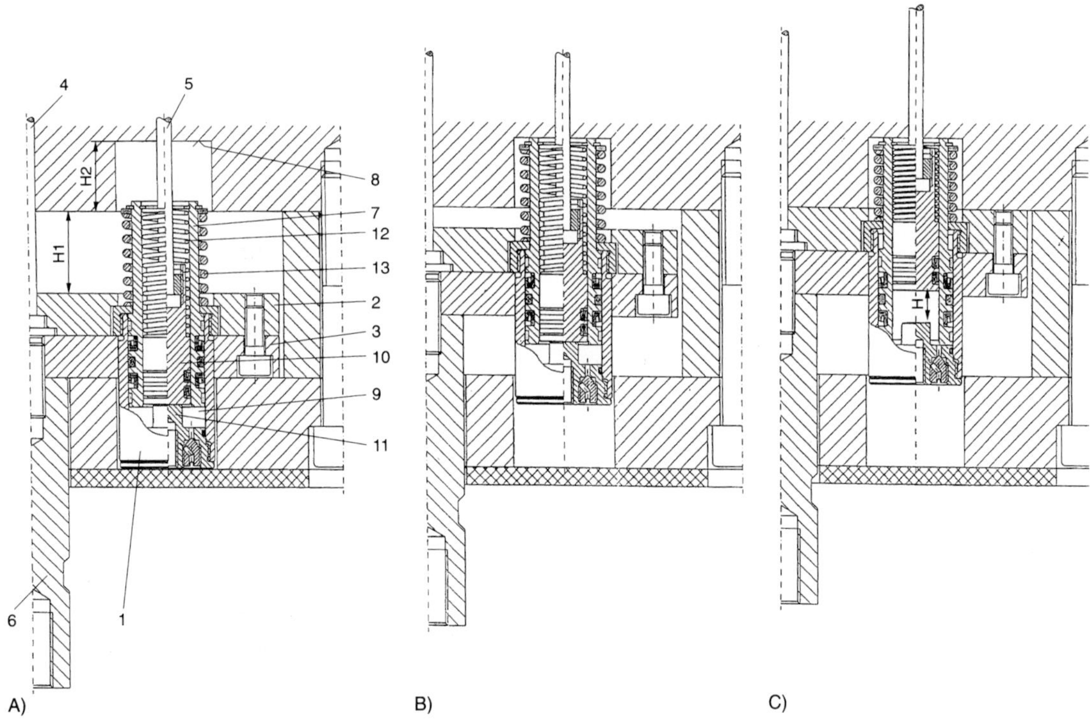

Figure 1 Hydromechanical ejector accelerators
1: ejector accelerator; 2, 3: ejection plates; 4, 5: ejection pins; 6: ejector bolt; 7: annular piston; 8: stopping face; 9: oil chamber; 10: central piston; 11: piston stop; 12, 13: restoring springs
A: ejector accelerator prior to ejecting; B: end of the synchronous movement of the ejector pins 4, 5 and beginning of the advance of pin 5; C: end of the ejection process
(Courtesy: Kühner Industrievertretungen GmbH, Reutlingen, Germany)

Table 1. Measures for the hydromechanical ejection accelerators

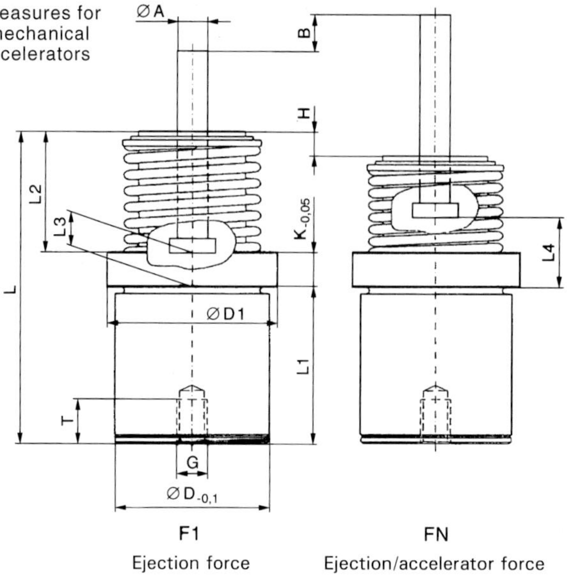

Typ	A	H	B	L	L1	L2	L3	L4	K	D	D1	G	T	F1	FN
size 1	1...4	6,0	8,4	70,0	33,5	28,5	7,0	15,4	8,0	23,0	26,0	-	-	175N	210N
size 2	2...8	8,0	12,5	105,5	52,5	41,0	11,5	24,0	12,0	40,0	44,0	M8	15	370N	430N

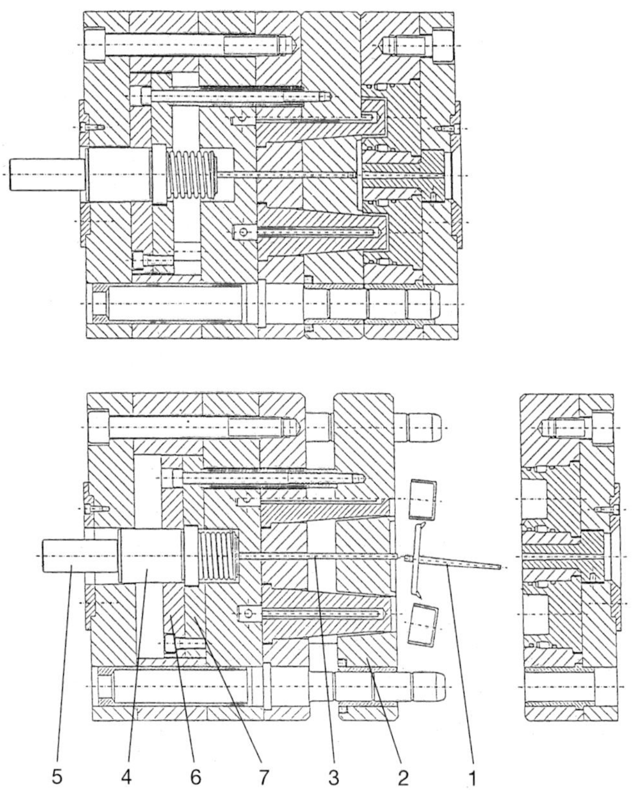

Figure 2 Two-cavity die for injection molding of caps
1: sprue; 2: stripper plate; 3: sprue ejector pin; 4: ejector accelerator; 5: ejector bolt; 6, 7: ejection plates

A pusher plate (2) is used to push the caps off the cores; this pusher plate also contains the runner and the boring for the gate extractor spigot.

When the mold is opened, the molded articles are being pulled out of their die cavities while the sprue is being pulled out of the sprue bush; the tunnel gate points are being severed and the sprue is snapped off the tunnel borings.

When the pusher plate (2) subsequently moves to the right, the caps dissociate from the cores and may fall off. The additional pusher stroke of the gate ejector (3) that is required to throw the sprue off the pusher plate (2) is provided by the ejector accelerator (4).

In this example, the ejector accelerator (4) is mounted in the center of the mold. The ejector bolt (5) is positioned in a thread boring on the bottom of the ejector accelerator which receives the two ejection plates (6) and (7).

...and of Casing Parts

The mold shown in Fig. 3 is used to manufacture casing parts which have projecting parts or undercuts along their edges. These are molded with the shaped pins (1, 2). The shaped pins (1, 2), the sprue ejector pin (3), the return pins (4), and the ejector accelerators (5) are mounted in the ejection plates (6, 7). During ejection, the shaped pins and the ejector accelerators initially push the parts away from the cores. The flatter casing part (on the left-hand side inside the mold), however, still adheres to the shaped pins for a while but is finally pressed out of these. The left-hand side of the photograph viewing the cavities from above (Fig. 3, below) demonstrates that three ejector accelerators are required for this purpose. Using only one or two ejector accelerators would cause tilting, so that one side of the casing part could get caught. Only the arrangement in one plane (triangle) avoids this danger. Release from the mold is therefore accelerated, since this technique obviates the need for the commonly used but time-consuming method of ejector jolting.

A coil spring (8) is sticking on the return pin (4); this coil spring enters a boring (9) as the ejector sheets are moving forward. This has the following reason: Table 1 lists two forces F1 and FN for each of the two ejector accelerator sizes. These are the forces exerted by the ejector accelerator immediately after touching the stopping face (8) (Fig. 1) or after a complete stroke of the ejection plates. These forces are effective even if there are no molded articles for the ejector pins (4, 5) to push (Fig. 1). If – as in this case – the ejector accelerators are positioned unsymmetrically relative to the ejector bolt (10) (Fig. 3), then an overturning moment is created, which may cause jamming of the ejection plates in their guideways. This overturning moment is almost entirely compensated by the two springs (8) sitting on the two return pins (4) opposite the ejector accelerators.

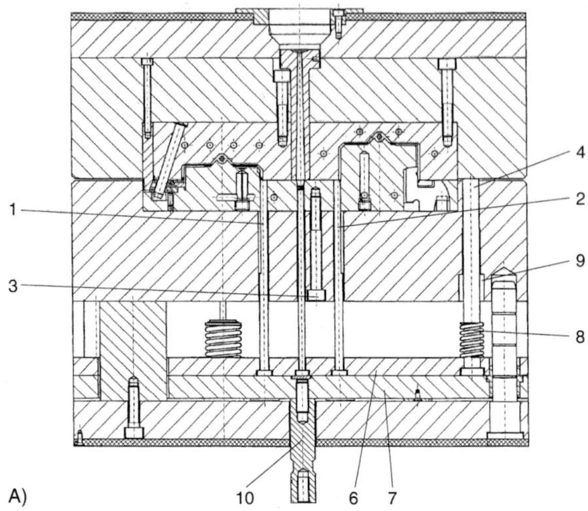

A)

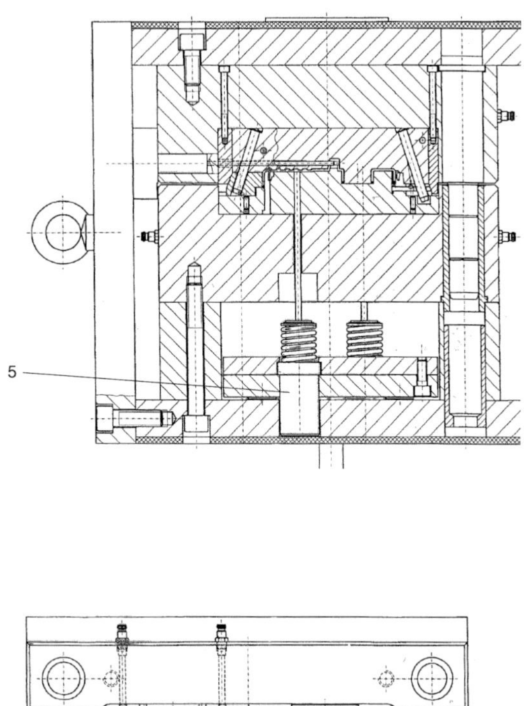

B)

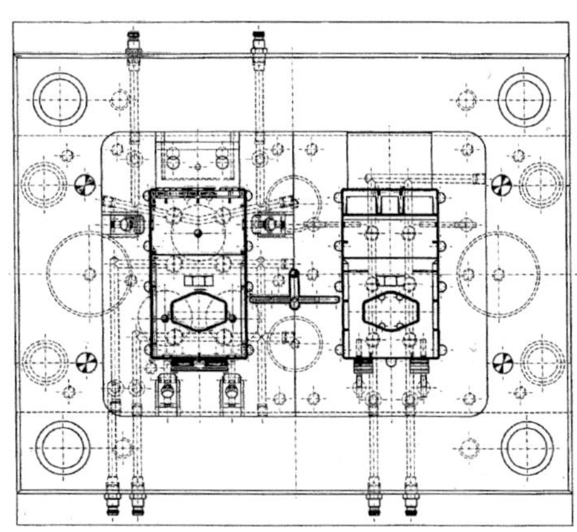

C)

Figure 3 Single injection mold for the upper and lower part of a casing made of polycarbonate
A: section; B: section turned by 90°; C: top view
1, 2: shaped pin; 3: sprue ejector pin; 4: return pin; 5: ejector accelerator; 6, 7: ejection plates; 8: coil spring; 9: boring; 10: ejector bolt
(Courtesy: Roth Plastik & Form, Balingen-Weilstetten, Germany)

Example 32, Injection Mold for a Polypropylene Container with a Threaded Neck

This is the description of a three-plate injection mold for a polypropylene container that does not show in the visual range any witness lines from slides. Sprue and ejectors are designed for automatic operation.

Specifications

When the development of a household appliance had been concluded, the task of constructing a tool for the container of the appliance arose. A molded part drawing served as the basis on which the geometrical shape, dimensions, permissible tolerances and polypropylene as material had been firmly established. The container concerned is a largely rotationally symmetrical high-quality molded part with relatively high demands on the surface, on dimensional accuracy (required for assembly and operation) and on the mechanical loading of the polypropylene.

Mold Design

The design shown in Figs. 1 to 4 is a three-plate mold (parting line I and II) with stripper plate (*p*). When the injection molding machine opens, the mold is forced to open at I by the latches (*b*). Through this movement the two thread-forming slides (*c*) are pulled out of the thread by the cam pins (*d*), thereby releasing the undercuts formed by the thread. The latches are guided by the cam strips (*e*) and they release after 28 mm of travel. Mold plate (*f*) stops. The mold opens in the main parting line (II). The cavity retainer plate is guided by four additional guide pins (*g*). This plate is secured by end-stops (*h*).

Runner System

The melt is fed into the cavity via sprue, conventional runners and gates (*i*). Despite the relatively high wall thickness of the molded part (3.5 mm) attempts were successfully made to inject via four submarine gates (*i*) of 1.8 mm diameter each. The gates are sheared off during mold opening (I). The gates are in the upper mold cavity. The choice of this gating system and its position offer the following advantages:

- Central gating. This ensures even mold filling
- Position of the gate is outside the functional and visible areas of the molded part
- Feasibility of automatic production without post-molding finishing of the gate area.

Temperature Control

Uniform mold temperature control to 50°C (122°F) is necessary to satisfy the quality requirements. Effective temperature control had also been asked for in order to achieve short cycle times. The mold was therefore equipped with five temperature control systems TS, independent of one another:
System 1: Temperature control for the main core (*k*)
System 2: Temperature control for the cavity (insert) (*l*)
System 3 + 4: Temperature control for the thread-forming slides (*c*)
System 5: Temperature control for the contour-forming sprue bushing (*m*) and the auxiliary core (*n*). The effectiveness of temperature control in the auxiliary core is increased by a copper pin (*o*).

Ejector Systems

The molded part is stripped off the core by the stripper plate (*p*). This movement is effected by a hydraulically operated ejector on the machine. The stroke of the ejector plate (*p*) is limited by two bolts (*q*) with end stops (*r*). The molding bead (*s*) in the ejector plate could cause it to stick on demolding. For that reason the air blast (*t*) has been provided, which is charged with compressed air at the end of the ejection process. This ensures certain discharge of the molded part. The sprue is pushed out of the retaining undercut and ejected by a separate ejector system (*u*). The movement of this ejector is actuated by contact of the bolt head (*v*). This bolt serves to couple the machine ejector with the ejector plate (*w*).

Materials Used in the Mold

The following materials were chosen for the individual parts of the mold:

- Part-forming inserts: case-hardening steel, case hardened
- Cavity retainer plate and stripper plate: nitrided annealed steel
- Mold base: tool steel, unalloyed

Fig. 1

Fig. 2

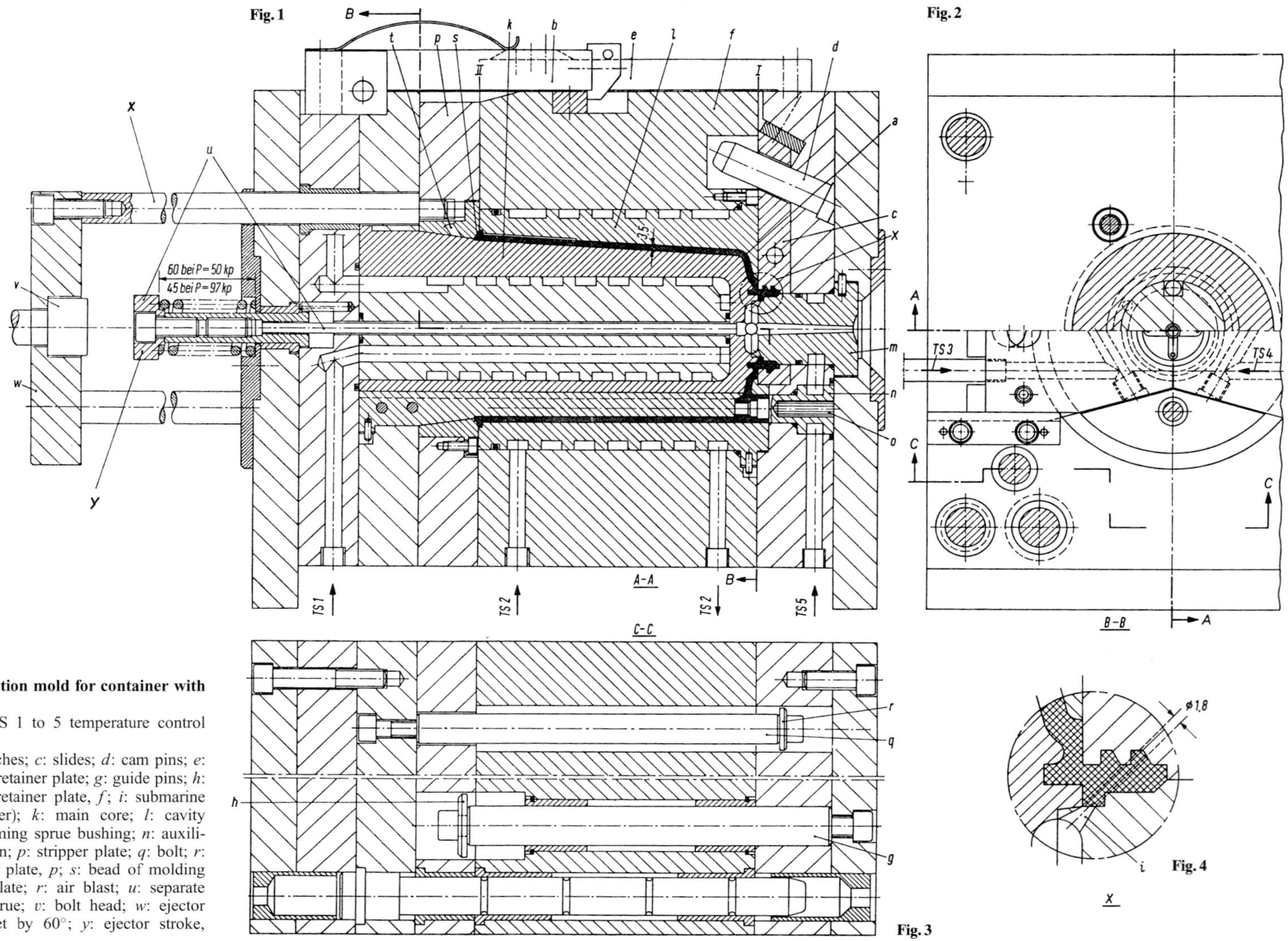

Fig. 3

Fig. 4

Figures 1 to 4 Injection mold for container with threaded neck
I, II: parting lines, TS 1 to 5 temperature control systems
a: molded part; *b*: latches; *c*: slides; *d*: cam pins; *e*: guide cams; *f*: cavity retainer plate; *g*: guide pins; *h*: end stops for cavity retainer plate, *f*; *i*: submarine gate (1.8 mm diameter); *k*: main core; *l*: cavity (insert); *m*: shape-forming sprue bushing; *n*: auxiliary core; *o*: copper pin; *p*: stripper plate; *q*: bolt; *r*: end stops for stripper plate, *p*; *s*: bead of molding running in stripper plate; *r*: air blast; *u*: separate ejector system for sprue; *v*: bolt head; *w*: ejector plate; *x*: shown offset by 60°; *y*: ejector stroke, 15 mm maximum

Example 33, Three-Plate Injection Mold with Stripping Device for a Precision Magazine

Integrated circuits (IC) are generally mounted automatically during production of electronic devices. This requires that they be stored in magazines. The magazine shown in Fig. 1 has an undercut U-shaped groove to hold the ICs and a recessed bottom that is reinforced by transverse ribs. Snap fits and mounting holes are located at either end of the magazine.

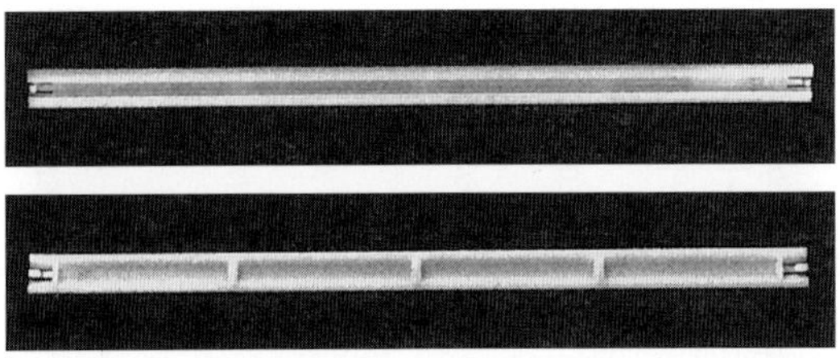

Figure 1 Precision magazine to hold 50 integrated circuits

top: view from above with undercut groove to hold the ICs; bottom; view from below showing the stiffening ribs by which the magazine is pulled off the groove-forming core

Mold

In the mold (Figs. 2 to 11), the longitudinal axis of the molded part is transverse to the opening direction of the mold. The ribbed bottom faces the runner system. The cavity is filled via two pinpoint gates from a runner system and a heated sprue bushing (16).

Mold Operation

Parting Line I
After a partial opening stroke of approx. 7 mm, the spring-loaded bolts (69) cause the gate to shear off. The sucker pins (22) hold the runner in the runner plate (5), and the snap fits as well as the mounting holes of the molded part are released by retraction of the slides (33). Mold plates (3) and (4) remain closed during this time.

Parting Line II (mold opening)
Parting line II (between plates 3 and 4) is opened during further opening of the injection mold. This phase of opening is concluded after a distance of 80 mm. During this opening motion, the ejector pin (36) is pushed into the eject position by the lever (29) with the aid of a stripper bolt (38). This lifts the hook-shaped detent on the molded part so that it can subsequently be stripped off the mold core.

Parting Line III
After a further opening stroke of 60 mm, parting line I opens completely through the action of stripper bolt (67). Two further stripper bolts (66) engage mold plate (4), pulling the runner plate (5) forward. The complete runner system is now pulled off the sucker pins (22) and away from the heated sprue bushing (16), dropping out of the mold. The spring-loaded strippers (72) prevent the runner from sticking to the runner plate (5). Shoulder bolts (68) prevent the mold plates from being pulled off the guide pins completely. They do not, however, serve as a mechanical stop. This is found on the molding machine.

Ejection of the Molded Part

During the opening sequence of the mold, the stripping device, which is mounted on the moving half of the mold, has moved into the ejection position. Only in this position will the limit switch initiate operation of the stripping device.
During the inward movement, the guide blocks (56) glide under the guide strips (31), the carefully adjusted latches (54) slide over the transverse stiffening ribs of the molded part and, once the stripping device is fully inserted, engage behind two of these ribs. A limit switch (58) is actuated simultaneously in this position to initiate the stripping operation. The molded part is now stripped off the entire length of the mold core (27), which is 162 mm long, and then drops out of the mold. Once the stripping device has returned to its starting position, limit switch (59) is actuated and another cycle can start if the infrared mold safety has also cleared the start.
The infrared mold safety monitors the lower mold half and checks whether the molded part and stripping device have cleared the tool area. Only when both conditions have been fulfilled are the machine controls able to initiate another molding cycle.

Figures 2 to 11 Three-plate injection mold with stripping device for a precision magazine
1: movable-side mold base plate; 2: spacer plate; 3, 4: mold plate; 5: runner plate; 6: cam pin retainer plate; 7: stationary-side mold base plate; 8, 9: leader pins; 10, 11, 12, 13, 14, 15: guide bushings; 16: heated sprue bushing; 17: movable-side locating ring; 18: insulating plate; 19: junction box for heated sprue bushing; 20: locating sleeve; 21: cam pin; 22: sucker pins; 23: spring; 24: mold core; 25, 26: core insert; 27: core; 28: wedge; 29: lever; 30: cavity end insert; 31: guide strip; 32: slide guide; 33: slide; 34: core; 35: articulated head and guide for ejector pin 36; 36: ejector pin; 37: fulcrum for lever 29; 38: stripper bolt; 39: spring-loaded bolt; 40: spring; 41: pivot pin; 42: housing for sensor; 43: sensor; 44: microswitch; 45: housing for microswitch 44; 46: mechanical stop for slide 33; 47: spring; 48: stripper head; 49: base plate for stripping device; 50: bearing support for guide; 51: sheet metal reinforcement; 52: support plate for stripping device; 53: outboard bearing for guide; 54: latch; 55: latch spring; 56: guide block; 57: mechanical limit for inward movement; 58, 59: limit switches; 60: cylinder connection; 61: spring; 62: hydraulic cylinder; 63: linear bearing; 64: guide rod; 65: circlip; 66, 67, 68: stripper bolts; 69: spring-loaded bolt; 70: runner ejector; 71: adjustment screw; 72: stripper; 73: stationary-side locating ring ▶

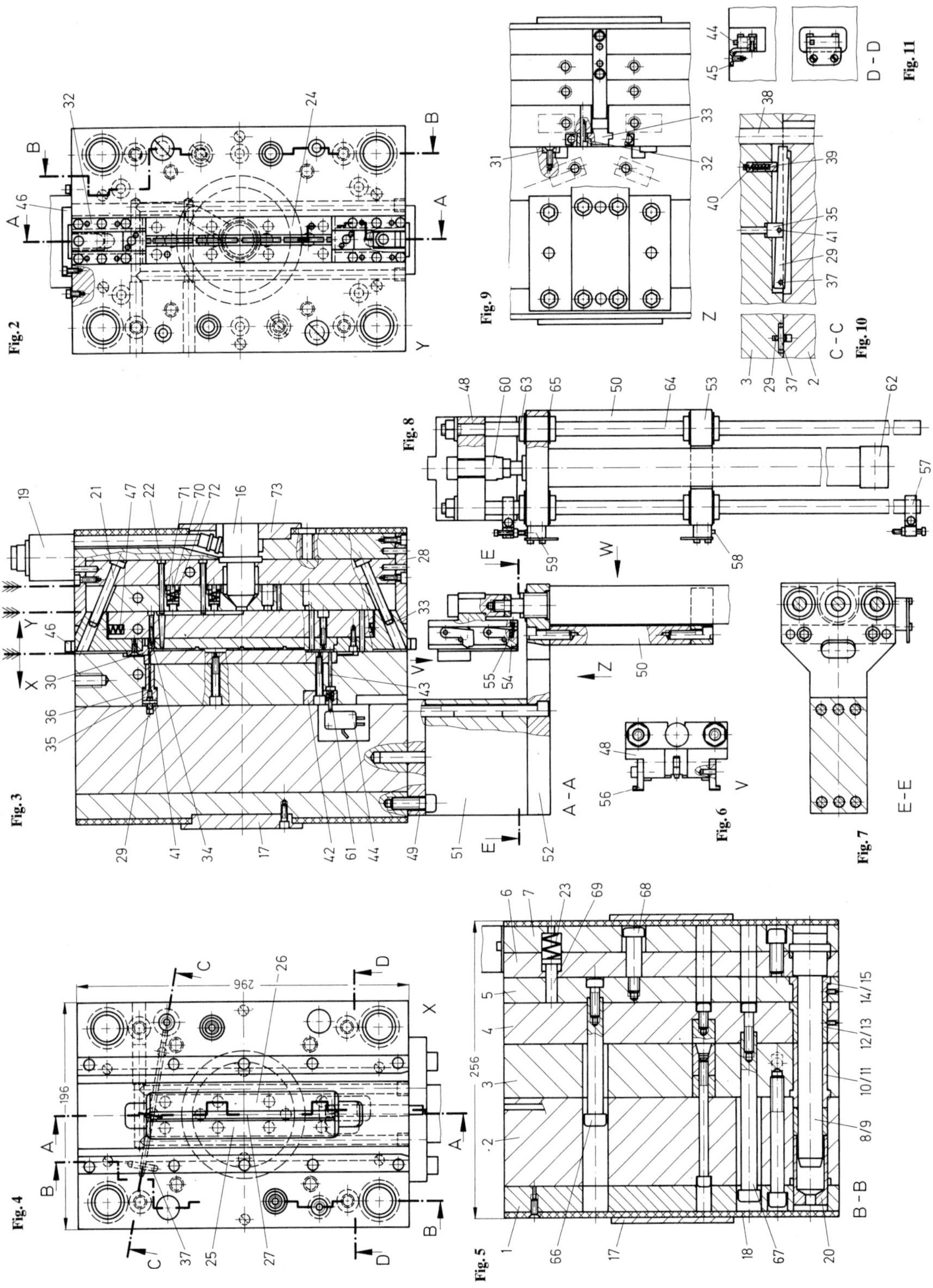

Example 34, 6-Cavity Metal-Powder Injection Mold (MIM) for Transport Fasteners

To produce threaded transport fasteners in MIM technology, a standard mold (see also DIN ISO 12165:2002-06, Fig. 11) was selected. The six-cavity mold is designed mainly with Z standard elements (Hasco). For reasons of wear and temperature, no. 1.2344 (X40 Cr Mo V 5-1) hot-work steel hardened to 56-58 HRC was used for the inserts. The molded parts are produced from the powdered molding compound, Catamold FN02 [1], [2]. Leaving aside the thermoplastic matrix, this material consists of $< 0.1\%$ C, 1.9 to 2.2% Ni, the rest being Fe. In hardened condition, approx. 55 HRC hardness is achieved.

Several peculiarities have to be considered before processing this compound. Standard injection molding machines are, in principle, suited for processing powdered compounds, however the plastifying unit should be as wear-resistant as the mold. If "normal" thermoplastics are processed on the same machine, the screw, non-return valve, and cylinder have to be purged (not just "pumped empty"!) after retooling in order to eliminate negative consequences for molded part properties during subsequent sintering. Compared to unfilled thermoplastics, this powdered compound has significantly higher temperature and thermal conductivity (approx. $10\times$ greater). In consequence, for example, early gate closing can cause cavity underfilling. For this reason, it is necessary to design flow channels generally as wide as possible. This is doubly necessary, since melt viscosity and therefore pressure loss, are high. A tendency to jetting must also be considered. These problematic issues can be kept under control by stepping the injection rate profile.

It should be observed that the processing window of the powdered compound is significantly smaller compared to standard products, so that considerable experience is required. To produce the transport fastenings with a combined molded part weight of 6×4.9g and a sprue weight of 17g, the following characteristic processing parameters were chosen:

Injection rate	$80 cm^3/s$
Injection pressure	1500 bar
Cooling time	25 s
Cycle time	38 s

Mold

This mold cooling system is relatively complex. Mold wall temperature is 130 °C. In addition to temperature control using pressure water – given a temperature of 130 °C, the pressure is approx. 3 bar – a regulated electrical heating system is provided. The outer mold surfaces are completely covered by insulating sheets against heat loss. Together with temperature regulation, they provide the thermal balance that is a prime requirement for uniform molded part quality (Fig. 1). Manifold and gate dimensions are given in Fig. 2. Jetting is not to be expected, since injection is directed against a core. Polished venting gaps of approx. 0.03 mm are ground into the parting planes of the mold. After the mold opens, the still fragile ("green") parts are demolded gently together with their runners, picked up by a removal device and set down. The runner is mechanically separated, regranulated and remixed with the original material for reprocessing without loss of properties. It is possible in principle to process powdered compounds on hot-runner systems; however, the increased sophistication and increased equipment requirements often, as in the case at hand, lead to the selection of a standard mold.

Subsequent Treatment of the Green Parts

Subsequent to injection, the green parts are catalytically debindered in a separate operating procedure (the polymer matrix is removed under specified conditions) and sintered. Debindering takes place in a nitrogen atmosphere with small traces of gaseous nitric acid at approx. 140°C in a drying furnace. The polymer matrix is thereby entirely removed. Subsequent sintering is also performed in nitrogen atmosphere at 1200°C. Molded parts produced by this method require no further treatment.

Literature

1. Data sheet Catamold FN02 D/Ca 006e Oct. 2001, BASF
2. *Hickmann*, Th., *Klemp*, E.: Metal Injection Molding – MIM: Schnell und günstig zu Präzisionsteilen, Kunststoffe 94 (2004) 11, p. 62–65

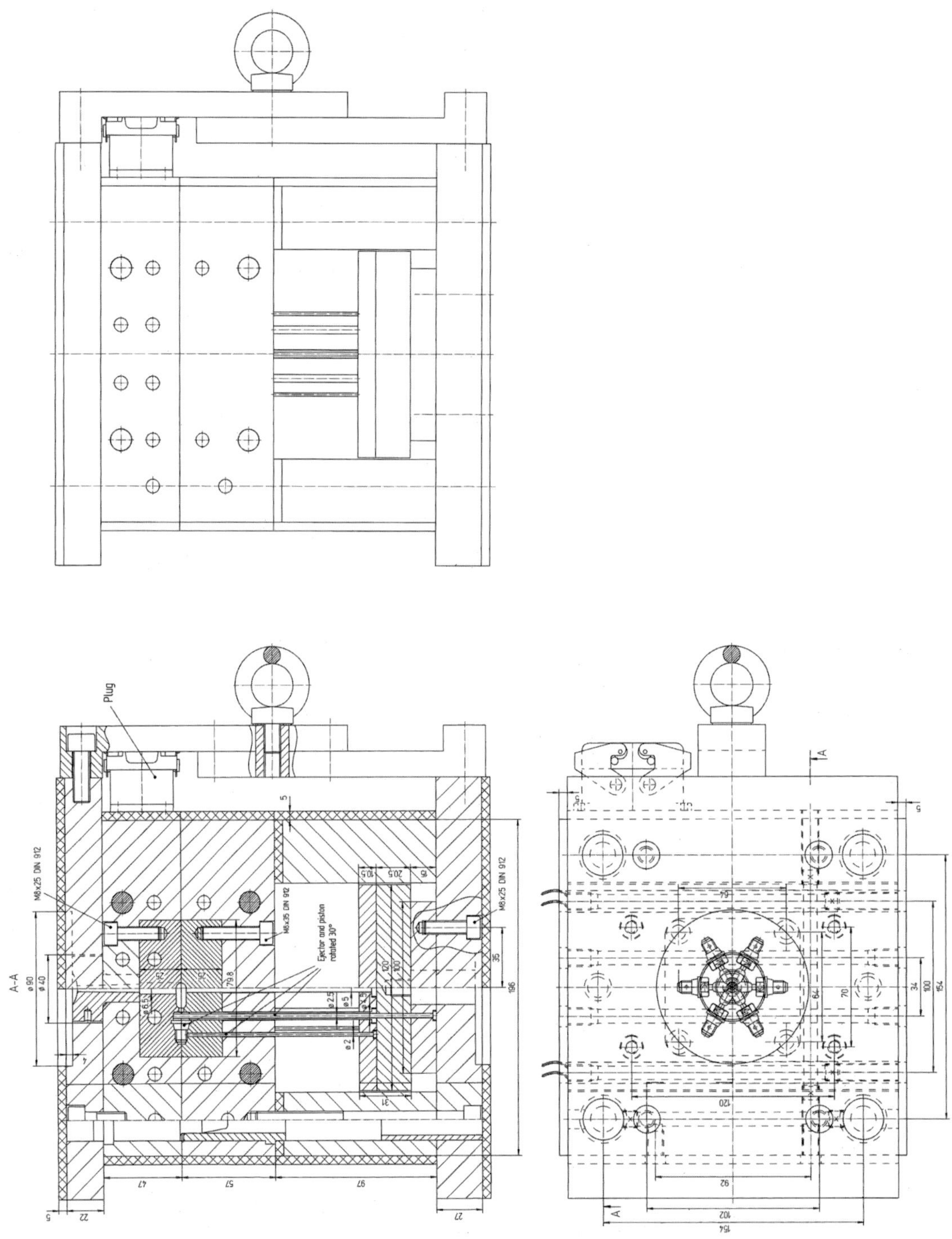

Figure 1 Six-cavity metal-powder injection mold (MIM) for transport fastener

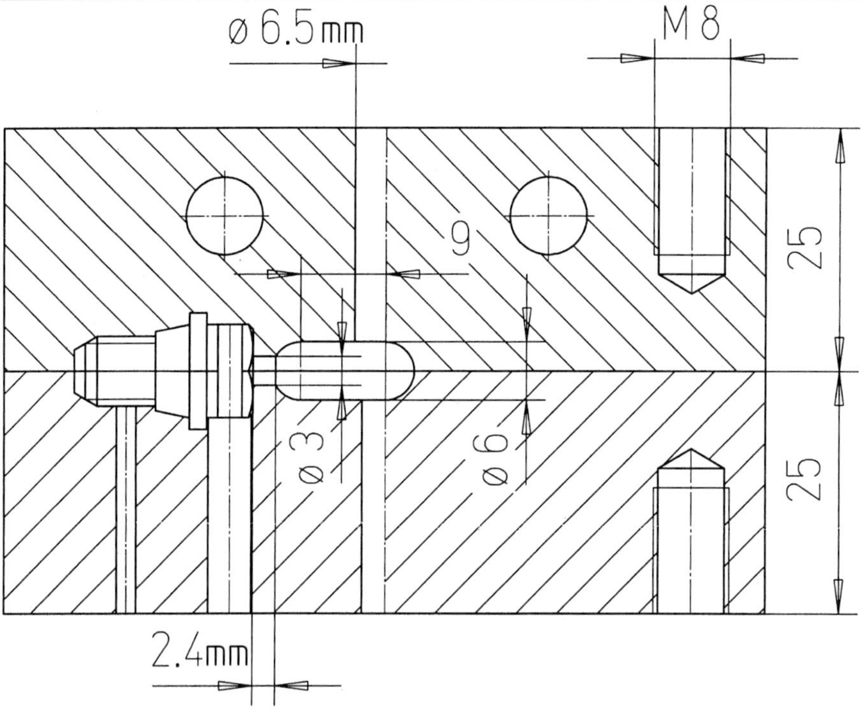

Figure 2 Configuration of runner and gate
(Courtesy: Eisenhuth, Osterode)

Example 35, Mold for a Polyamide V-Belt Pulley

Internal and external undercuts on injection molded parts cause the costs for both mold making and mold maintenance to increase considerably, since slides and their actuating mechanisms become necessary. Moreover, with increasing use of the mold, the reliability drops due to wear of precisely these additional components.

The polyamide V-belt pulley shown in Fig. 1 is an example of a practical design showing how slides in the mold can be avoided. If the V-belt pulley were of a one-piece design, two or three slides per cavity would be required to form the V-belt groove in the pulley. Any flash formed in the groove with this mold design could be removed only with a great deal of effort. If the flash were not carefully removed, there would be a danger of damaging the V-belt. By dividing the pulley across the axis of rotation, the mold design shown in Fig. 2 becomes possible. Furthermore, the snap fit required for assembly may also be achieved without the use of slides. Both halves of the pulley were designed to be identical, so that the mold inserts are identical and any two pulley halves may be assembled together.

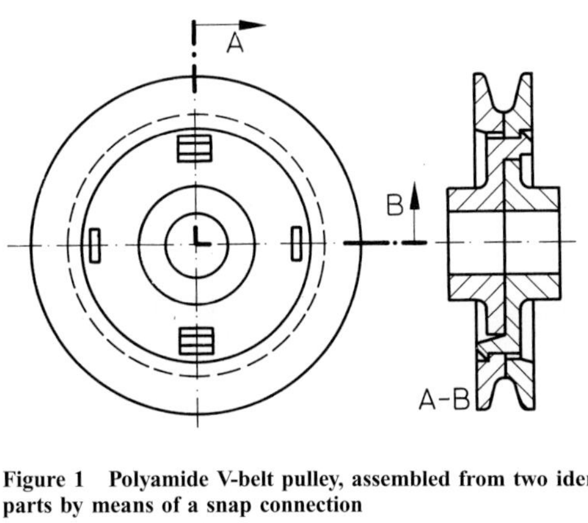

Figure 1 Polyamide V-belt pulley, assembled from two identical parts by means of a snap connection

The three-plate mold (Fig. 2) operates fully automatically. Opening begins at parting line I, since plates (3) and (4) are held together by pin (25) and latch (24) until the bar (23) releases the latch (24) via the adjustment screw (27). Further opening movement of (3) and (4) is prevented by stop bolts, so that parting line II also opens, and the molded parts may be ejected by the ejector pins (17). After removing the diaphragm gates, the molded parts may be snapped together to form the V-belt pulleys.

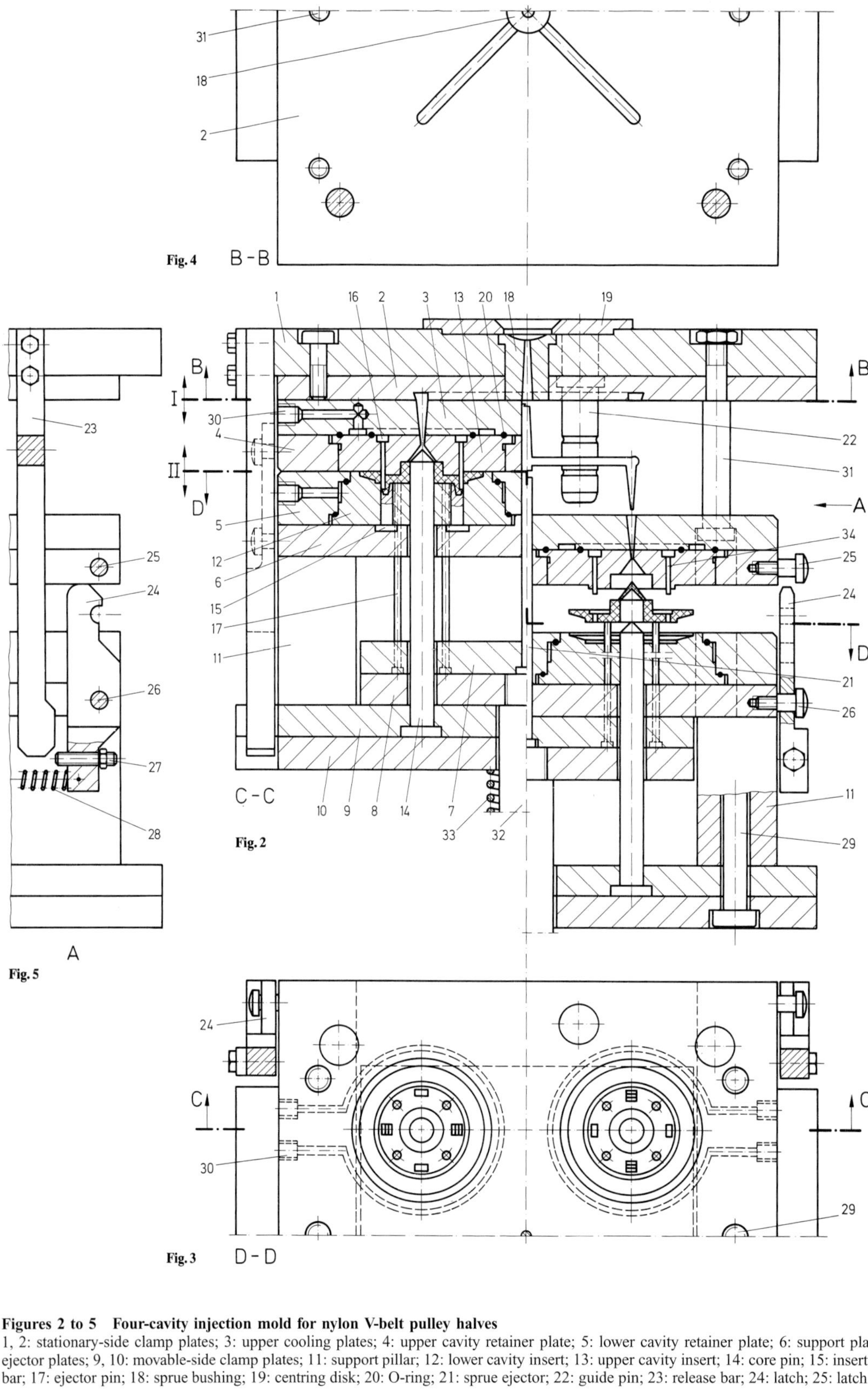

Figures 2 to 5 Four-cavity injection mold for nylon V-belt pulley halves

1, 2: stationary-side clamp plates; 3: upper cooling plates; 4: upper cavity retainer plate; 5: lower cavity retainer plate; 6: support plate; 7, 8: ejector plates; 9, 10: movable-side clamp plates; 11: support pillar; 12: lower cavity insert; 13: upper cavity insert; 14: core pin; 15: insert; 16: flat bar; 17: ejector pin; 18: sprue bushing; 19: centring disk; 20: O-ring; 21: sprue ejector; 22: guide pin; 23: release bar; 24: latch; 25: latch pin; 26:

Example 36, 2 × 8-Cavity Hot-Runner Stack Mold for Yoghurt Cups Made from Polypropylene

Stack molds for thin-walled cups must be built extremely ruggedly and precisely in order to avoid variations in wall thickness. Air ejection eliminates the need for mechanical ejector mechanisms and involves little wear.

Stack molds are employed whenever parts with low weight, minimal wall thickness and large projected area have to be produced in large numbers.

These polypropylene yoghurt cups weigh 13.4 g and have a wall thickness of 0.63 mm. A unique feature of these cups is the bottom rim (raised bottom), which requires special release techniques. Depending on the molding material used, yoghurt cups have wall thicknesses of 0.4 to 0.65 mm. The wall thickness within the cavity requires great accuracy, i.e. positioning of core and cavity must be extremely precise, so that the melt does not advance down one side, displacing the core in relation to the cavity. Should this be the case, it will be impossible to obtain properly filled parts.

Mold

Figures 1 and 2 show how ruggedly the entire mold, including core and cavity, has been dimensioned. Each core (1) and its allotted cavity (2) are aligned by the ring (3). All part-forming components are hardened, while the mold plates are nickel-plated. The mold has been designed with the aid of a CAD system and its components have been produced by means of CAM techniques.

Mold Temperature Control

The minimal wall thickness of the molded parts allows for rapid cooling if the cooling system in the mold has been laid out correspondingly. As is customary with such cup molds, the cooling channels in core and cavity lie close together just below the mold surface. The cores are capped (8) with Cu–Be.

The paths of the cooling channels are shown in Fig. 3.

Fig. 1

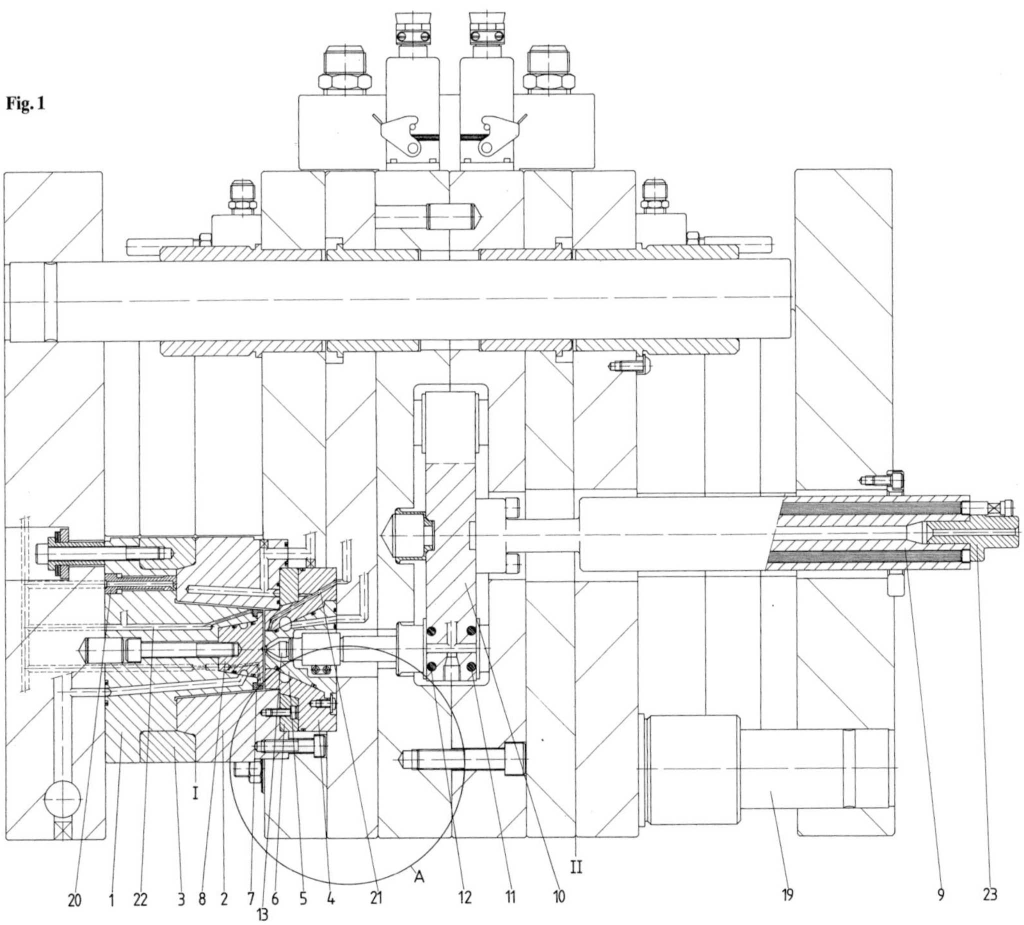

Fig. 2

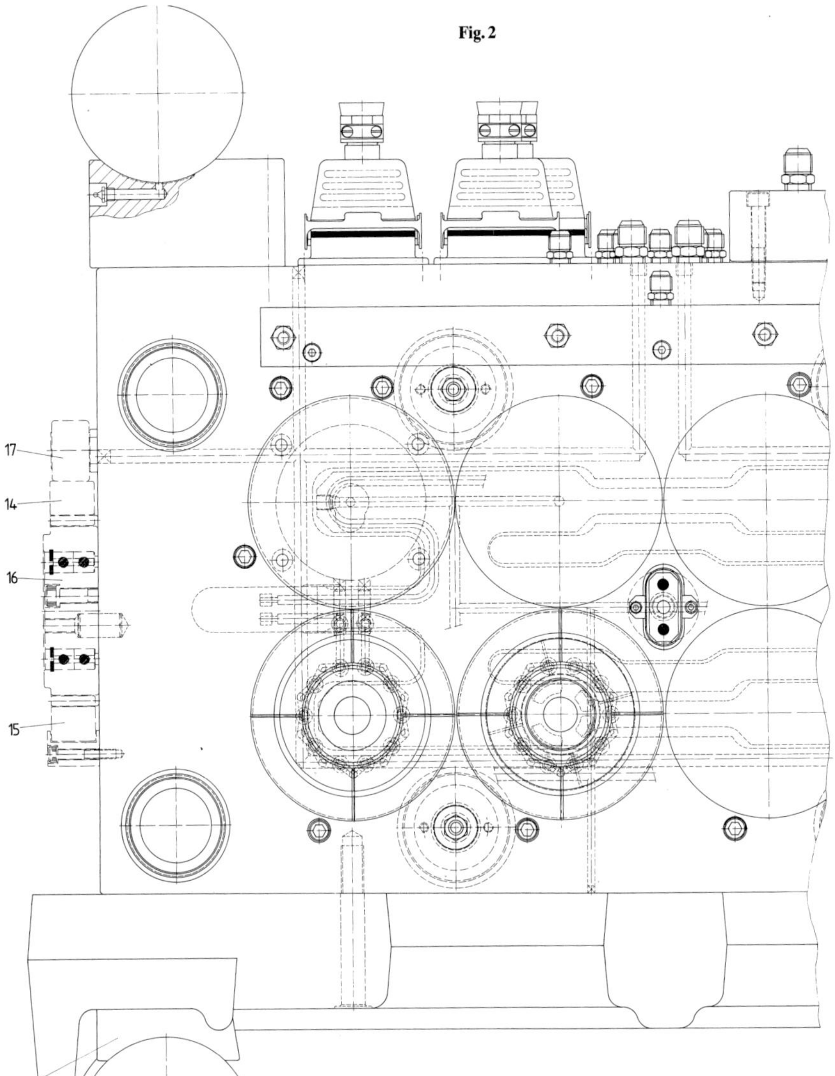

Figures 1 and 2 2 × 8-cavity hot-runner stack mold
1: core; 2: cavity; 3: locating ring; 4: cavity; 5: nozzle bushing; 6: locating ring; 7: core head ring; 8: core cap; 9: heated sprue bushing; 10: manifold; 11, 12: heating coils; 13: heated nozzle; 14, 15: gear racks; 16: pinion; 17: support cover; 18: support shoe; 19: guide rod; 20: air jet; 21: air channel for cavity bottom; 22: air channel for core; 23: sliding adapter; 24: piston

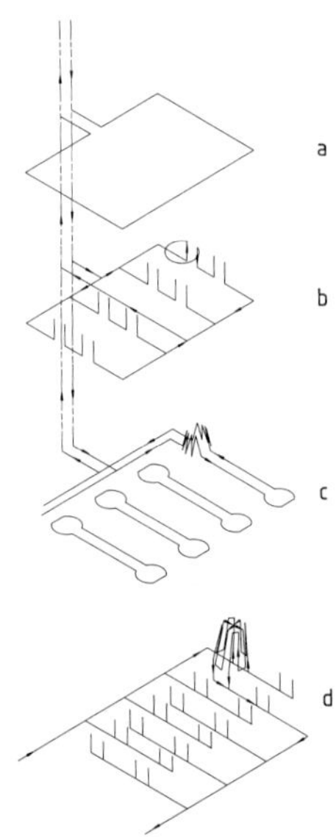

Figure 3 Schematic cooling diagram for the 2 × 8-cavity hot-runner stack mold for yoghurt cups
a: cooling for hot-runner plate; *b*: temperature control for gate insert plate; *c*: cavity temperature control; *d*: core temperature control

Detail A

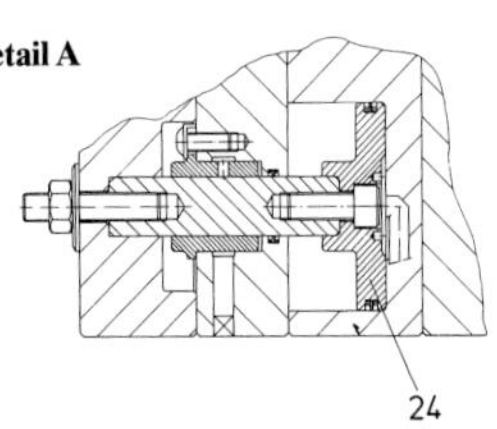

Figure 4 Air cylinder for releasing the bottom rim

Runner System/Gating

The melt to be injected flows to the cavities through a heated sprue bushing (9), which is bolted to the manifold (10). The manifold itself is heated by two heating coils (11) and (12) embedded in it. Eight healed nozzles (13) lead to the cavities from the manifold (10).
There is a sliding adapter piece (23) where the melt enters the heated sprue bushing (9). When the molding machine's nozzle lifts off after injection, the adapter piece follows the nozzle, thereby increasing the volume of the manifold. The melt in it is allowed to expand, thus stopping the sprue bushing from drooling.

Guides and Supports

A pair of racks (14, 15) is mounted on either side of the mold, with a pinion (16) in between. They ensure the synchronous movement of the two parting lines when the mold opens or closes. Support pillars (17) above and below the racks absorb the radial forces transmitted by the meshed teeth when in operation.

The mold weighs 2330 kg and is supported on the machine base by shoes (18). There are also four guide rods inside the mold (19).

Part Release/Ejection

The rim on the bottom of the cup can only be safely released if the cups are freed from the cavity bottom before the main parting line I opens. This is achieved with compressed-air pistons (24), which push the mold apart first at parting line II on mold opening (Fig. 4). In order to prevent suction from occurring at the cup bottom, air is blown in. The cups now remain between the cavity insert (2) and core (1) until parting line I opens. The molded cups are pulled out of their cavities and then blown off the cores by compressed air. This is achieved via two annular gaps between core head ring (7) and the adjacent components (1) or (8). There is a moving air jet (20) at the base of each core. As the mold opens, compressed air enters the air channel (21), the air jet (20) moves forward and an air blast blows the falling cups in a downward direction out of the mold.

Example 37: 2 × 2-Cavity Stack Mold for Covers Made from Polypropylene

The cover for a coffee maker has a diameter of 135 mm and is 13 mm high. Two small depressions are located on the outside edge, while two snap fits are found on the inside. In addition, an "ear" is found on the side of the cover.

Mold

The mold has dimensions of 646 mm × 390 mm and a shut height of 736 mm, with a weight of 1000 kg. The part-forming inserts are made of through-hardened steel (material no. 1.2767).
For such a cover (large projected area, minimal height, minimal weight), it makes sense to use a stack mold. The four cavity inserts (2) are arranged in opposite pairs in the center plate (1) of the mold.

Gating

The center plate also holds the four externally heated hot-runner nozzles (4) and the hot-runner manifold (6), that is heated with heater cartridges (5). Melt is conveyed to the manifold via the heated sprue bushing (7). The heated sprue bushing (7) follows the motions of the center block and is enclosed in a stationary protective tube (8) that prevents any melt from drooling into the stationary-side ejector housing from the bushing plate (9).
A short runner (10) connects each of the four hot-runner nozzles (4) to the "ear" of the cover via a submarine gate.

Sucker pins (11) pull the solidified runners away from the hot-runner nozzles and out of the tunnel gates upon mold opening and then eject them.

Temperature Control

The cores and cavity inserts are provided with a system of cooling channels that cover their respective surfaces. Additional cooling lines are located in the center plate (1) to remove the unavoidable heat radiated by the hot-runner system.

Demolding

To release the outer depressions, each cavity has two slides (13) actuated by cam pins (12) attached to the cavity insert and guided along the core. The inner snap fits are released by lifters (14).
Ejector pins (15) are used to eject the molded part from the core. Three moveable core pins (17) position the cover during ejection. After a stroke "X", the cover is stripped off the core pin by a sleeve ejector (16).
The stationary-side ejector plate (19) is operated by hydraulic cylinders (21), while the moveable-side ejector plate (20) is operated by the machine's ejector. Both ejector plates run in ball guides (22).
Two racks (23) connected to each other by means of a pinion (24) are provided on the two narrow sides of the mold to ensure synchronous opening and closing of both parting lines.

Fig.2

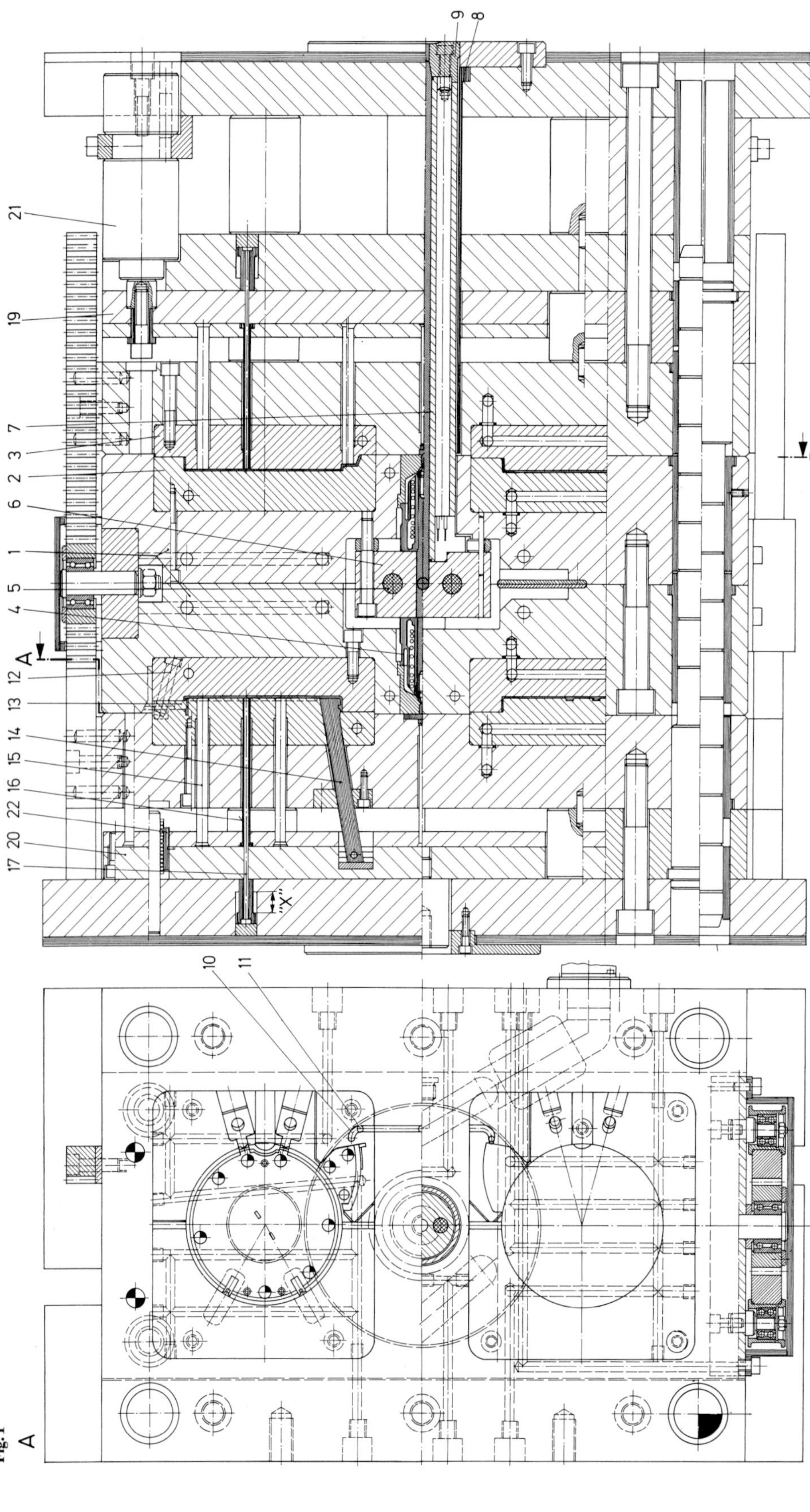

Fig.1

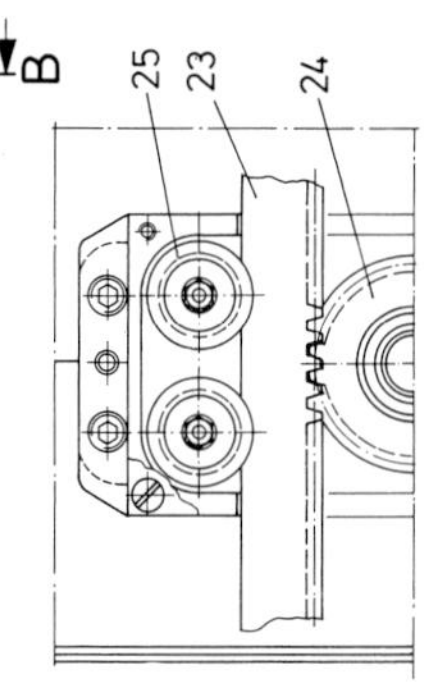

Fig.3

Figures 1 to 3 2 × 2-cavity stack mold for a coffee maker cover of polypropylene
1: center block; 2: cavity insert; 3: core; 4: hot-runner nozzle; 5: heater cartridge; 6: hot-runner manifold; 7: heated sprue bushing; 8: protective tube; 9: bushing plate; 10: runner; 11: sucker pin; 12: cam pin; 13: slide; 14: lifter; 15: ejector pin; 16: sleeve ejector; 17: core pin; 19, 20: ejector plates; 21: hydraulic cylinder; 22: ball guide; 23: rack; 24: pinion; 25: support pillar

Example 38, 2 × 5-Cavity Stack Mold for Cases Made from Polypropylene

For a case in which the base and lid are connected by an integral hinge (Fig. 1), a stack mold has been designed with five mold cavities in each of the two parting lines. This means that each shot produces 10 complete cases. Due to the surface quality specified for the outside, the molded parts are gated on the inside surface; base and lid each require a separate gate.

Figure 1 Case in which the base and lid are connected by an integral hinge (left opened, right closed)

Mold

The mold mounting dimensions have been so selected that it can be used on two different injection molding machines with different distances between tie bars. Because of the mold's weight, it is also supported by shoes that fit over the tie bars of the injection molding machine in addition to its own four guide columns. The mold shown in Figs. 2 to 5 can be adapted to either tie bar spacing with the aid of the reversible adaptor (10) and the two semi-circular bearings (28) contained therein.
The moving central section of the mold accommodates the hot-runner manifold (5), which is heated by heater elements (63) cast in aluminum. The melt flows from the hot-runner manifold through a heated nozzle (42) to the respective gate in the base or lid of the case. The individual cavities are formed by the inserts (17) and (18). For appearance reasons, the gates are positioned asymmetrically. The hot-runner manifold (5) is fed by the machine's nozzle through a sprue bushing (20) which is heated by the tubular heater (64).
The total installed heating capacity amounts to about 13.5 kW. It is divided up as follows: 20 × 300 W for the hot-runner probes, 7000 W for the hot-runner manifold and 450 W for the tubular heater used around the sprue bushing. When in operation, the connected load is about 7.5 kW. The hot-runner probes can be independently controlled; their temperatures are monitored by built-in thermocouples. Four control circuits are provided for the hot-runner manifold. The temperature of the sprue bushing is controlled by the injection molding machine's closed-loop control in the same way as is the nozzle on the machine.

Mold Temperature Control

Twenty circuits connected to a water manifold by quick disconnect couplings have been provided for cooling the molded parts.

Part Release/Ejection

To eject the cooled parts, the mold opens with synchronous separation of the two parting lines ensured by the action of the racks (6) and pinion (7). During the opening motion, the molded parts are retained on the central section until the racks actuate the ejector plates (3A) and (3B), thereby ejecting the molded cases and separating them from the gates. During mold closing, pushback pins return the ejector plates to their original positions again.

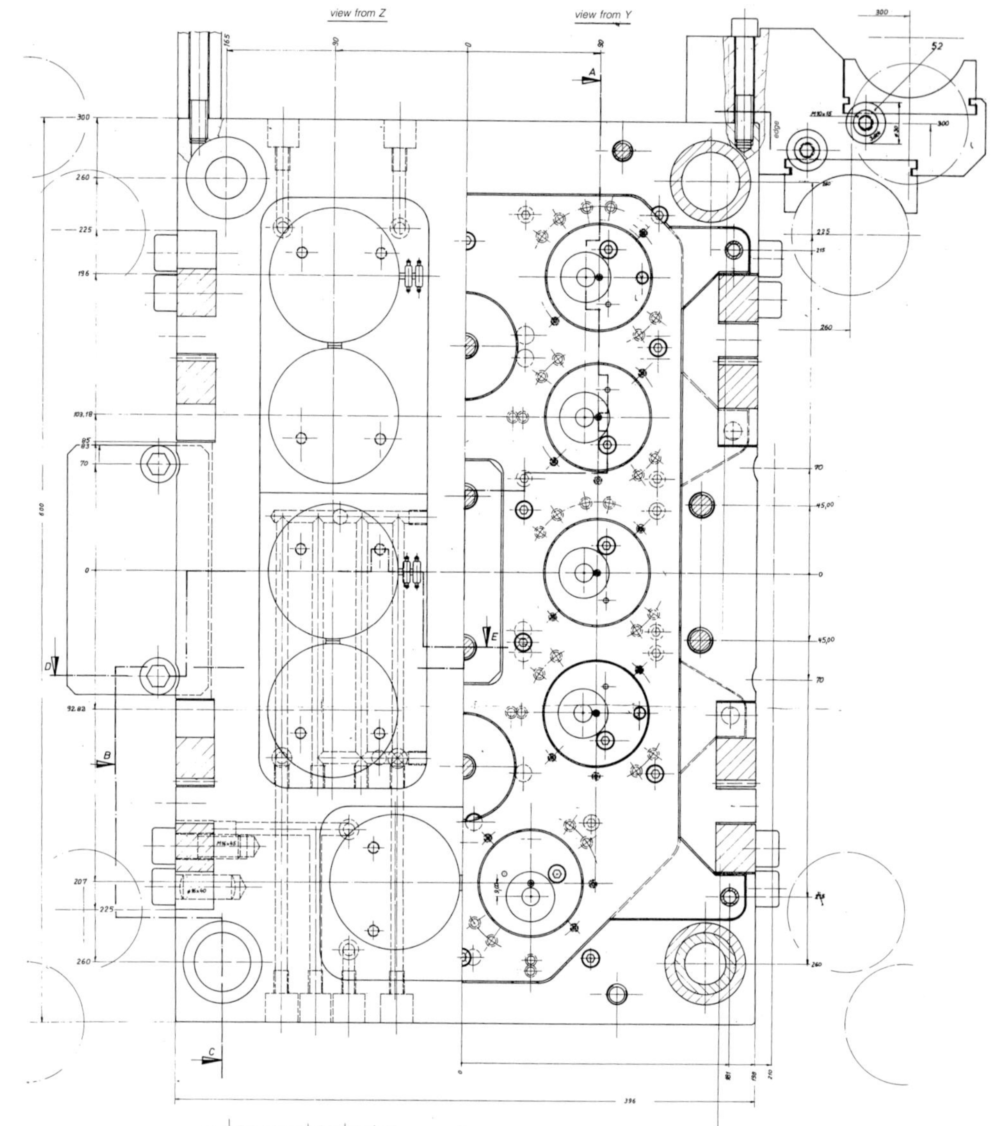

Fig. 2

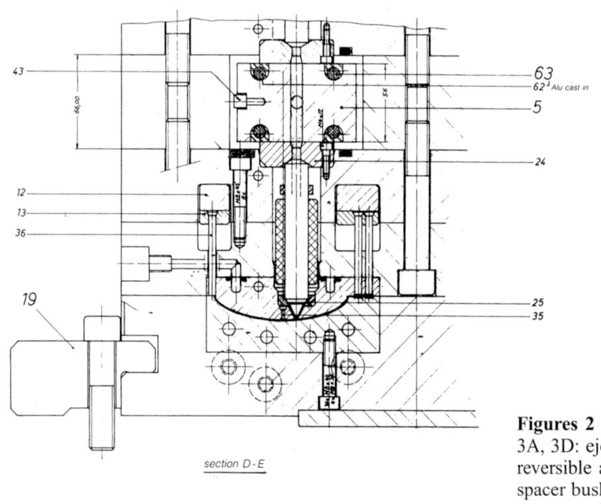

Fig. 3

Figures 2 to 5 2 × 5-cavity stack mold for cases
3A, 3D: ejector plates; 5: hot-runner manifold; 6: rack; 7: pinion; 10: reversible adaptor; 17, 18: mold inserts; 19: mold clamp; 20: heated spacer bushing with filter element; 28: semi-circular bearing; 42: hot-runner probe; 47: locating ring; 63: heat element; 64: tubular heater

Fig. 4

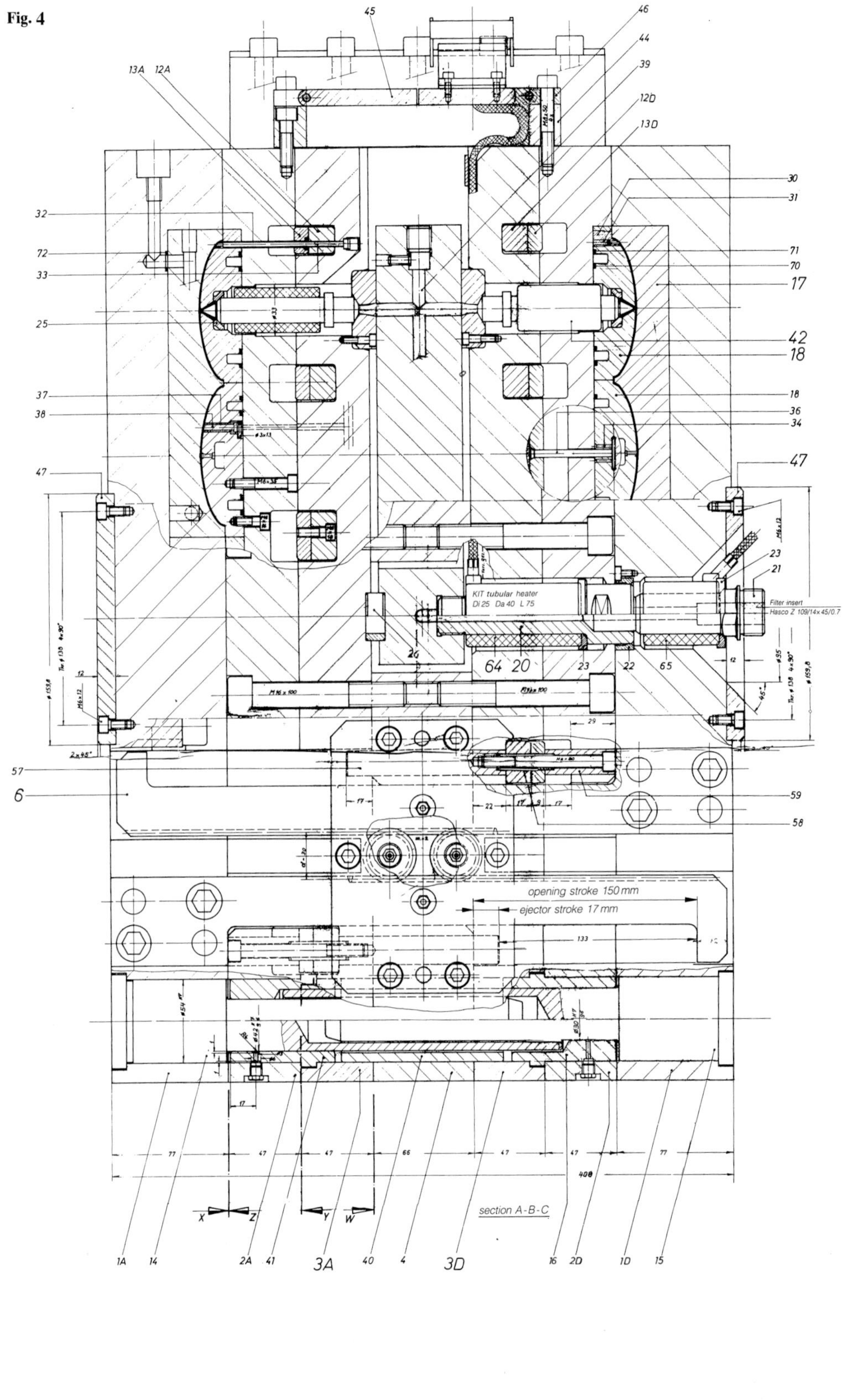
KIT tubular heater
Di 25 Da 40 L 75
Filter insert
Hasco Z 109/14×45/0.7
opening stroke 150 mm
ejector stroke 17 mm
section A-B-C

Fig. 6

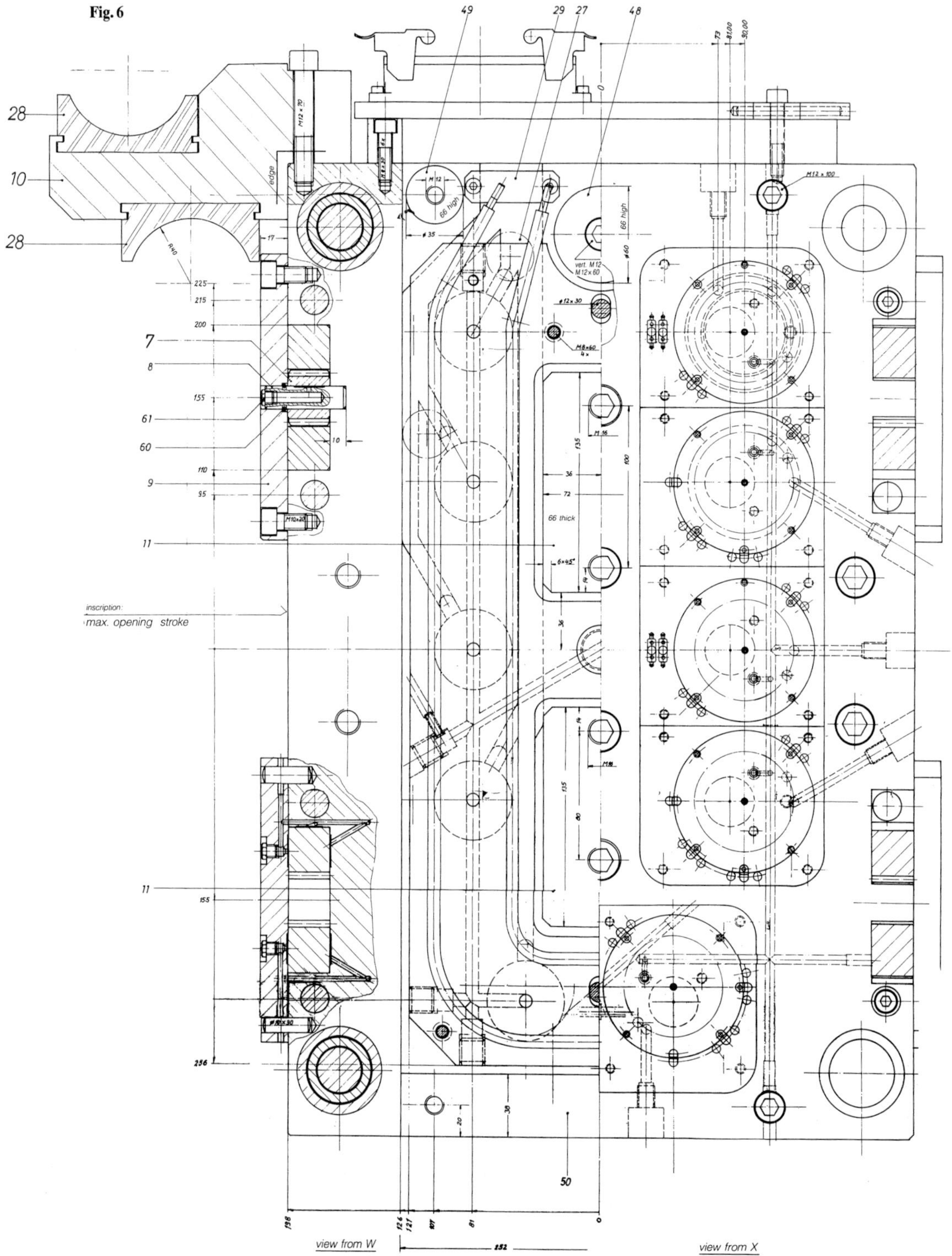

Figures 2 to 5 2 × 5-cavity stack mold for cases
3A, 3D: ejector plates; 5: hot-runner manifold; 6: rack; 7: pinion; 10: reversible adaptor; 17, 18: mold inserts; 19: mold clamp; 20: heated spacer bushing with filter element; 28: semi-circular bearing; 42: hot-runner probe; 47: locating ring; 63: heat element; 64: tubular heater

Example 39, 16-Cavity Hot-Runner Mold for Packaging of Medical Parts made from Polypropylene

The tubular medicinal packaging parts are internally threaded on both sides separately and divided in the middle by a thin wall. They are produced from an easily flowing polypropylene known for being ropy. To keep this effect from causing problems in production and application, suitable measures had to be taken with regard to the hot-runner system as well as to thermal control. In addition, it was also necessary to comply with the strict requirements of cleanroom technology, which involved consequences for the choice of a suitable hot-runner system.

Mold

For the mold, an individually regulated 16-cavity model with externally heated hot-runner manifold and open, externally heated gating nozzles *with* tips (Figs. 1 and 2) was selected. Special difficulty was presented by the necessity to locate each of the gating nozzles between the unscrewing spindles (not adequately illustrated on the drawings) on both the nozzle as well as the closing side. The solution arranged the gating nozzles horizontally to gate the molded parts directly from the side. Lateral configuration *without* tips would be at least problematical. Depending on cycle time, materials, and temperature, the gate in such a system tends to freeze; any resulting "cold plug" would then be injected into the cavity at the next cycle and could cause unacceptable surface defects, for example. These problems can be solved with inside mounted heat-conducting (torpedo) tips in conjunction with optimum mold cooling. To keep the melt from dripping into the cavity, melt decompression (pressure release, e.g., by screw retraction) is required. Each nozzle body is radially sealed and centered in the gating area by titanium alloy sealing rings chosen for their low thermal conductivity factor (to avoid heat loss in the gating area). The head mount of the gating nozzle at the hot-runner manifold block is a sliding seal face. The reactive force resulting from both injection pressure and pressure-stressed difference surfaces has a desired sealing effect by generating surface pressure between the nozzle head and hot-runner manifold block. The lateral forces are absorbed by the moveable cavity plate. The mold inserts and gating nozzles with four cavities each can be dismounted, e.g., for maintenance, on the machine when the mold is open. The hot-runner manifold block with a width of only 10mm in the nozzle area is a very compact design. The drilled runner channels are *naturally balanced*. The narrow, "sword-shaped" area of the hot-runner manifold block is indirectly heated. Thanks to the twofold heat source (from hot-runner and nozzle), temperature difference should be negligible.

Cooling

Effective cooling is ensured by 2×4 separately connectible cooling circuits.

Demolding

De-spindling takes place before the mold opens, whereby the molded parts remain on the ejector side. Subsequent to opening, the clamping side is de-spindled. Each of the molded parts is ejected by an ejector pin.

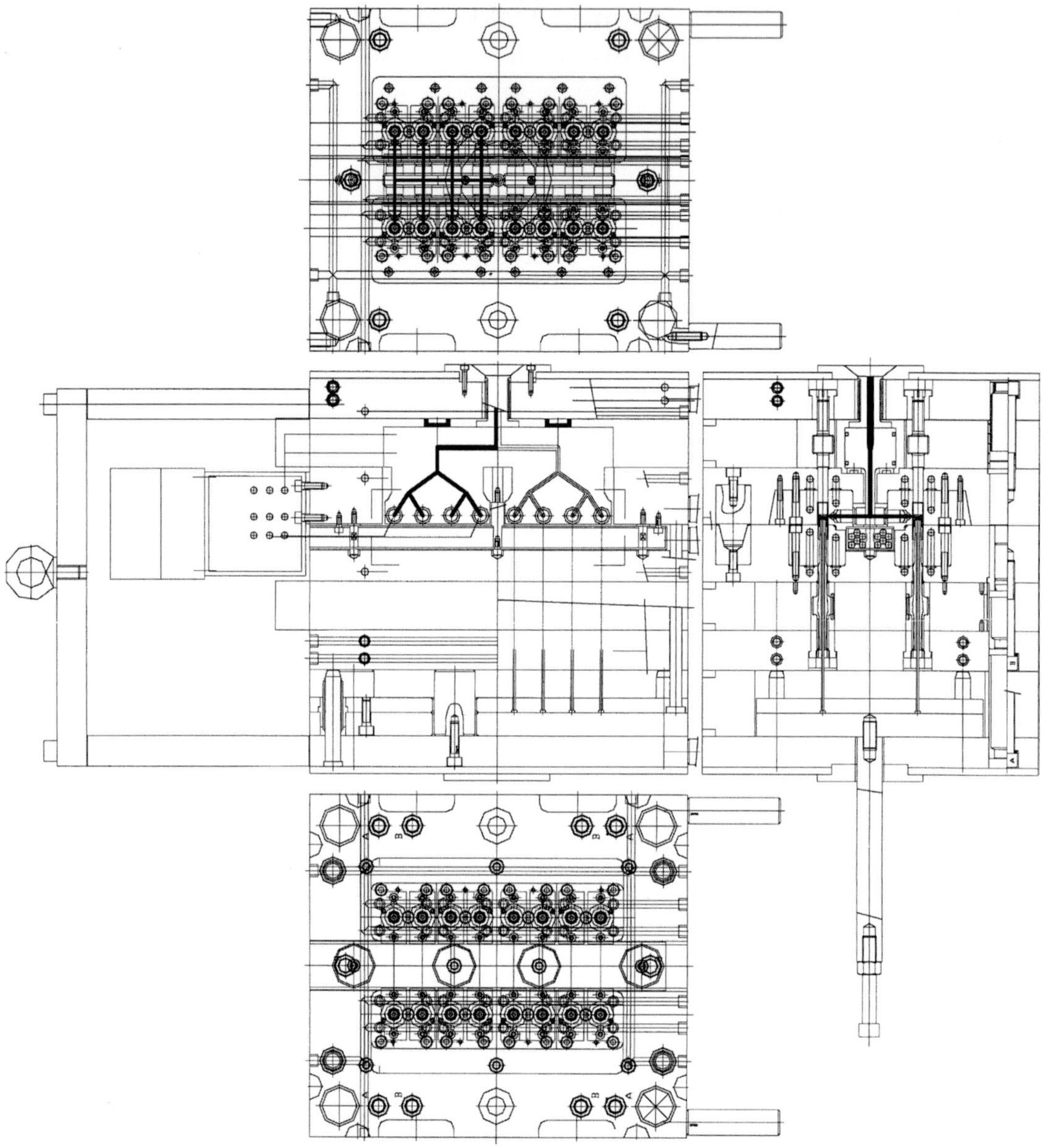

Figure 1 16-cavity hot-runner mold
with so-called sword distributor and horizontally arranged, heated gating nozzles with tips

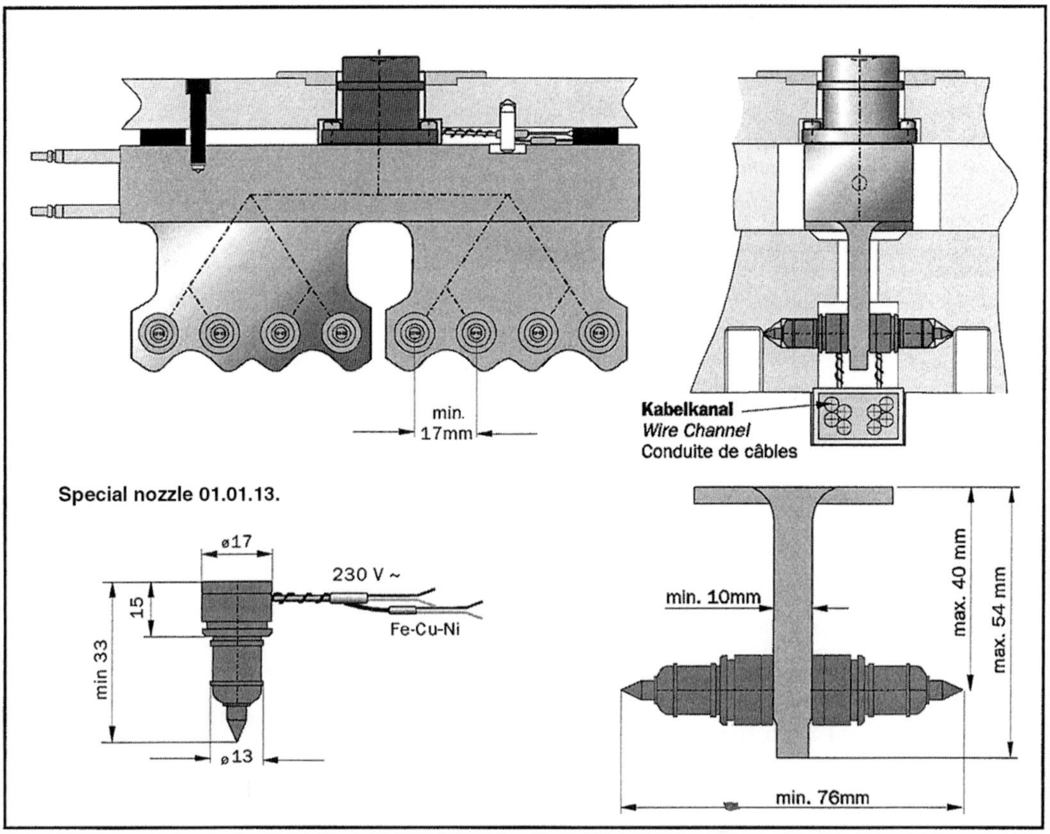

Figure 2 Sword-shaped hot-runner manifold block
(Courtesy: Heitec, Burgwald, Germany)

Example 40, Hot-Runner Stack Mold for a Water Distribution Block Made from Polypropylene

The water distribution block for a dishwasher (Fig. 1) consists essentially of two flat parts of polypropylene that are welded together so as to be watertight. Because of the postmolding shrinkage of polypropylene, it is necessary that the two halves be welded together either after a minimum of two days' storage or, better yet, immediately after molding. The latter approach is especially simple if both parts (which differ in weight and projected area) are produced together in a single mold.

The decision was made in favor of a stack mold to produce both parts. With regard to this decision, the following points had to be considered:

In a stack mold, the hot-runner manifold is always located between the two parting lines that accommodate the mold cavities. As a rule, the melt is conveyed to the manifold via a feed pipe which is coaxial with the injection and clamping units and passes through the first parting line of the mold. This is not possible here, however, because the molded part covers the entire mold surface of this parting line.

Another possibility would be to supply the melt to the manifold from the side. This requires that the injection unit of the molding machine be repositioned by 90 degrees (L position) so that the direction of injection lies in the plane formed by the runners. For plant-related reasons, this design could not be considered.

Figure 1 Water distribution block for a dishwasher

Mold

A design was selected (Fig. 2) in which the melt coming from the machine nozzle is directed from the sprue bushing (314) around the first parting line (20, 21, 22) to the hot-runner manifold (1). Such redirection of the melt flow results in an increased pressure loss during injection (Fig. 3). The arrangement shown at the bottom of Fig. 3 was finally selected.

Another unique aspect of this arrangement is that both the thickness of plate (22) as well as the opening stroke of the first parting line affect the shut height of the mold. The entire melt-conveying system (20, 21, 22) is fastened to the hot-runner manifold located in the center section of the mold and executes the same opening and closing motions. The distance H (Fig. 2) must be provided for these motions. The hot-runner plate (22) is supported against the central section (11, 12) by means of pillars (3).

A simplification of this arrangement would be possible with an injection molding machine where the injection unit is raised by the amount E (Fig. 2) so that the machine is coaxial with the melt-conveying bridge pipe (21). The shut height of the mold would then be reduced and the pressure loss in the melt-conveying system less.

Runner System/Gating

Analyses of the filling pattern (Fig. 4) resulted in the channel dimensions given in Fig. 5 for the hot-runner manifold. By varying the runner channel dimensions, it was possible to compensate for the different flow path lengths and volumes of melt in the runner system so that the 8 hot-runner nozzles (Fig. 6) could remain identical.

Part Release/Ejection

The part farther away from the machine nozzle is ejected by means of the machine's ejector system: the part closer to the nozzle is ejected by means of hydraulic cylinders (not shown) that actuate ejector plates (6 and 7).

Heating the Hot Runner System

As the result of thermal expansion, the bridge pipe (21) also increases in length. In order to avoid thermal stresses in the melt-conveying system and

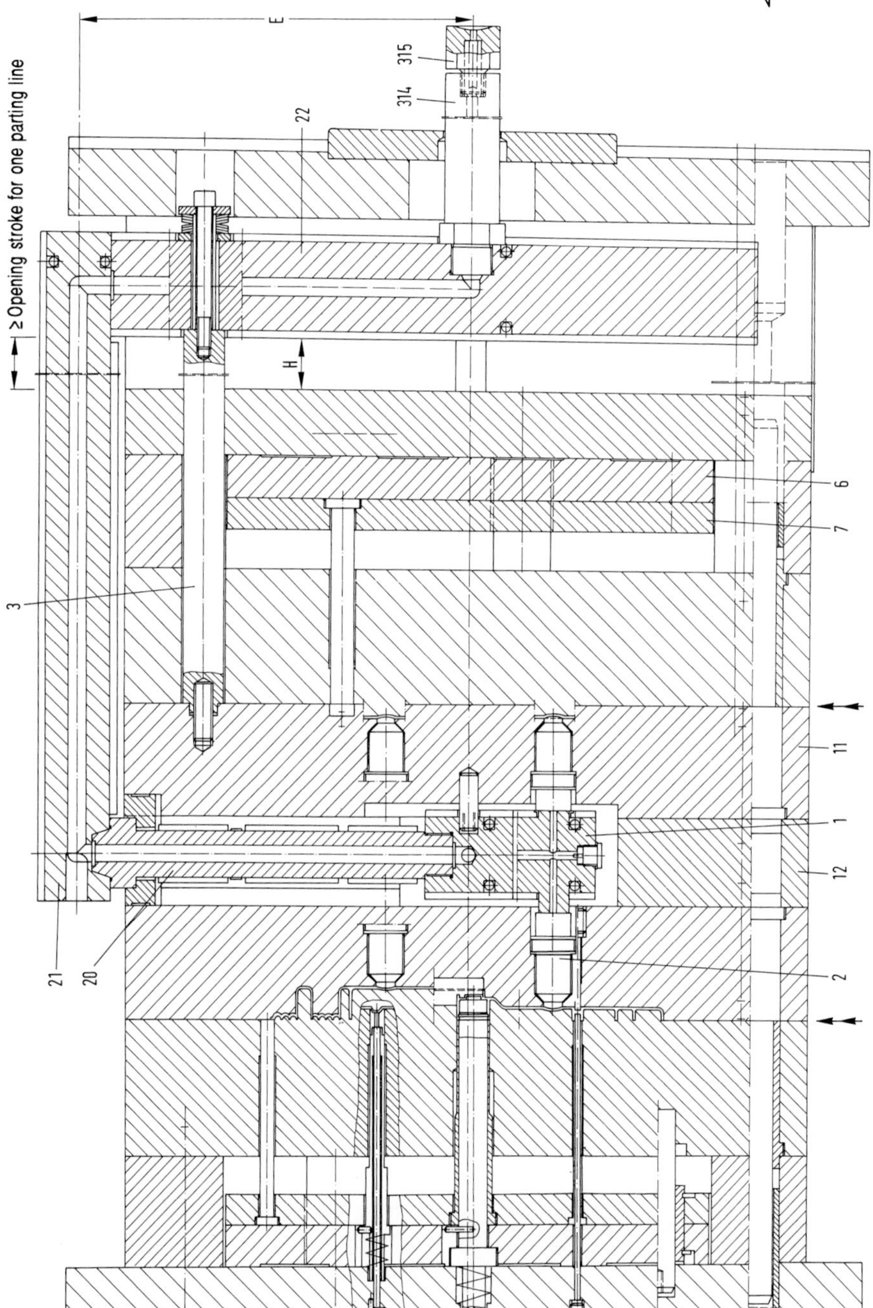

Figure 2 Connection of the mold and its central hot-runner manifold plate with the machine nozzle
1: hot-runner manifold; 2: hot-runner nozzle; 3: support pillar; 6, 7: ejector plates; 11: mold plate; 12: spacer plate; 20: feed pipe; 21: bridge pipe; 22: hot-runner, (314) heated sprue bushing, (315) adaptor

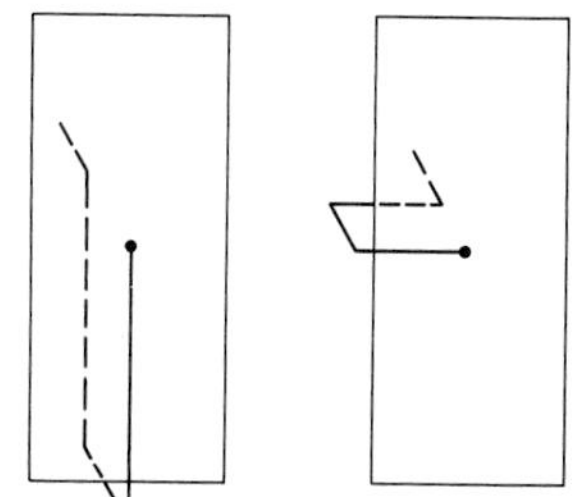

Figure 3 Comparison of possible runner system arrangements, top: around the narrow side of the mold; bottom: around the wide side of the mold, with less pressure loss (injection time; 5 s, material; PP)

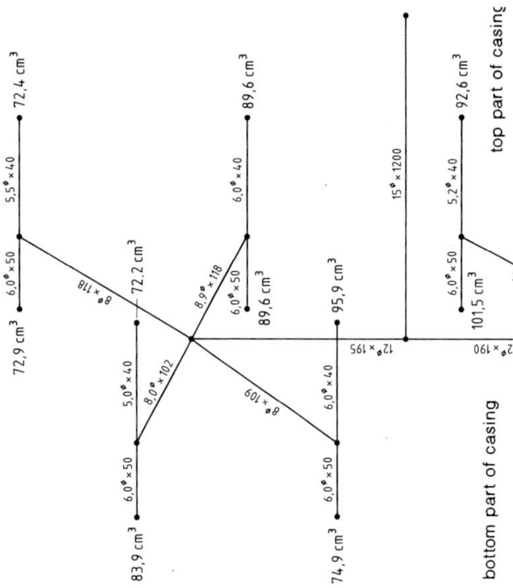

Figure 5 Final proposal for the hot-runner manifold (with runner dimensions and volumetric flows)

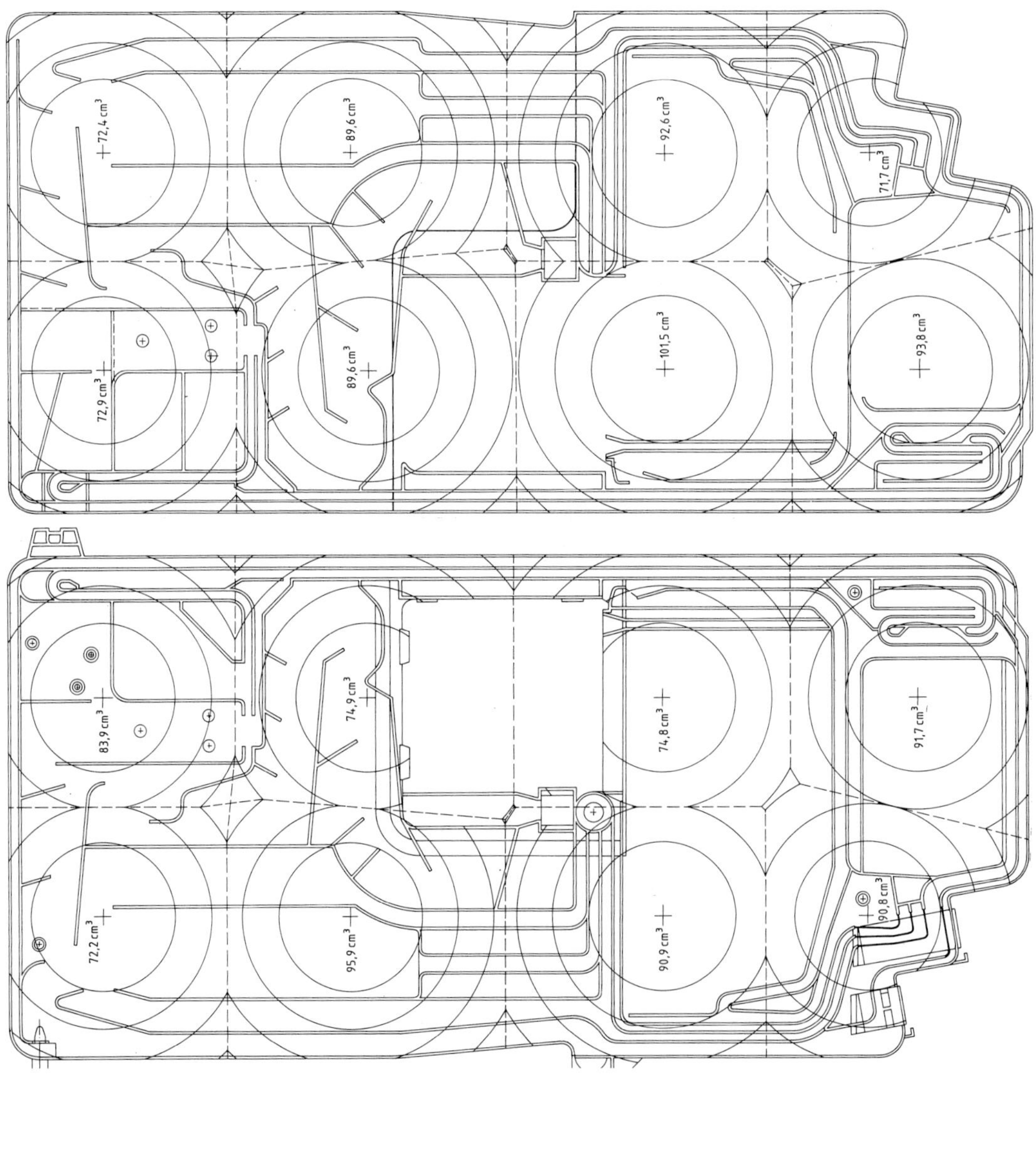

Figure 4 Lower half (left) and upper half (right) of the water distribution block for a dishwasher; gates marked, with indication of the volumetric flows

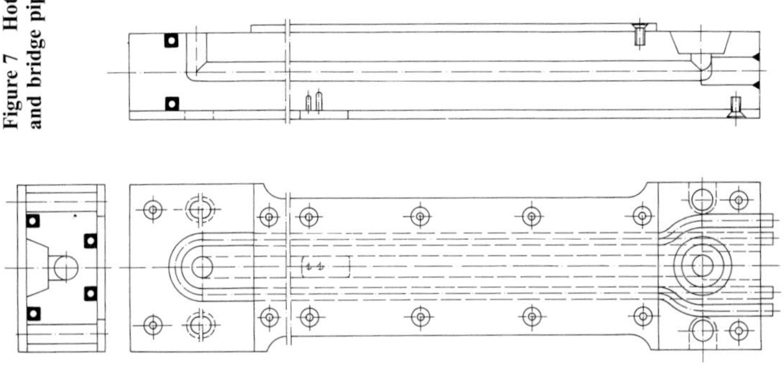
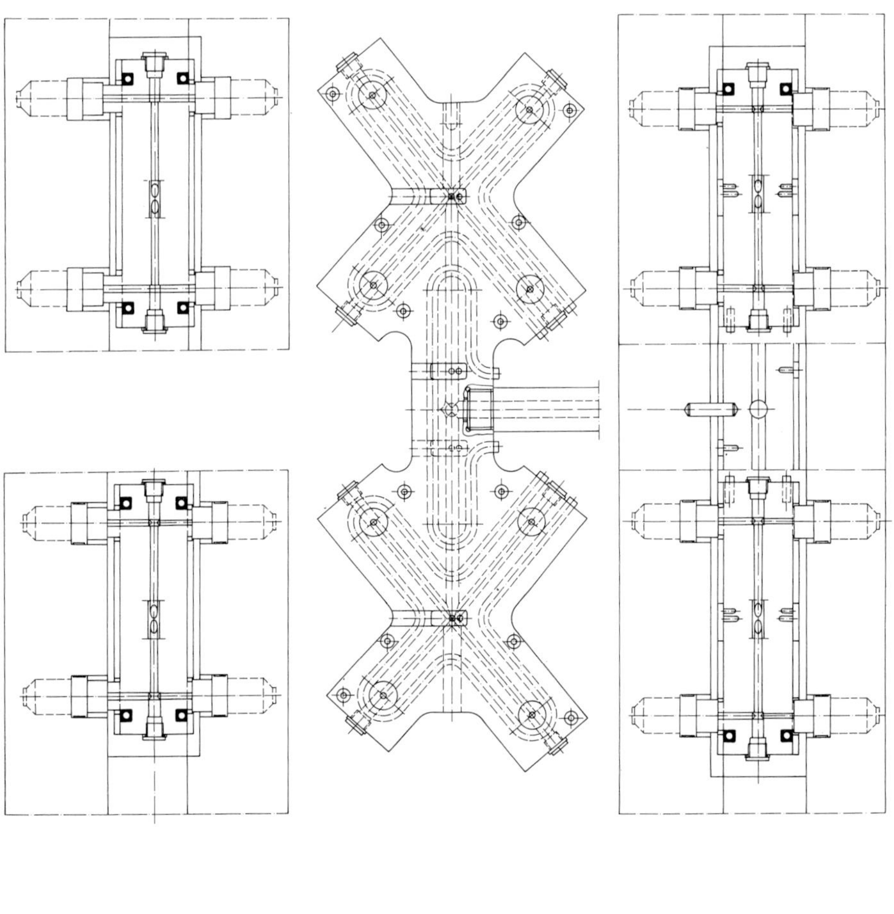

Figure 7 Hot-runner manifold with cast-in tubular heaters (left) and bridge pipe (right)

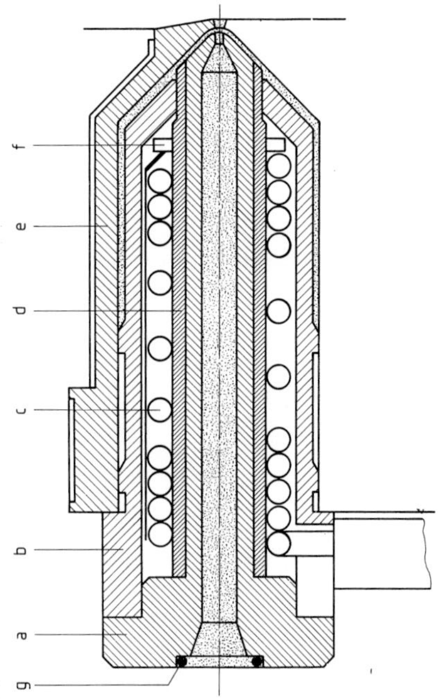

Figure 6 Externally heated nozzle selected for the gates; top: for installation with an additional sleeve (e); bottom: for direct installation in the hot-runner plate

a: nozzle body; *b*: housing; *c*: heating coil (220 V); *d* high-conductivity BeCu sleeve; *e*: sleeve; *f*: thermocouple; *g*: O-ring (Courtesy: PSG)

hot-runner manifold, and thus any possible leakage, the mounting screws between the bridge pipe (21) and hot-runner plate (22) can be tightened only after reaching the operating temperature. These screws must be loosened again before turning off the heating system.

Operation

Opening and closing of the two parting lines is accomplished with the aid of two racks connected via a pinion on each side of the mold (Fig. 8). Both parting lines thus open and close synchronously.

Figure 8 Stack mold with racks for synchronous opening of the two parting planes

Example 41, 2 × 8-Cavity Stack Mold for Lozenge Box Made from Polystyrene

A mold was to be designed for a transparent, thin-walled lozenge-box bottom in polystyrene. The2 manufacturing costs had to be kept in an acceptable relation to the production costs. An injection molding machine of sufficient daylight opening was available, hence the decision for an 2 × 8-cavity stack mold (Figs. 1 to 5). This mold consists of three plate assemblies with interspaced ejector plates (3) and (9). In the central plate assembly (4 to 8) are installed the hot-runner manifold (15) with the heated sprue bushing (19), the hot-runner nozzles (14) and the mold cavities in plates (4) and (8). The latter can be cooled intensively, as can be the cores (12) fixed in the outer plate assemblies.

The H-shaped hot runner manifold (15) is heated by four cartridge heaters (30). A cartridge heater (18) installed in the torpedo (16) ensures uniform heating of the melt in the sprue bushing (19). The specially designed nozzle from the machine's plasticizing unit extends into the sprue bushing (20). The hot-runner nozzle (14) projects its tapered tip up to the shape-forming surface of the mold cavity, so that "runnerless" injection is feasible. The gate insert sleeve (21) required with indirectly heated thermally conductive nozzles for thermotechnical reasons had to be shortened, because a continuous gate insert sleeve would have left an unacceptable marking on the molding.

Opening and closing of the two mold parting lines are coordinated by two laterally fitted angle levers (32, 33) that are linked to the plate assemblies. When the mold opens, the central plate assembly (4 to 8) is held centered between the two outer plate assemblies (1, 2) and (10, 11) by this lever arrangement. The ejector plates (3) and (9) are set in motion simultaneously by the levers (34), so that the molded parts are pushed off the cores (12) by the stripper plates (13) during the continued opening movement.

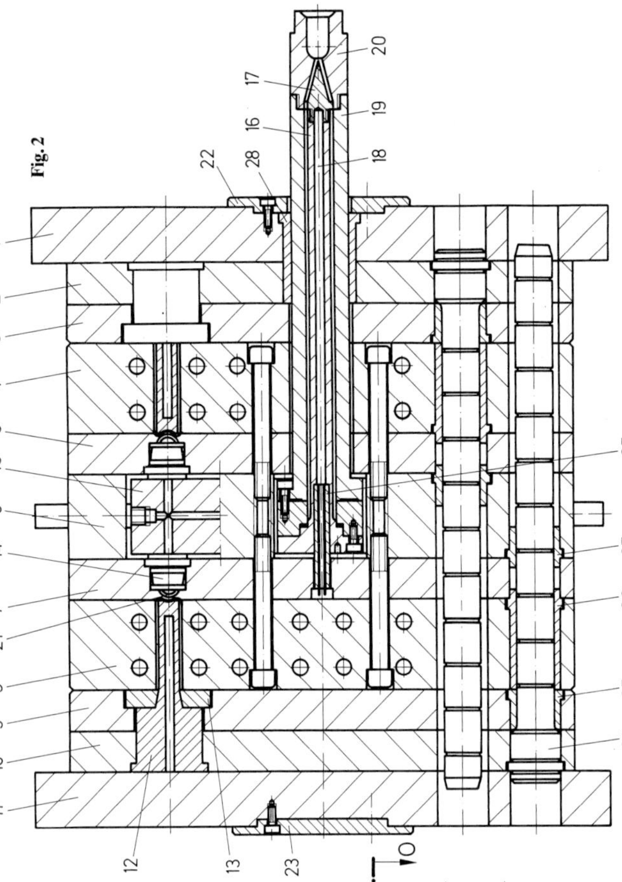

Fig. 2

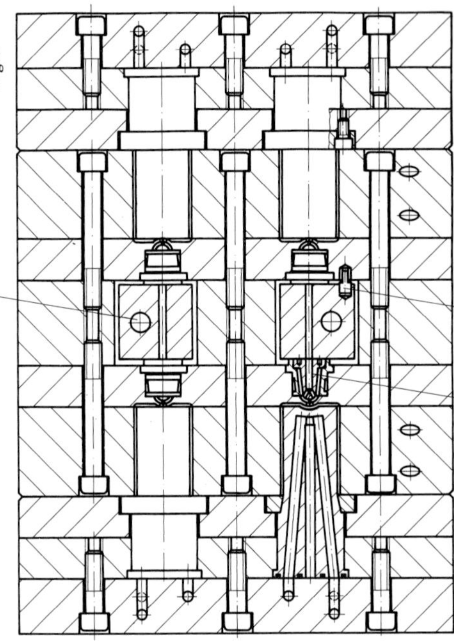

Fig. 3

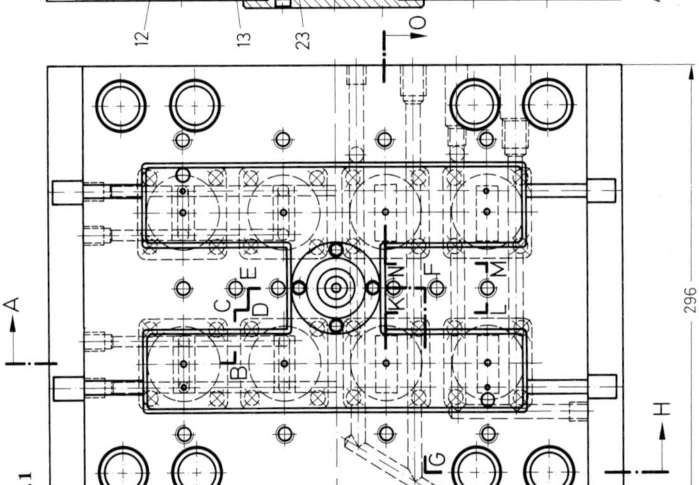

Fig. 1

Fig. 4

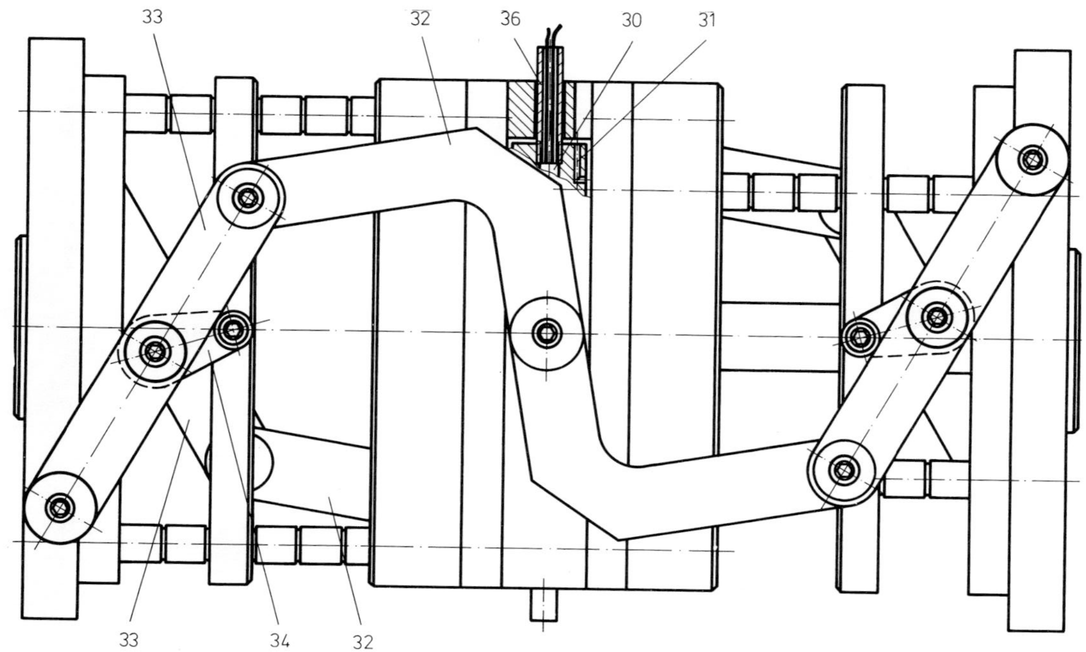

Fig. 5

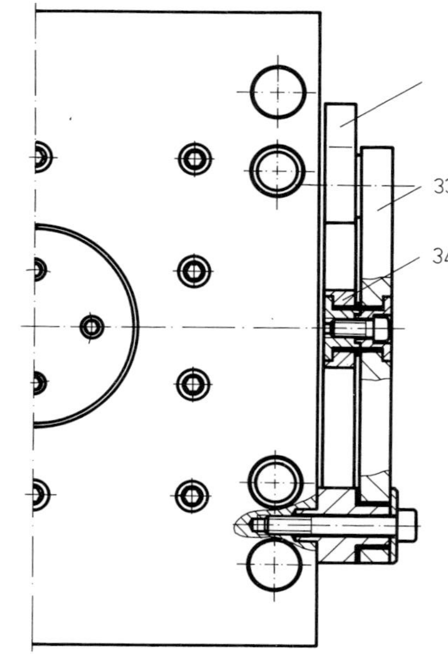

Figures 1 to 5 2 × 8-cavity stack mold for lozenge-box bottoms in polystyrene
1: fixed-half mold mounting plate; 2: fixed core retaining plate; 3: fixed-half stripper plate; 4: fixed-half mold cavity plate; 5: fixed-half nozzle plate; 6: hot-runner plate; 7: moving-half nozzle plate; 8: moving-half mold cavity plate; 9: moving-half stripper plate; 10: moving core retaining plate; 11: moving-half mounting plate; 12: core; 13: stripper insert; 14: nozzle; 15: hot-runner block; 16: torpedo; 17: torpedo tip; 18: cartridge heater; 19: heated sprue bushing; 20: sprue bushing; 21: hot-runner gate insert sleeve; 22, 23: locating ring; 24: leader pin; 25, 26, 27: guide bushing; 28: guide bushing; 29: locating pin; 30: cartridge heater; 31: thermocouple; 32: angle lever; 33: connecting lever; 34: pilot link; 35, 36: cartridge heater housing
The cores 12: are not shown in Fig. 4

Example 42, Two-Cavity Injection Mold for a Tail Light Housing Made from ABS

Car tail light housings have decorative wall areas which form part of the vehicle exterior surface and areas which have to meet the optical and technical requirements of the lights. The high degree of integration of the injection molding process makes it possible to fulfill these requirements in a single injection molded part.

Figure 1 shows an automobile tail light housing (weight about 570 g) side as viewed from the back of the vehicle. The six chambers are specially designed with parabolic and spherical surfaces to act as reflectors and light bundling areas. These optical surfaces are later coated with light-reflecting paint before the entire casing side is welded to the colored lenses made from PMMA. The circular openings in the chambers take the bulbs.

Figure 2 shows the inside of the housing with four insert molded screws to fix the casing to the vehicle body and a central receiving element for the bayonet closure that secures the cap with the printed circuits and bulb holders. In Fig. 2, the casing is already welded to the colored lenses, which can be seen through the light openings. On the left edge of the casing is a wall area which forms part of the vehicle exterior surface.

If metal parts such as screws are insert molded, accumulations of material occur at the fixing points. These may cause sink marks in the part surface and also require a relatively long cooling time. One of the special tasks in designing the mold was to avoid these material accumulations in order particularly not to interfere with the geometries of the reflector surfaces. The housings are produced in two different designs (for EU and USA). It was therefore necessary to provide two readily interchangeable inserts for each cavity. The distinctions between the two designs (trapezoidal area, Fig. 1 left) can be seen in Figs. 1 and 2.

Mold

The mold with the dimensions B × H × L: 896 × 796 × 902 mm^3 is designed as a two-cavity mold for a pair of housings (right, left). Figure 3 is a view of the parting line on the fixed mold half. The parting line on the moving mold half is shown in Fig. 4. Figure 5 is a vertical longitudinal section through a cavity at the height of the gates. Figure 6 is a vertical longitudinal section through the mold at the height of the two lower fixing screws visible in Fig. 2. Figure 7 is a horizontal longitudinal section which shows the situation below the bayonet closure for the cap and through a fixing screw. Figures 7 and 8 are sections through various core regions and show the path of the cooling channels within them.

The contour-forming areas of the mold are constructed from high-alloy, pretempered steel. The inserts, cavity parts and openings are produced by spark and hot wire erosion. The high degree of pretempering permits mold proving and production of short runs even before final hardening.

Centering and Guidance

The mold inserts are fixed into their receiving plates with wedges, for example (22) and (24) (Fig. 5). In addition to the four guide pins 61 (Fig. 5), there are four guide bars (64) above the parting line I (Fig. 7). These are arranged in each case in the center of the longitudinal and transverse sides of the mold. The guide bars have a closer fit than the pins because their accuracy of fit is independent of temperature differences in the two mold halves.

As can be seen from Fig. 2, in the walls of the lower three light chambers, there are openings formed by mold surfaces contacting each other at the core and

Figure 1 Housing: outside view

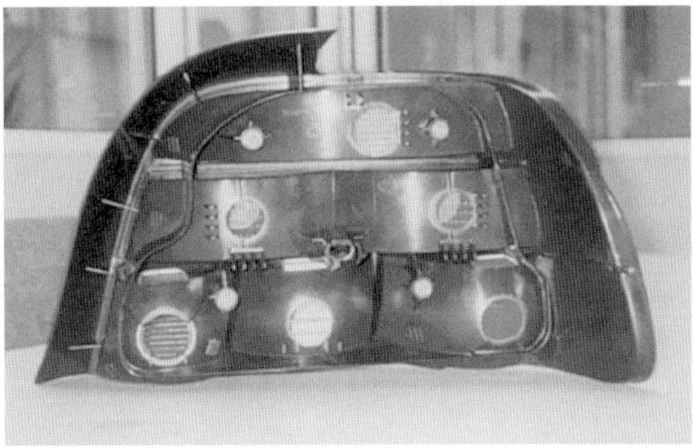

Figure 2 Housing: inside view

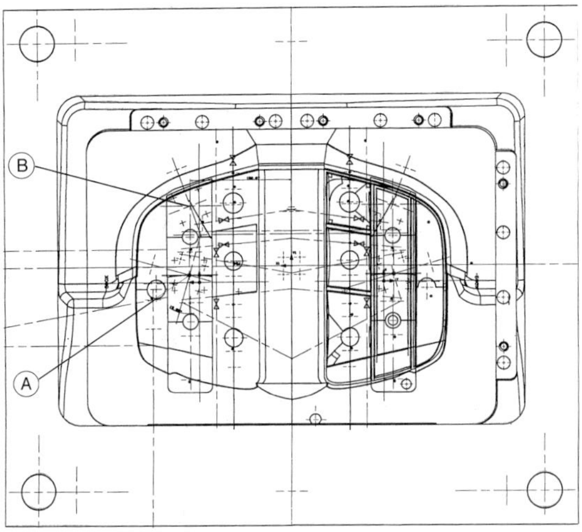

Figure 3 Injection mold, fixed mold half; A: needle shutoff nozzle; B: nozzle with edge gate

cavity plate. When parting line I is closed, this exact guidance of the mold is required to prevent the bedding-down surfaces from damaging each other. This exact guidance is even more important when the mold is opened. It prevents so-called sagging of the two mold halves and hence the formation of scratch and drag marks on the grained, decorative surfaces of the molded parts at points with lower draft. The final fit in the closed mold is effected by the tapered surfaces (117) (Fig. 5) and (116) (Fig. 7).

Gate

Each of the two parts is gated in the center via a hot runner needle shutoff nozzle A and also via a nozzle located on the side wall with an edge gate (hot edge)

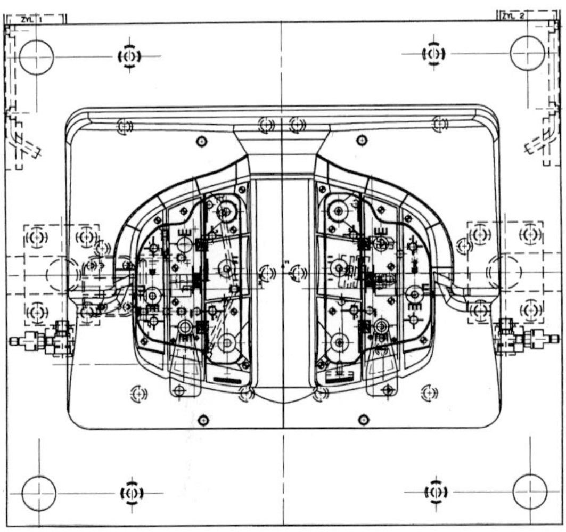

Figure 4 Moving mold half

B (Figs. 3 and 5). The nozzle needle is actuated with compressed air and controlled by the injection molding machine.

Mold Temperature Control

For mold temperature control, 20 different water circuits are provided. The connections for these are grouped on coupling plates for the fixed and moving mold halves. The complicated shape of the molding necessitates a complex temperature control channel system which in many cases had to be constructed by deep-hole drilling work. To facilitate insert changing on the machine, the interchangeable mold inserts do not have cooling medium flowing directly through them. They have bores into which long slender cooling probes (46) (Fig. 5) seated in the mold plates are inserted. To improve heat transfer, thermally conducting pastes are applied when installing over the inserts.

Demolding

The mold has a series of submarine splits with which recesses transverse to the demolding direction are demolded and material accumulations displaced. To actuate these elements, a second opening of the mold in parting line II (Fig. 7) is provided. After the main parting line I has opened, parting line II then opens and the following movements take place:

- the control bar (36) (Fig. 6) pulls with its two top tapered guides the two displacing splits (11) from the cavities below two inserted screws.
- the displacing split (13) (Fig. 7) is pulled by the control bar (38).
- the split (37) which is laterally movable in the plate (56) (Fig. 7) pulls with its top tapered guide the submarine split (12) from the cavity below the bayonet closure.

Parting line II is opened by a pair of cylinders 80 via coupling bolts (74) (Fig. 7). The ejector plate (5, 6) (Fig. 6) should not be actuated until parting line II has been opened in order to prevent the ejectors damaging the molded part. When the mold is closed, the ejectors must be withdrawn before the two parting lines close.

For correct timing of these movements, two electronically controlled limit switches are installed in each of the two block cylinders (80). These send signals to the machine and ejector control systems. Since faults in the machine and mold control systems cannot be completely ruled out and unintentionally faulty operation is likely during mold installation and machine setting, a series of mechanical safeguards are provided to prevent mold damage:

- During mold opening: in parting line II a control bolt (118) acts together with a control slide (119) (Fig. 6) to ensure that the bolt (120) seated in the ejector plate (5, 6) cannot move to the right until

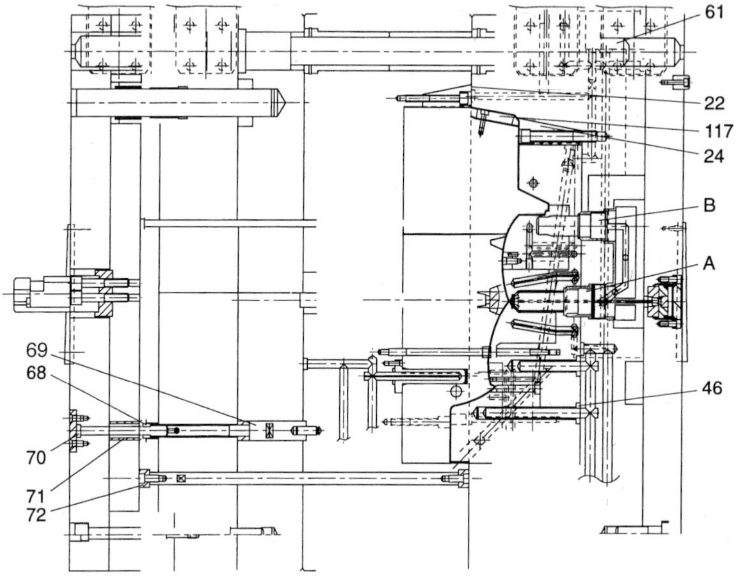

Figure 5 A: needle shutoff nozzle; B: nozzle with edge gate; 22, 24: wedges; 46: cooling probe; 61: guide pin; 68–71: ejector retraction device; 72: return bar; 117; tapered surface

the parting line is completely opened. The ejector retraction device (68–71) (Fig. 5) operates with a similar purpose.

– During mold closing: the return bars (72) (Fig. 5) press the ejector plates back so that the mold cavity on the fixed mold half is not damaged by projecting ejector pins. If during mold closing parting line II closes before the ejectors are retracted, the bolt (120) (Fig. 6) blocks further movement of the control slide (119) and the control bolt (118).

When the two hydraulic cylinders (80) open parting line II, the return bars (72) (Fig. 5) take the ejector plates with them commensurately with the amount that the parting line opens. The coupling of the plates with the central machine ejector has a correspondingly long idle stroke so that the machine ejector is not moved. As a result of this idle stroke,

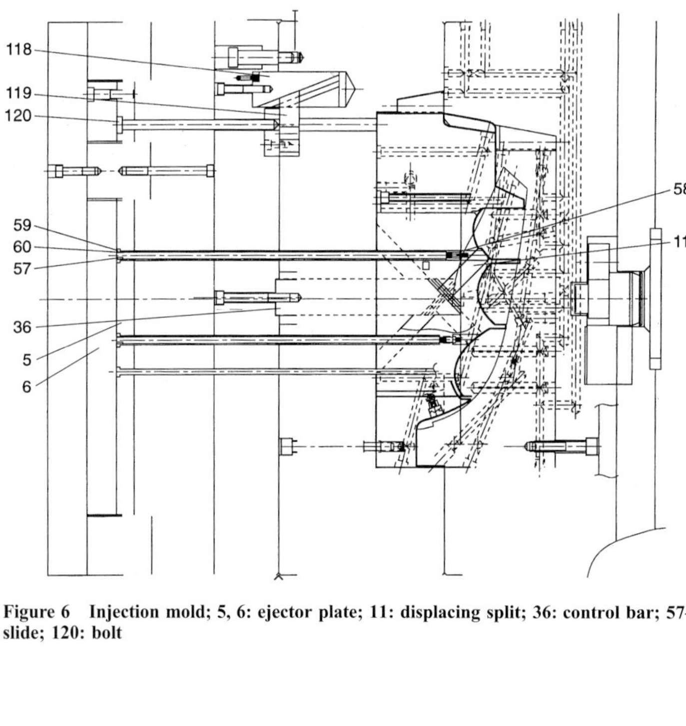

Figure 6 Injection mold; 5, 6: ejector plate; 11: displacing split; 36: control bar; 57–60: sleeve ejector; 118: control bolt; 119: control slide; 120: bolt

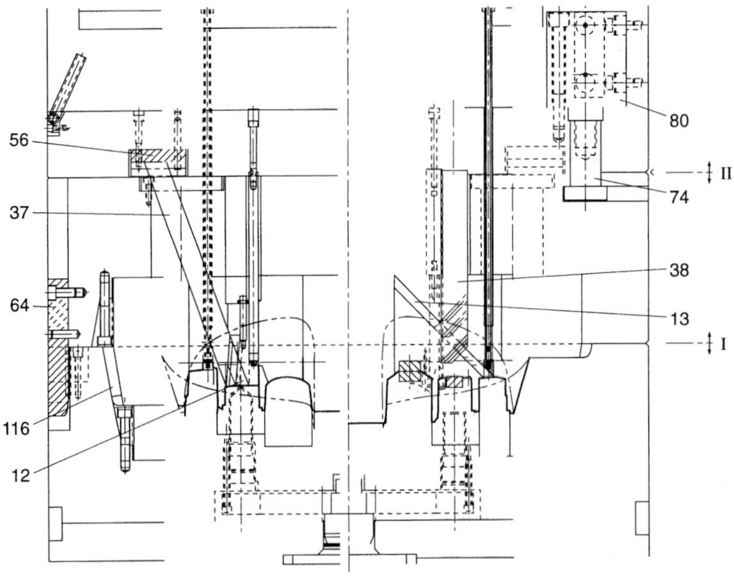

Figure 7 Injection mold; 12: submarine split; 13: displacing split; 32: lifters; 38: control bar; 64: guide bar; 74: coupling bolt; 80: hydraulic block cylinder; 116: tapered surface

the machine ejector cannot completely retract the ejector plates during mold closing. This function is carried out by the retraction device (68–71). The return bars ensure the correct final positioning of the ejectors. Mold operation is fully automated. The inserted screws are automatically placed into the sleeve ejector (57–60) (Fig. 6). When the ejectors have advanced, the moldings with their screws are lodged in the sleeve ejectors from where they are removed by demolding robots.

The complex system of safeguards described for split and ejector movement has proved very successful in practice and mold breakage and production breakdowns have been avoided.

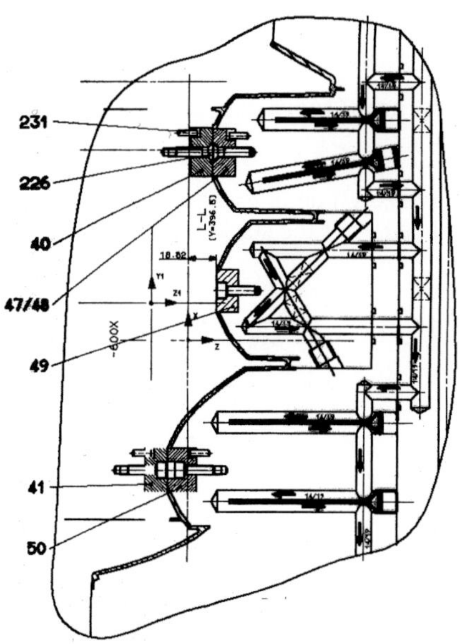

Figure 8 Injection mold for tail light housing, part section showing cooling channel arrangement

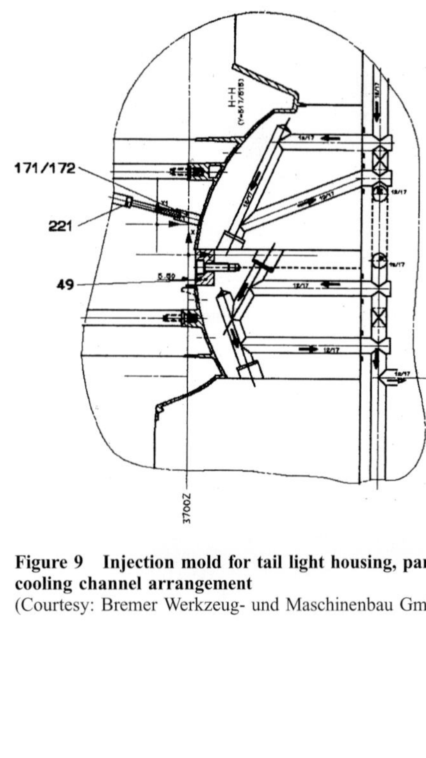

Figure 9 Injection mold for tail light housing, part section with cooling channel arrangement
(Courtesy: Bremer Werkzeug- und Maschinenbau GmbH, Germany)

Example 43, 2 × 2-Cavity Stack Mold with a Hot-Runner System for Runnerless Molding of Polystyrene Container Lids Using Direct Edge Gating

In selecting a suitable injection molding machine, the necessary clamping force, shot volume, mold height and mold opening stroke must be in a balanced ratio to each other. This, however, is achieved only partially in the production of relatively flat and thin-section parts unless the number of cavities in a mold is doubled by adopting the multidaylight design. This increases injection volume, mold height and necessary mold opening stroke while the necessary locking force remains unchanged.

Stack Molds for Container Lids

The task was to produce polystyrene lids (Fig. 2) for a polystyrene container (Fig. 1) using a stack mold (Figs. 3 to 5, see pp. 120 and 121) so as to make better use of the machine. The outer surface of the lid must not, however, show any gate mark, so that the lids can only be gated on the inside or externally from the side.
Gating on the inside of the lid is not possible since it would be too difficult and complicated to design a hot-runner system passing through the mold core and the necessary ejector system. This means that the only solution is to provide for direct edge gating of the lid, the gate being situated on an external side wall surface.

Figure 1 Polystyrene packaging container

Figure 2 Lid for the container shown in Fig. 1

Although standard hot-runner nozzles specially developed for the purpose are now on the market for runnerless, direct edge gating, the shape of the article will dictate whether such a hot-runner nozzle can be used. As Fig. 6A shows, the space for hot-runner nozzles used for direct edge gating should be far enough away from the mold cavity for the cavity wall lying in between to be able to absorb the stress produced during injection. On the other hand, the thinner the cavity wall and the shorter therefore the gate land, the smaller will be the residue remaining inside the gate until the next injection molding cycle. Under no circumstances must the residue be longer than the component wall thickness.
In the present container lid this kind of runnerless edge gating of the article with a hot runner nozzle cannot be realized since this would make the gate disproportionately long (Fig. 6B) because of the angle of the side wall relative to the lid surface.
To gate the component directly on the side wall nevertheless, the hot-runner system of the stack mold was equipped with hot-runner nozzles. In contrast to the generally used direction of installation (along the longitudinal axis of the mold), installation in this case was at right angles to the longitudinal mold axis. This hot-runner nozzle is of pointed conical shape at the front end, which fits into a conically shaped gates insert so that the nozzle tip can be flush with the cavity wall. In this way the formation of a gate vestige, which could prevent release of the component, is prevented (Fig. 7).

Construction and Operation of the Stack Mold

Two container lids (C) lie in each of the two parting surfaces (A) and (B) of the stack mold (Fig. 5). The

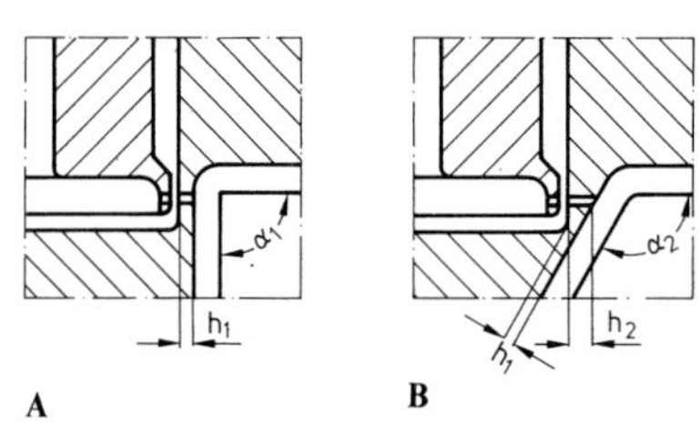

Figure 6 Dependence of gate height on article wall thickness for the minimum distance of the gate insert from the cavity (governed by strength considerations) for hot-runner nozzles used for direct edge gating
h_1: gate height for right angled position of article side wall to base and minimum distance of mold cavity from the antechamber of the hot-runner nozzle; h_2: gate height for non-rectangular position of article side wall to base ($a_2 \geqslant 95°$). In this case $h_2 > h_1$

Figure 7 Smoothed-out gate mark of container lid

mold consists of three plates (1, 2, 3), the cavities being in the center plate (2), formed by the cavity plates (4a, 4b, 5a, 5b) and core inserts (7) and (8). The core inserts are attached to plates (9) and (10), which form part of the plate assemblies (1) and (3). The plate assemblies (1) and (3) are guided via leader pins (11) and guide bushings (12) (Fig. 4). The position of the three plate assemblies relative to one another is ensured by means of further centering units (13), which lie in the parting surfaces (A) and (B).

The container lids (C) are injected via heated hot-runner nozzles (14) in the center of a longitudinal side at a distance of 10 mm from the lid bottom. Each mold cavity is filled through the annular gap of about 0.3 mm between hot-runner nozzle and gate. To keep the heat requirements for the two heated hot-runner nozzles as low as possible compared with those of the cooled mold, the nozzles are surrounded by thermally insulating gate inserts (15). These center the nozzles and at the same time support them relative to the cavity plates. Each gate insert lies centrally in an insert well and the gate linked to it. This ensures that the nozzle's tip is exactly centered in the gate.

At the center of the stack mold there is a hot-runner (16). This is rectangular; only near the band heater (650 W) (17) and the centering collar is it round. In the rectangular part of the hot-runner the cartridge heaters are accommodated by means of bars fixed to it (18). Two high-capacity cartridge heaters (19) are incorporated, each with a heating capacity of 800 W (Fig. 3).

On the outside of the mold there are clamps (20) that pull the cavity plates (4a) and (4b) as well as (5a) and (5b) toward the hot-runner manifold over the hot-runner nozzle when the center plate assembly is being assembled. These parts are thus clamped in such a way that there is no risk of leakage between the end of the nozzle and hot-runner manifold. The surface contact pressure between nozzles and manifolds is further increased during operation because of thermal expansion.

A flangelike thickening at the hot-runner manifold is clamped between cavity plates (4a, 4b) and (5a, 5b), so that the injection unit's nozzle contact force acting axially on the manifolds is absorbed. The manifold is centered in the intermediate plate (6) as well as by the two two-part centering pieces (23) and (24).

The mold cavities are fed with melt through the central feed channel (21) and the four runners (22) lying at right angles to it and the four hot-runner nozzles. At the front side the hot-runner manifold is closed by a sliding shutoff nozzle (25) when the machine nozzle moves away, so as to prevent molding compound escaping, which would inevitably cause production problems after cooling and solidifying. Because of the axial displacement of the torpedo (26) when the hot-runner manifold moves away from the machine nozzle, the melt compressed in the hot-runner system during the injection process can expand in the resultant space of the channel (21). This prevents the melt escaping through the gates when the mold opens.

The claddings (27) and (28) protect the hot-runner manifold from major heat loss when the mold is opened and at the same time serve to protect the operator against accidental contact with the hot-runner.

When the mold is opened, the plate assembly (3) is pulled toward the left by the mold clamping plate (29), which is fixed to the moving platen of the injection molding machine, thereby opening parting line (B). During this operation, a synchronous opening movement of the two parting lines is achieved via a rack and pinion drive (30), which lies diagonally on the front and back surface of the mold (Fig. 3).

Ejection of the container lids takes place in the parting line (A) via the ejector mechanism (31), which is operated via two pneumatic cylinders (32). These lie on two opposite outer mold surfaces diagonally relative to each other. The ejector movement for the articles in parting surface (B) is carried out, as usual, via the ejector mechanism (33), which is actuated via the ejector rod (34).

The hinge holes on the component are produced by the core pins (35), which lie in the slides (36). The slides' movement at right angles to the direction of demolding is achieved by cam pins (37) (Fig. 4).

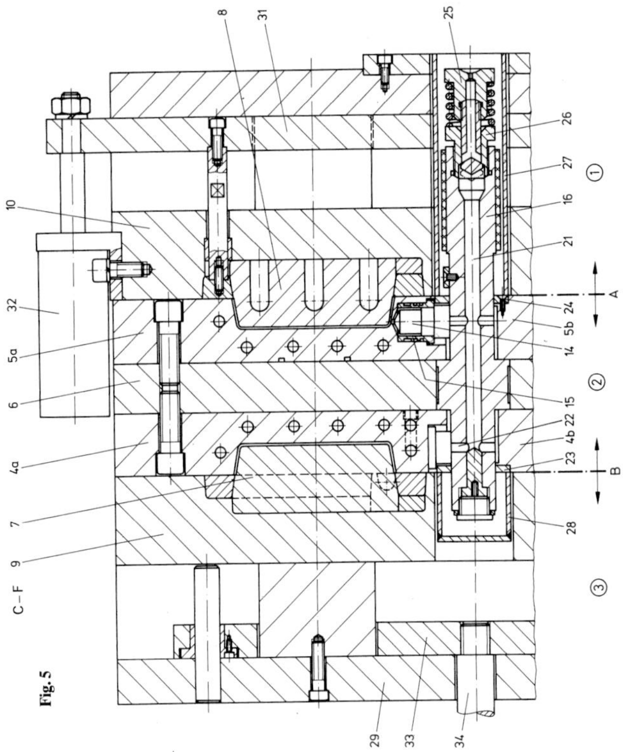

Fig. 5

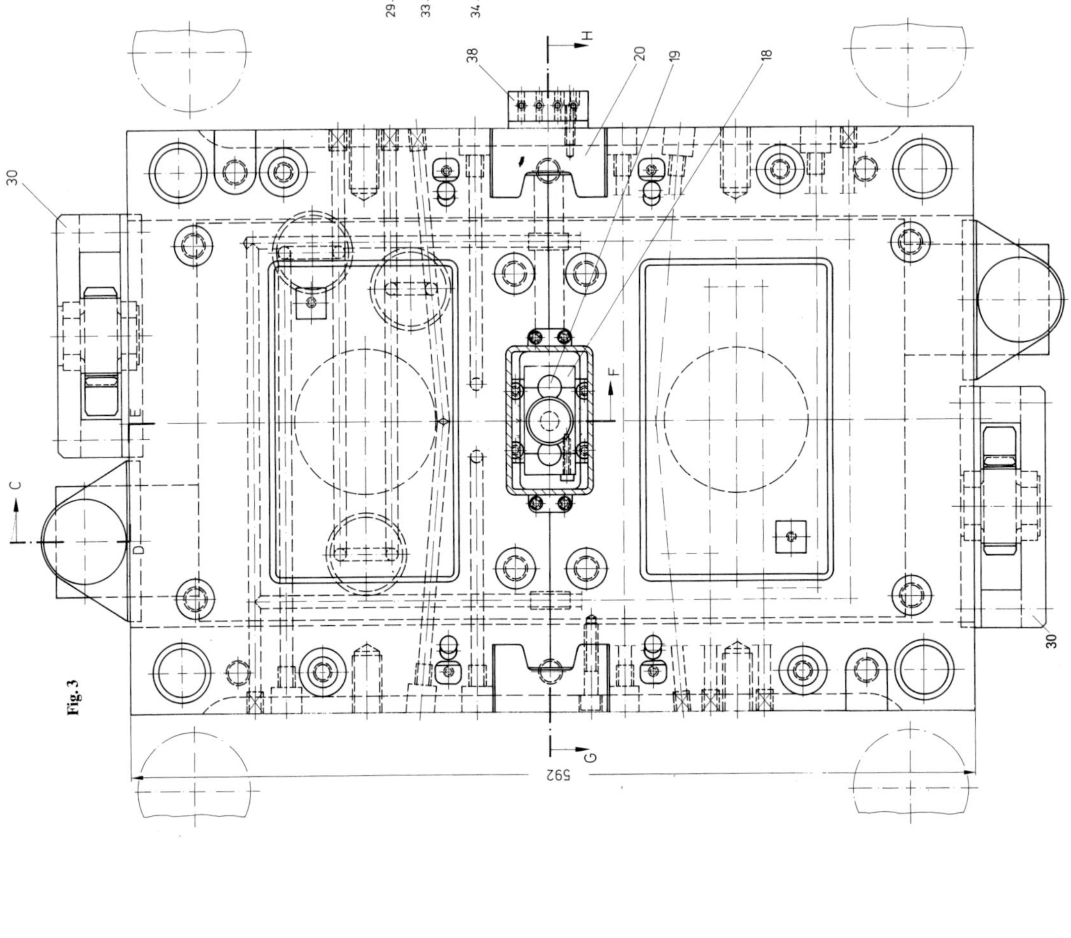

Fig. 3

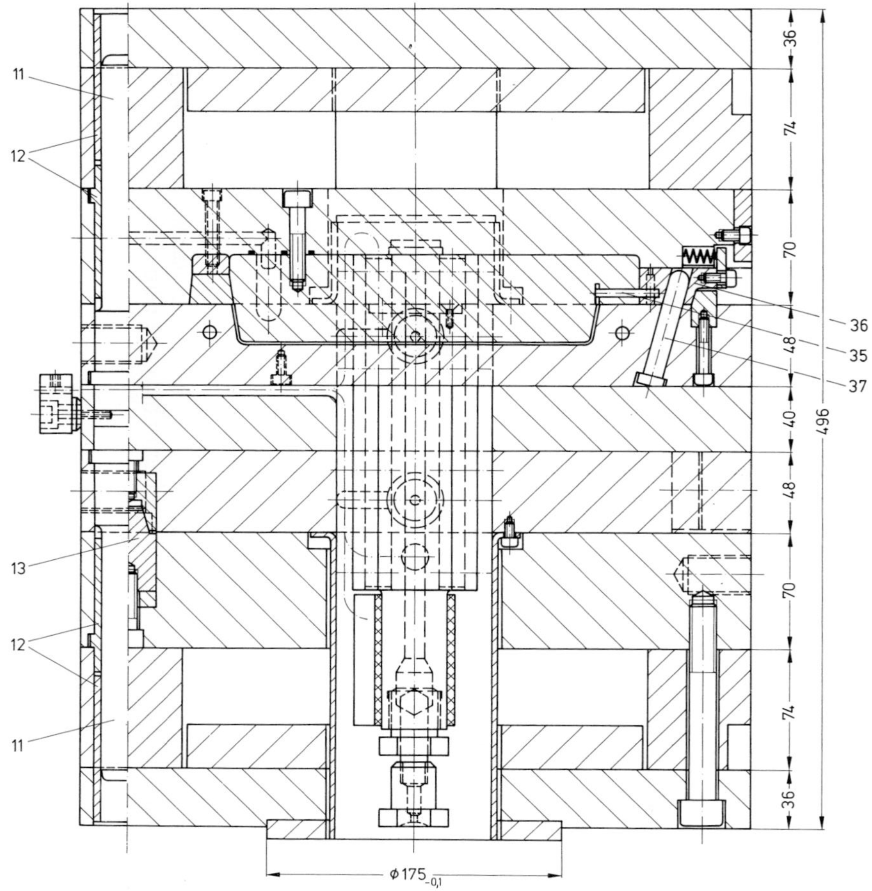

Fig. 4 G–H

Figures 3 to 5 2 + 2-cavity stack mold with hot-runner system for direct runnerless edge gating of polystyrene container lids

1, 2, 3: plate assembly; 4a, 4b, 5a, 5b: cavity plates; 6: intermediate plate; 7, 8: core inserts; 9, 10: support plates; 11: leader pin; 12: guide bushing; 13: centering unit; 14: hot-runner nozzle; 15: gate insert; 16: hot-runner; 17: band heater (650 W); 18: cover strip; 19: high-capacity cartridge heater; 20: centering clamp; 21: principal hot-runner; 22: secondary hot-runner; 23, 24: centering piece, double-shell; 25: sliding shutoff nozzle unit; 26: torpedo; 27, 28: sheet metal cladding; 29: mold clamping plate; 30: rack-and-pinion drive; 31: ejector mechanism; 32: pneumatic cam pin; 38: electrical terminal block for hot-runner nozzles

Example 44, 2 × 4-Cavity Hot-Runner Stack Mold for Dessert Cups Made from Polypropylene

The mold described here is used to produce polypropylene dessert cups with an average diameter of 60 mm, a height of 85 mm and a wall thickness of 0.55 mm. The cups weigh 7.5 g. The cup is footed so that an undercut that must be released by means of slides is formed between the foot and body of the cup.

Mold

The mold weighs approx. 2200 kg, has a shut height of approx. 700 mm and is designed as a 2 × 4-cavity stack mold. As is customary with stack molds, it consists essentially of three sections, namely the two end sections each of which has a clamping plate (1, 11) with approx. dimensions of 540 mm × 800 mm and a core retainer plate (2, 12). The center section (5) with the two bottom retainer plates (3, 13) holds the hot-runner manifold (4).
Two cavity retainer plates (6, 16) that accommodate the cavity inserts (7) are located between the two end sections and the center section.
The core inserts (8) are held by the core retainer plates (2, 12). Rugged locating rings (9) with conical surfaces that engage and center the cavity inserts (7) are fitted over the core inserts.
Guide strips (10) in which the slides (14) move sideways are bolted to the cavity retainer plates (6, 16). Each slide forms half of the outer shape of one foot for two neighboring cavities. Cam pins (15) mounted in the center section serve to actuate the slides. The opening stroke of the mold is 2 × 200 mm.

Runner System/Gating

Melt flows from the sprue bushing (17) with an attached sliding shutoff (18) into the feed pipe (19) and from there to the hot-runner manifold (4), which is heated by four heater rods (20). The directly heated hot-runner nozzles (21), the tips of which extend to the gate openings in the bottom inserts (22), are attached to the hot-runner manifold.
The heater bands for the feed pipe are enclosed by a protective tube (23), since the feed pipe is exposed via parting lines (I A) and (II A) when the mold is opened. The total installed heating capacity is approx. 6 kW.
The sliding shutoff prevents leakage of melt when the feed pipe is pulled in to the stationary section of the mold during mold opening. In addition, the increase in volume of the hot-runner system produced by the sliding shutoff when the machine nozzle backs away prevents drooling from the hot-runner nozzles.

Mold Venting

Vent gaps (32) and vent channels to remove the air displaced by the melt entering the cavity are provided at the ends of the flow paths around the rim of the cup and at the foot.

Temperature Control

Thin-wall parts such as these cups transfer heat quickly to mold surfaces, so that increased outlay on the cooling system is worthwhile in molds for such parts. Cooling of the core should be given special attention. A beryllium/copper cap (24) with six radial cooling channels is placed on the core insert (8). These cooling channels require drilling of the center tube (25) leading to the cap. This drilling weakens the center tube and there is the risk of rupture if the mounting nuts (26) are tightened excessively. Compression springs (27) are provided to permit an exactly defined tightening torque.
Coolant is supplied to the cooling channels in the slides via tubes (28) threaded into the slides. Slots (29) in the guide strips (10) allow these tubes to follow the motion of the slides.

Part Release/Ejection

Prior to mold opening, the hydraulic cylinders (30) are pressurized, so that the parting lines (I A, II B) open first. The undersides of the feet are released, the slides separate. Ball detents (33) secure the opened slides in their end positions.
Once the piston in the hydraulic cylinders has completed its full stroke, the mold opens at (I A, I B), and the cups, still retained on the cores, are withdrawn from the cavity inserts (7).
Finally, compressed air is introduced into the annular gap between the core insert (8) and core cap (24) via the channels (31). The molded parts are now blown off. A rack and pinion arrangement not shown in the drawing is used to ensure synchronous opening of the mold parting lines.

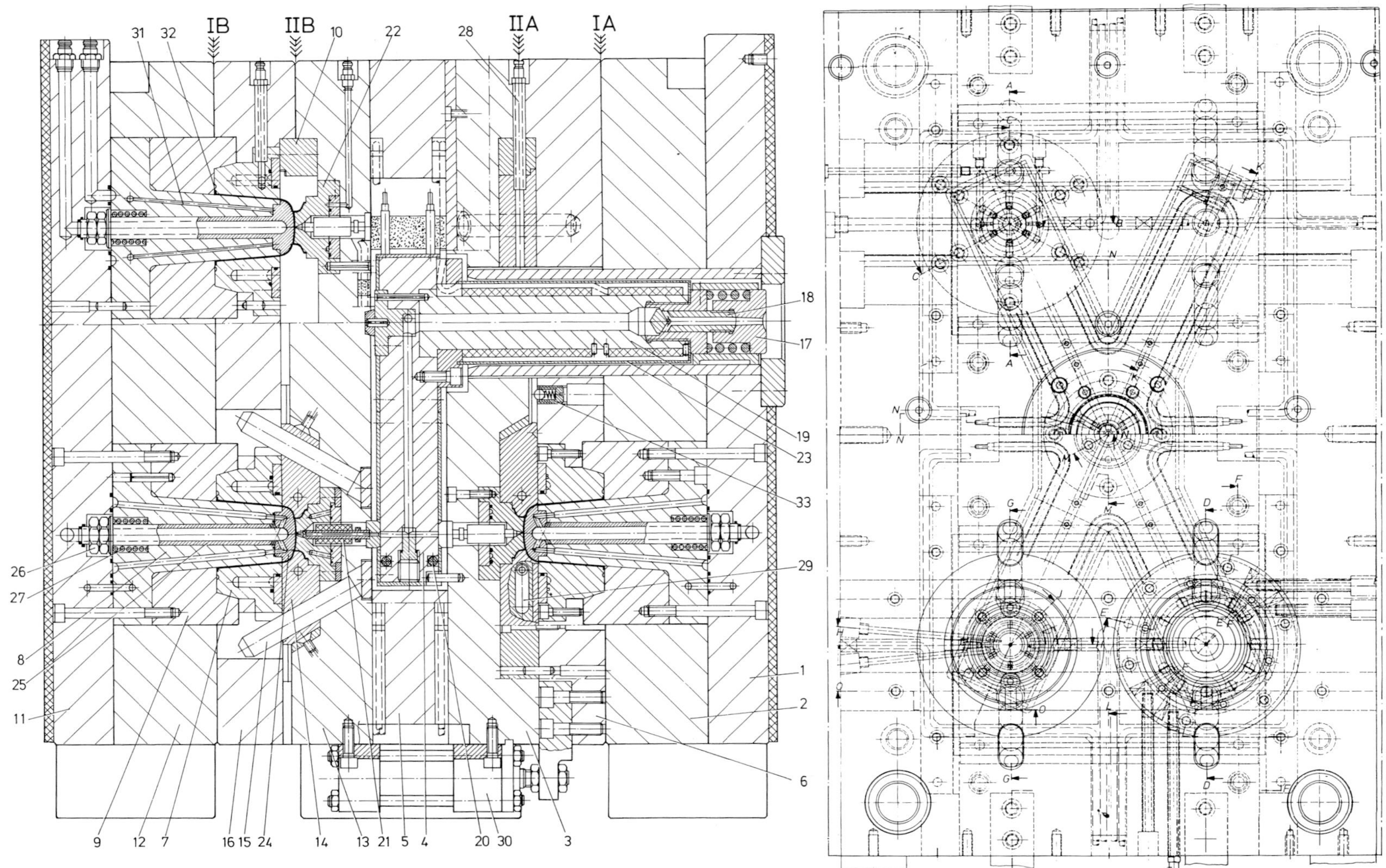

Figures 1 and 2 4 + 4-cavity hot-runner stack mold for dessert cups of polypropylene
1: clamping plate; 2: core retainer plate; 3: bottom retainer plate; 4: hot-runner manifold; 5: center plate; 6: cavity retainer plate; 7: cavity insert; 8: core insert; 9: locating ring; 10: guide strip; 11: clamping plate; 12: core retainer plate; 13: bottom retainer plate; 14: slide; 15: cam pin; 16: cavity retainer plate; 17: sprue bushing; 18: sliding shutoff; 19: feed pipe; 20: heater rod; 21: hot-runner nozzle; 22: bottom insert; 23: protective tube; 24: core cap; 25: center tube; 26: mounting nut; 27: compression spring; 28: cooling water connection; 29: slot; 30: hydraulic cylinder; 31: air channel; 32: vent gap and vent channel; 33: ball detent

Example 45, Hot-Runner Mold for Bumper Fascia Made from Thermoplastic Elastomer

Bumper fascias of TPE (thermoplastic elastomer) can be found today on most automobiles. To protect the vehicle, the sides of the bumper fascia wrap around to the side by a significant amount, so that in conjunction with numerous stiffening ribs, openings and mounting elements, very large molds with rather intricate part release are necessary.
The bumper has an overall width of approx. 1750 mm. With its wrap-around sides, it forms a U with a depth of 750 mm. Numerous ribs are located on the inside, and the side sections have transverse and longitudinal depressions that form undercuts in the direction of draw. The lower surface of the front section contains holes.

Mold (Figs. 1 and 2)

The mold has dimensions of 2800 mm × 1500 mm, with a shut height of 1740 mm and a weight of 32 t. To facilitate machining and handling, the cavity and core blocks are built up from a number of parts. The cavity block (1) is bolted to the filling pieces (2). The core half consists of the core retainer plate (3) and the core proper (4). These two parts of the core are fitted together with the aid of wear strips (6) and wedges (7). When the mold is closed, the cavity and core are centered with respect to one another by means of taper locks and attached wear plates (5).
To guide the core and cavity, four guide blocks (8) are provided, one at the center of each side of the mold. In contrast to the usual round leader pins, such guide blocks permit the core and cavity to be operated at different temperatures without binding. In addition, subsequent corrections in the event of uneven wall thicknesses are possible.
The part-forming components of the mold are made from polishable steel (material no. 1.2311) heat treated to a strength of 1100 to 1200 N/mm^2. For the remaining components, material no. 1.2312 is employed because of its better machinability. The mold clamping plates are made of material no. 1.1730; the wear plates are made of material no. 1.2162 and case-hardened. Bronze is employed for sliding pieces and guides. The movable slide inserts in the core are also made of bronze, in part because of the better thermal conductivity. Lifters (12) that are actuated by push rods (13) are used to release the undercuts on the inside of the front surface. The push rods are movably mounted in the ejector plate (14). With these lifters, the short U-shaped sections on the inside of the top surface of the bumper fascia can be released.
The inside surface of the two wrap-around side sections is released by internal slides (15) that are also actuated via push rods attached to the ejector plate (14).
The outer surface of the wrap-around side sections is located in hydraulically (cylinder 17) actuated external slides (16). Recesses with holes on the bottom surface of the bumper fascia are formed by core pins (18). They are operated by hydraulic wedge gate (19) located along the adaptor plate area of the mold.

Runner System/Gating

The part is filled from a hot-runner manifold (9) with two nozzles (11) with external heater bands (10). Each nozzle fills a short sprue and runner with a film gate. The two sprues, runners and film gates are removed from the molded part in a subsequent operation.

Temperature Control

The front surface of the molded part is cooled via cooling lines in the cavity, while the outer surfaces of the wrap-around side sections are cooled by cooling lines in the external slide (16).
Cooling lines in the lifters (12) and the internal slides (15) serve to cool the inside of the molded part. Supply and return of the coolant takes place via channels in the push rods (13). Space permitting, cooling lines are also located in the stationary core components.

Part Release/Ejection

The core pins (18) are pulled prior to mold opening. During opening, the cylinders (17) push the two external slides (16) in the open direction. The molded part is released from the stationary cavity surfaces as well as from the slides (16); the sprues are pulled out of the tapered orifices of the hot-runner nozzles.
After the part has been withdrawn from the cavity half, the ejector plate (14) is advanced by the cylinders (20). This actuates all lifters (12) as well as the internal slides (15) and the sprue pullers (29). The molded part is pushed off the core; the internal undercuts are released. It must be also be ensured during ejection that the wrap-around side sections of the fascia do not become caught by the shape of the internal slides. Blocks (21, 22) are provided for this purpose. With the aid of guides (23, 24), they ensure that the wrap-around side sections do not follow the sideways motion of the slides.
To release the molded part from the lifters that have advanced along with it, the ejector plate (25) is now actuated by hydraulic cylinder (26). With ejector plate (14) stationary, the molded part is pushed off by the ejector rods (27) and thrust blocks (28).

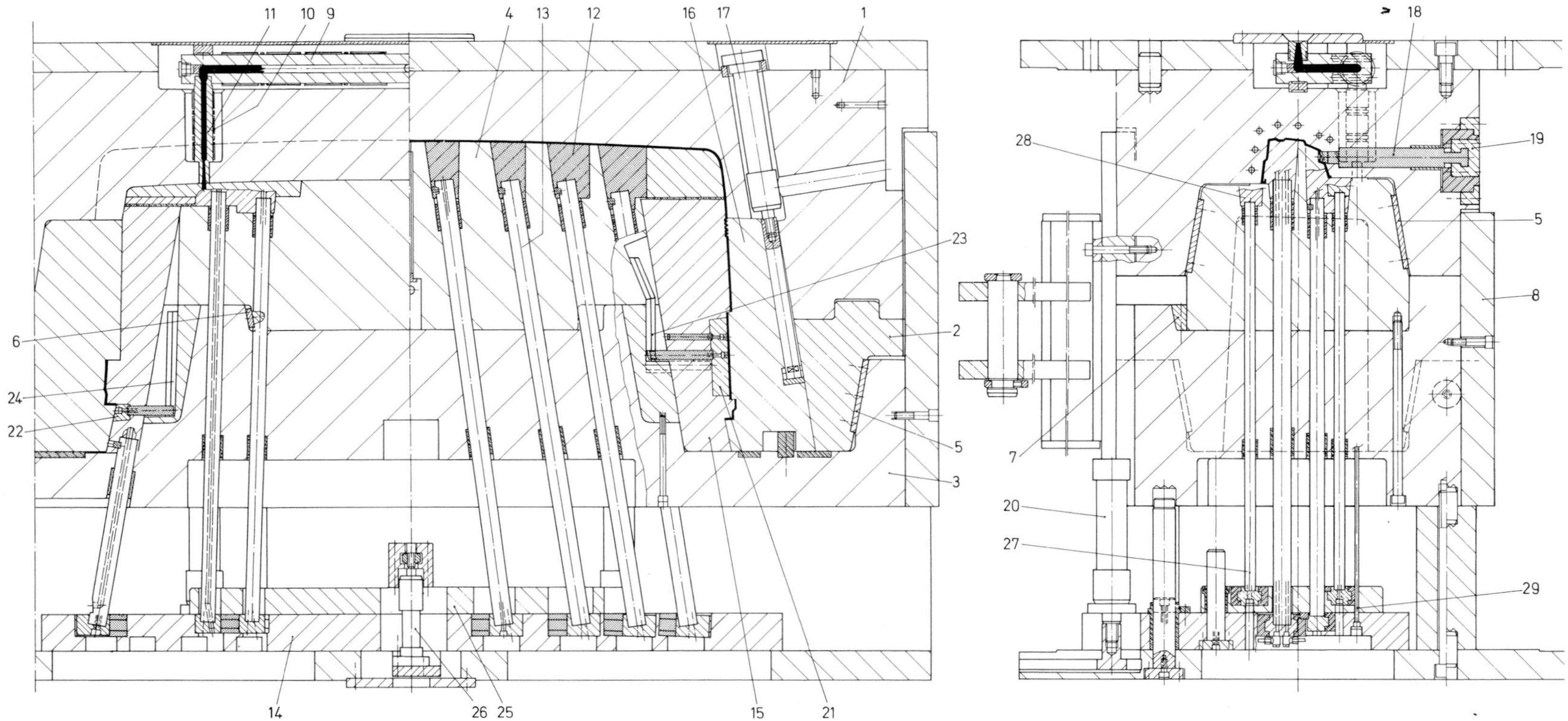

Figures 1 and 2 Hot-runner mold for a bumper fascia
1: cavity block; 2: filling piece; 3: core retainer plate; 4: core assembly; 5: wear plate; 6: wear strip; 7: wedge; 8: guide block; 9: hot-runner manifold; 10: heater band; 11: hot-runner nozzle; 12: lifter; 13: push rod; 14: ejector plate; 15: internal slide; 16: external slide; 17: hydraulic cylinder; 18: core pin; 19: wedge gate; 20: hydraulic cylinder; 21, 22: block; 23, 24: guides; 25: ejector plate; 26: hydraulic cylinder; 27: ejector rod; 28: thrust block; 29: sprue puller

Example 46, Four-Cavity Hot-Runner Mold for Threaded Covers Made from SAN

The appearance of cosmetic containers must, as a rule, meet very high standards. Thus, no gate marks are permissable on the appearance surface of the cover for a cream jar (60 mm diameter, 15 mm high). Gating on the outside either at the center or at the edge via a submarine gate, for instance, is not allowed.

It is thus necessary to gate the part through the core that forms the threads. In such a case, it would be possible to keep the cores stationary and rotate the cavities for unscrewing. The unscrewing mechanism would be simpler; the flow paths shorter. This is not possible, here, however, because – as mentioned – the external surface of the cover must be completely smooth so that no elevations or depressions to assist in unscrewing can be present.

Mold

As shown in Figs. 1 to 4, the unscrewing mechanism was thus located on the injection side.

The core inserts (1) are placed in threaded sleeves (2) that run in guide bushings (3) and are driven by drive shaft (5).

A hollow core (6) with helical cooling channel is located within the core insert (1) and accommodates within its 22 mm diameter a hot-runner nozzle (7) with a length of 150 mm.

The hot-runner system employed here is described in greater detail in Example 50 (toothpaste dispenser).

Radial grooves (i.e. ribs on the inside surface of the cover) that prevent the cover from turning during unscrewing are located on the part-forming surfaces of the core insert (1) and hollow core (6).

A drive shaft (5) extends through the movable-side mold clamping plate (8). The unscrewing motor does not follow the opening motion; the guide bushing (9) slides back and forth on the shaft (5) during opening and closing of the mold.

Mold Temperature Control

Cooling water reaches the mold cores through the hollow cores (6). Cooling lines are provided in the cavity plate (10) and the stripper plate (11). Channels in the core retainer plate (12) supply the hollow cores (6) with cooling water.

Part Release/Ejection

Unscrewing of the threaded sleeves is initiated upon mold opening. The molded parts are firmly held by the ribs on the core insert (1) and hollow core (6) until the latch (13) (Fig. 3) engages the stripper plate (11) and ejects the molded parts. The stripping motion is limited by the mechanical stop (14) (Fig. 4).

Fig. 1

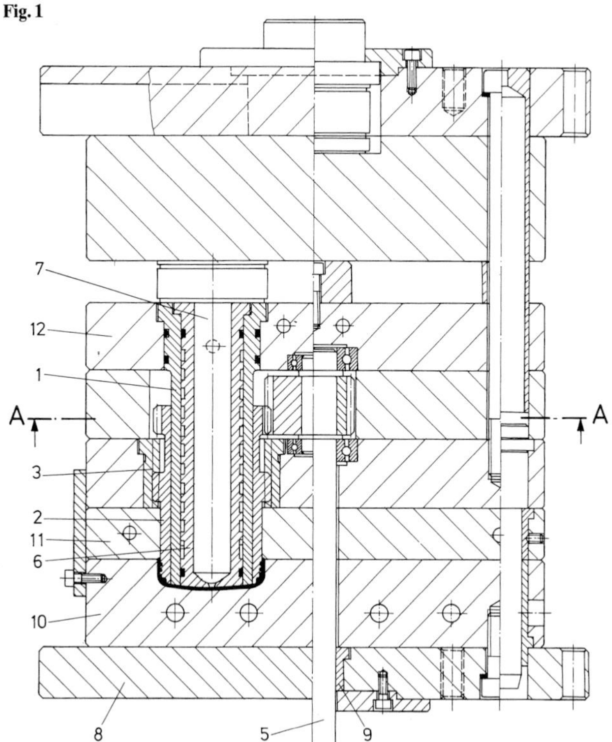

Fig. 2

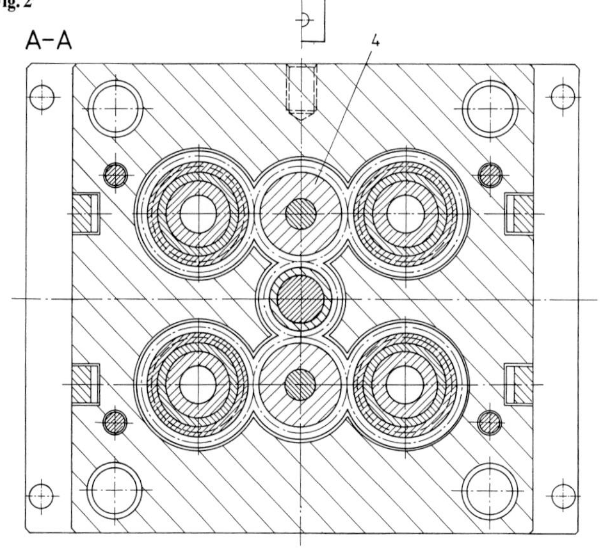

Fig. 3

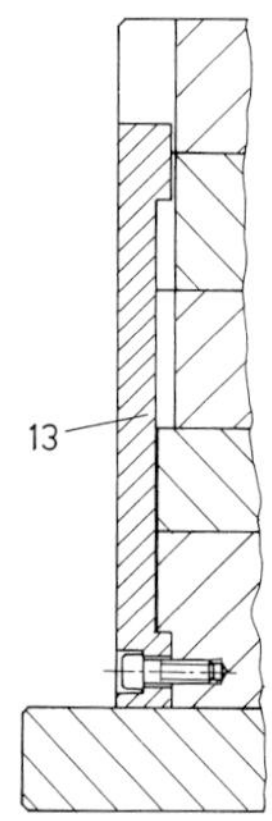

Fig. 4

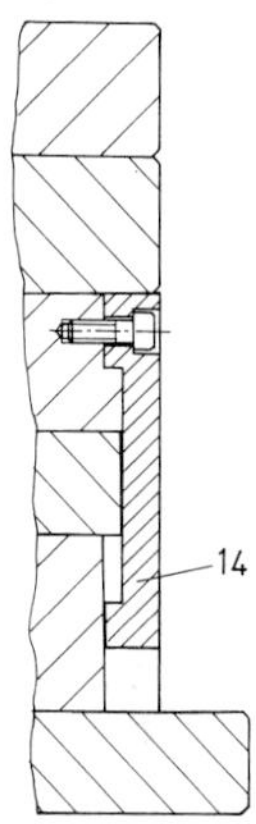

Figures 1 to 4 Four-cavity hot-runner mold for threaded covers
1: core insert; 2: threaded sleeve; 3: guide bushing; 4: gear; 5: drive shaft; 6: hollow core; 7: hot-runner nozzle; 8: mold clamping plate; 9: guide bushing; 10: cavity plate; 11: stripper plate; 12: core retainer plate; 13: latch; 14: mechanical stop
(Courtesy: Günther Heißkanaltechnik, Frankenberg, Germany)

Example 47, Two-Cavity Hot-Runner Mold for Trim Bezels Made from ABS

The two trim bezels (Fig. 1) have outside dimensions of 150 mm × 155 mm × 30 mm and are to be chrome-plated. They are installed in motor vehicles in pairs. For installation, each part is provided with eight snap hooks on the installation side to snap into the vehicle body.

Figure 1 Automotive trim bezel

Mold

The mold contains a pair of parts (Figs. 2 to 4). The distance between the two mold cavities is determined by the slide (15) that must be placed between them to release the snap hooks found there. The remaining hooks are released by slides (12 to 14). Mold inserts (16, 17) attached to the slides form the part-forming surface for the snap hooks. The slides are operated by cam pins (24, 25, 33). When the mold is closed, the slides are secured by heel blocks (22, 23) and the bracket (18). The ejector pins are secured against turning (pin 32) since their ends are shaped to the part-forming surface.
The mold has dimensions of 596 mm × 396 mm with a shut height of 482 mm and a weight of 725 kg.

Runner System/Gating

Each of the parts is filled via two submarine gates at either end of a runner positioned in the opening diagonally between two snap hooks. Melt reaches the two mold cavities through a heated sprue bushing (4), a hot-runner manifold (20) and two attached hot-runner nozzles (21). The hot-runner manifold contains two heating coils (19). The manifold block is supported against the opening force resulting from the injection pressure by support pads (26) of a high-strength thermally insulating material.
A transducer (28) to measure the melt pressure in the runner is located behind the ejector pin (27).

Mold Temperature Control

Each cavity is provided with two cooling circuits on the stationary side and one circuit on the movable side. The cooling circuits are formed by channels drilled to follow the shape of the molded parts. Thermocouples (29) to provide information on temperature changes of the coolant in the mold are provided at the inlet and outlet of the cooling circuits.

Part Release/Ejection

Upon mold opening, the parts and solidified runners are retained on the movable mold half, since the snap hooks and runners are still held in the slides and sprue pullers as well as submarine gates respectively.
After the snap hooks are released by the slides, the molded parts and runners are ejected by the ejector pins. This also shears off the submarine gates.
During mold opening, the slides disengage from the cam pins required to operate them. Coil springs (30) hold the slides in the opened position. In this way, the cam pins can reenter the slides during mold closing without suffering any damage. The ejectors are retracted during closing by means of pushback pins (31).

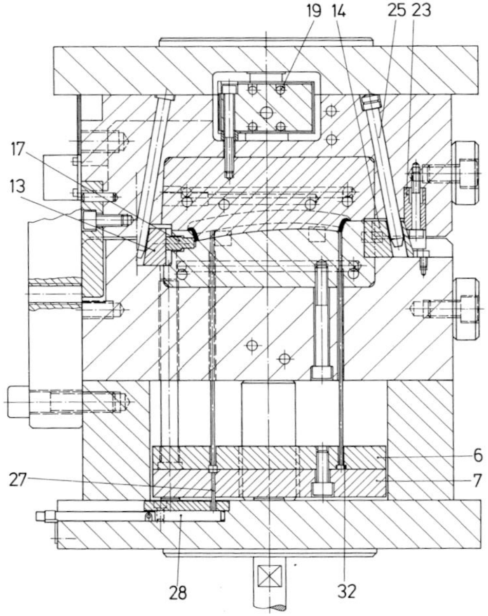

Fig. 2

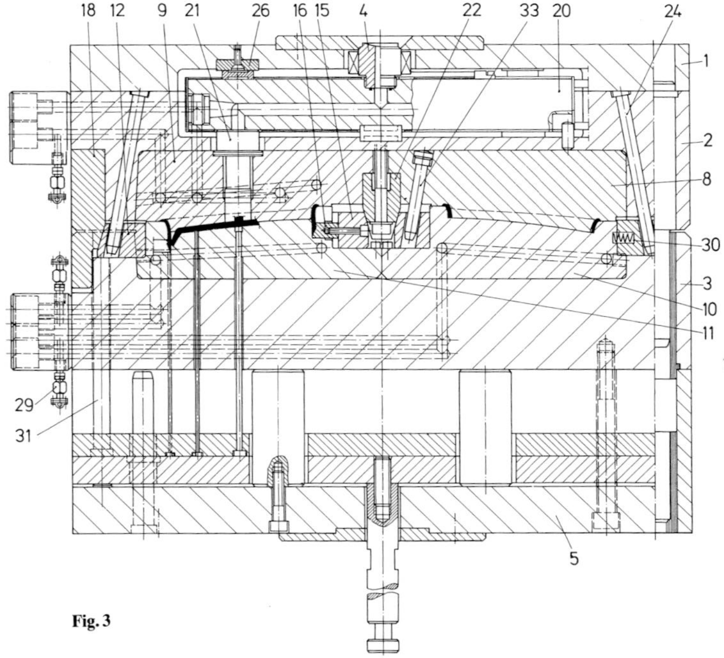

Fig. 3

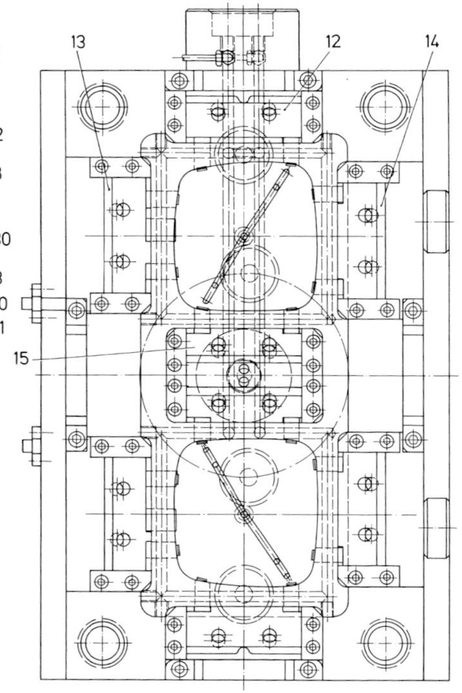

Fig. 4

Figures 2 to 4 Two-cavity injection mold for automotive trim bezels
1: mold clamping plate; 2, 3: mold plate; 4: sprue bushing; 5: mold clamping plate; 6, 7: ejector plates; 8, 9, 10, 11: mold inserts; 12, 13, 14, 15: slides; 16, 17: mold inserts; 18: locking bracket; 19: heating coils; 20: hot-runner manifold; 21: hot-runner nozzle; 22, 23: heel block; 24, 25: cam pin; 26: insulating support pad; 27: ejector; 28: transducer; 29: thermocouple; 30: coil spring; 31: pushback pin; 32: securing pin; 33: cam pin

Example 48, Four-Cavity Hot-Runner Mold for Control Flap Made from Polyacetal Copolymer

The control flaps (Fig. 1) are installed in pairs in the flush valve of a toilet cistern and permit water-saving interruption of flushing.
The parts have approximate overall dimensions of 55 mm × 65 mm × 55 mm and consist essentially of a cup-shaped float chamber that functions also as a valve body and a number of attached spring levers.

Figure 1 Toilet cistern flush valve of polyacetal (POM) copolymer

Mold

The mold has dimensions of 496 mm × 316 mm with a shut height of 427 mm and contains four cavities (Figs. 2 to 5). The four cavities are arranged in a line so that the spring levers attached to the float chambers can be molded together in a single slide (21). The slide runs in guide strips (22) and on wear strips (25) and is actuated by two cam pins (45). Wear plates (24) hold the slide in position when the mold is closed.
Four mold inserts (20) are attached to the slide. In addition, four ejector plates (27, 28) with ejector pins (54) and pushback pins (53) are located in the slide.
The cavities of the float chambers are formed by cores (23).
The ejector assembly (4) containing the ejector pins (51), blade ejectors (52) and pushback pins (47) runs in ball guides (48).

Runner System/Gating

The mold halves are aligned as usual by leader pins (11) and guide bushings (12, 13). Locating strips (29) ensure proper final alignment. The mold inserts (16, 17, 20) and the slide (21) are made of hardened steel (material no. 1.2767). The cores (23) are made of Cu–Be.
Melt flows from the sprue bushing (39) through a filter insert (62) to the hot-runner manifold (30), which is heated by four heater cartridges (64) with a heating capacity of 800 W each. From there it flows to the four gate chambers where it is kept warm by the indirectly heated thermally conducting torpedoes (34). The torpedo tips extend into the gate openings so that the gate separates cleanly from the molded parts.

Mold Temperature Control

The slide (21) and mold inserts (16, 17) contain cooling lines and bubblers with baffles (33) to direct the cooling water. The Cu–Be cores (23) transmit the heat they absorb to the surrounding, directly cooled components via conduction.

Part Release/Ejection

As soon as the mold opens, the slide (21) moves sideways away from the molded parts, allowing the ejector plates (27, 28) with the attached ejectors (54) to push the spring levers out of the recesses in the mold inserts (20) through the action of the compression springs (57). The slide is secured in the opened position by spring-loaded ball detents (59). Pushback pins (53) return the ejectors (54) to the molding position as the mold closes.
The ejector pins (51, 52) eject the molded parts from the cores (23) and from the recesses in the inserts (17).
Figure 6 gives a view of the ejectors in the open mold. The mold inserts (20) in the slide (21) can be seen at the right, a core (23) is visible at the top left and a molded part being ejected by the ejector pins (51, 52) can be seen at the lower left.

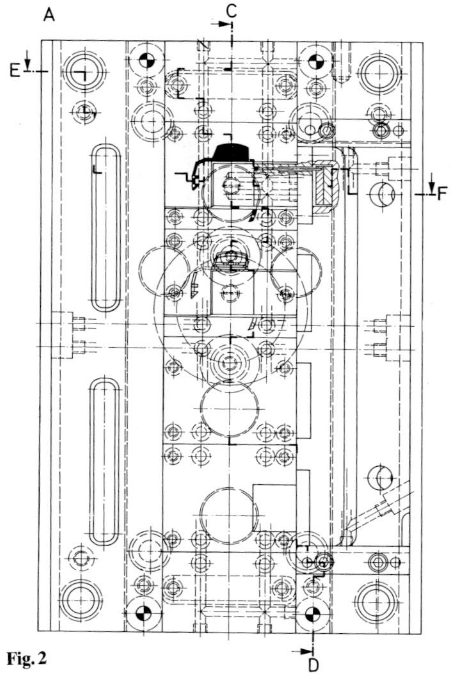

Fig. 2

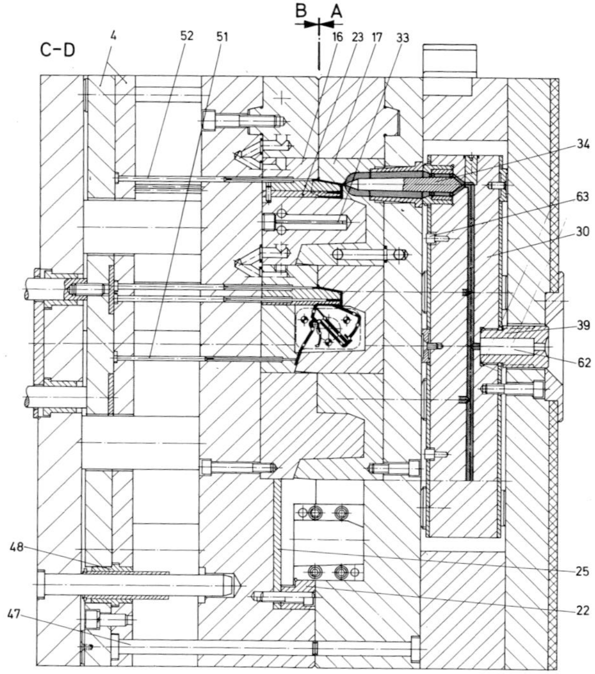

Fig. 3

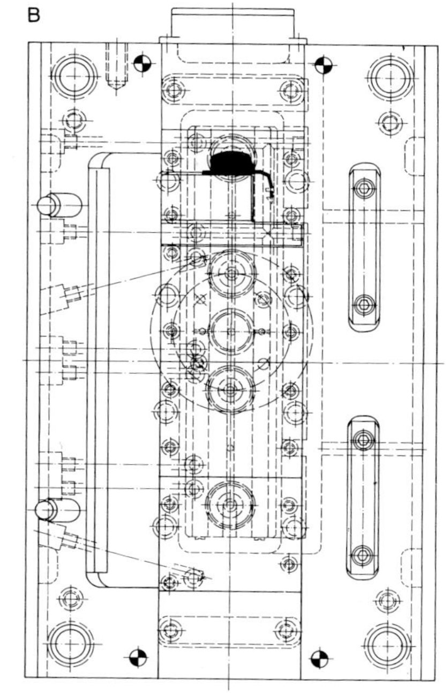

Fig. 4

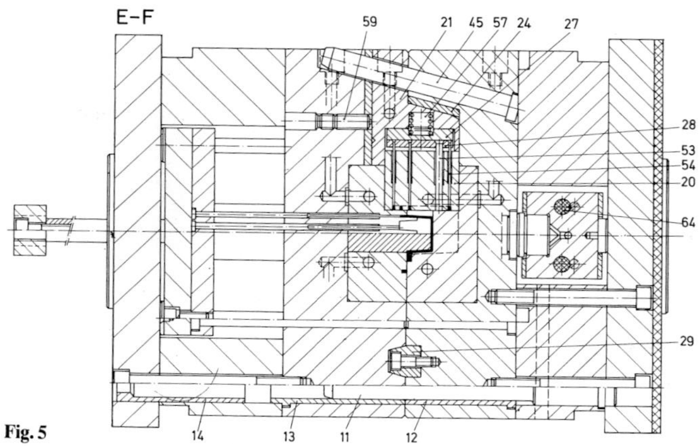

Fig. 5

Figures 2 to 5 Four-cavity hot-runner mold for a control flap of polyacetal copolymer for a toilet cistern flush valve
4: ejector assembly; 11: leader pin; 12, 13: guide bushings; 16, 17: mold inserts; 20: insert; 21: slide; 22: guide strip; 23: core; 24, 25: wear strips; 27, 28: ejector plates; 29: locating strip; 30: hot-runner manifold; 33: baffle; 34: thermally conducting torpedo; 39: sprue bushing; 45: cam pin; 47: pushback pin; 48: ball guide; 51: ejector pin; 52: blade ejector; 53: pushback pin; 54: ejector pin; 57: compression spring; 59: ball detent; 62: filter insert; 63: thermocouple; 64: cartridge heater

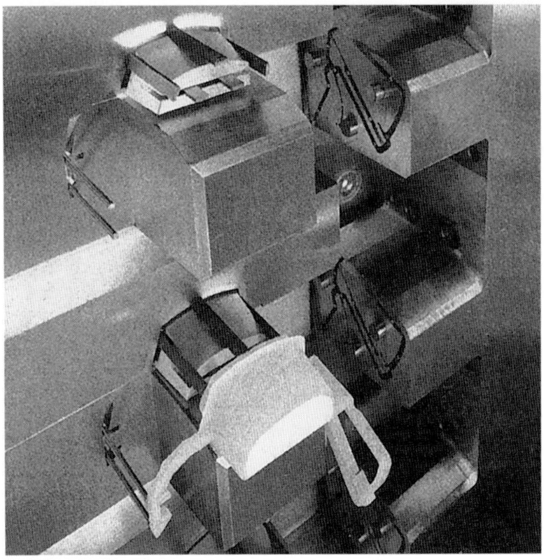

Figure 6 Ejection of the control flap

Example 49, 64-Cavity Hot-Runner Mold for Seals Made from Thermoplastic Elastomer (TPE)

Seals for disposable injection syringes (Fig. 1) are increasingly being produced from thermoplastic elastomers (TPE), whose processability by the injection molding method has advantages over the rubber hitherto employed. In the mold introduced here 64 seals of 14 mm diameter, 8 mm high are produced in a runnerless manner. The cycle time is about 20 s.
The external mold dimensions are 740 mm × 550 mm, and the mold height is 463 mm. The 64 cavities have been arranged in four blocks of 16. They are supplied with melt through a hot-runner system. The cavity inserts (22), cores (23) and ejector sleeves (24) are identical and interchangeable (Fig. 2 and 3).

Figure 1 Seal for disposable syringes

Runner System/Gating

The hot-runner manifold block is of two-storey construction, so that the runners leading to the mold cavities can all be of equal length. Thus a natural balancing of the flow resistances in the manifold is achieved.
The melt arriving from the machine's nozzle enters the manifold block A (12) through the sprue bushing (25). The manifold block is in the shape of a St. Andrew's cross, guiding the melt through four channels of equal length into the center of the four distributors B (13). From there bores – also of equal length – lead to individual heated nozzles (14) of the spear torpedo type. Steel O-rings (17) serve as seals between the manifold blocks, the heated nozzles and the sprue bush. The two manifold blocks are heated by cartridges (18, 19). Every heating zone is controlled within itself.
The torpedoes have two different heating zones. Whereas heating (20) in the torpedo shaft has a constant effect, heating in the torpedo tips (21) is switched ON and OFF in such a manner during the injection cycle that thermal opening and closing of the gates is achieved. By closed-loop controlling the shaft-heating it is possible to achieve fine-tuning of the melt volume entering individual mold cavities. This has the advantage that these changes in temperature have no influence on the opening and closing of the gate passages.

In order to achieve a clean break at the gates, the diameter of the gate orifice must not be larger than 0.5 mm maximum and the discharge opening must have very sharp edges.

Mold Temperature Control

The cavity side has been provided with numerous cooling channels (29) for removing heat from the molded parts and hot runner system dissipation. The cores are cooled by central bubbler tubes (28) housed inside them.

Part Release/Ejection

When opening the mold in the parting line, molded parts and cores are withdrawn from the cavities. The machine's ejectors then push against the bushings (27), moving the ejector plate (7) together with the core plate (8) forward, so that the ejector sleeves (24) force the seals off the undercuts on the cores. With the start of the closing movement the coil springs (26) return the ejector plate to its starting position.

Fig. 2

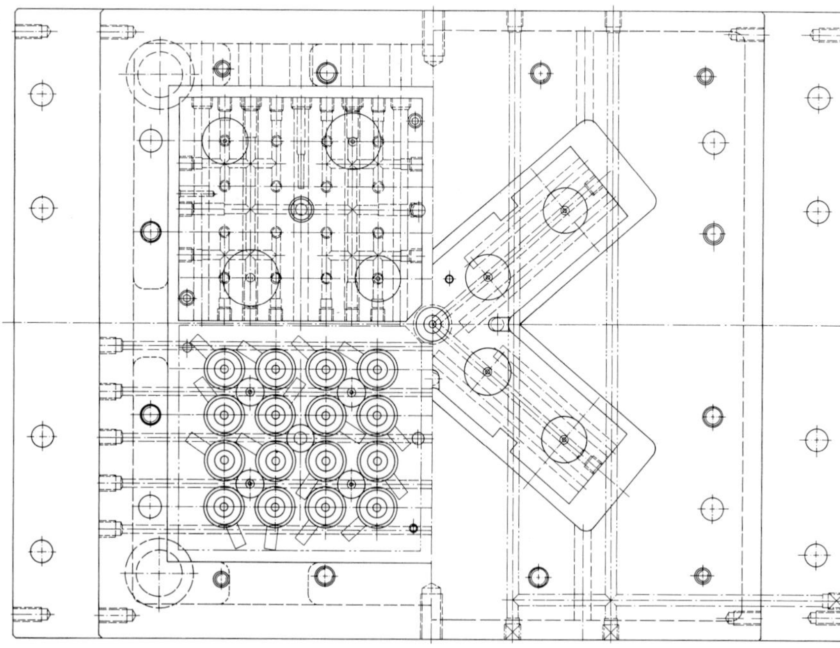

Fig. 3

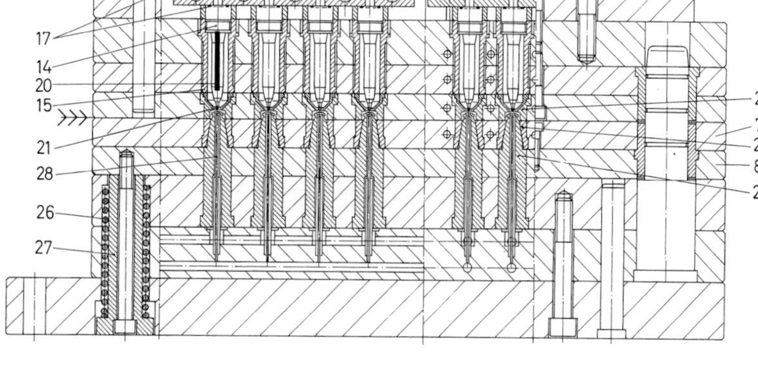

Figures 2 and 3 64-cavity hot-runner mold for seals of thermoplastic elastomer
7: ejector plate; 8: ejector plate; 12: manifold block A; 13: distributor block B; 14: heated hot-runner nozzle, system spear, seiki; 15: insulating bushing; 17: O-ring (steel); 18, 19: cartridge heaters; 20: torpedo shaft heating; 21: torpedo tip heating; 22: cavity insert; 23: core; 24: ejector sleeve; 25: sprue bushing; 26: coil spring; 27: shoulder bushing; 28: bubbler; 29: cooling channel

Example 50, Eight-Cavity Hot-Runner Mold for PP Toothpaste Dispenser

The polypropylene toothpaste dispenser (Fig. 1) is a cylindrical, 146 mm long article, essentially consisting of two tubular sections, the first of which is of 36 mm internal diameter and 26 mm long. The second part is 120 mm long with an internal diameter of 38 mm. There is a partition between the two sections, equipped with various functional components.

Injection of Long Tubular Moldings Straight Through the Core

If one were to gate this molding at one point on the outside, the long core would curve due to the unilaterally entering melt: the article would not fill uniformly.

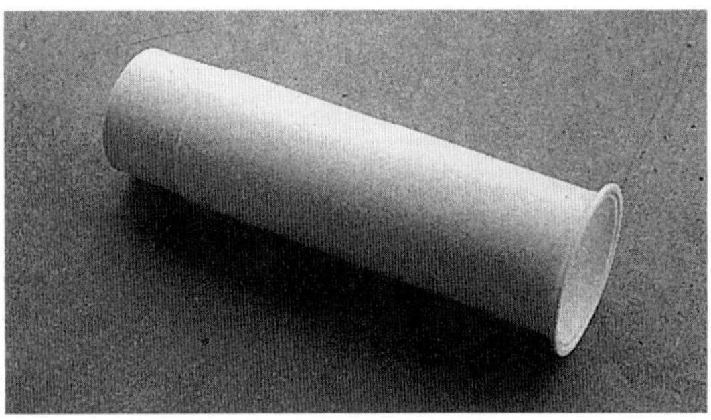

Figure 1 Polypropylene toothpaste dispenser

A number of gates distributed around the periphery, although preventing the core from being displaced, would leave gate marks and mean a large percentage of wasted material in the runner.
Internal gating would be a possibility with a three-plate mold and a break-away pinpoint gate. However, even greater material loss in the form of sprue would be unavoidable, as the specific shape of the partition only allows gating through the long tubular section.
An externally heated, temperature-controlled hot-runner nozzle (Fig. 2) can be employed here to advantage, with the melt being carried in a tube electrically heated to the required processing temperature at low voltage (3 to 5 V) and effectively insulated against heat losses. As Fig. 3 illustrates, it is thereby possible to employ a cooling spiral equipped hollow core (9) inside the long core for the tubular section of 38 mm internal diameter, which accommodates the 200 mm long hot-runner nozzle in a bore of 22 mm diameter.
This nozzle is equipped with a cone-shaped heated tip entering the gating point. Thus a small ring-shaped gate is created that ensures clean article separation. There is no interference between cooling system and hot-runner nozzle; the article is therefore allowed to cool rapidly.
The melt conveying system inside the manifold (Fig. 6) is designed differently from that in the nozzle. The melt-carrying heating tube a is electrically insulated and surrounded by a supporting tube f. This tube system is enclosed in a stationary layer of rigid material d and rests in a bore of the manifold plate (4). The "frozen" material layer d acts as heat insulation, so that the manifold plate (4) is allowed to make full surface contact with the two adjacent mold plates without requiring any other form of heat containment. A hot-runner system so designed combines the advantages of externally heated hot-runners (thermally homogeneous melt) with those of the internally heated systems (simple manifold construction, good thermal insulation against the environment).
The bores in the manifold plate (12) (Figs. 3, 4, 5) have been arranged in two layers, one above the other. Thus the channels serving the eight cavities – grouped in two rows in the mold – can be of equal

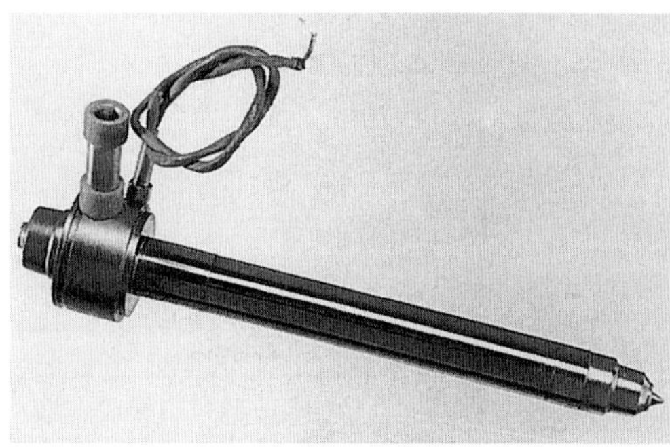

Figure 2 Hot-runner nozzle for injecting through the long tubular section

length, so that a natural balancing of the flow resistances can be achieved between sprue bushing and cavities.

Mold Temperature Control

The short cores (2) (Fig. 3) are accommodated in the moving mold half. They have been equipped with an effective spiral cooling system (8). The mold cavities are formed by two cylindrical sleeves each (3, 4), around which spiral cooling grooves have been arranged. Even the stripper rings (16) have a grooved ring for cooling.

Part Release Ejection

The mold opens at parting line I. Plates (5, 6) are retained by the latches mounted on the fixed half of

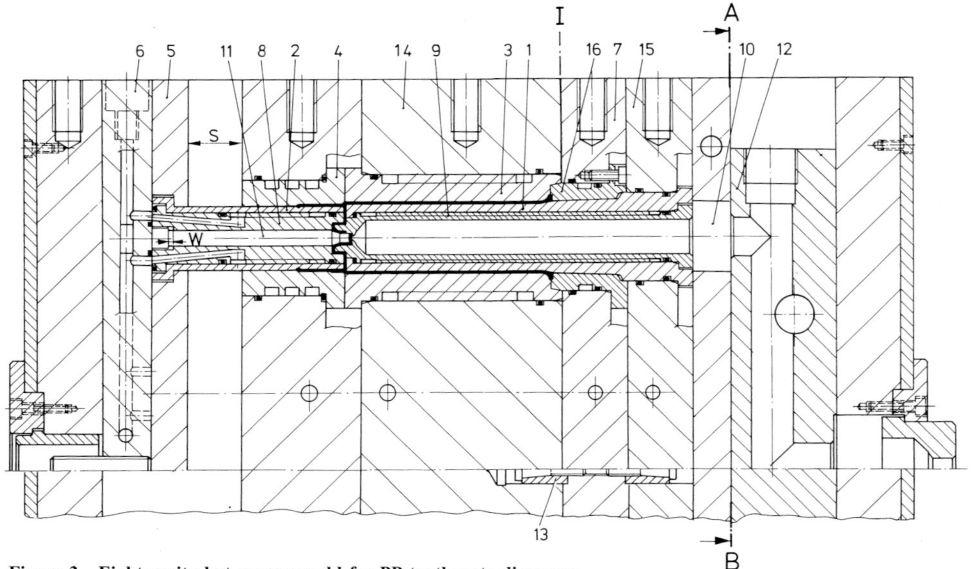

Figure 3 Eight-cavity hot-runner mold for PP toothpaste dispenser
1: long core; 2: short core; 3: cavity sleeve, long; 4: cavity sleeve, short; 5, 6: ejector plate; 7: stripper plate; 8, 9: hollow cores; 10: hot-runner nozzle; 11: sliding core; 12: manifold housing; 13: cushioning device; 14: cavity plate; 15: core retainer plate; 16: stripper ring
(Courtesy: Günther)

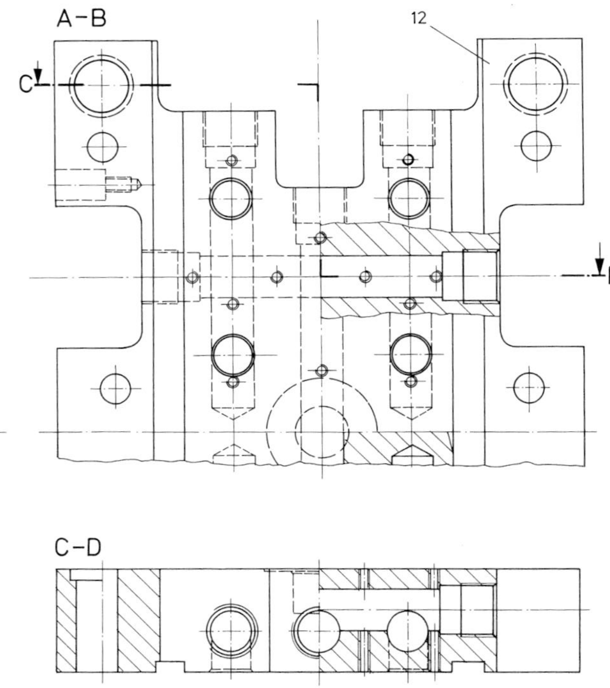

Figures 4 and 5 Manifold plate for the mold shown in Fig. 3

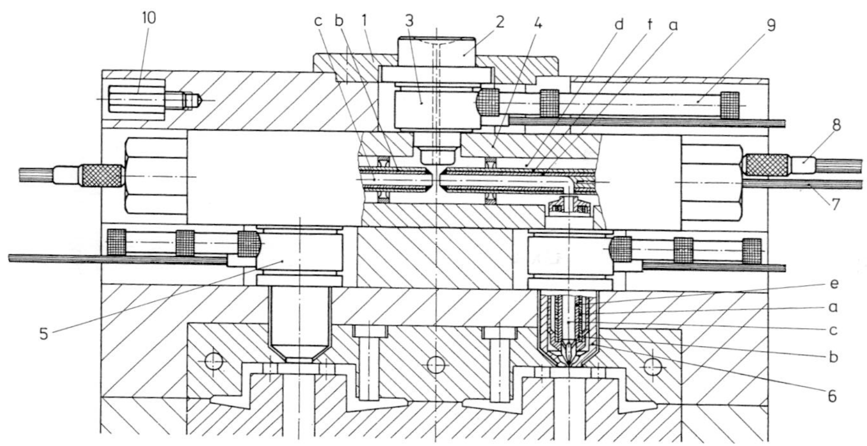

Figure 6 Hot-runner system for injecting through the long tubular section (see text for explanations)

the mold, so that the moldings remain on the two cores (1, 2) to start with. They are only displaced in relation to the cavity (3, 4).
After distance S, the two platens (5, 6) continue the opening movements, the molded part remains on the long core, the short core is pulled and the undercut on the thin end of the part is demolded.
One sliding core (11) each is housed in the center of the cores (2) with a displacement stroke W. When the articles are released from the short cores (2), each core (11) travels with the departing molding for the distance of stroke W. Having reached the end of the stroke the core – which had given shape to the molding's partition – is only then released from it. Finally, the stripper plate (7) pushes the moldings off the long cores (1).
A cushioning device (13) has been fitted in the stripper plate (7). Its two ends, which protrude beyond the plate (7) enter bores in the cavity plate (14) and the core retaining plate (15) with a friction fit. This prevents the mold plates from chattering when being pushed together during mold closing. Core (1) returns the sliding core (11) to its starting position with the closing movement.

Example 51, 2-Cavity Hot-Runner Mold for Polyethylene Jars

This jar with a diameter of 50 mm and a height of 28 mm (Fig. 1) requires a smooth, clean gate point. For this reason, valve-gating was chosen. The universal bulge on the inner edge of the jar forms an undercut, and the jar edge has four small high spots. In order to ensure that the molded part falls out of the mold (Figs. 4 to 7), the demolding sequence was divided into two steps.

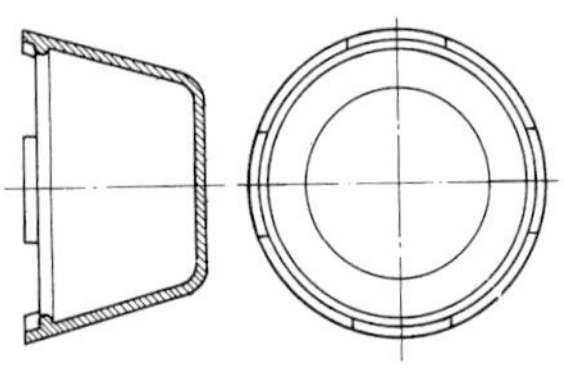

Figure 1 Jar made from PE

Gating System

Melt flows through the sprue bush (43), hot-runner manifold (56) and two heated sprue nozzles (37) into the cavity. The hot-runner block is heated by four heater cartridges (39) inserted in pairs at each end. The heater cartridges and their mounting bores are conical; this simplifies their installation and removal, as well as ensuring good heat transfer. A pneumatic valve needle (41) runs the length of each sprue nozzle. In addition to the advantage of leaving smooth gating areas on the molded part, valve gates generally require less injection pressure than pin-point gates since the opening they provide during cavity filling is larger. They are also not sensitive to impurities (granules) in the injected material. There is a thermal insulation sheet (14) on the nozzle side of the mold clamping plate (1) to minimize the heating effect from the hot-runner manifold on the machine plate.

Cooling

Inside the bored-out core (59) there is a standardized spiral core (46) with bi-directional loops for coolant circulation. The counter-sunk insert (58) has a universal cooling grove; its two O-rings (49, 50) for sealing deserve special mention. One ring (49) is smaller in diameter than the other (50) and the space for it. This avoids the smaller ring (49) being damaged as it is thrust past the bores K when the countersunk insert is installed. The gating area, especially the area of core 59 directly across from the gate, is a design-related hot spot. Gloss variations, waviness or noses (due to surface particles sticking and ripping out of the part at increased temperatures) are quite likely to occur with this kind of cooling. Heat transfer can be improved by using steel/copper pins (e.g., Hasco Z 494/ . . . steel/copper core pins, Fig. 2). The combination of external hardened steel jacket *and* a copper core can improve heat transfer, since this material combination significantly increases thermal conductivity. Depending on the type of heat transfer – in series, or parallel with two different materials in the composite – different equivalent thermal conductivity coefficients are obtained [1]. For a prismatic component composed of different materials, the equivalent thermal conductivity coefficient is λ_R or λ_p:

Series Conduction, λ_R

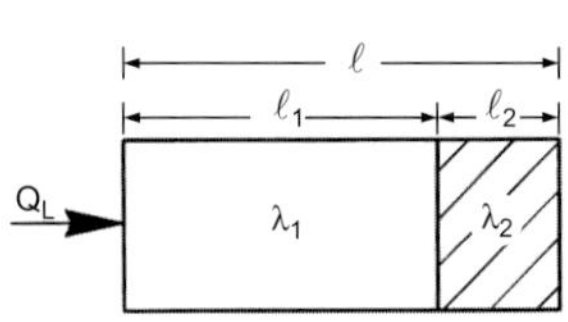

$$\text{((drawing))}\ \lambda_R = \frac{1}{\frac{\phi_1}{\lambda_1} + \frac{\phi_2}{\lambda_2}} \left[\frac{W}{m \cdot K}\right]$$

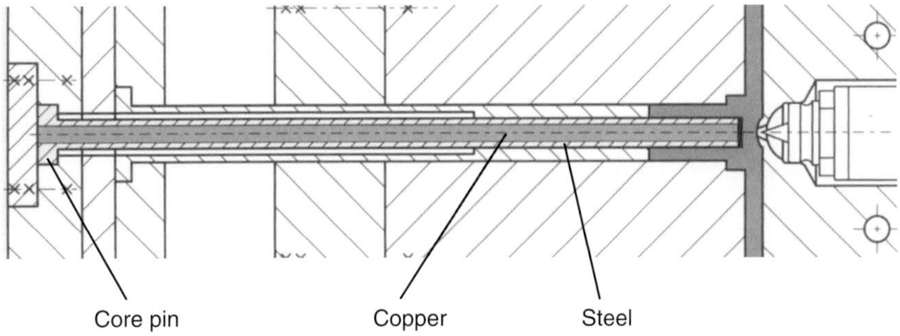

Figure 2 Steel/copper core pin Z 494/… instead of ejector pin (51 in Fig. 7) (Courtesy: Hasco)

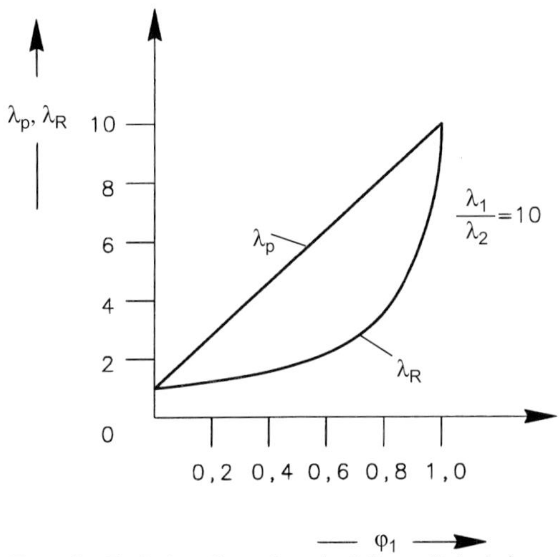

Figure 3 Equivalent thermal conductivity coefficients λ_R and λ_P as a function of volume portion Φ_1 of component 1. Assumption: $\lambda_1/\lambda_2 = 10$, which corresponds to the ratio $\lambda_{copper}/\lambda_{steel}$

Parallel conduction, λ_p

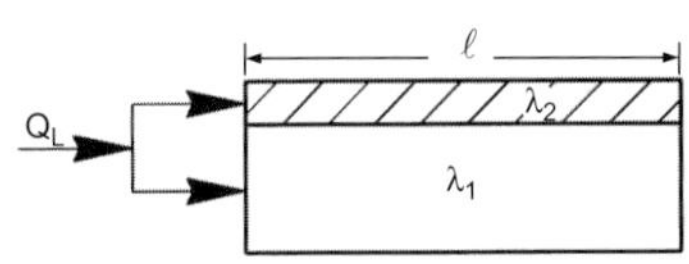

$$((\text{drawing}))\ \lambda_p = \phi_1 \cdot \lambda_1 + \phi_2 \cdot \lambda\left[\frac{\text{W}}{m \cdot K}\right]$$

$$\phi_1 = \frac{V_i}{V};\ \phi_1 + \phi_2 = 1$$

where ϕ_1 Volume portions
V_i Individual volumes
V Total volume.

For parallel conduction, the largest equivalent thermal conductivity coefficient results. By contrast, for series conduction the smallest equivalent thermal conductivity coefficient results (Fig. 3).

As a result, axial heat transfer is more effective than radial. However, the design in the example at hand enables heat transfer almost exclusively in radial direction. This problem can be solved by using steel/copper core pins. Such core pins are available as standard parts. The composite material exhibits thermal expansion different from that of its individual components: its equivalent thermal expansion coefficient a_{total} is greater than that of steel, but smaller than that of copper. It can be calculated according to the following relation:

$$\alpha_{total} = \frac{\alpha_1 \cdot E_1 \cdot \vartheta_1 + \alpha_2 \cdot E_2 \cdot \vartheta_2}{E_1 \cdot \vartheta_1 + E_2 \cdot \vartheta_2}$$

α_i thermal expansion coefficient of individual components [1/K] and E_i modulus of elasticity of the indivitual components [N/mm^2].

Demolding

The mold opens at I and the molded part is pulled from the conical hole by the core. When the standardized two-stage ejector (53) strikes the machine stop, it moves the ejector plates (10, 11, 12) with it. At the same time, one plate (10) moves the two stripper plates (5, 6) (opening at II) and the push-off ring (60) via a sleeve (23). The other plates (11, 12) also push the ejector pins (51) forward. After the molded part has been snapped off the undercuts at the core, the first plate (10) and push-off ring (60) stand still. The ejector (51) moves on and releases the part from the push-off ring contour. The part falls off. When the mold closes, the return pin (52) pushes the ejector system back into injection position. A pressure transducer (54) under one of the ejector pins (51) monitors internal mold pressure.

Reference

1. Unger, P., Hot Runner Technology, Hanser Publishers 2006

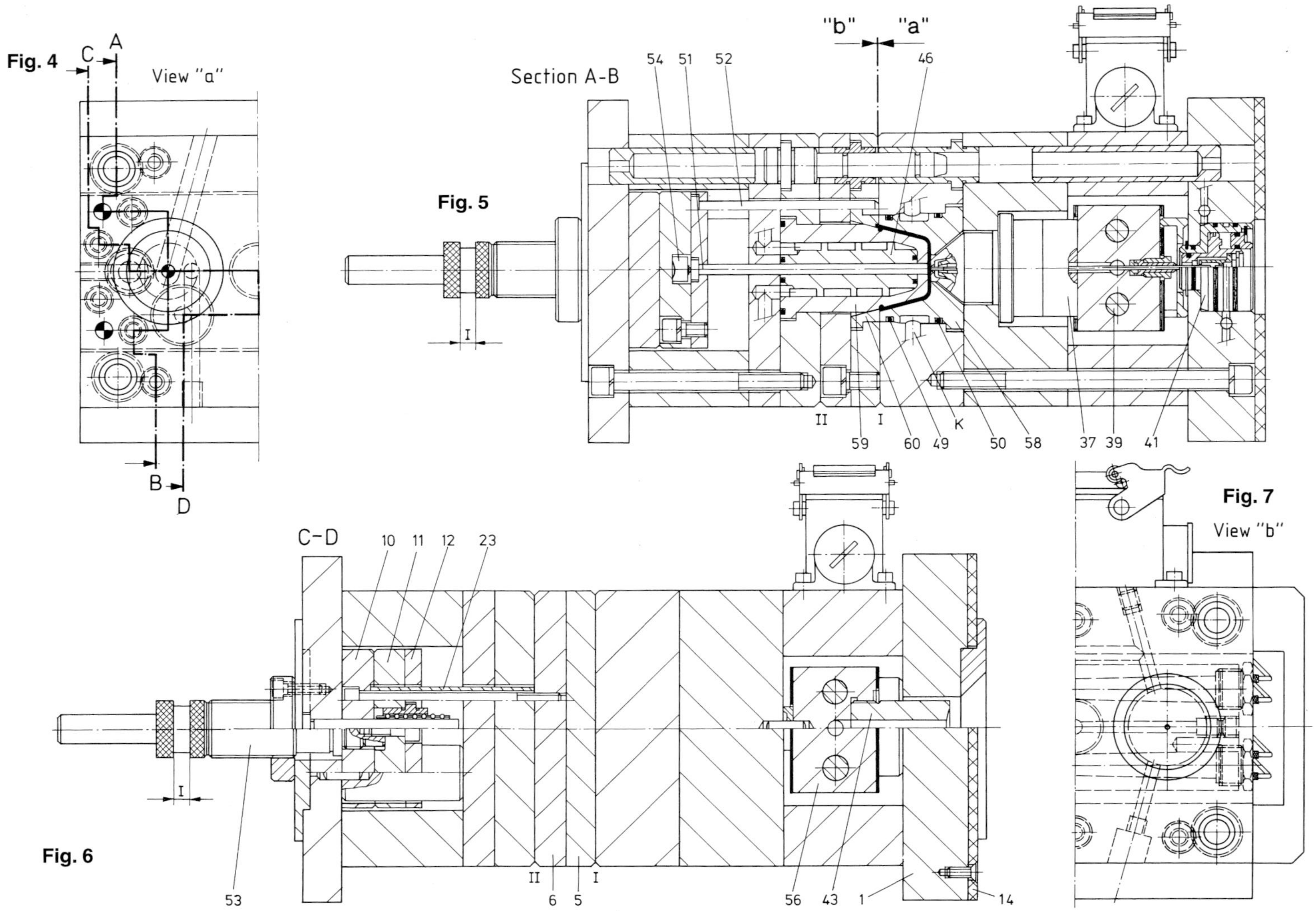

Figures 4 to 7 Two-cavity hot-runner mold for producing tubs of PE
1: mold clamping plate; 5, 6: strinpper plates; 10, 11, 12: ejector plates; 14: insulating plate; sleeve; 37: hot-runner nozzle; 39: hot runner nozzle; 39: cartridge heater; 41: pneumatically operated needle valve; 43: sprue bushing; 46: spiral core; 49, 50: Vision O-rings; 51: ejector pin: 52: pushback pin; 53: two-stage ejector; 54: transducer; 56: hot-runner manifold; 58: cavity insert; 59: core; 60: stripper ring

Example 52, Two-Cavity Hot-Runner Mold for Production of Connectors Made from Polycarbonate

A connector shell (Fig. 1) was supposed to be molded in glass-fiber-reinforced, flame-retardant polycarbonate.

The two different parts (upper and lower halves) were to be produced in a single mold (Fig. 1). Prior to the start of production, the question arose as to whether a second gate might not be required to fill the part. To avoid weld lines and entrapped air, however, it might also be necessary to use only a single gate. This flexibility as to gating was supposed to be possible simply by switching a hot-runner nozzle on or off. A further difficulty resulted from the small amount of space in which the gates were to be located on the face of the molded part. All of these requirements were satisfied through the particular arrangement of the conductive, internally heated hot-runner system.

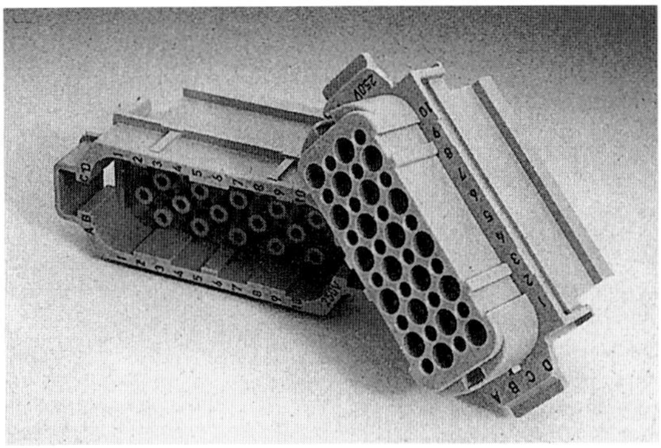

Figure 1 Connector shell (upper and lower halves) of glass-fiber-reinforced prolycarbonate

Mold

Figure 2 shows the basic construction of the mold. It consists of a heated sprue bushing (2), a hot-runner manifold (1) and four hot-runner nozzles (3). The hot-runner nozzles have not heen installed parallel to the longitudinal axis of the mold, but rather at an angle. In spite of the unfavorable geometric relationships, it is possible to gate each part on its face in this manner (Fig. 3). This arrangement is possible, because the solidified melt in the outer regions of the hot-runner manifold and nozzle channels precludes any possibility of leakage. Each of the hot-runner nozzles is individually controlled and can thus be switched on or off as required. It is thus possible to fill the part through either one or two gates. These measures alone, however, would not have been adequate to vary the gating possibilities. Attention also had to be given to the hot-runner manifold to ensure that no melt stagnated at continuously high temperatures in the runner channels when the various gating possibilites were being employed. The installation of four manifold end pieces (1, 2, 4, 5) in addition the central heating element (3) in the hot-runner manifold solved the problem (Fig. 4). Each of these manifold end pieces is individually controlled so that switching sections of the hot-runner manifold on or off is possible without subjecting the material to thermal loads in regions where it is not flowing. Switching the hot-runner nozzles on or off produces an H-, U- or Z-shaped runner in the manifold. Activating the control circuits (I, II, III, IV, V) produces an H-shaped runner. Activation of control circuits (III, IV) and (V) results in a U-shaped runner, while control circuits (I), (III) and (V) yield a Z-shaped runner. The H-shaped runner is a basic prerequisite for molding of both parts. The U-shaped runner permits one part to be gated at two locations. The Z-shaped runner permits both parts to be gated at only a single location. For variants U and Z, however, the corresponding hot-runner nozzles must also be switched off.

Fig. 2

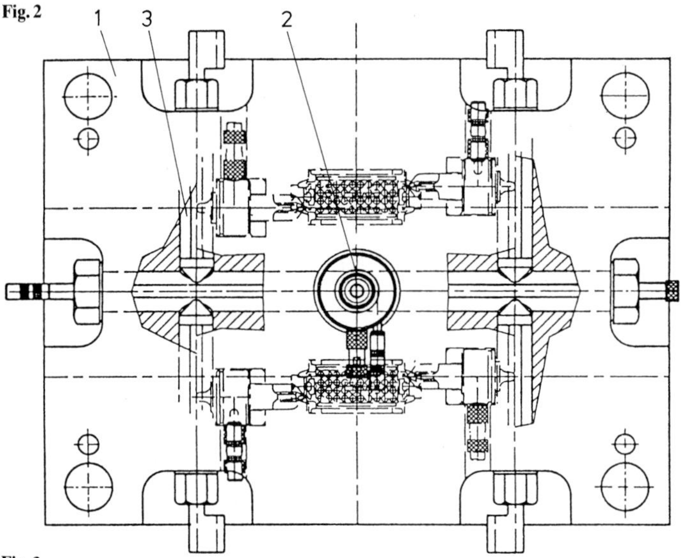

Figure 2 Two-cavity injection mold for producing a connector shell
1: hot-runner manifold; 2: heated sprue bushing; 3: hot-runner nozzles

Fig. 3

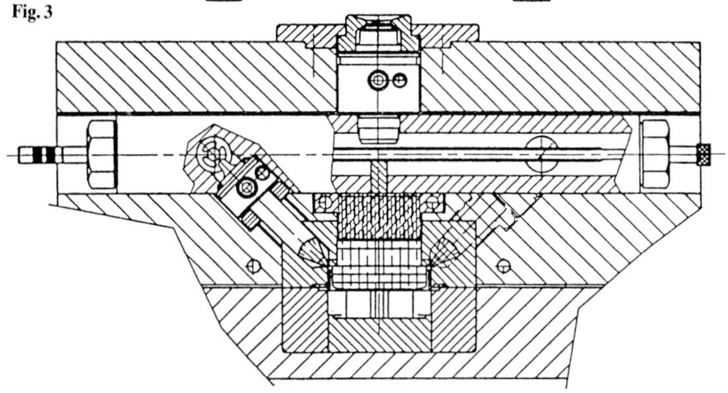

Fig. 4

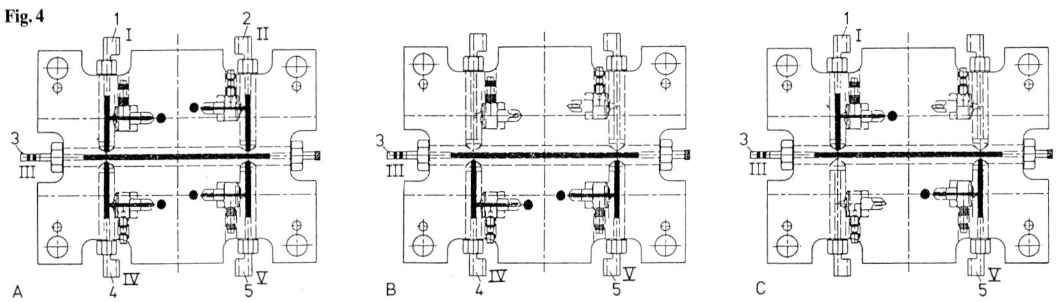

Figure 4 Runner variations in the manifold
A: H-shaped runner; B: U-shaped runner; C: Z-shaped runner
1, 2, 4, 5: manifold end pieces; 3: central heating

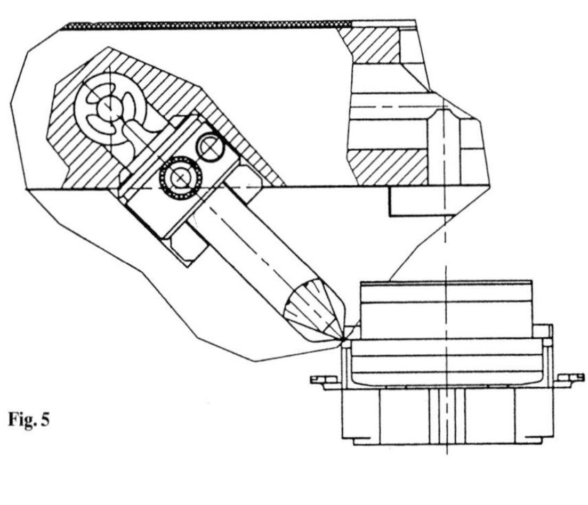

Fig. 5

Figures 3 and 5 Nozzle arrangement in the injection mold shown in Fig. 2
(Courtesy: Ewikon)

Example 53, Four-Cavity Hot-Runner Unscrewing Mold for Cap Nuts Made from Polyacetal (POM)

The four cavities of the mold are arranged in line because this results in an especially space-saving arrangement for the drive mechanism for the unscrewing cores by means of a hydraulically actuated rack.

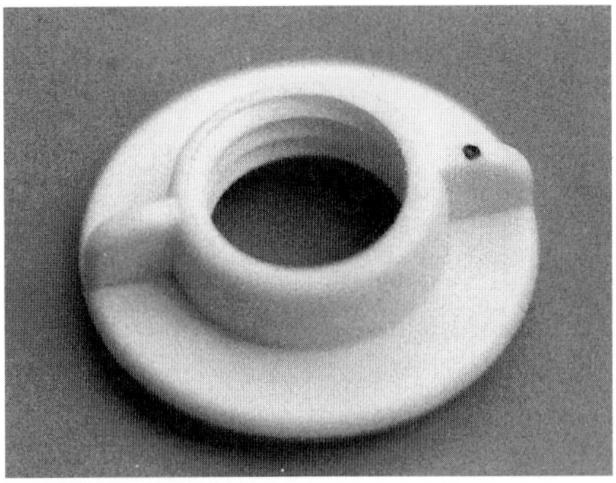

Figure 1 POM cap nut produced in a 4-cavity hot-runner mold

The unscrewing cores are of a multi-piece design and consist of the gears (33) with journals and lead threads and the part-forming threaded cores (31). The threaded cores must be unscrewed while the mold is still closed, since the ejector side of the cap nuts is flat and offers no means of preventing rotation. A hydraulically actuated gear rack (36) drives the threaded cores. The gear forces are absorbed by the guide plate (32) and threaded bushings (34).
After unscrewing has been completed, the mold opens and the ejector sleeves (30) eject the molded parts.
The replaceable mold inserts (13), (14) are made of the polishable, through-hardened steel (material no. 1.2767); mold plates (6), (7) are made of pre-hardened steel (material no. 1.2312).
The cap nuts are gated on their face along one of the two ribs via an off-center pinpoint gate.

An externally heated hot-runner system is employed. As a result of identical flow cross-sections and flow path lengths to the individual cavities, balanced and uniform filling is ensured.
The melt flows from the heated sprue bushing (46) into the manifold (49) via a central cylindrical insert. The manifold pipe is made of hardened steel and is encased in a heating element. The steel pipe and heating element are bonded together and thermally insulated. Attached cylindrical inserts (48) direct the melt to the gates.
The slip fits permit unobstructed movement of the manifold pipe in the cylindrical inserts as it expands during heating.
From the cylindrical inserts the melt flows to the four hot-runner nozzles (45) where it is directed to the gates. The heated nozzles (Fig. 5) have a central tip that extends into the gate orifice and ensures clean separation of the gate from the molded part. The cylindrical inserts and nozzles are thermally insulated from the surrounding mold components by air gaps. The manifold pipe and nozzles are fitted with replaceable thermocouples for closed-loop temperature control.

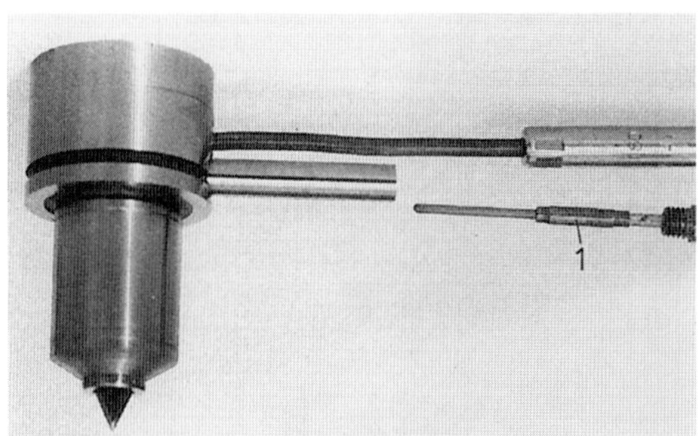

Figure 5 Hot-runner nozzle with replaceable thermocouple (1)
(Courtesy: Heitec)

Fig. 2

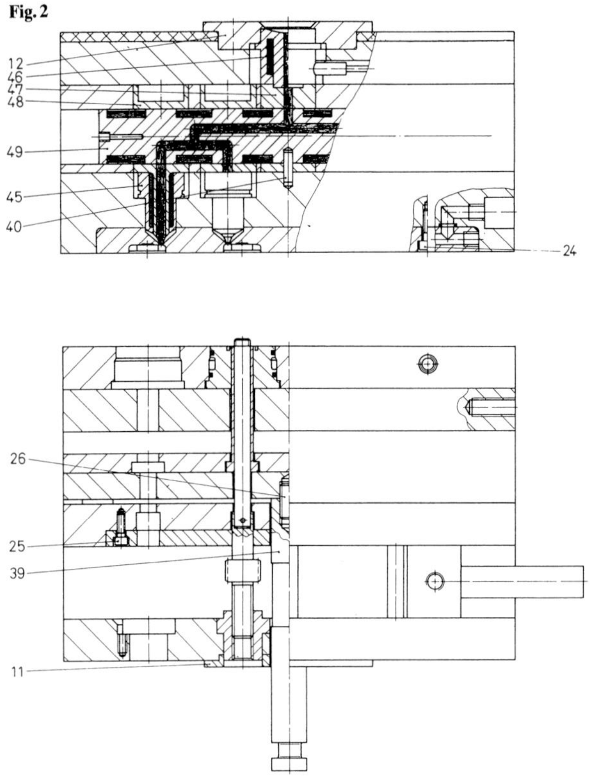

Fig. 3

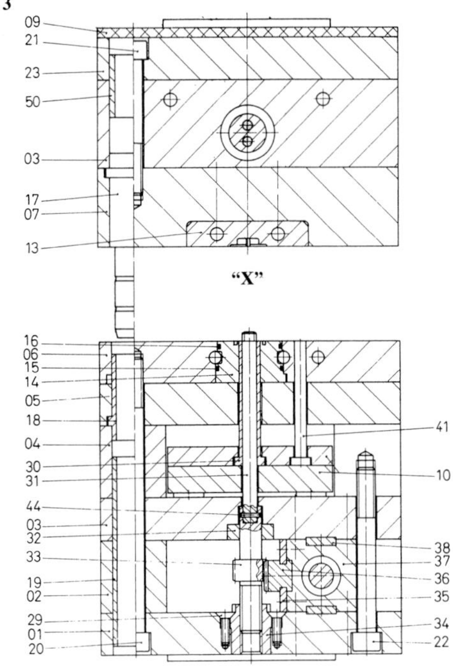

Item	Qty.	Designation	Dimensions	Material
50	4	locating bushing	Ø 30 × 60	
49	1	manifold pipe	Ø 35 × 250	
48	4	cylindrical insert	01.01 K	
47	1	cylindrical insert	01.01 K	
46	1	heated sprue bushing	01.09.06.40	
45	4	hot-runner nozzles	01.01.20.30 K	
44	4	straight pin	Ø 3 × 14	
43	1	flange	25/1	
42	8	cylinder head screw	M 5 × 30	
41	4	push back pin	Ø 8 × 80	2516
40	1	straight pin	Ø 6 × 20	
39	1	ejector rod	Ø 18 × 200	
38	4	guide plate	25	
37	1	hydraulic unscrewing mechanism	25/300	
36	1	gear rack	25/600/1.25	
35	2	guide strip	25/300	
34	4	threaded bushing	Ø 34 × 32	Bronze
33	4	gear	16/1.25	2767
32	1	guide plate	30 × 240 × 10	Bronze
31	4	threaded core	Ø 8 × 115	E130K
30	4	ejector sleeve	Ø 8 × 68	
29	8	flat head bolt	M 6 × 20	
28	1	cylinder head screw	M 8 × 20	10.9
27	1	cylinder head screw	M 8 × 30	10.9
26	1	set screw	M 10 × 20	
25	6	cylinder head screw	M 5 × 16	10.9
24	8	cylinder head screw	M 6 × 20	10.9
23	1	clamping plate D	196 × 296 × 27	1730
22	2	cylinder head screw	M 12 × 100	10.9
21	4	cylinder head screw	M 12 × 50	10.9
20	4	cylinder head screw	M 12 × 170	10.9
19	2	locating bushing	Ø 30 × 120	
18	3+1	guide bushing	46/22/24	
17	3+1	leader pin	46/22/24 × 65	
16	4	O-ring	36/3	
15	4	O-ring	36/3	
14	4	round core	Ø 48 × 27	2767
13	1	insert D	80 × 16 × 250	2767
12	1	locating ring D	Ø 110 × 13	
11	1	locating ring A	Ø 110 × 8	
10	1	ejector plates	196 × 296	1730
09	1	insulating plate	196 × 296/5/90	
08	1	manifold plate	196 × 296/46	1730
07	1	mold plate D	196 × 296/46	2312
06	1	mold plate A	196 × 296/27	2312
05	1	spacer plate	196 × 296/27	1730
04	2	strip	196 × 296/46	1730
03	1	spacer plate	196 × 296/27	1730
02	1	strip	196 × 296/46	1730
01	1	clamping plate A	196 × 296/27	1730

Fig. 4

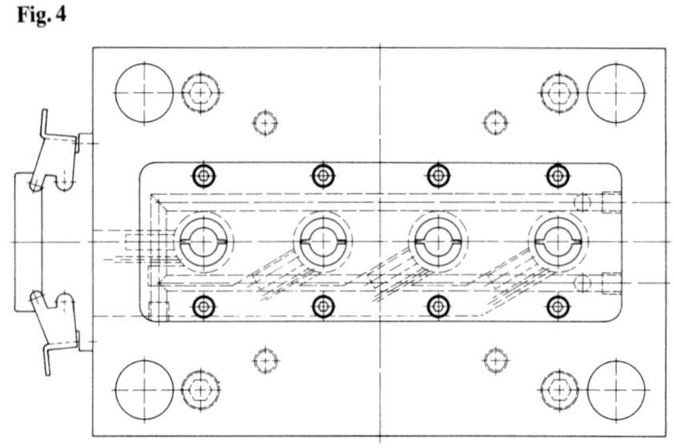

view "X"

Figures 2 to 4 Four-cavity hot-runner unscrewing mold for production of cap nuts
(Courtesy: Heitec)

Example 54, Four-Cavity Hot-Runner Mold with a Special Ejector System for a Retainer Made from Polypropylene

A retainer for insulating material having a total length of 168 mm and a weight of 9.5 g (Fig. 1) is to be produced in polypropylene. In the axial direction

Figure 1 View of injection molded polypropylene retainers for insulating material, complete (top) and in section (bottom)

the projected area of one part amounts to 33.7 cm^2. An injection molding machine of 1300 kN clamping force has been chosen from the available machinery for the intended four-cavity tool. This machine however has insufficient mold mounting height available and possesses too short an ejection stroke to be able to demold the injection molded article by a conventional ejection system. In order to make it possible to produce these parts on this machine despite these shortcomings, the hot-runner mold described below was designed with a special ejection system.

Four-Cavity Hot-Runner Mold

The molded part is injected at its tip via a hot-runner pinpoint gate. As can be seen in Fig. 2 the system employed is that of the "indirectly heated thermally conductive torpedo" [1, 2]. The torpedoes were surface-treated by electroless nickel plating. Heating is by two tubular heating elements of several bends (19) which are embedded in the hot-runner manifold by heat conducting cement. The hot-runner manifold is covered by heat protection plates to reduce heat losses and the contact areas of the nozzle bodies (9), pressure pads (12) and (18) have been kept as small as possible.

The melt compressed in the hot-runner manifold must not be allowed to expand into the cavity at the start of mold opening, as the core (64) would otherwise be damaged on mold closing. To prevent this from happening, the manifold is relieved of pressure immediately after the injection process is finished by withdrawing the machine nozzle from the recessed sprue bushing (14).

Cooling

Figure 3 shows the arrangement of the cooling circuits as well as specific cooling data. Blind holes for housing the thermocouples (34) have been drilled in various places so that the temperatures of the mold plates can be checked during production.

The cores (64) had originally not been equipped with cooling pins (39). It was found during the first molding trials, however, that just the high core temperature alone can prevent a faster production cycle being achieved. New cores were produced with bores to house the cooling pin [3]. A 23% reduction in cycle time was achieved by this action. The heat transfer between cooling pin and temperature control medium would be improved still further by flowing water directly onto the cooling pin. However, this would require extensive reconstruction of the mold.

The frontal area of the core is additionally air-cooled during ejection. The air is supplied via three channels in the center ejector (27) for a duration of 5 s. (Fig. 2, section Y b).

Figure 2 Four-cavity hot-runner mold for insulating material retainers ▶

1: mold clamping plate; 2a, 2b: spacer strip; 3: stationary-side mold plate; 4, 5: stationary-side mold insert; 6: sealing ring; 7: hot-runner manifold; 8: end plug; 9: nozzle body; 10: torpedo; 11: locating insert; 12: pressure pad; 13: sealing ring; 14: recessed sprue bushing; 15: heater band; 16: thermocouple; 17: locating ring; 18: pressure pad; 19: tubular heater; 20: spacer sleeve; 21: plug plate; 22a, 22b: plug-in connection; 23: movable-side mold plate; 24: actuating cam; 25: actuating bar; 26: pneumatic valve; 27: central ejector; 28: stop bushing; 29: core carrier plate; 30a, 30b, 30c: stop strip; 31: pneumatic valve; 32: sealing ring; 33: guide bushing; 34: thermocouple; 35: sealing ring; 36: movable-side mold insert; 37: threaded plug; 38: locating ring; 39: cooling pin; 40: clamping plate; 41: ejector bar; 42, 43: sealing ring; 44: water connection; 45: cover plate; 46: sealing ring; 47: water connection; 48: piston; 49: circlip; 50: air ejector; 51: support bolt; 52: leader pin; 53: guide bushing; 54: gear; 55: cover plate; 56: gear; 57: washer; 58: shaft; 59: bearing plate; 60: draw plate; 61: nut; 62: threaded pin; 63: guide; 64: core; 65: clamping piece; 66, 67: spacer plate; 68: clamping piece; 69: guide; 70: draw plate; 71, 72: gear rack; 73: strips; 74: housing for infrared light barrier; 75: transmitter/receiver

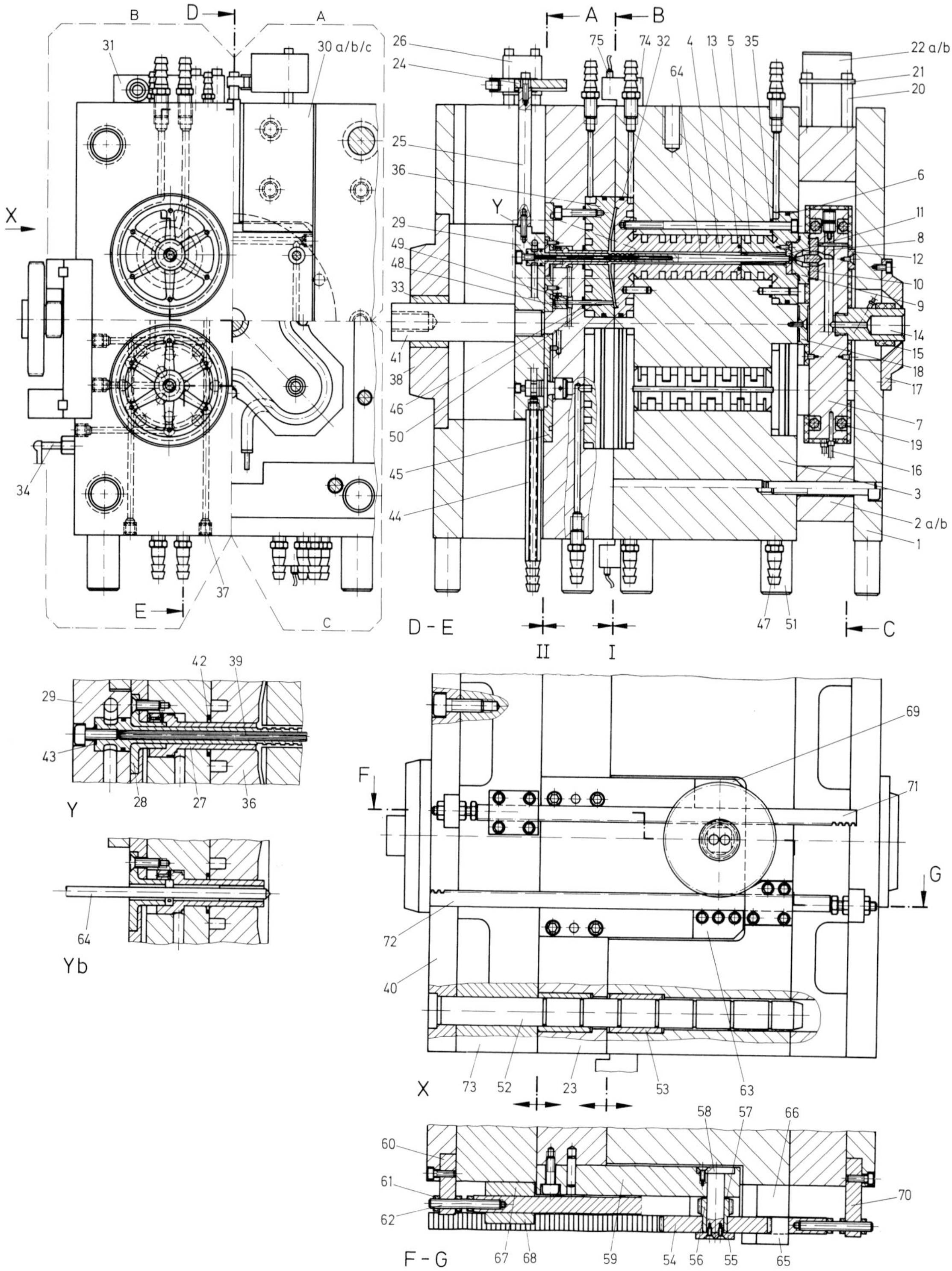
A
B
C
D
E
X
D - E
II
I
Y
Yb
F
G
X
F - G

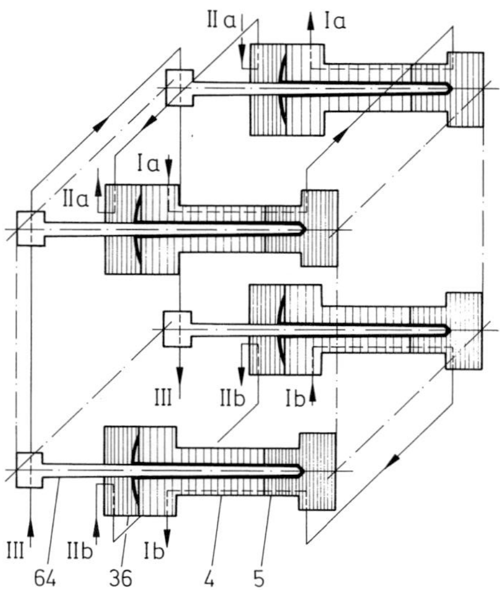

cooling circuit no	cooling medium	mean temperature of the coolant [°C]	volumetric flow [m³/h]	flow rate [m/s]	Reynolds number	notes
Ia & Ib	Water	40	0.5	3.3	35000	heating/cooling unit
IIa & IIb	"	40	0.7	4	50000	" "
III	"	5.5	0.6	9.5	32000	chiller

Figure 3 Cooling diagram
Numbers: Position numbers according to Fig. 2; I, II, III: cooling circuits

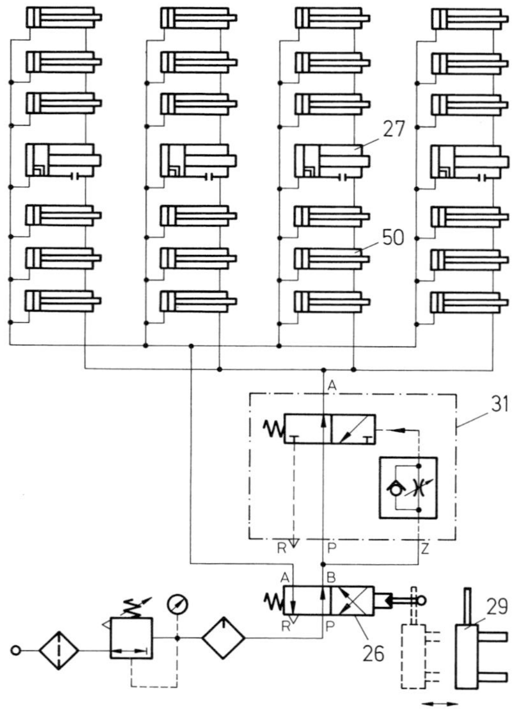

Figure 5 Pneumatic diagram
Numbers: Position numbers according to Fig. 2; A, B, P, R, Z: connecting markings of the pneumatic components

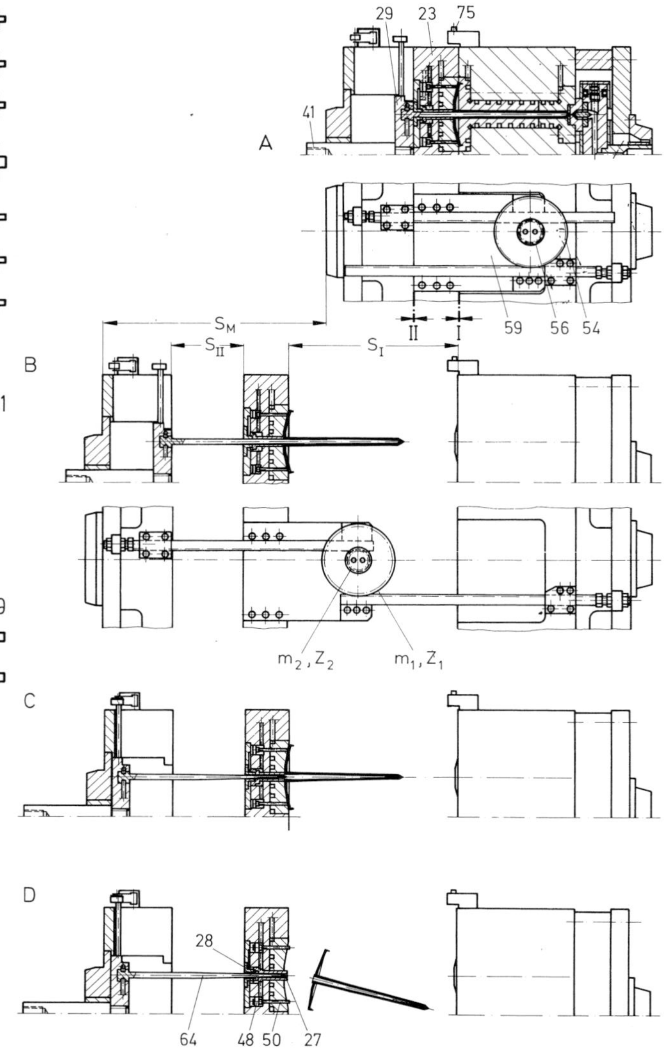

Figure 4 Ejection sequence
Numbers: Position numbers according to Fig. 2; m: gearing modulus; S: distances; Z: number of teeth; I, II: parting lines; A: injecting; B: mold opening; C: retract core; D: ejector

Part Release/Ejection

The core carrier plate (29) is connected to the hydraulic ejector of the injection molding machine via the ejector bar (41). The core carrier plate rests against the mold plate (23) in the injection position, i.e. the hydraulic ejector has moved forward. The force created by the hydraulic ejector suffices to hold the core in position during injection.
One gear drive each is situated in each mold half. The two gears (54) and (56) which cannot be turned against each other are housed in the bearing plate (59) and connected by this to the mold plate (23).
When mold opening starts, the parting lines I and II open simultaneously but at different speeds and for different distances because of the different pitch diameters of the gears.
The molded part is pulled out of the fixed-half cavity by its adhesion to the core, additionally aided and increased by indentations on the latter. Partial stripping from the core takes place simultaneously through the opening of parting line II (Fig. 4B). At the conclusion of the machine opening stroke the hydraulic ejector retracts. A pneumatic control is activated at the end of the stroke so that each part is ejected from the moving mold half by six air ejectors (50) and a central ejector (27) per mold cavity. The three holes in the central ejector leave the zone of the stop bushing (28) during the ejection stroke. This allows air to pass from the pressurized cylinder area through these channels for additional core cooling (64) (Fig. 2 section Y b).
Air cooling of the cores terminates with the return stroke of the central and the air ejectors, activated by pneumatic controls as soon as the core carrier plate (29) moves from its rear position.
The pistons (48) and the central ejector (27) are not sealed against the bores. DIN – fit H_6/g_6 was chosen as the tolerance between bores and pistons. This results in leakages. The supply lines and valves have therefore been dimensioned to sufficiently large nominal sizes to ensure that the pressure in the cylinder interior is high enough despite the losses through leakage. Figure 5 shows the pneumatic controls. Ejection stroke and return stroke are initiated by the core carrier plate (29) via the valve at (26). To keep leakage losses low, the supply line is shut off by valve (31) soon after the return stroke.

Literature

1. Unger, P.: Kunststoffe 70 (1980) p. 730/737
2. Unger, P.; Hörburger, A.: Kunststoffe 71 (1981) p. 855/861
3. Wübken, G.: Kunststoffe 71 (1981) p. 850/854

Example 55, 2 × 16-Cavity Two-Component Injection Mold for Microswitch Covers Made from Polyamide and Thermoplastic Elastomer

The individual cavities in injection molds are placed as close together as possible so as to make optimum use of the given mold surface. If small-area moldings are to be made by sprueless gating with hot runner nozzles, the distances between the cavities are often determined not by the dimensions of the parts, but by the size of the nozzles. Parts which can be side-fed by means of submarine gates can be moved closer together by feeding the melt with hot runner nozzles to individual groups of cavities, each with a small, cold sub-runner. But this solution does not work in the case of molded parts which have to be gated on their surface.

Figure 1 Microswitch covers produced in a 2 × 16-cavity mold

For this reason, attempts are being made to design hot runner nozzles in such a way that their projection onto the mold clamping surface takes up a minimum of space. If needle shutoff mechanisms have to be fitted into the nozzles, the need to take into account the needle to be housed in the nozzle aperture also influences the size of the outside diameter of the nozzle. These nozzles are shown in Fig. 2. They are designed either as a needle valve nozzle (1), with a continuous aperture, or as an open nozzle (2) with a central point extending into the gate. If the inside diameter of the nozzle tube is 4.8 mm, the outside diameter at the nozzle shaft is 15 mm and at the input end 27.5 mm. Distances of 28 mm between the

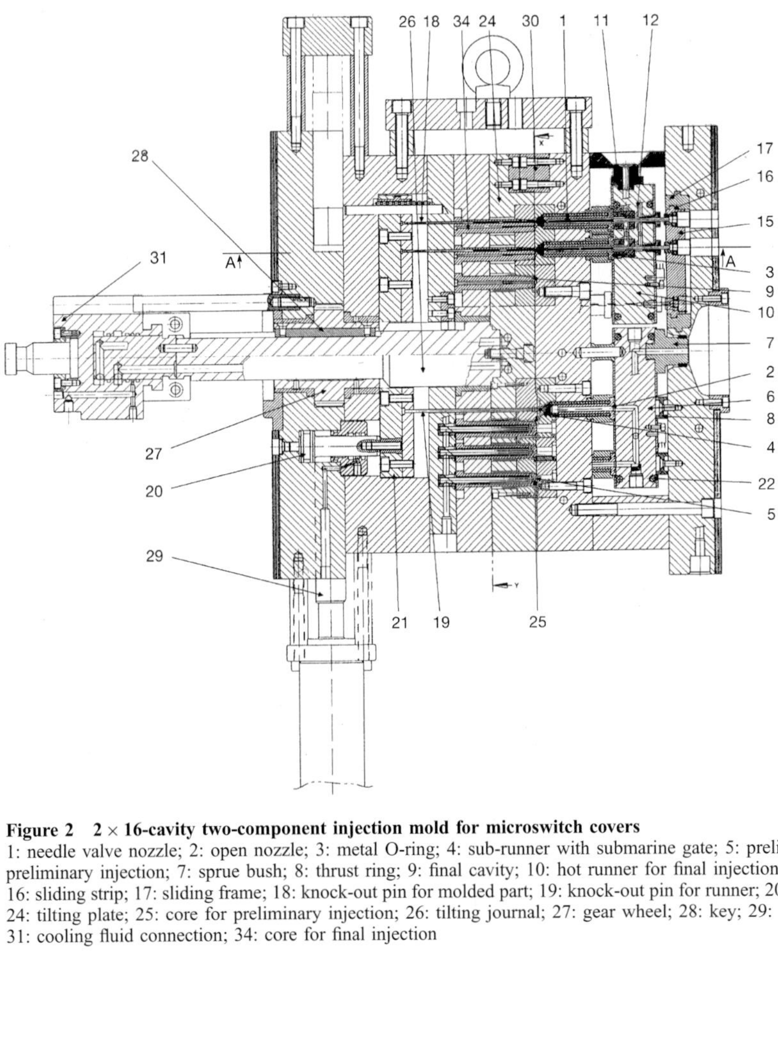

Figure 2 2 × 16-cavity two-component injection mold for microswitch covers
1: needle valve nozzle; 2: open nozzle; 3: metal O-ring; 4: sub-runner with submarine gate; 5: preliminary injection cavity; 6: hot runner for preliminary injection; 7: sprue bush; 8: thrust ring; 9: final cavity; 10: hot runner for final injection; 11: sprue bush; 12: manifold; 15: plate; 16: sliding strip; 17: sliding frame; 18: knock-out pin for molded part; 19: knock-out pin for runner; 20: piston; 21: ejector plate; 22: heating rod; 24: tilting plate; 25: core for preliminary injection; 26: tilting journal; 27: gear wheel; 28: key; 29: hydraulic cylinder; 30: centering element; 31: cooling fluid connection; 34: core for final injection

cavities are, therefore, possible even in a needle shutoff mechanism.
The nozzle heating input in zones with a higher heat requirement near the gate, at the mating point in the cavity and at the center of the shaft (near the cooling channel) should be boosted to the extent necessary to keep the runner at a uniform temperature. The heater voltage is 24 V. The nozzle has only one connecting line; the current is returned via the die. A thermocouple near the gate at the nozzle point permits accurate temperature control. The transition point from the hot runner manifold to the nozzle is sealed with a metal O-ring (3).

Mold

The mold, which is shown in Figs. 2 to 7, is a 16-cavity two-component injection mold with dimensions W × L × H = 269 × 446 × 453 mm. It is used to manufacture small flat covers for microswitch elements (photo). They consist of a square frame made from glassfiber-reinforced polyamide PA 6.6 + 35% GF, W × L = 12 mm, with a hole 8.5 mm in diameter, a wall thickness of 0.8 mm and a weight of 0.1 g. An elastic disc made from TPE (Thermoflex TF 60) with a wall thickness of 0.3 mm and a weight of 0.08 g is injected across this frame.

Feed Side

The four preliminary injection nozzles (2) are open sprue nozzles with a central point. They lead into four small cold runners (4), each weighing 0.45 g, which conduct the melt via submarine gates to four sets of four preliminary injection cavities (5). The preliminary injection nozzles are located alongside a

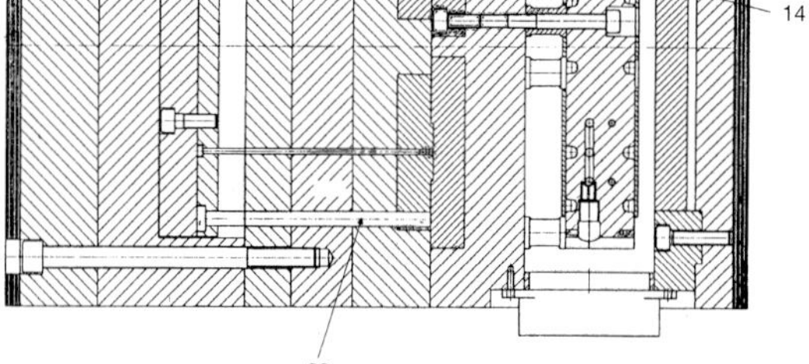

Figure 3 Section A-A (Fig. 2); bolt
13: screw; 14: needle shutoff mechanism; 23: pawl; 32: hydraulic cylinder; 33: ejector plate return pin

Figure 4 View of sliding frame, frame drawn toward the outside (needles open)

hot runner manifold (6) which receives its melt via the sprue bush (7) on the longitudinal mold axis. Thrust rings support this runner and its nozzles against the buoyancy force arising during the injection process. Each of the 16 final cavities (9) is fed via a hot runner nozzle with needle valve (1).
The melt passes to the needle valve nozzles (1) via a hot runner (10) after it has entered the mold via the sprue bush (11) vertical to the mold opening direction. This runner is connected by bolts (Fig. 3, 13) directly to the rear of the cavity plate and braced with the needle valve nozzles, as the needle drive has to be mounted on the rear of the runner. Because of the small distance of 28 mm between the needles, and the large number of needle shutoff mechanisms, it was not practical to actuate each needle individually, for example hydraulically or pneumatically. The needles (Fig. 3, 14) are instead attached to a plate (15). On two sides of this plate are located angled sliding strips (16) which engage in mating grooves of a sliding frame (17). To open and close the needles, this sliding frame is moved back and forth by a hydraulic cylinder. This produces a needle stroke of 5 mm (Fig. 4).
The needle gate leaves a mark 0.8 mm in diameter and 0.2 mm deep on the molding. The melt channels in the two hot runners are naturally balanced, i.e. the distances from the sprue bushes to the respective gates are of equal length and the corresponding channel cross-sections are of the same size. This ensures that the cavities are uniformly filled. Above the melt channels in the needle runner (10) are manifolds (12) in which melt leakage escaping to the rear at the needle shafts is collected and removed.

Figure 5 View of the two hot runners
top: runner for preliminary injection with sprue bush at the mold center; bottom: runner for final injection with lateral sprue bush and 16 guide bushes for the needle shutoff mechanisms (Courtesy: Alwa, Deißlingen, Germany)

Heating

Each nozzle has a heating input of 200 W. The preliminary injection runner (6) is heated with 1370 W and the final injection runner (10) with 1970 W. Heating rods (22) embedded in grooves and cast in brass are used to heat the runners.
The low shot weights of 3.4 g during preliminary injection and 1.5 g during final injection mean that the two melts are subjected to long residence times in the hot runners. The melts withstand this stress only because the respective temperature levels are very uniform and dead spots were avoided in the channels.

Ejector Side

On the ejector side of the mold is also located the tilting device. After each shot, the mold opens at "x" in the main parting line. The secondary parting line at "y" is initially held closed by a pawl (Fig. 3, 23). In the process, the preliminary and final moldings move out of their cavities on the feed side and the cold submarine runners (4) from their feed channels.
The ejector plate (21) now moves to the right with the aid of the pistons (20). The molded parts are ejected by means of the knock-out pins (18) and the cold runners (4) by the pins (19). The preliminary moldings remain in their ejector-side cavities in the tilting plate (24). After the pawl (Fig. 3, 23) is released, the tilting plate (24) comes to a stop and the mold opens at "y".
The opening stroke at "y" is so large that both the cores and the knock-out pins (19) and the ejector plate return pins (Fig. 3, 33) move out of their holes in the plate (24). The plate (24) attached to a tilting journal (26) is tilted through 180° so that the preliminary moldings in it reach the final station. The tilting journal (26) is driven by a gear wheel (27), a gear rack and two hydraulic cylinders. The two ends-of-travel of the rack are signalled by limit

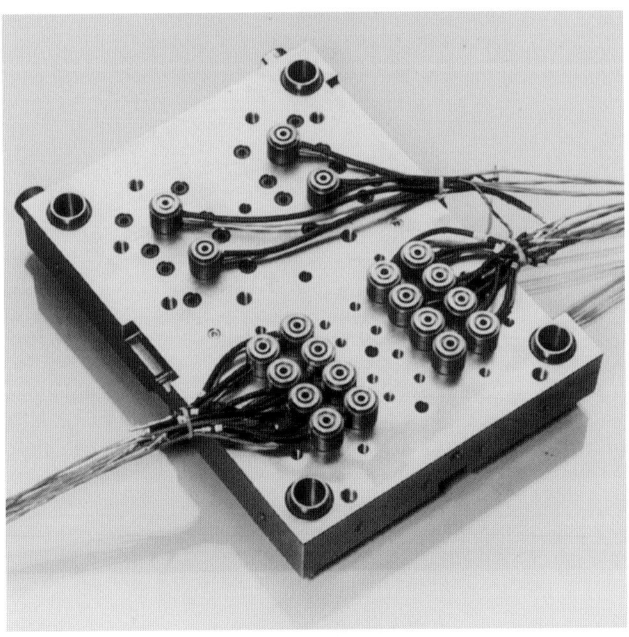

Figure 6 View of the inlet apertures of the hot runner nozzles and the nozzle wiring
top: 4 nozzles for preliminary injection; bottom: 16 nozzles for final injection (Courtesy: Günther Heißkanaltechnik, Frankenberg, Germany)

Figure 7 View of the feed-side parting line
top: preliminary injection cavities with cold runners; bottom: final cavities (Courtesy: Günther Heißkanaltechnik, Frankenberg, Germany)

switches. The longitudinal adjustability of the gear wheel (27) is ensured by a key (28). When the mold closes, the parting line "x" closes first, so that the preliminary moldings are held securely when the cores and the knock-out pins and ejector plate return pins move into the tilting plate and assume their molding position.

Centering

Along with four guide pillars, four wedge-shaped centering elements ensure that the cavities are accurately positioned.

Temperature Control

Holes in the two mold plates and in the preliminary injection cores (25) serve to control the temperature of the mold. The feeding of the cooling medium to the tilting plate and its removal are performed via a rotatable connection. As is shown by this example of a mold, the nozzles introduced here permit the space-saving accomodation even of small cavities on a limited area and reductions in mold volume and weight. Thus it is possible to operate with a small, economical injection molding machine.

Example 56, 32-Cavity Hot-Runner Mold for Production of Packings Made from Polyethylene

A two-plate mold with a conventional runner system (Figs. 1 and 2) was used as the production mold for packings for atomizer pumps (the piston pump principle). In this mold, ejection of the molded parts and runner system took place separately via synchronized ejector mechanisms (12, 13) and (14, 15), the latter being actuated subsequent to extension of ejector mechanisms (12) and (13) which severs the molded parts from the submarine gates. To provide the most economical production possible, the mold was designed with 32 cavities. Each part is molded via a single submarine gate with a diameter of 0.8 mm. After extensive mold trials, PE-LD was selected as the suitable material for the packings (Fig. 5), which had to be produced with high precision. With a total weight of 11.2 g (= 32 × 0.35 g) for the molded parts, the weight of the runner system in this mold design was 10.03 g (Fig. 6). The ratio of part to runner volume was 1.1 : 1.

As part of a campaign to improve production efficiency, the mold was supposed to be redesigned at the lowest possible cost and in the least amount of time to reduce the volume of the runner system and shorten the cycle time, if possible, through use of a suitable hot-runner system. After detailed study of various hot-runner systems, a system utilizing indirectly heated probes (torpedoes) was selected for reasons of

- Simple and problem-free conversion
- Lowest temperature control requirements
- Low space requirement
- Little susceptibility to trouble
- Low maintenance requirements
- Low cost

The design is shown in Figs. 3 and 4. For space reasons, the hot-runner manifold (H-pattern) (21) was designed with eight copper probes (22) (E-Cu F 37; DIN 40 500). The probes were electroless hard-nickel-plated by the Kanisil technique. To ensure adequate integrity of the probe-locating bushing in the limited space available, the probe shank was made as long as possible (18 mm).

With a weight of 12 kg for the hot-runner manifold block, a heating circuit consisting of two tubular heating elements (23) with a total heating capacity of 3.5 kW was provided. This corresponds to a specific heating capacity of about 300 W/kg of hot-runner manifold. To provide good heat transmission the tubular heating elements are embedded in thermally conductive cement. To reduce heat losses due to radiation, the surfaces of the hot-runner manifold are covered with bright-rolled aluminum plates (24). The hot-runner manifold is controlled with the aid of only one temperature controller.

As a result of the conversion of the mold, the weight of the runner system was reduced to 4.15 g (Fig. 7). This corresponds to a volume reduction of 59%. The ratio of part to runner volume changed to 2.7 : 1. Through the reduction of recovery time and shot volume, it was possible to reduce the cycle time by 25%.

Use of the hot-runner system in conjunction with a shortening of the flow paths led to noticeably lower pressure drop along the runner compared with that of the original mold design. This resulted in greater part density and better dimensional accuracy.

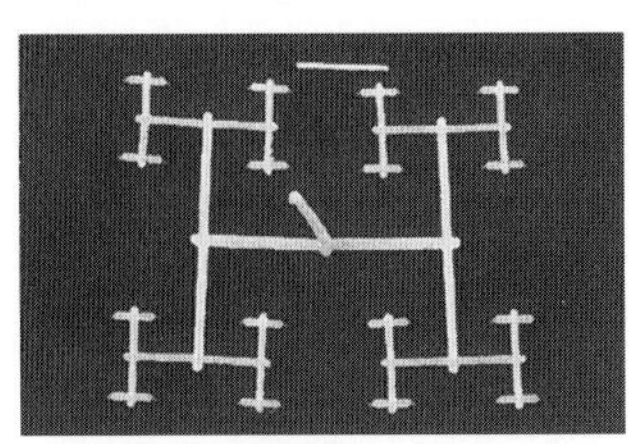

Figure 6 Runner system before mold conversion

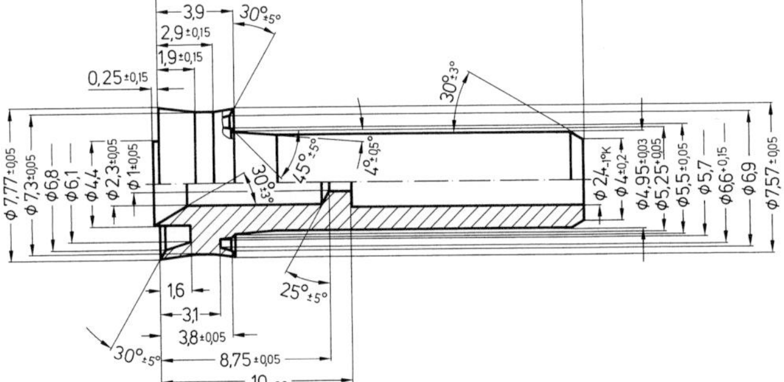

Figure 5 PE-LD plunger packing for atomizer pump

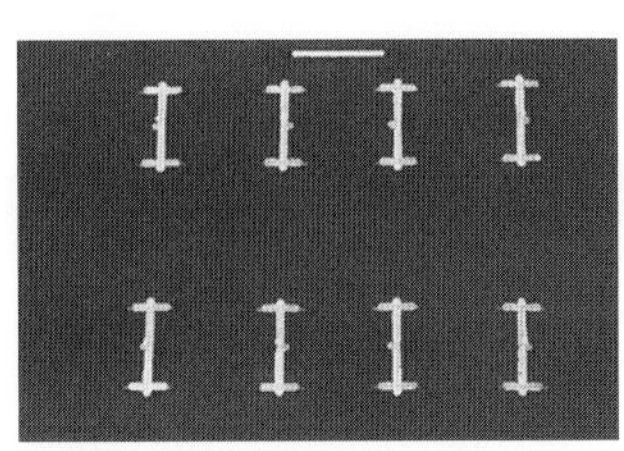

Figure 7 Runners between sprue nozzles and molded parts after mold conversion

Fig. 1

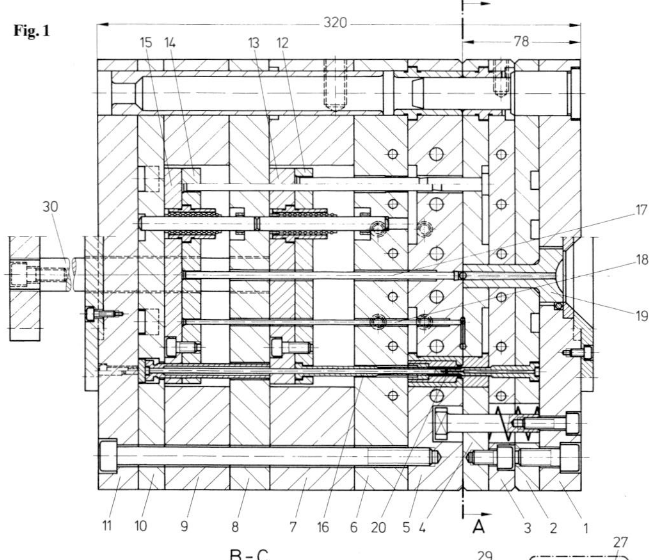

Fig. 2

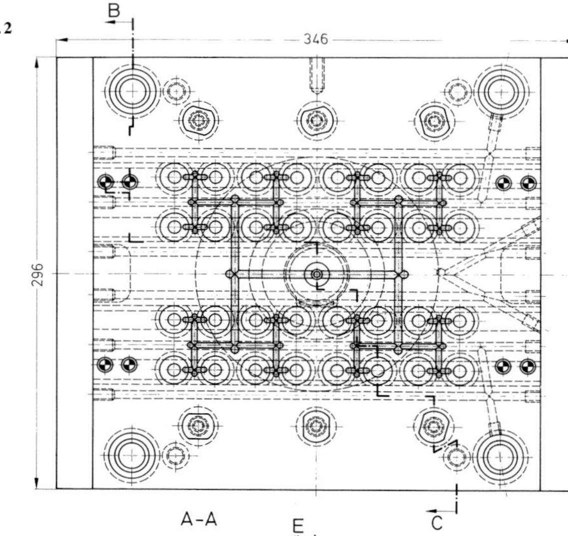

Fig. 3

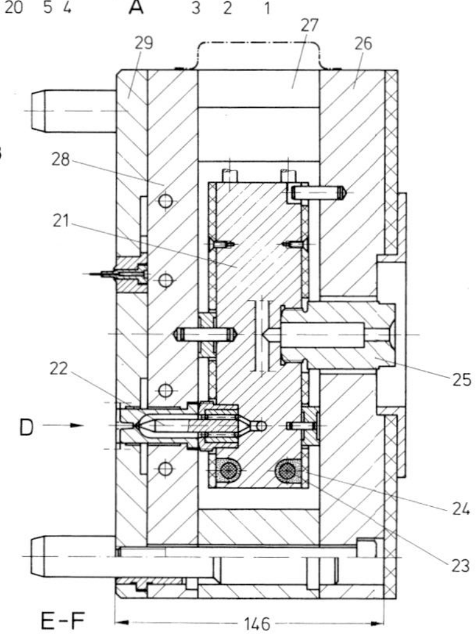

Fig. 4

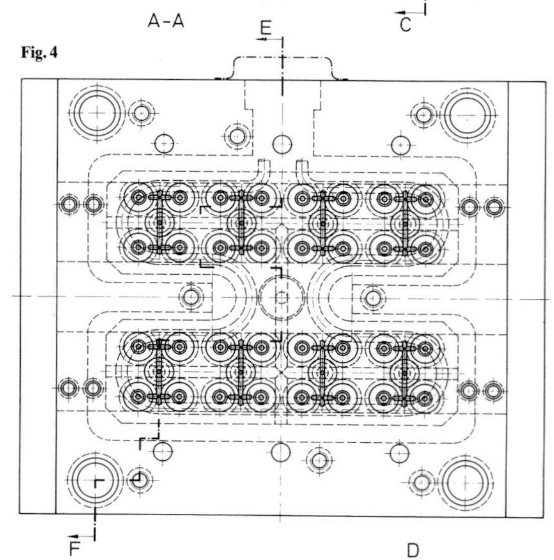

Figures 1 to 4 32-cavity injection mold for atomizer pump plunger packings. Above, mold with conventional runner system before conversion; below, hot-runner mold after conversion (the movable mold half remains unchanged and for this reason is not shown)
1, 2, 3, 4: stationary-side mold plates; 5, 6, 7, 8, 9, 10, 11: movable-side mold plates; 12, 13: ejector retainer plate and ejector plate for the molded parts; 14, 15: ejector retainer plate and ejector plate for the runner system; 16: ejector for the molded part; 17, 18: ejector for the runner system; 19: sprue bushing; 20: cavity; 21: hot-runner manifold; 22: probe (torpedo); 23: tubular heating element; 24: aluminum plate; 25: sprue bushing for hot-runner manifold; 26, 27, 28, 29: stationary-side mold plates; 30: ejector rod
(Courtesy: Calwar Albert)

Example 57, 12-Cavity Hot-Runner Mold with Edge Gates for Bushings Made from Polyacetal Copolymer

This 12-cavity hot-runner mold was fitted with indirectly heated probes (thermally conductive torpedoes) to produce bushings of polyacetal copolymer.

Figure 1 Edge-gated injection molded bushings of polyacetal copolymer

Design Features of the Mold

The hot-runner block (1) is located between the mold clamping plate (2) and cavity plate (4) in the surrounding spacer plate (3) such that there is no heat loss by the so-called "chimney effect". The block is supported by bushings (5) and backup washes (6) which are pressed into the locating holes and ground flat on a surface grinding machine to ensure a uniform reference surface, which is necessary to ensure a leak-free system. The backup washes are overdimensioned by 0.03 + 0.01 mm. The end plug (8) is longer than in prior versions. This excludes the possibility that the mold maker, by chamfering the face, could unknowingly introduce an undercut into which melt would penetrate and, after a sufficiently long residence time, degrade and lead to rejects. The hot-runner manifold (1) is connected to the mold clamping plate (2) by dowel pins (13) and bolts (14), and thus centered and secured against falling out when the mold is disassembled. To reduce radiation losses, the hot-runner manifold (1) is clad with bright aluminum plate (9). The block is heated by three cartridge heaters (10) (diameter 12.5 mm, length 160 mm, 1000 W each). At 3000 W, this gives a specific heating capacity of 250 W/kg of hot-runner block. Good heat transmission from the cartridge heaters (10) to the hot-runner manifold (1) is ensured if the cartridge heaters are inserted into polished wells with heat-conducting paste (tolerance H 7). The thermocouple (11) is located above the central cartridge heater and seated so deeply in the block that the measuring position lies between two gates. This guarantees that the requirements for temperature measurement and control will be met.

The probes (12) are generally of copper (E-Cu 2.0060). It is recommended that probes of smaller diameter be made of CuCrZr 2.1293 because of its greater hardness, particularly at higher processing temperatures.

Compared to standard probes, these probes have a flattened tip with two or more conical protrusions at the side to conduct heat directly to the gates. The rear end of the probe is inserted into the support bushing (5) with the tolerance H 7/m 6. This guarantees uniformly effective heat transfer from the hot-runner manifold (1) through the support bushing to the probe (12). The normal dimensioning rules are applicable for calculating the main probe dimensions in advance.

Upon mold opening, the gates shear off the molded parts. The thin gate residue remaining in the gate orifice is typically not melted during the following shot. The molded parts are ejected by means of the ejector plates (15) and the ejector sleeves (16). To protect the probes when assembling the mold, the leader pins are located on the injection side and dimensioned such that the injection-side cavity plate slides on the leader pins before reaching the tips of the probes.

Usually, the mold clamping plate (2), the spacer plate (3) and the cavity plate (4) are bolted together. To permit rapid separation of the cavity plate and spacer plate, they are held together by two clamps (18) in the present design. This is of advantage for quick color changes or when processing regrind. If a contaminant plugs a gate or the color is to be changed, the clamps are released from the mold cavity plate (4) and the spacer plate (3) and used to hold mold cavity plate (4) and mold plate (19) together. On starting up the injection molding machine, the gate side of the mold opens between plates (3 and 4). The insulating material is stripped from the exposed hot probes (12), the mold closed and the clamps returned to their original position. After a few minutes, the mold is ready again for use.

Fig. 2

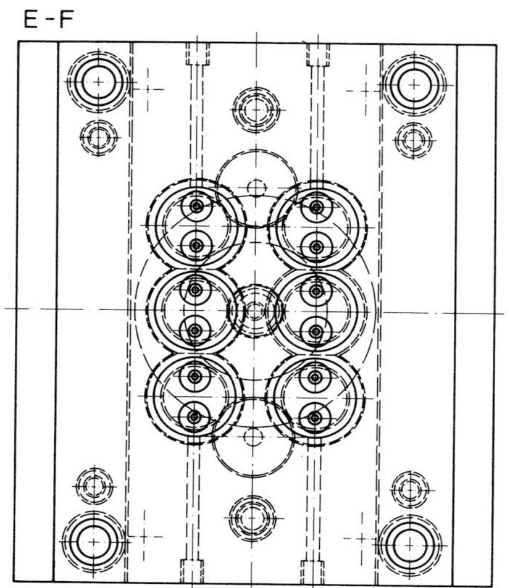

Fig. 3

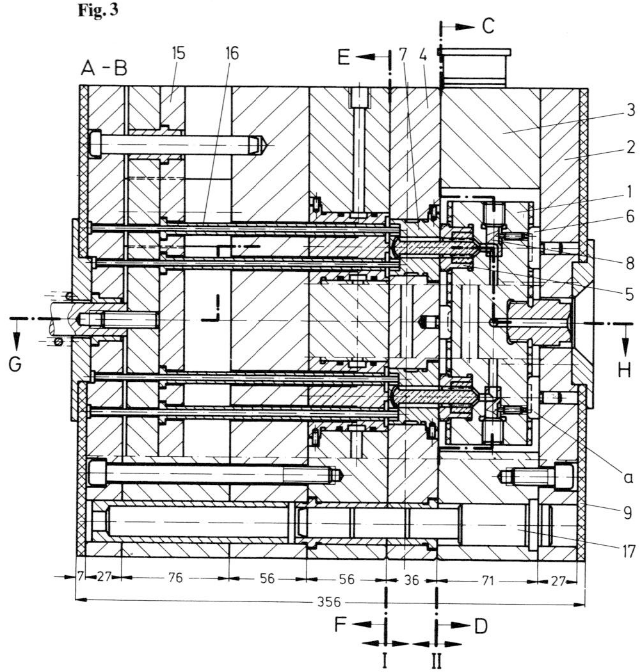

Fig. 4

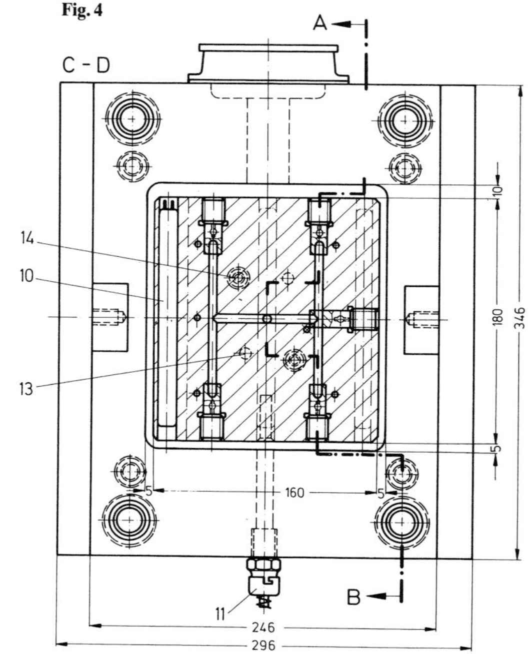

Fig. 5

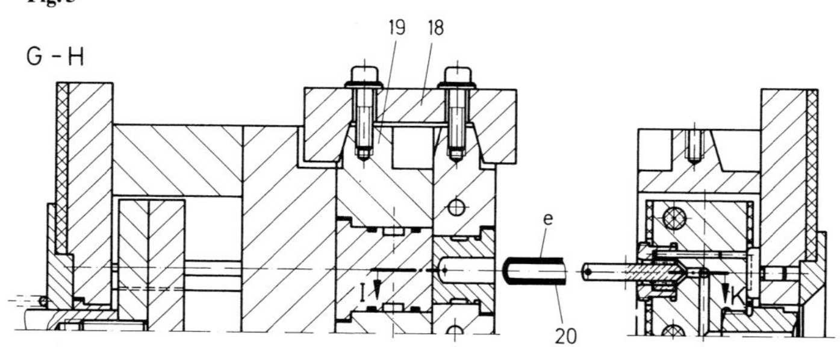

Fig. 6

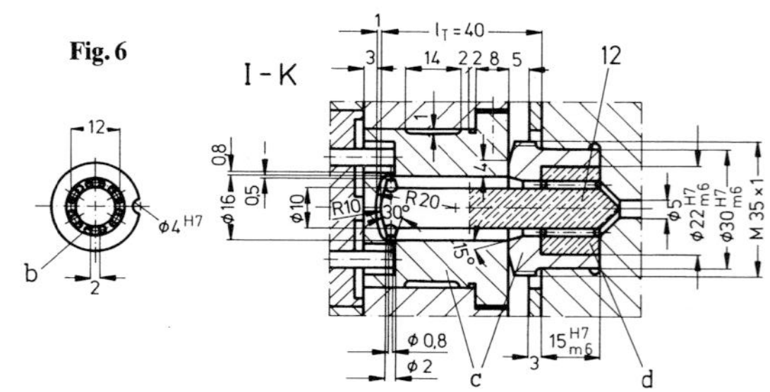

Figures 2 to 6 12-cavity hot-runner mold for the production of bushings
1: hot-runner manifold; 2: mold clamping plate; 3: spacer plate; 4: cavity plate; 5: support bushing; 6: support pad; 7: gate insert; 8: end plug; 9: aluminum plate; 10: cartridge heater; 11: thermocouple; 12: probe; 13: dowel pin; 14: socket-head cap screw; 15: ejector plate; 16: ejector sleeve; 17: leader pin; 18: clamp; 19: cavity plate; 20: insulating material
a: interference for pressure pad 0.03 + 0.01 mm, 1.2767 hardened; *b*: transitions rounded; *c*: 1.2767 hardened; *d*: E-Cu, 2.0060 or CuCrZr, 2.1293; *e*: insulating material stripped, pre-hardened tool steel 1.2311, 1.2312 was used for plates and hot-runner manifold
I: operating parting line; II: parting line for cleaning; section G-H/
parting line II opened
(Courtesy: Hoechst AG)

Example 58, Single Injection Mold for Sleeves Made from Glass-Fiber and Talcum Reinforced PA 66

The frame-shaped molded part, dimensions 20 mm × 107 mm × 5.3 mm, has openings with varying diameters (Fig. 1). For its functionality as a technical component, especially to meet the quality specification of zero warping, dimensional stability and minimal molded part tolerance have to be fulfilled. Due to the, in part, considerable differences in wall thickness, these specifications can only be achieved by intensive mold cooling with separate cooling circuits. The molded part weighs 5 g.

Mold

This is essentially a two-plate mold with inserts (1) and (2) made from annealed 1.2343 hot-work steel on nozzle and ejector side. The mold inserts are fitted into pockets and screw-fastened (Fig. 2). Processing shrinkage was calculated at 0.4%. To create the lateral contours of the molded part which cannot be demolded in the direction of mold opening, contoured mold splits (3 and 4) are located on both sides of the part. The choice of the gating position is the reason for the off-center cavity position. The mold dimensions are 246 mm × 296 mm. It is built up mainly from standardized parts like plates, splits guides, etc.

The size of splits required is based on the article contour surfaces (undercuts, recesses, lateral embossments), the sealing surfaces on the core inserts, angular pin placement and space required for splits cooling. Splits size and opening path determine the space requirement and the mounting room available on the working surface of the cavity plates. Both cavity plates were machined for the splits function.

The cavity plate (5) manufactured from 1.2764 case-hardened steel holds the non-positively and form-fit mounted locking heel (7) and angular pillars (8). To adjust the sleeve system at the sealing surface, hardened pressure plates (9) are fastened to the slanted surfaces of the locking device. The cavity plate on the ejector side (6) has had slide paths machined out of it to accommodate splits (3) and (4), splits slide plate (10) and guide bar (11). To secure the splits in open position, spring-loaded spring plungers (12) have been incorporated. In addition, ejector pins (13) are used for retaining them. The projecting clamping plates (16) and (17) are equipped with threads and drill holes on their front ends for transport and clamping; thermal insulation sheets (14) and (15) are screwed on to them. Four support pillars (18) are mounted in the ejector box to stabilize the mold. The ejector

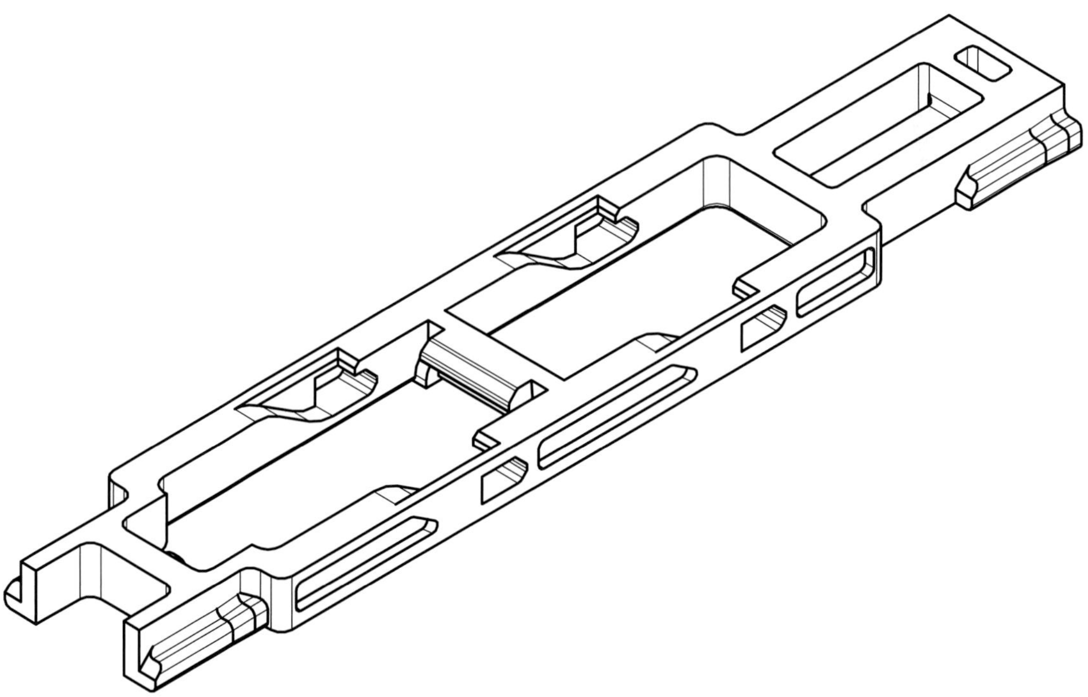

Figure 1 Sleeve made from polyamide, diagram

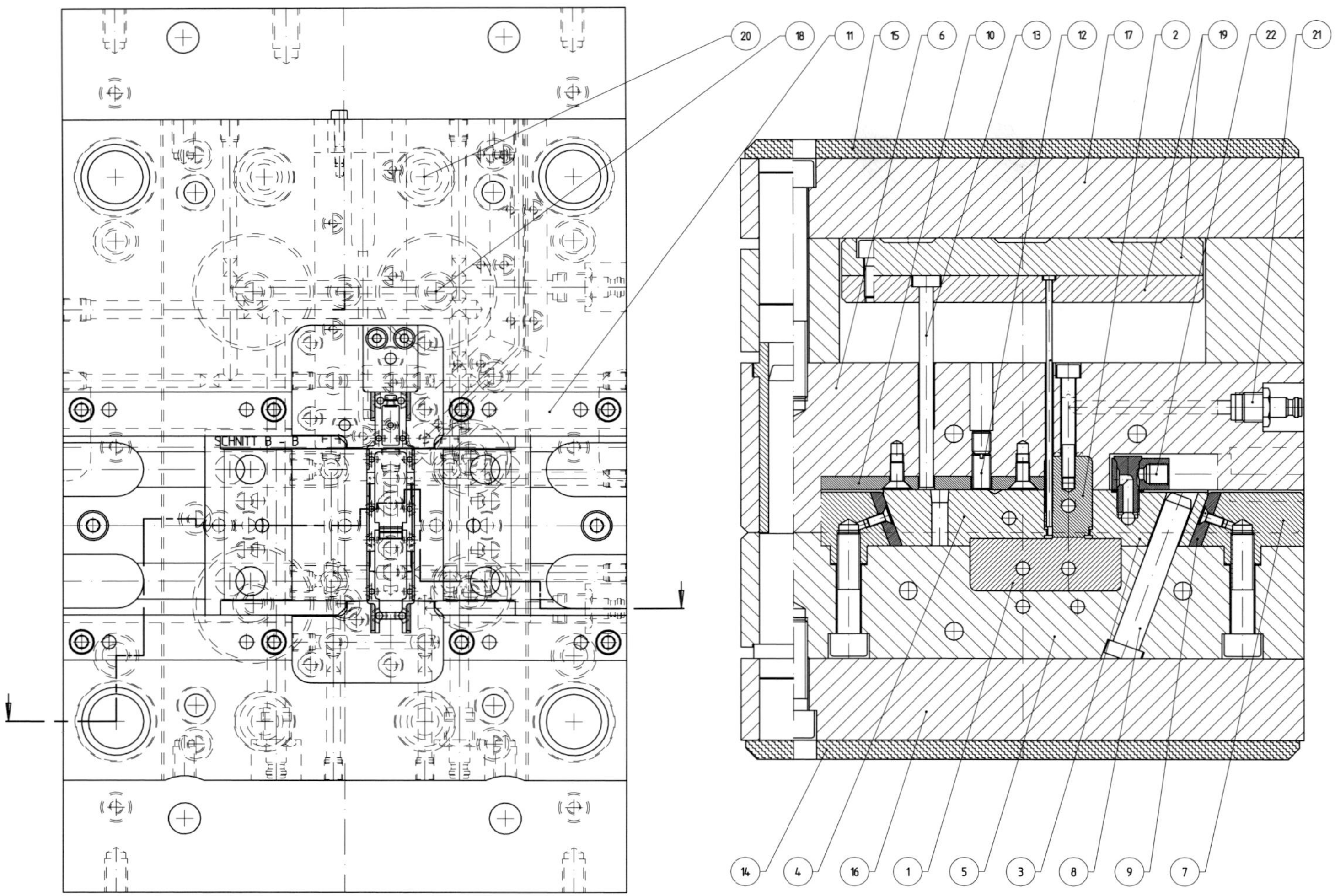

Figure 2 Single injection mold for sleeves made from polyamide
1: mold insert FS, 2: mold insert FS, 3: mold split large, 4: mold split, small, 5: cavity plate FS, 6: cavity plate BS, 7: splits locking heel, 8: angular pillars, 9: pressure plate, 10: slide plate, 11: guide bar, 12: spring-loaded thrust element, 13: ejector pin, 14, 15: thermal insulation sheet, 16, 17: clamping plate, 18: support roller, 19: ejector assembly, 21: nipple, 22: connector
(Courtesy: Hasco, Lüdenscheid; Möller, Bad Ems)

assembly (19) guided with four ball-bearing guides (20).

Cooling

The very effectively lay out of the cooling system of the mold uses parallel and serial cooling circuits in the cavity plates, mold inserts and splits. Coolant flows into and returns from the cavity plates and mold inserts via nipples (21) on the exterior of the mold. To supply the splits, standardized connectors (22) are used. They are mounted underneath the splits and fastened in a position pointing outward. The cavity plate FS (6) has cut outs out in this area. By using nipple extensions, the connections can be taken outward and are thus more accessible.
Mold temperature is measured and controlled near the cavity; internal mold pressure is monitored by a pressure transducer.

Example 59, Two-Component Injection Mold for Drink Can Holders Made from Polypropylene and Ethylene-Propylene Terpolymer

The 10 g can holder (Fig. 1) is part of a special automobile interior fitting designed to hold drinks cans securely. The holder itself consists of polypropylene, while the insert-molded inner ring is made of EPDM elastomer (ethylene-propylene terpolymer). The dimensions of the mold are 295 mm × 460 mm × 415 mm (length without rotary shaft and coupling). Figure 2 shows a simplified longitudinal section through the 1 + 1-cavity mold. To produce the two-component holder, the preinjection molded part made from the first component is placed in the second station (final injection molding station) where the second component is injected into a space formed between the mold cavity and parison. Placement of the preinjection molding is carried out by an integrated rotary device.
In the open mold the entire mold plate (1) with the mold inserts (3, 4) and the runners for the first station (5) are retracted by a rotary shaft (2). The rotary shaft is moved out by the hydraulic ejector system of the injection molding machine. The mold cores of the first and second station (6, 7) therefore remain unchanged in their position. The retracting mold plate (1) and the inserts (3, 4) assume the additional function of a stripper plate. The difference in thermal expansion resulting from the different injection temperatures of the plastics is taken into account in the design. A common cooling system for the inserts of the two stations (fixed and moving

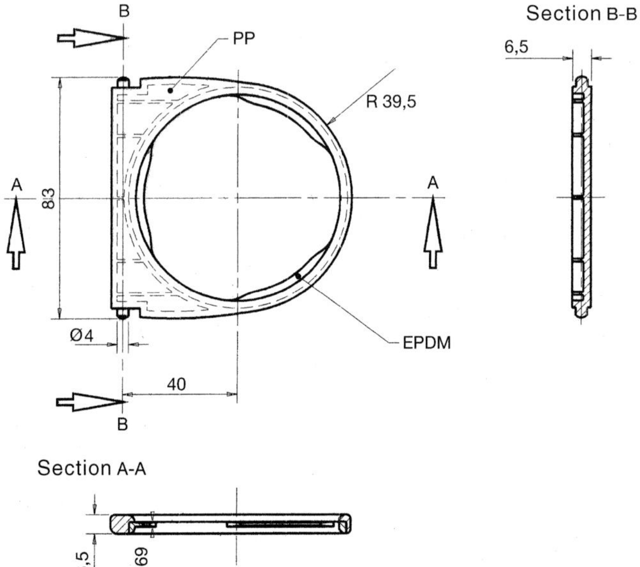

Figure 1 Swivel-mounted can holder made from PP with retaining lip made from EPDM

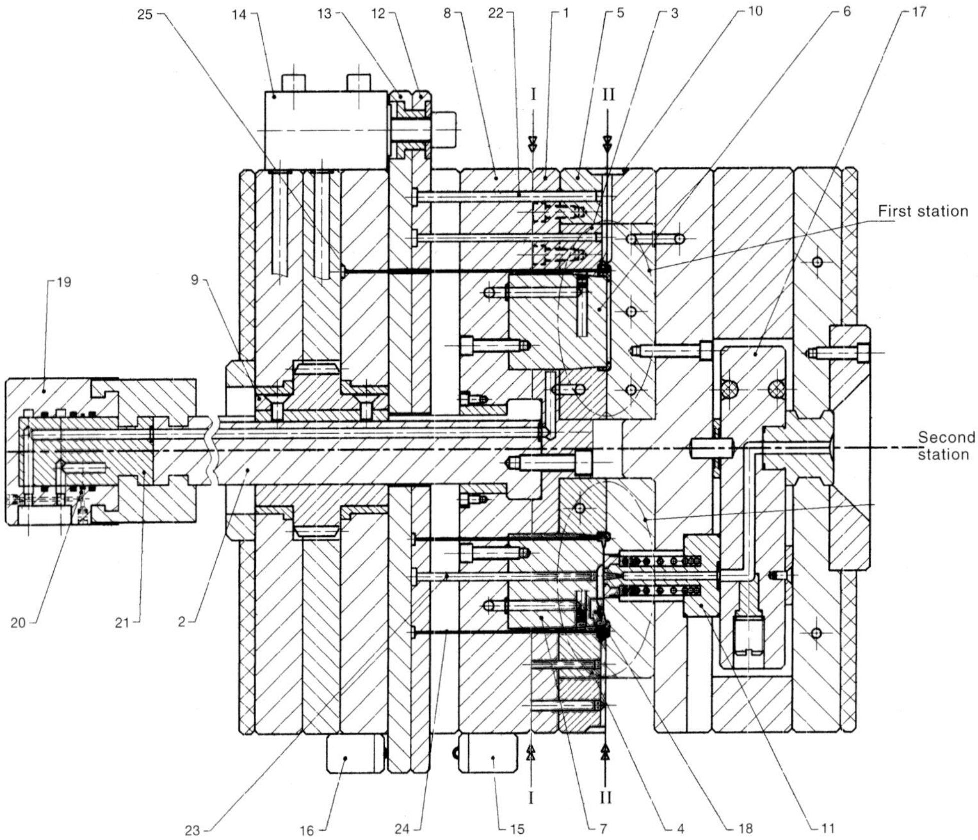

Figure 2 Simplified section through the closed mold
1: mold plate; 2: rotary shaft; 3, 4: mold insert; 5: split sprue bushing; 6, 7: core; 8: mold plate; 9: rack and pinion; 10: nozzle system; 11: heated sprue nozzle; 12, 13: ejector plate; 14: block cylinder; 15, 16: limit switch; 17: hot runner; 18: tunnel gate; 19: rotary coupling; 20: leakage line; 21: separable shaft coupling; 22 to 24: ejector pin; 25: pin

mold halves in each case) compensates for the expansion differences accordingly.

Cycle Sequence

- Injection of PP in the first station via the parting line (II), and injection of EPDM in the second station via a hot runner nozzle (11).
- Mold opening 1: two rubber springs built into the parting line I (Fig. 3) hold the mold plate (1) with the mold inserts (3, 4) on the moving mold half closed for a distance of 4 mm. This ensures that the mold cores (6, 7) release the injection molding on the inner ring; this prevents disortion during retraction and ejection.
- Mold opening 2: the mold halves travel about 95 mm apart, the ejector plates (12, 12) are moved forward and backward by two block cylinders (14) and eject the finished molding from the second station. Limit switches (15, 16) safeguard the operation.
- When the limit switch (16) is actuated again, the hydraulic ejector mechanism of the injection molding machine moves out the entire rotary device (rotary shaft (2), mold plate on the moving mold half (1), runner on the moving mold half (5) and mold inserts (3, 4)).
- A hydraulic cylinder on the mold (25) turns the rotary device through 180° via the rack and pinion (9). End stops limit the travel path of the rack and limit switches protect the area of the racks.
- The hydraulic ejector mechanism of the machine retracts the rotary device. The injection molding from the first station is now in the second final injection molding station.
- The mold closes and the cycle starts again.

Gate

The first station is gated via the parting line II with a tunnel gate. The sprue breaks off when the molding is moved. The sprue is held by an undercut in the ejector bore. The second station is gated via a hot runner (17), a hot runner nozzle (11) and an elbow-shaped tunnel gate (18). Here again, the sprue breaks off when the injection molding is ejected. The sprue remains suspended on the moving mold half by the elbow tunnel.

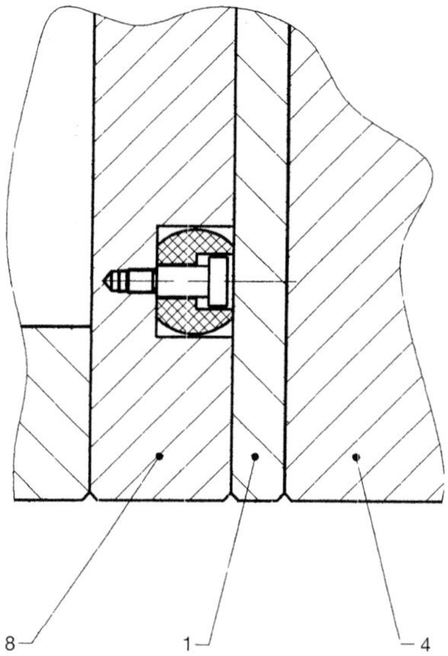

Figure 3 Simplified section through a rubber spring

Mold Temperature Control

Four cooling-water circuits control the temperature of different mold areas (Fig. 4, right):

- the mold plate on the moving mold half (1) and inserts (3, 4) via the rotary coupling and rotary shaft, the mold plate (8) and mold cores (6, 7),
- the mold plate and mold inserts on the fixed mold half,
- the fixed platen.

The mold plate on the moving mold half with the mold inserts is cooled by two longitudinal bores in the rotary shaft (Fig. 2). The cooling water connections are in the rotary coupling (19), which at the same time provides the connection to the ejector hydraulic mechanism of the injection molding machine. The coupling is sealed on the shaft by two O-rings. In addition, it is provided with a leakage line (29) which is intended to carry away any leaking water before it escapes from the coupling.

To counter premature wear in the corrosion-prone contact zone between the shaft and coupling, this area of the shaft is protected by a special surface treatment. The actual rotary shaft (2) is flange-connected to an additional shaft end (21) with a separable coupling for easier mold installation. The rotary coupling with the shaft end is first mounted on the injection molding machine and connected to the ejector hydraulic system.

Temperature control of the mold cores on the moving mold half (6, 7) is effected uniformly around the periphery to ensure that the molding materials cools at a constant rate and to prevent warpage of the molding.

The cooling curve for the mold inserts on the fixed mold halt and fixed platen is shown in Fig. 4 (left). An FeCuNi thermocouple measures the temperature of the mold insert on the fixed mold half in the first station.

Demolding

The sprues of the first and second stations are ejected by two ejector pins (22 and 23). To prevent the injection molding being distorted as it is moved or ejected, two rubber springs push the two mold cores (6, 7) backward out of the way before the mold halves are opened. The finished injection molding is ejected in the second station by eleven pins (24) on the periphery. The mold inserts (3, 4) act as the shaping elements for the first and second station, alternating after each cycle. The ejector bores must therefore be closed off for the preinjection molding by ejector pins (25), which are retracted when the mold opens:

- the mold opens, the pins (25) for the first station are retracted,
- the ejector plates (12, 13) are moved out by two block cylinders (14) and the finished part and sprues of the first and second station are ejected,
- when the limit switch (15) is actuated the block cylinders move in again, and
- limit switch (16) gives the signal to move out the rotary unit.

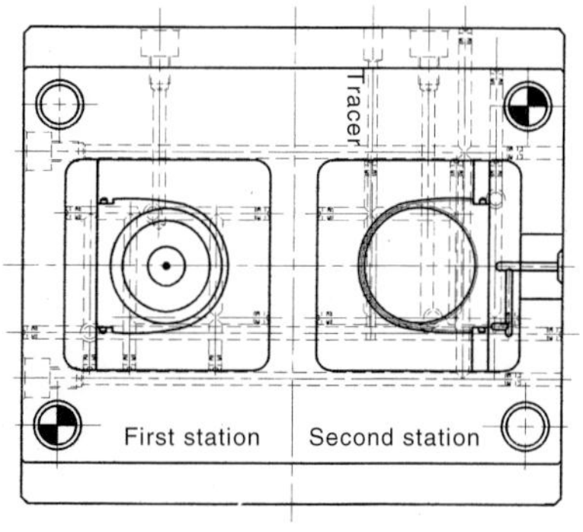

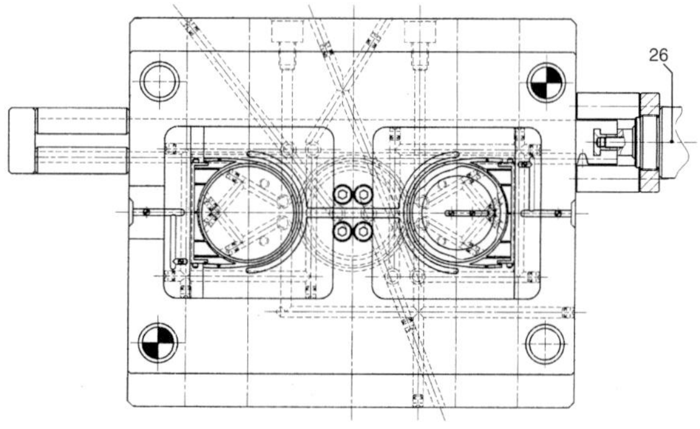

Figure 4 Simplified view of the fixed (left) and moving (right) mold halves
26: hydraulic cylinder
(Courtesy: Braun Formenbau GmbH, Bahlingen, Germany)

Example 60, Hot-Runner Mold for Polypropylene Clamping Ring with Internal Undercut around the Circumference

Profile clamping rings are used to join two components, such as a container and lid, with an interlocking connection that can be detached as often as desired. The ring is placed around the surrounding rim at the container and lid and tightened by a tangential screw. To receive the clamping screw, the ring has two external bored lugs (Fig. 1).

Mold, Sprue and Temperature Control

A single-cavity mold is used (Fig. 2), which is made mainly of standardized components (mold unit, slide bar actuator, ejector system, sprue system). The ring is gated at the two lugs by means of a hot-runner block (HK in Fig. 3) and a heated nozzle. Holes for the coolant are located in the two mold inserts and in the platens below them. The slides consist of a conductive metal, which rapidly transfer the heat to the adjacent, directly cooled mold area.

Demolding

The holes for the clamping screws are demolded by two slides (Fig. 2). The internal undercut is demolded by means of 2 × 4 slide elements S, AS (Fig. 2). Ultimate ejection is accomplished by ejector pins AW (Fig. 5).

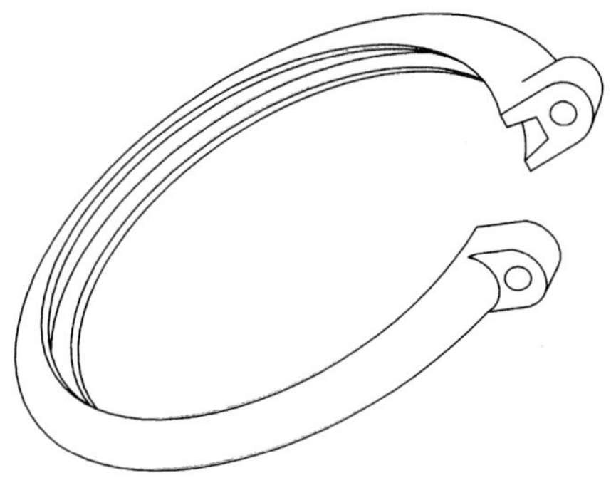

Figure 1 Clamping ring of polypropylene

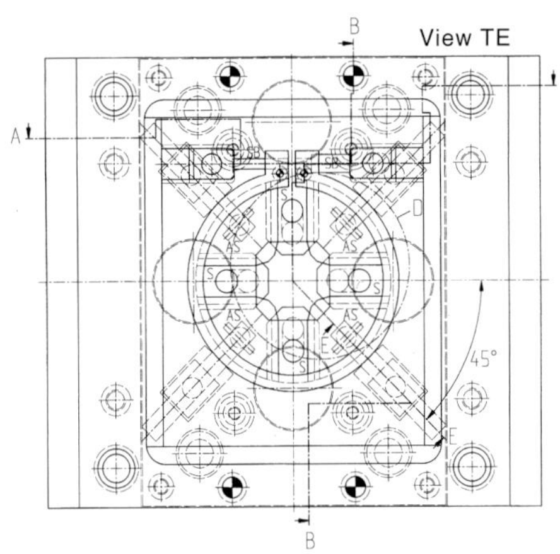

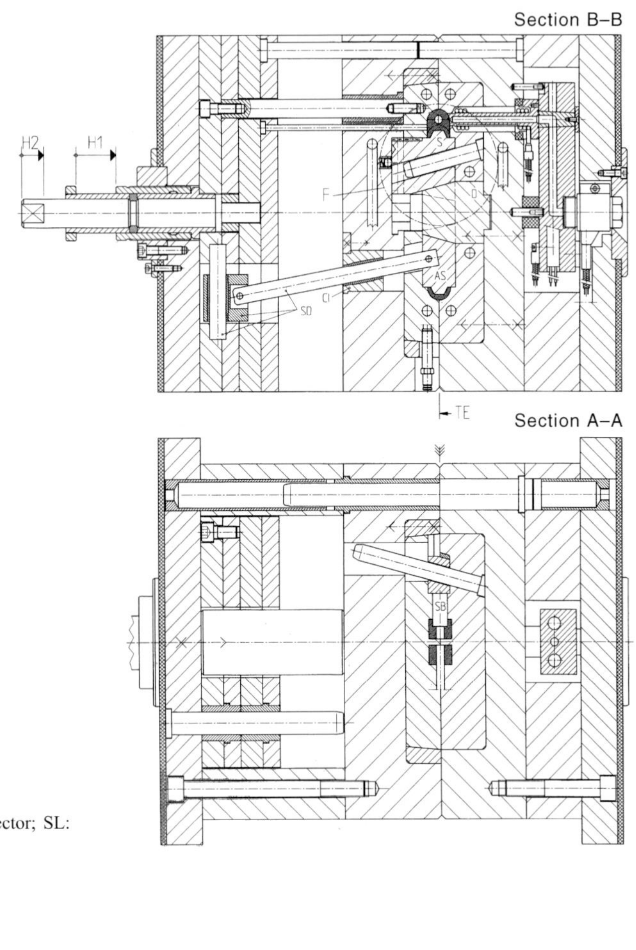

Figure 2 Injection mold for clamping ring
D: pad; S, AS: slides for internal undercut; SD: angled ejector; SL: ejector guide; TE: mold parting line

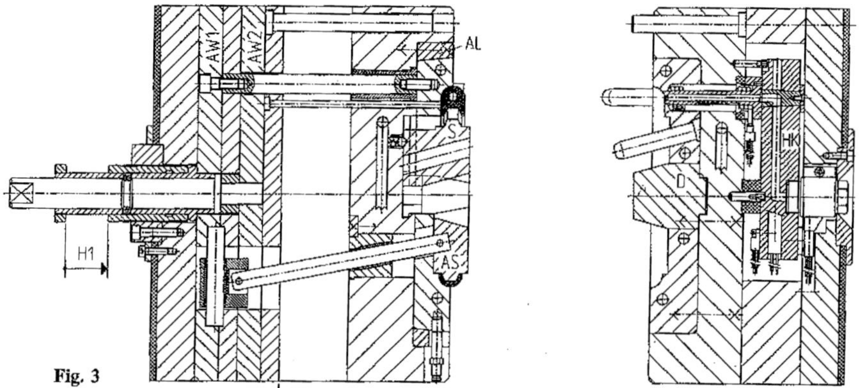

Fig. 3

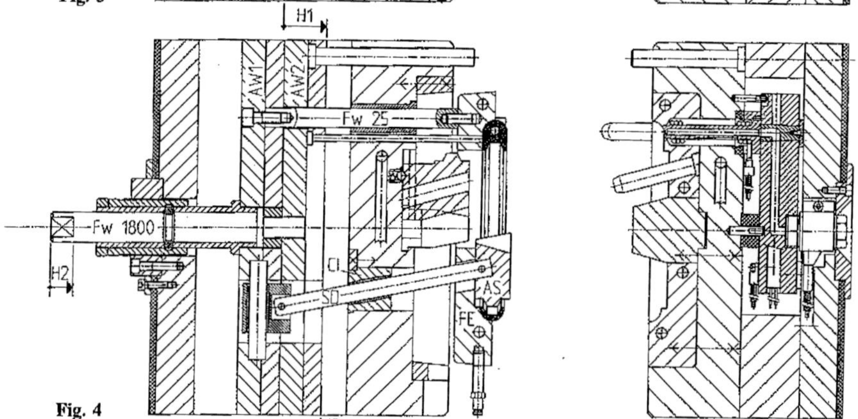

Fig. 4

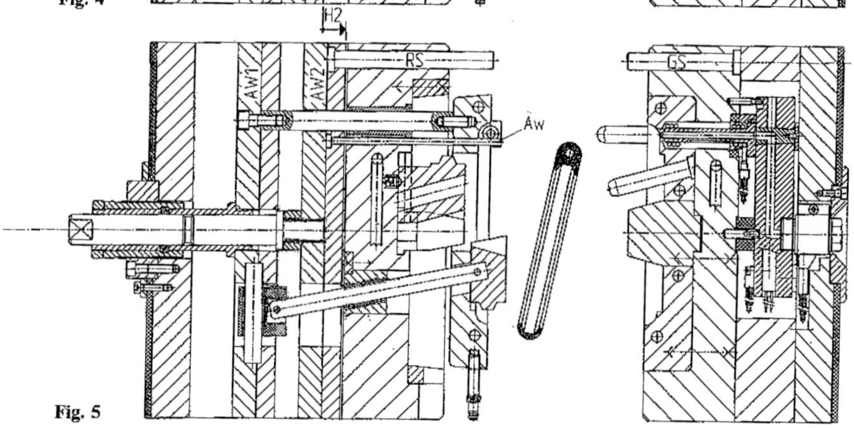

Fig. 5

Figure 3 Demolding the inner contour of the clamping ring, 1st stage
AW1, AW2: ejector plates; AL: centering strip for mold insert; HK: hot-runner system with heated nozzle

Figure 4 Demolding the inner contour, 2nd stage
Fw 25: ejector bolt; Fw 1800: two-stage ejector; FE: mold insert

Figure 5 Ejecting the clamping ring from the mold insert
AW: ejector pin; RS: back-pressure pin; GS: counter-pressure pin

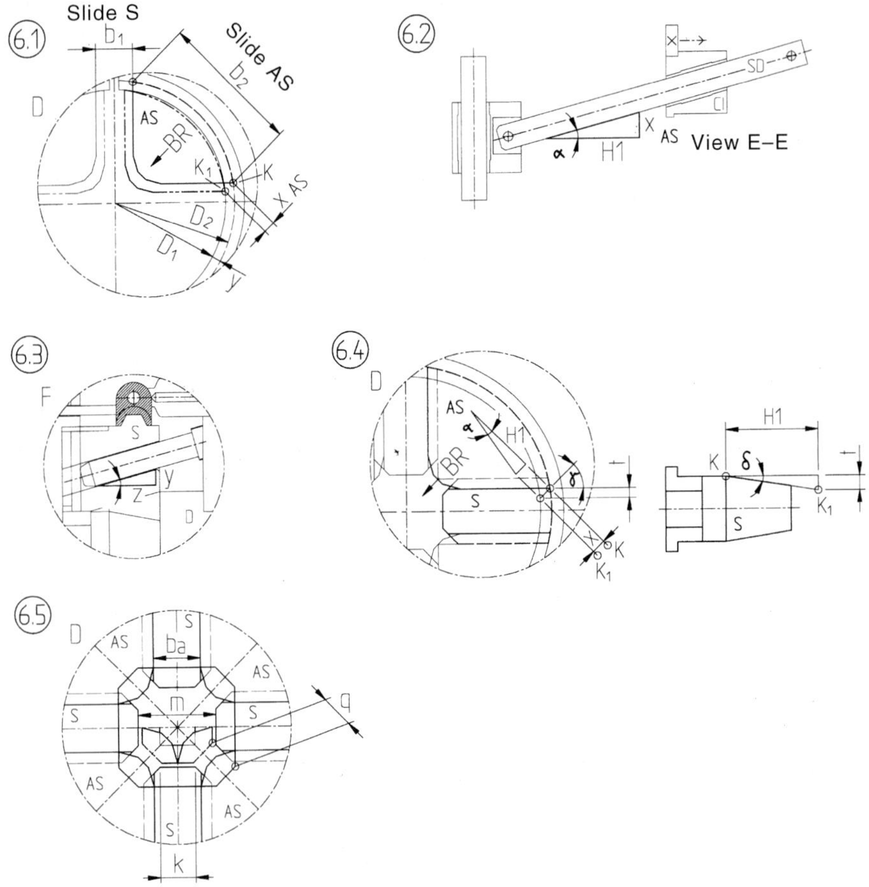

Figure 6 Angles and paths of the slide elements
(Courtesy: EOC, now DME)

The Demolding Sequence

As soon as the mold opens (Fig. 3), the two slides SB (Fig. 2) for demolding the tensioning-screw holes are drawn apart by means of angled columns. This process releases inner slides S and AS, which are locked by the angled flanks of the pad D when the mold is closed. The four slides S are moved inward by the angled columns. The slides SB and S are guided in T-grooves on the moveable mold half. Their position when the mold is open is secured by ball detents. During these demolding operations, the molding remains held in its original position on the moveable mold half by the slides Ap. The subsequent demolding sequence is carried out by means of the ejector device, acting via the two ejector plate assemblies AW1 and AW2 (Fig. 4), which are connected to the machine ejector system via the two-stage ejector Fw 1800. First the two ejector plates advance together through the stroke H1. The mold insert FE is attached to AW1 via the ejector pin Fw 25. Also in AW1 are the four angled ejectors SD, which in conjunction with the ejector slides AS travel out to free up the remaining areas of the internal undercut. After the stroke H1, the plate assembly AW1 with the mold insert FE and the slides AS remain stationary. The plate assembly AW2 advances further by the stroke H2 (Fig. 5), and the ejectors AW push the molding out of the mold insert FE. During the stroke H2, the assembly AW1 is locked in position via the two-stage ejector Fw 1800.

Clamping Operation

Before the mold is closed, the two-stage ejector is drawn back by the machine ejector. The ejector assemblies AW1 and AW2 with the ejectors AW, slide elements AS/SD and the mold insert FE return to their injection position. The moveable and shaping components of the mold are protected against damage in the event of accidental closing of the mold halves (during mold mounting or removal, storage and maintenance) by backpressure pins RS and counterpressure pins GS, which forcibly reset the ejector assemblies AW1 and AW2.

Angles and Paths of the Slide Elements

To demold the annular undercut of depth y, the slide AD has to move in the direction BR, covering a path of x_{AS} (Fig. 6.1). Its corner K, originally at diameter D_2 must pass through the circle formed by D_1.

$$x_{AS} = \tfrac{1}{2}\left(\sqrt{D_2^2 - b_2^2} - \sqrt{D_1^2 - b_2^2}\right).$$

If, instead of b_2, the width b_1 of slide S is used, the equation yields its path x_S. The relationships between the slide stroke H1 and angle α of the inclined guide of the slide AS are shown in Fig. 6.2 $x_{AS} = H1 \tan\alpha$. The angle φ and stroke Z for slide S (Fig. 6.3) can be determined in the same way. When slides AS are actuated, they move with respect to slides S, against which they were tightly abutted during the injection molding process. The mutually contacting surfaces must have a particular minimum angle of inclination so as to prevent jamming. This angle of inclination δ depends on the angles α and γ (Fig. 6.4). $\tan\delta = \tan\alpha \sin\gamma$. Since in the present case $\gamma = 45°$, then

$$\tan\delta = \frac{\tan\alpha}{\sqrt{2}}.$$

The angle δ at the slide S acc. to Fig. 6.4 must be greater than or equal to this value. The minimum distance of slide S (Fig. 6.5) must be at least twice the path x_S of this slide, plus the width of its head k. The slide path x_S must be equal to or greater than

$$q \geq \frac{ba}{\sqrt{2}}.$$

Example 61, Injection Mold for Compact Discs Made from Polycarbonate

A Compact Disc (CD) contains information in the form of small "pits" on one side. This side is made reflective by means of a metal coating. When the CD is played, a laser beam is transmitted through the opposite, non-reflective side and reflected from the inside of the reflective, information-bearing surface. It thus passes twice through the disk body. This process is only reliable if the disk has maximum precision of flatness, thickness, waviness and surface quality.
Because of these high requirements which CDs have to meet, they are injection molded in single-cavity molds with a sprue cone and annular gate. The mold cavity has to meet tolerances of the order of a few micrometers. The mold surfaces are polished to a mirror finish. The shaping parts of the mold are of steel 1.2083 with a Rockwell-C hardness of 50.

Mold

The mold has an overall height of 168 mm and a diameter of 235 mm and is specially stiffened. The two cavity plates are bonded over their full surface by a special hard soldering or diffusion welding process to plates (12, 16) which enclose the cooling channels. As a result, the plate assembly has high stiffness and the cooling channels can be arranged close to the cavity and uniformly below the molding surfaces. These two plates are centered with respect to one another by means of conical surfaces in plate (3). Bolt (8) fixes the relative position of the two mold halves. The funnel-shaped sprue bush (18) is also in full contact with the sleeve surrounding it. These two parts also enclose a cooling channel. At the moving mold side, ejector (11) is in full contact with the sleeve surrounding it, enclosing a cooling channel.
The exchangeable, 0.3 mm thick cavity plate bearing the information is placed on the ejector side of the mold plate and held in place by a vacuum applied to its reverse side via the gap between the bush (15.2) and the hole in plate (12).

The Process

After the cooling process, the machine nozzle is lifted from the sprue bush (18). Then, with the mold still closed, the ejector (11) is moved 0.5 mm to the right by compressed air. The sprue bush moves the same distance to the right so that the annular gate in the bore of the CD is sheared off. Then the mold opens, the sprue cone, which is held in the undercut in ejector (11) is pulled out by the sprue bush and ejected by ejector pin (11). Finally, bush (14.1) pushes the molding toward the handling unit, which has entered the mold.
The mold is operated with a mold temperature of 60°C (140°F); the cycle time is less than 4 s.

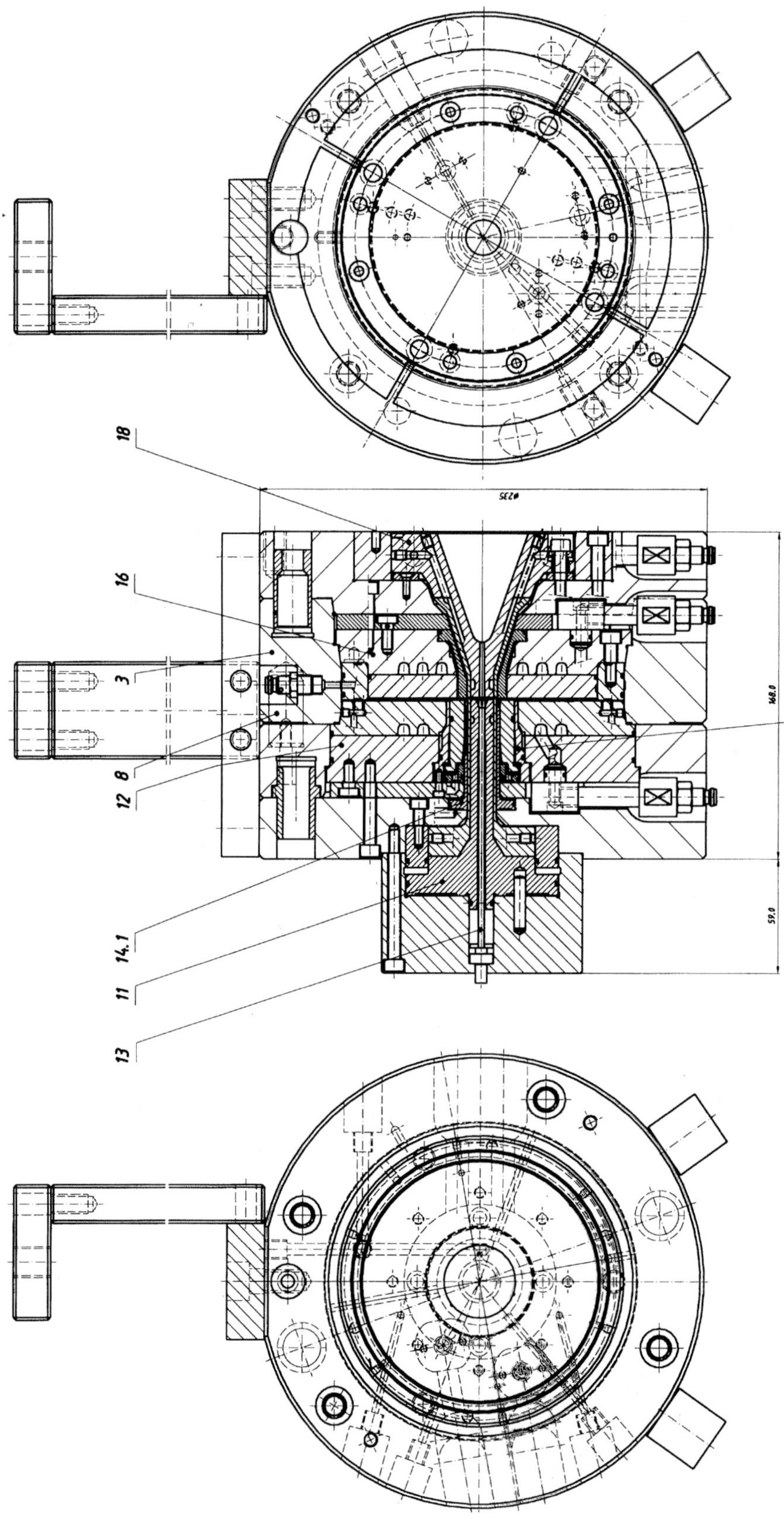

Figure 1 Single-cavity mold for polycarbonate compact discs
3: plate; 8: bolt; 11: ejector; 12, 16: plate; 13: ejector pin; 14.1, 15.2: bush; 18: sprue bush
(Courtesy: Krallmann GmbH & Co. KG, Hiddenhausen)

Example 62, Single-Cavity Injection Compression Mold for a Cover Plate Made from Unsaturated Polyester Resin

When injection molding thermosetting resins, undesired fiber orientation in the molded part can be largely reduced by employing injection compression. If side action is also needed to release the molded part, the drive mechanism for this side action must take into account the compression movement.
The cover plate (Fig. 1) is produced from a free-flowing thermosetting resin and has a dovetail-shaped slot that must be released by means of a slide.

Mold (Fig. 2)

The cavity is formed between the core insert (1) and cavity insert (2). The core fits into the cavity recess; the lateral shear surfaces have a slight taper to facilitate entry. The slide (3), which is attached to the piston rod (5) of a hydraulic cylinder by means of the slide retainer (4), is located in the cavity. A lock (6) fits into an opening in the slide retainer (4) to hold the slide in position. The lock fits against the wear plates (7).

Runner System/Gating

The molding material enters the mold via the jacketed sprue bushing (8). A system of cooling channels (10) in the sprue bushing keeps the molding compound within at a temperature of 90 to 100°C (194 to 212°F) to prevent curing. The insulating gap (9) ensures thermal separation between the heated mold (approx. 180°C (356°F)) and the sprue bushing (8).

Heating

Heating of the mold is accomplished with the aid of high-capacity cartridge heaters (11) that are divided into 4 heating circuits. Each heating circuit is provided with a thermocouple for individual temperature control. The power and thermocouple leads are brought to a junction box (16) in accordance with the appropriate electrical codes (VDE 0100).

Mold Steels

The mold is constructed of standard mold components. The part-forming components (core, cavity and slide) are made of hardened steel (material no. 1.2083). The slide retainer and wear plates are made of case-hardened steel (material no. 1.2764).

Operation

Prior to mold closing, the slide is hydraulically set in the cavity so that the lock (6) enters the opening in the slide retainer (4) before the core (1) enters the cavity (2). The mold is not completely closed during injection of the molding material. The exactly metered shot volume initially fills the gap in the runner region and a portion of the cavity. During the subsequent closing motion (compression phase), molding compound fills the entire cavity and cures there under the action of heat.
During the compression stroke, the lock (6) prevents the slide (3) from being displaced outward by the molding pressure.
The molding material in the runner region also cures. The boundary between cured and uncured material in the sprue bushing is located approximately at the cavity end of the cooling channel (10). The sprue puller (13) and ejector (14) remove any remaining cured runner material. The molded part is ejected by means of four ejectors not described here.

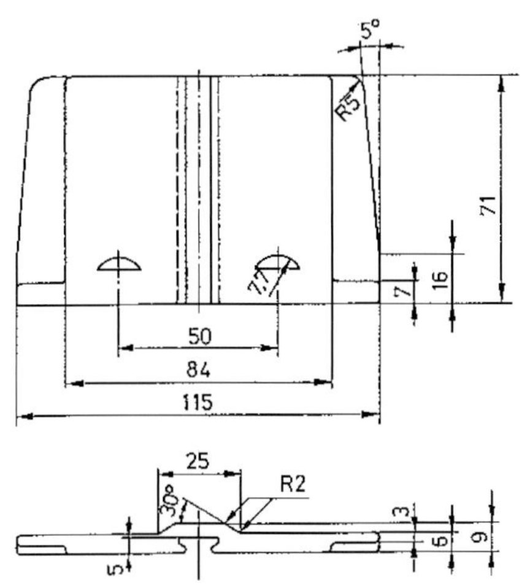

Figure 1 Cover plate

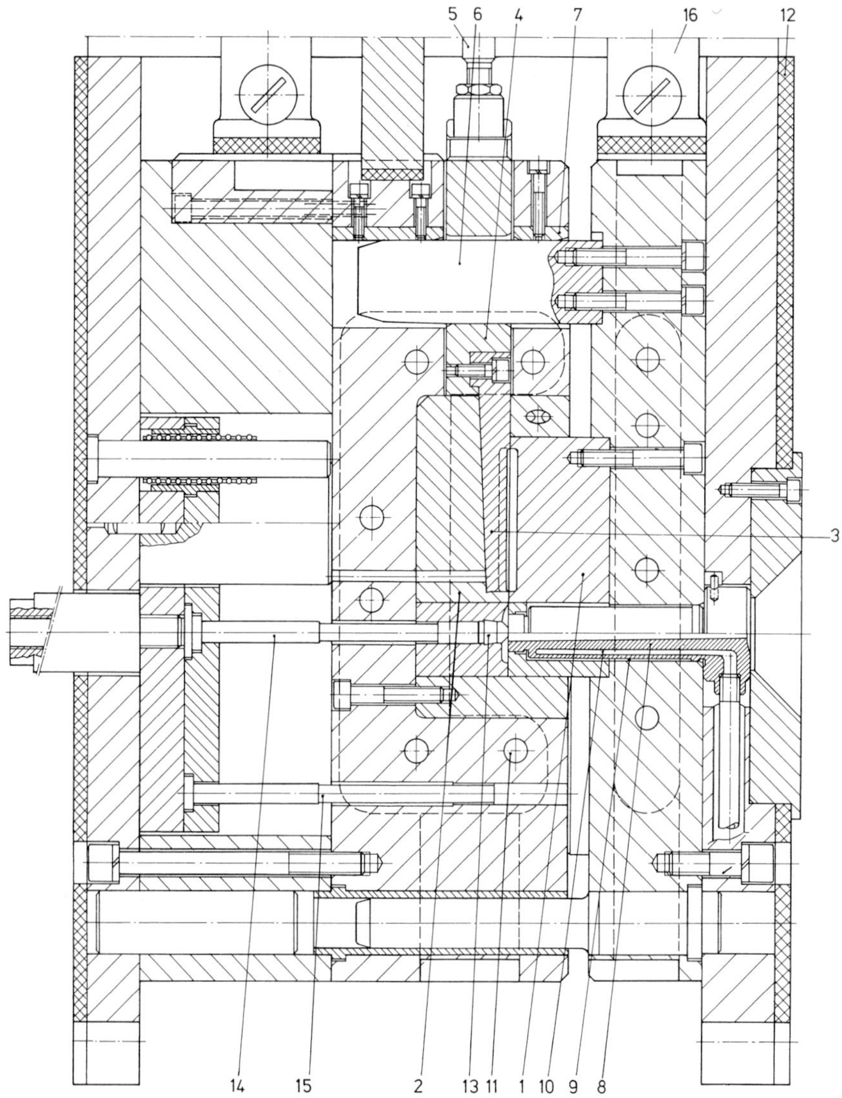

Figure 2 Single-cavity injection compression mold
1: core insert; 2: cavity insert; 3: slide; 4: slide retainer; 5: piston rod; 6: lock; 7: wear plate; 8: jacketed runner (cold runner); 9: insulating gap; 10: cooling channel; 11: cartridge heaters; 12: insulating plate; 13: sprue puller; 14: ejector; 15: pushback pin; 16: junction box (Courtesy: Hasco)

Example 63, Two-Cavity Injection Compression Mold for a Housing Component Made from a Thermosetting Resin

Fiber orientation, deflashing and lost runner material are problems that result in costs especially in the area of thermoset processing. The mold presented in this example shows how expenses for the above can be reduced. A device that permits more exact metering of the molding compound to the two mold cavities is described.
The molding material is injected into the partially opened two-cavity mold (Fig. 2 and 5).

Flow Divider

Distribution of the molding material to the two cavities is accomplished with the aid of a conical flow divider (1) with appropriately designed grooves. During injection, the flow divider is opposite the discharge opening of the sprue bushing (2). After injection, the molding compound lies in the common pocket (Bakelite system) at the mold parting line in the form of two approximately equal masses.

Compression Step

With final closing of the mold, the molding compound is forced into the two cavities (3, 4), where it cures under the action of the mold temperature (approx. 180°C (356°F)). As a result of the compression step, fiber orientation in the molded part is considerably less than would have been the case with injection into a closed mold.

Degating

The flow divider (1) protrudes into the sprue bushing (2) during compression and blocks it off from the parting line.
The standardized jacketed sprue bushing is provided with cooling channels (5), as a result of which the molding compound in the sprue bushing is held at a temperature of 90 to 100°C (194 to 212°F), so that it does not cure ("cold runner system"). Only the protruding tip of the flow divider is warmer – as a result of mold heating – and molding compound cures here. As a result, material lost in the form of a runner is limited to only the small amount of material in the grooves of the flow divider (1). An insulating gap (17) provides thermal separation between the sprue bushing and mold.

Flash

During the compression step, the molding compound flows past the projected area of the mold cavities and forms flash.
The mold cavities (3, 4) are provided with flash edges (7, 8) to ensure clean separation of the molded parts from the flash during ejection. Figure 3 shows the common pocket (9) with flash edges (7, 8) located on the movable side of the mold parting line. Figure 4 shows the two molded parts and the associated flash.

Common Pocket

The shear edge (12) defines the size of the common pocket. Details of the shear edge configuration and gap are shown in Fig. 5. The different edge radii (0.8/2.4 mm) impart increased stiffness to the flash rim (13) and give the numerous ejector pins located behind it a good means for ejecting the flash. A slight undercut (14) holds the flash on the movable side during mold opening.

Mold Steels

The mold is constructed largely of standard mold components. The part-forming inserts are made of steel (material no. 1.2767, hardened).

Heating

The mold is heated by means of high-capacity cartridge heaters divided into 6 control circuits. Six thermocouples control the mold temperature.

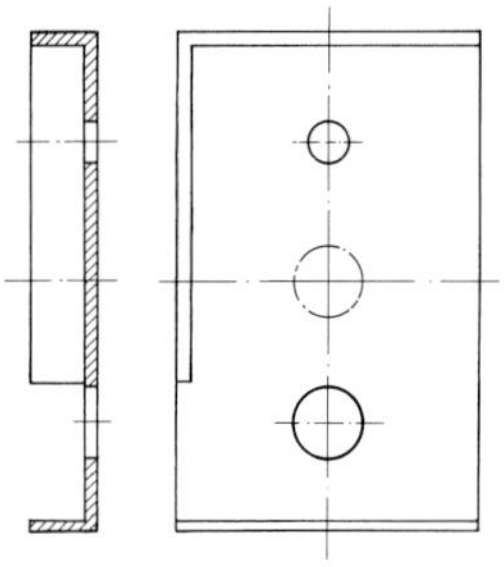

Figure 1 Housing component

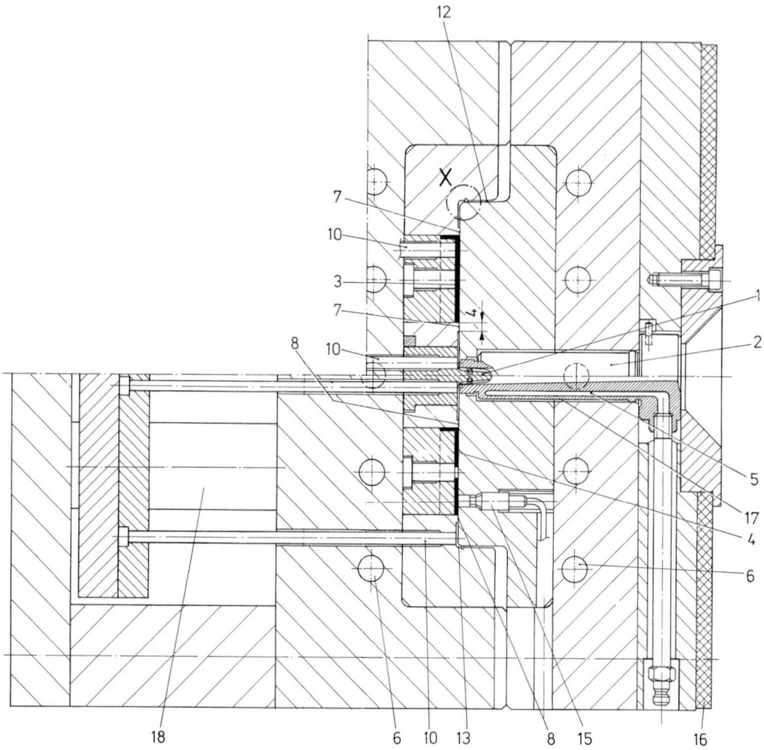

Figure 2 Two-cavity injection compression mold for a housing component
1: flow divider; 2: sprue bushing; 3, 4: mold cavity; 5: cooling channel; 6: cartridge heater; 7, 8: flash edge; 10: ejector; 12: shear edges; 13: flash rim; 15: pressure sensor; 16: insulating plate; 17: insulating gap; 18: support

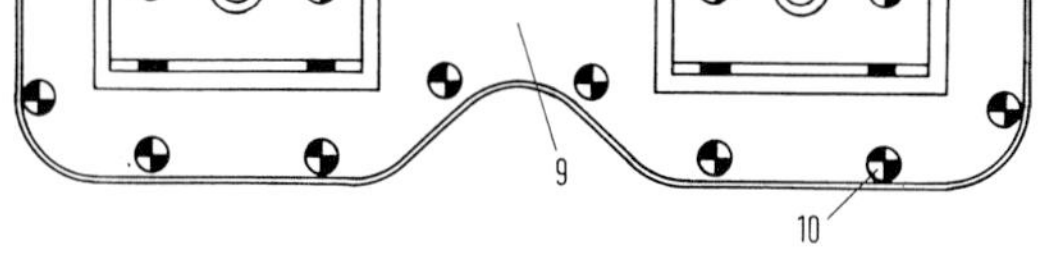

Figure 3 Common pocket
7, 8: flash edge around cavity; 9: common pocket; 10: ejector

Figure 4 Molded (bottom) parts with separated flash (top)

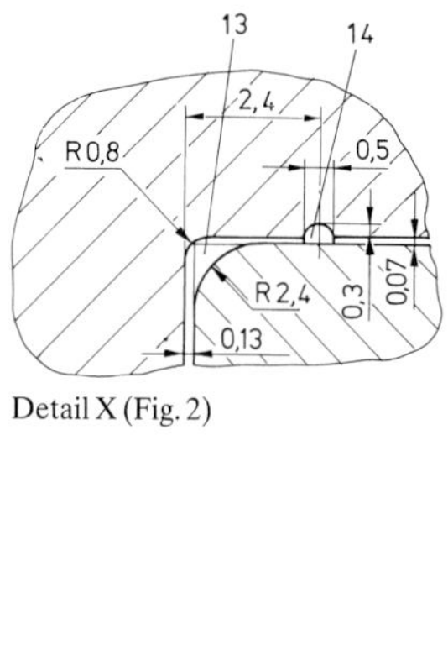

Figure 5 Shear edge
13: flash rim; 14: undercut

Example 64, Injection Compression Mold for a Plate Made from Melamine Resin

Plates, cups and a variety of household items are often made of melamine resin, type 152.7. In addition to the "classical" compression molding technique, injection molding machines are employed to mass-produce such parts by means of the injection compression technique. Figure 1 shows the mold in the three steps of production: injection (I), compression (II) and ejection (III).

The plate is molded using a pinpoint gate. When injection molding without subsequent compression with this type of gating, the melt would be subjected to severe orientation that could lead to molded-in stresses in the part and thus warpage or even cracks.

With the injection compression technique utilized here, the mold is closed until a gap of only 6 to 8 mm remains and then the molding compound is injected. After injection, the machine closes the mold and compresses the molding compound in the cavity. In this way, production of warp- and stress-free molded parts is ensured. To permit injection compression, the mold must have a shear edge, usually in the vicinity of the parting line. With this rotationally symmetrical part, a mold was selected in which the compression plate *c* passes through the mold plate *b* and forms the underside of the plate.

The mold operates as follows: the injection molding machine closes the mold until the two mold plates *a* and *b* contact one another and a compression gap *z* is formed between mold plate *b* and compression plate *c*. After injection of the carefully metered amount of molding compound, the mold is closed completely, compressing the material in the mold cavity. As the mold opens, the spring washers *t* initially cause plates *b* and *c* to separate by the amount of the compression gap *z*, which is limited by the stripper bolts *x*. Since the machine nozzle *d* is still in contact with the mold at this point in time, a vacuum that holds the molded plate against mold plate *b* is formed in the "molding chamber". After the mold has opened completely, the machine nozzle *d* retracts from the mold. Because of the undercut *h*, the cured sprue is pulled out of the sprue bushing and ejected from the nozzle with the aid of a pneumatically actuated device. With opening of the gate, the vacuum in the molding chamber *f* is released. The molded plate is ejected by means of a pneumatically actuated valve ejector *v*. During ejection, the molded parts are held by the suction cups on a part extractor and subsequently placed on a conveyor belt.

Cartridge heaters *k* heat the mold, while the sprue bushing is heated by a heater band *m*. Each heating circuit is individually controlled. The mold is separated from the machine platens by means of

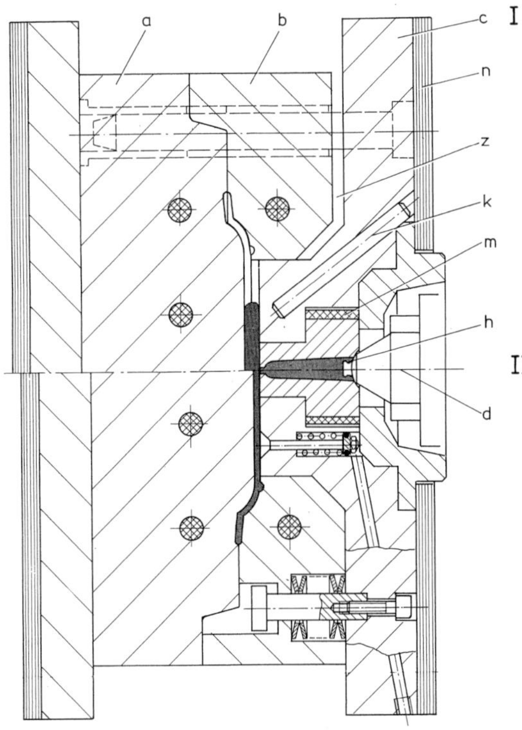

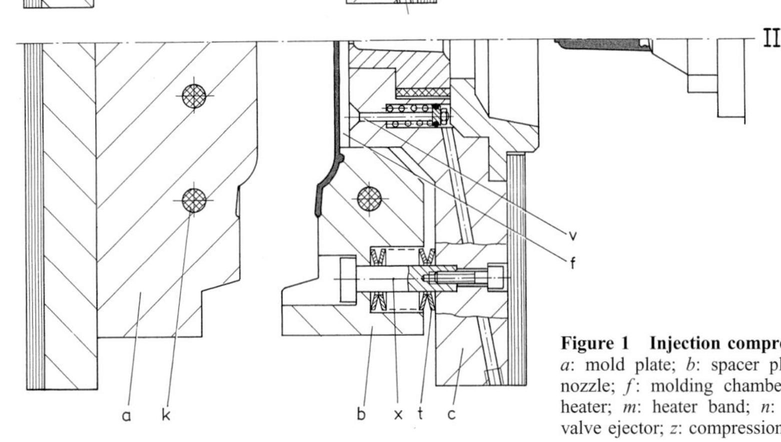

Figure 1 Injection compression mold for a plate
a: mold plate; *b*: spacer plate; *c*: compression plate; *d*: machine nozzle; *f*: molding chamber; *h*: undercut on nozzle; *k*: cartridge heater; *m*: heater band; *n*: insulating plate; *t*: spring washers; *v*: valve ejector; *z*: compression gap; *x*: stripper bolts

insulating plates *n*. The gate is so designed that upon retraction of the machine nozzle *d* only a relatively small gate vestige remains on the molded part after the sprue breaks away. This vestige is removed mechanically in a subsequent finishing operation.

Example 65, Five-Cavity Unscrewing Mold for Ball Knobs Made from a Phenolic Resin

Ball knobs of a thermoset resin, e.g. type 31, in a variety of diameters with and without internal threads are often employed for handles and levers on machinery and equipment. An alternative to compression molding as a means of producing these ball knobs is given by injection molding, which permits shorter cycle times and an automatic production cycle to be achieved. With the injection

Fig. 1

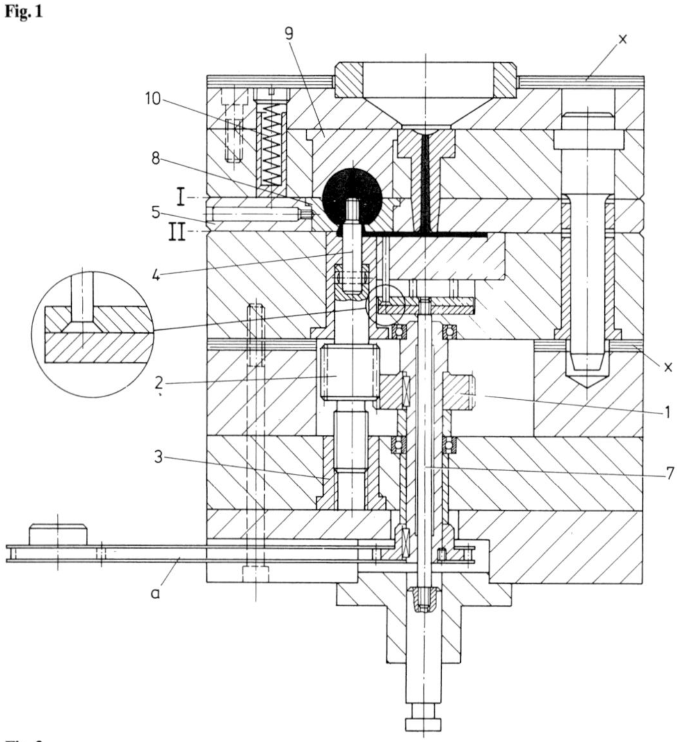

Fig. 2

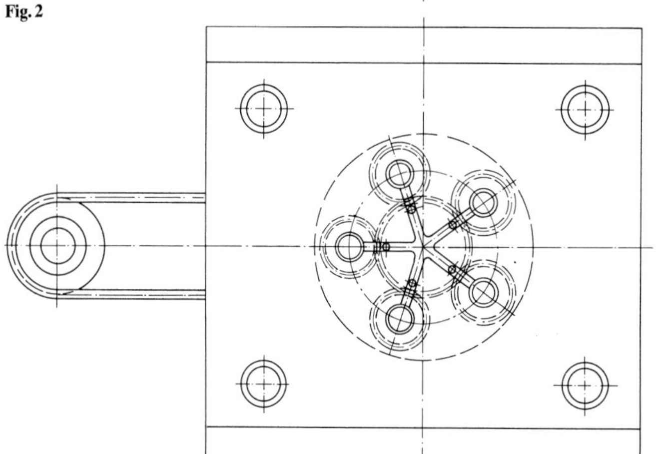

Fig. 3

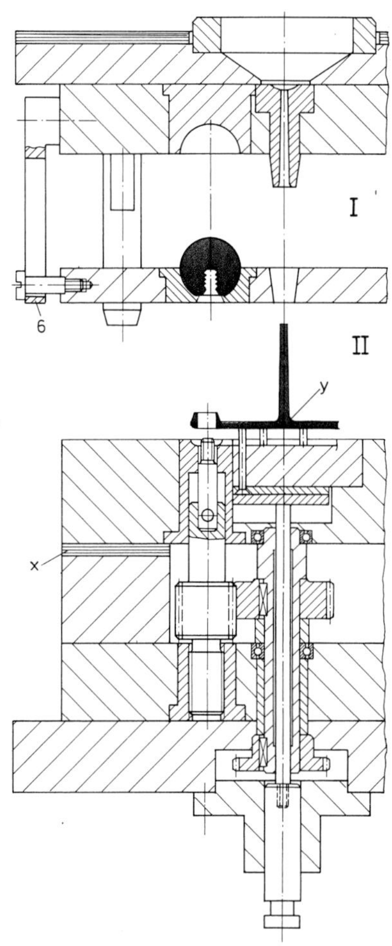

Figures 1 to 3 Five-cavity unscrewing mold for ball knobs of a thermoset resin
1: gear; 2: threaded spindle; 3: guide bushing; 4: threaded core; 5: center plate; 6: stop; 7: ejector rod; 8, 9: cavity inserts; 10: spring bolt; I, II: parting lines; *x*: insulating plate; *y*: runner; *a*: chain

mold shown schematically in Figs. 1 to 3, it is possible to produce ball knobs with different diameters and optionally with or without internal threads. Initially, molds were produced in which a film gate was located in the parting line on the periphery of the ball knobs. During degating, however, the molded parts were often damaged and could not be repaired even in a secondary finishing operation.

With conversion to a three-plate mold with two parting lines, it was possible to mold the ball knobs by means of a ring gate on the seating surface. Since the relatively clean gate mark after degating is not on a visible surface or functional area of the molded parts, subsequent finishing is not required. To permit production of ball knobs with different diameters, all part-forming components have been designed to be interchangeable (mold inserts (8, 9)). By replacing the threaded cores (4) with unthreaded core pins, ball knobs without internal threads can be produced. If threads with a different pitch are to be molded, the threaded spindles (2) and guide bushing (3) must also be replaced. The threads of the guide bushing (3) must always have the same pitch as the threads on the threaded cores (4). Only in this way is it possible to release the threads and ensure exact positioning of the threaded cores prior to injection.

The mold is heated by cartridge heaters located in the mold plates and insert retainer plates. The heating circuits are closed-loop controlled. Insulating plates *x* are provided to separate the mold from the machine platens and the drive mechanism.

The mold operates as follows: with the mold closed and the cores in the forward position, the molding compound is injected into the cavities via the ring gates. After the molded parts have cured, the threaded cores (4) are unscrewed from the ball knobs by a hydraulic motor that is controlled through an interface on the machine. To prevent the ball knobs from turning, unscrewing takes place while the mold is closed. The rotary motion is transmitted to the threaded spindles (2), which are displaced axially during unscrewing, by the chain *a* and the gear (1). Upon mold opening, the spring bolts (10) separate parting line I. Following this, plate (5) continues moving until it reaches the stop (6) after parting line II has also opened. Undercuts hold the runner on the movable half of the mold after the ring gates have separated from the ball knobs. Next, the runner *y* is ejected by the ejector rod (7) which is connected to the machine ejector. During mold closing, parting lines II and I close automatically. Following this, the threaded cores (4) are returned to the molding position by the hydraulic motor.

Example 66, Four-Cavity Injection Mold for a Thin-Walled Housing Made from a Phenolic Resin

The housing component shown in Figs. 1 to 3 was produced in a thermosetting resin by means of injection molding. The special features of this part are the thin wall sections of 0.7 mm, some of which taper down to 0.3 mm. As a result of the very slight shrinkage, there is no guarantee that the molded parts will remain on the core for ejection. It was not possible to provide undercuts to hold the molded part on the core. This means that ejection poses a particular problem. Since there was also no possibility to eject the part only by means of ejector pins because of the extremely thin wall sections, a three-plate mold was selected.

Fig. 1

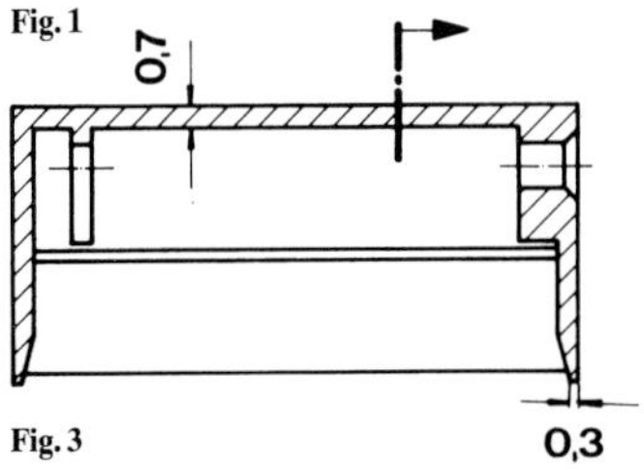

Fig. 2

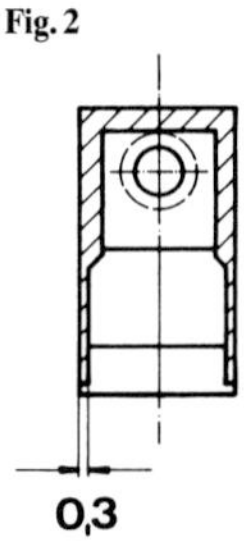

Fig. 3

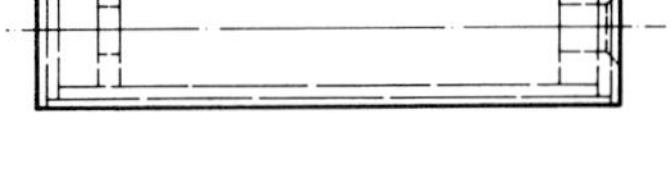

Figures 1 to 3 Thin-walled housing component of a thermoset resin

The four-cavity injection mold shown in Figs. 4 to 10 operates as follows: after the housings have been molded via the sprue (4) and runner system and the molding compound has cured, the mold opens at parting line I through the action of the spring-loaded inserts (3). This pulls the sprue (4) out of the sprue bushing, since an undercut is provided in the guide bore for the somewhat recessed center ejector. Simultaneously, the slide (5), which forms the holes in the side of the housing is pulled by the cam pin (6) and held in position by the spring-loaded detent (7). Parting line I now opens until mold plate (8) is stopped by latch (9), whereupon parting line II opens. This pulls the core (10) out of the housing.

Fig. 4

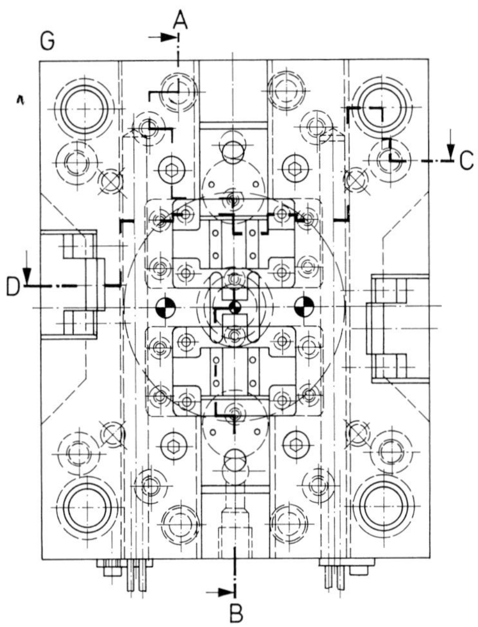

Fig. 5

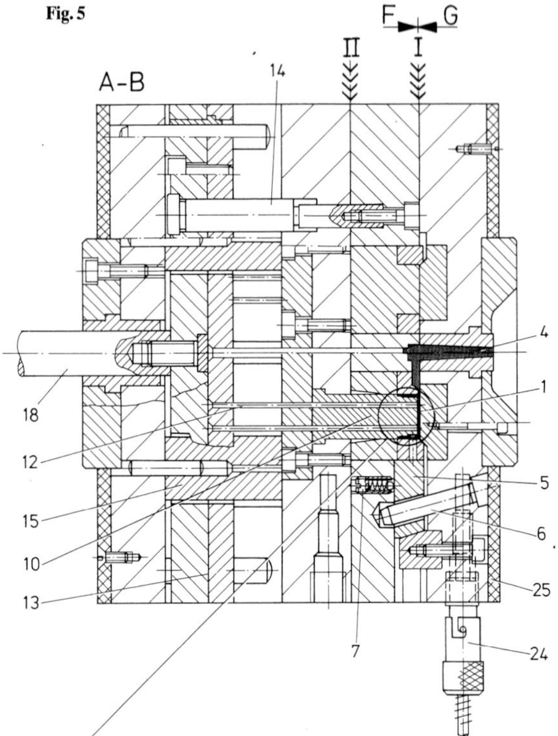

Fig. 7

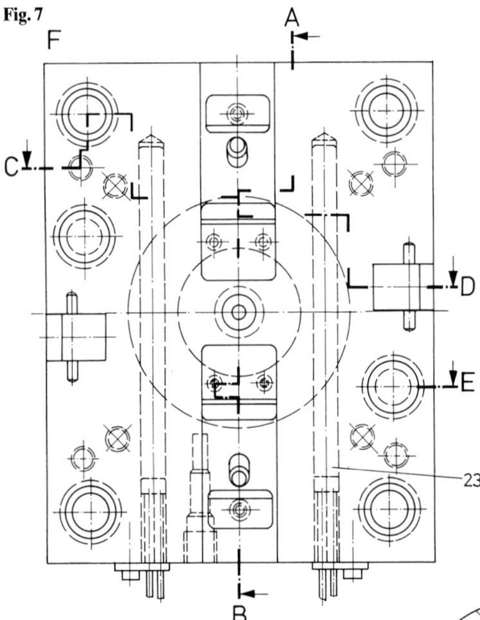

Fig. 8

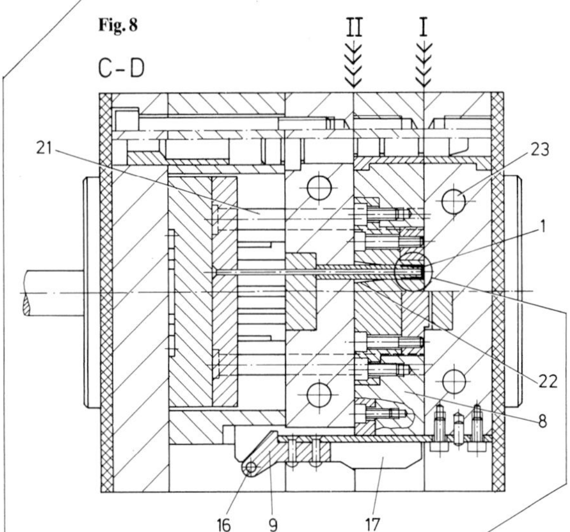

Fig. 10

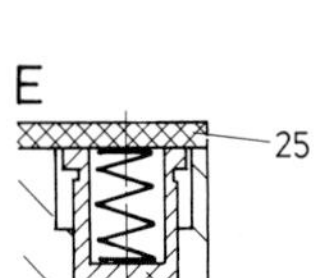

Fig. 6

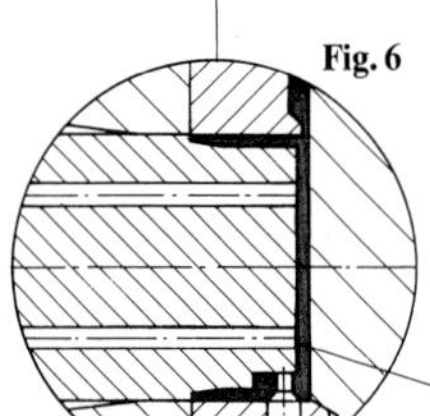

Fig. 9

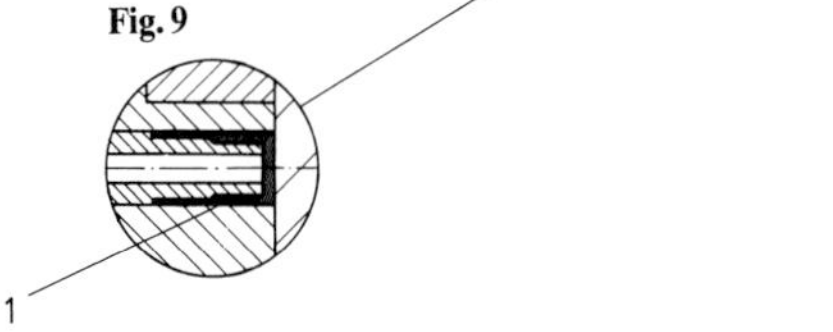

Figures 4 to 10 Four-cavity injection mold for a thin-walled housing component of a thermoset resin
1: housing component; 3: spring-loaded insert; 4: sprue; 5: slide; 6: cam pin; 7: spring-loaded detent; 8: plate; 9: latch; 10: core; 11: narrow side of housing; 12: ejector; 13: ejector plates; 14: stop bolts; 15: support pillar; 16: pin; 17: ejector plate guide; 18: ejector rod; 21: pushback pin; 22: relief on core; 23: cartridge heater; 24: thermocouple; 25: insulating plate

The molded part is supported by the two ejectors (12) during this motion. The ejector plate (13) is connected to mold plate (8) by stop bolts (14) so that the ejectors (12) do not change their position with respect to the molded part during opening of parting line II. As the mold opens further, pin (16) releases latch (9) so that the movable half can now retract completely. Ejector rod (18), which is connected to the hydraulic machine ejector, now advances the ejector plates (13) so that the ejector pins (12) eject the housings from the cavities in plate (8) along with the runner system. Advancing and retracting the ejector plates several times ensures that the molded parts do not stick on the ejector pins. This pulsating ejection also clears the ejector guide bores of any slight flash that might impair venting of the cavities and operation of the mold. In the present case, the parting line around the core (10) provides a good means for venting. After a short guiding surface, plate (8) is relieved (22). In addition to functioning as a vent, this relief acts as a discharge for any thin residual flash that could otherwise cause a malfunction. The mold is heated by high-capacity cartridge heaters (23); the temperature is controlled with the aid of thermocouples (24). The insulating plates (25) prevent heat transfer to the machine platen, thereby saving energy and ensuring a more accurate temperature in the mold.

Example 67, Thermoset Injection Mold for a Bearing Cover Made from Phenolic Resin

The bearing cover shown in Fig. 1 (dimensions: 50 mm × 70 mm × 25 mm) is to be injection molded in a glass-reinforced phenolic molding compound. Because of the production quantities expected, a 2-cavity mold was envisioned.

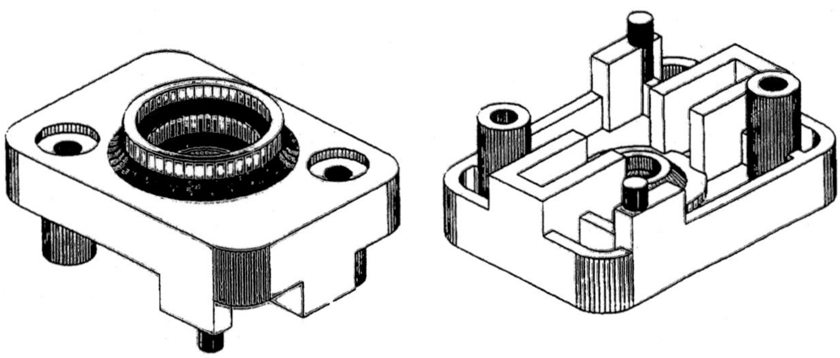

Figure 1 Thermoset injection molded bearing cover for an electric motor

Molds for processing of thermoset molding compounds are, in principle, comparable to those employed for processing of thermoplastics, with the understanding that there are certain material-specific considerations. Molds must be designed to be very rigid in order to prevent "breathing" and deformation, which contribute to the formation of flash. To monitor the injection pressure, which serves as the basis for mechanical design calculations, the design incorporates pressure sensors in the stationary and moving mold halves, for which blind plugs (35) are inserted as placeholders. The mold base utilizes standard mold components.
Steel grade 1.2767 is used for the mold inserts (39, 40), while grade 1.2312 (heat-treated to a strength of 1080 N/mm^2) is employed for mold plates (4, 5) as well as the ejector plate (9). Steel grade 1.1730 is used for the remaining plates and rails. Thermal insulating plates (19, 20), which are available in sizes to match the standard mold plates, serve to insulate the mold from the machine platens.
The molding compound enters the mold via the jacketed (temperature-controlled) sprue bushing (21). While the mold is heated to a temperature of about 170°C (338°F) by cartridge heaters (22, 23) to allow the molding compound to cure, the temperature of the material in the sprue bushing is kept below the cross-linking temperature, allowing it to be processed further. Material in close proximity to the gate cures. The interface between cured and uncured material in the sprue bushing is located at approximately the face of the sprue bushing (Fig. 2). A more recent version of this sprue bushing contains a restriction, or narrowing, in this region, which provides a well-defined interface. The fluid used for temperature control reaches the jacketed sprue bushing via extension nipples (33).
The molded part is ejected via knockout pins. In addition to providing for ejection, these pins serve to vent the cavity during filling. It is in part for this reason that the knockout pins are located beneath ribs and other deep sections of the part, where entrapped air is to be expected.
The mold filling pattern during injection of the molding compound as well as the mechanical and

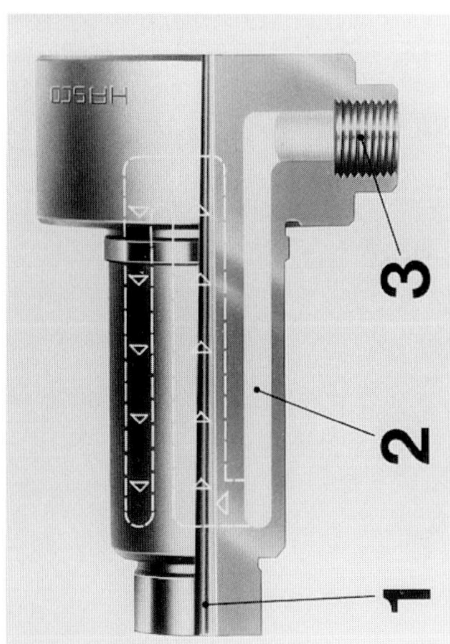

Figure 2 Standardized jacketed sprue bushing
1: sprue bushing bore; 2: jacket for temperature control; 3: connection threads

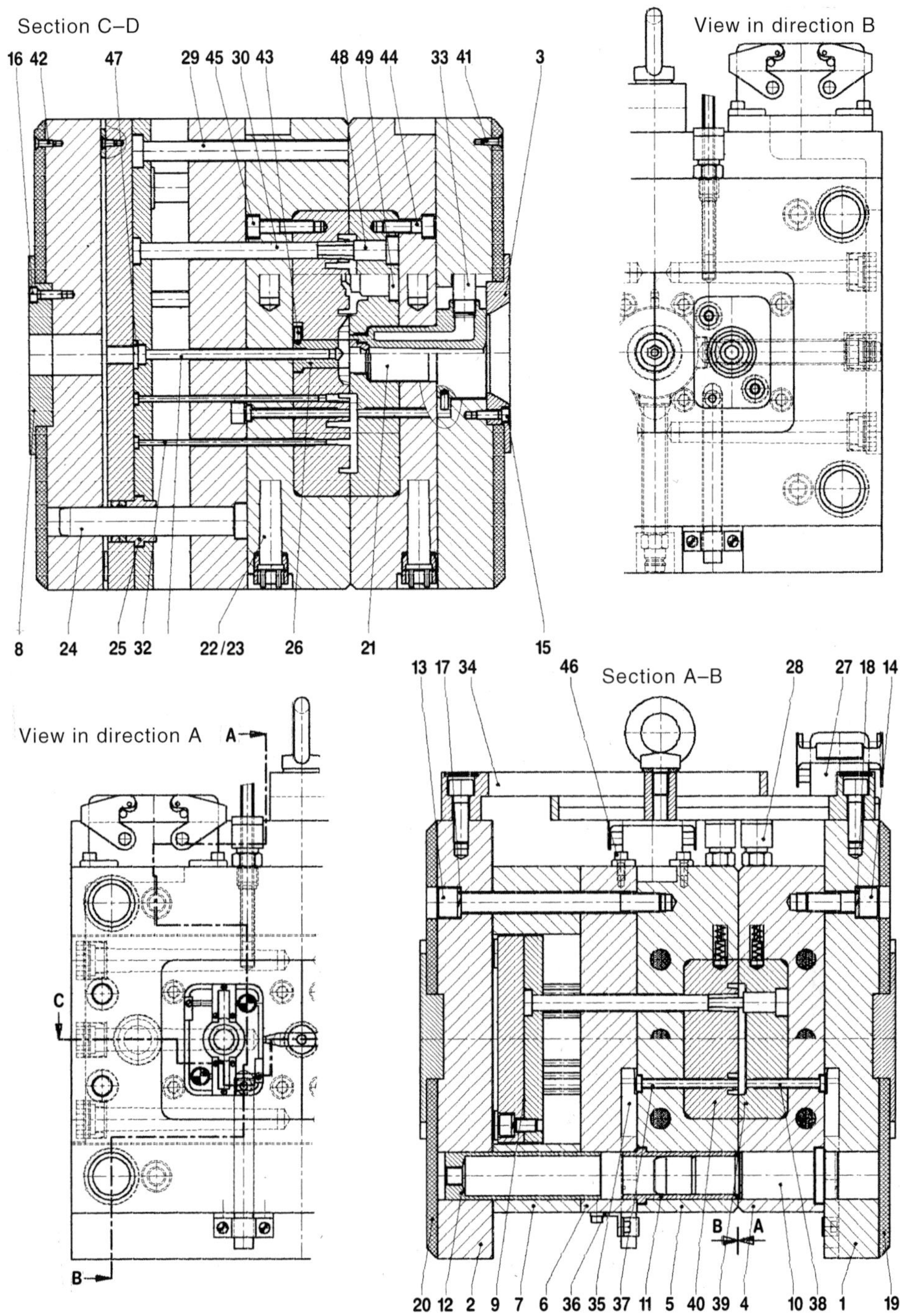

Figure 3 Thermoset injection mold for a bearing cover
1, 2: clamping plates; 4, 5: mold plates; 6: backing plate; 7: rails; 9: ejector plates; 19, 20: thermal insulating plates; 21: sprue bushing; 22, 23: high-performance cartridge heaters; 27: multi-pin connector; 29: pushback pin; 33: extension nipple; 34: transport strap; 35: blind plug; 36: retainer ring; 37: pressure sensor pin; 39, 40: mold inserts
(Courtesy: Hasco, Lüdenscheid, Germany)

thermal aspects of the mold design were simulated and established during the design phase with the aid of appropriate computer programs.
Thermoset parts exhibit very little shrinkage at the moment of ejection. Accordingly, appropriate measures must be taken to ensure that the parts remain in the ejector half of the mold. The mold is heated by tapered cartridge heaters (22, 23), which are distributed over four heating circuits. Each heating circuit is provided with a thermocouple and can thus be controlled independently. Power and thermocouple leads are terminated in junction box (27) in compliance with VDE guidelines (VDE 0100).

Example 68, 6-Cavity Hot-Runner Mold for Coffee Cup Covers Made from Polypropylene

Airtight sealing covers in various colors for coffee cups are injection molded in this mold from easily flowable polypropylene. The demands on quality are high for this molded part. To open or close the lock, a turn of < 30 angular degrees is required, i.e., segmentation is specified for the inside of the cap. For economical reasons, a hot-runner system with open, externally heated gating nozzles (PSG system, Fig. 1) was selected for the high level of production required.
To obtain efficient cycling times, but also to eliminate the drool commonly associated with PP processing, very effective temperature control is provided by a total of eight independent cooling circuits. Particularly the cavities and the threaded segments which are demolded by angular slides are cooled separately close to contour (Contura system [1], Figs. 2 and 3).
To exclude heat marks on the gate side of the molded part, 10 bonded copper cores provide very efficient thermal exchange in the gating area. Temperature at the hot-runner manifold and sprue nozzles is 240 °C. Depending on the color setting, cycle times of less than 12.5 s are achieved. Since the various pigments affect the dimensional behavior of molded parts differently during molding, correspondingly different mold wall temperatures have to be selected. This explains the variation in cycle times. The minimum input temperature of the coolant (water) is 15 °C.

Mold

This mold is a 6-cavity hot-runner system with open sprue nozzles and tips. The nozzles and the hot-runner manifold are each heated and temperature regulated externally. The intermediate gate with filter insert is unheated and equipped with an immersion nozzle for melt decompression. The hot-runner manifold support disks are composite structures equipped with a steel jacket for support and a ceramic core to minimize heat loss by conduction [2]. The sprue nozzles are connected to the hot-runner manifold non-positively by a sliding seal face. The thread segments arranged at 90° angles to each other are demolded on the core side by lifters. The double-wall design with a distance between walls of 5mm cannot be temperature regulated by conventional systems. To this end, the slides were equipped with bonded copper cores whose front end is in contact with the coolant (Fig. 3). To avoid heat marks, the insert on the nozzle side is equipped with a coolant channel system that follows the contour and ten additional copper cores for efficient heat removal from the gate area (Fig. 3). In all, the mold has over 36 (!) cores, kufters, etc., manufactured by System Contura in order to optimize the thermal conditions with the ultimate goal of reducing cycle time.
The tool steel used – material no. 1.2343 ESU – has a hardness of $50 + 4$ HRC and is partly nitrated to ensure wear and slideability.

Demolding

The thread segments are released by angular slides after the mold opens. The parts are demolded to free-fall with compressed-air assist (see detail 16, Fig. 1).

References

1. Contura Mold Temperature Control GmbH, Menden, Germany
2. Unger, P.: Hot Runner Technology, 2006, Hanser Publishers, Munich

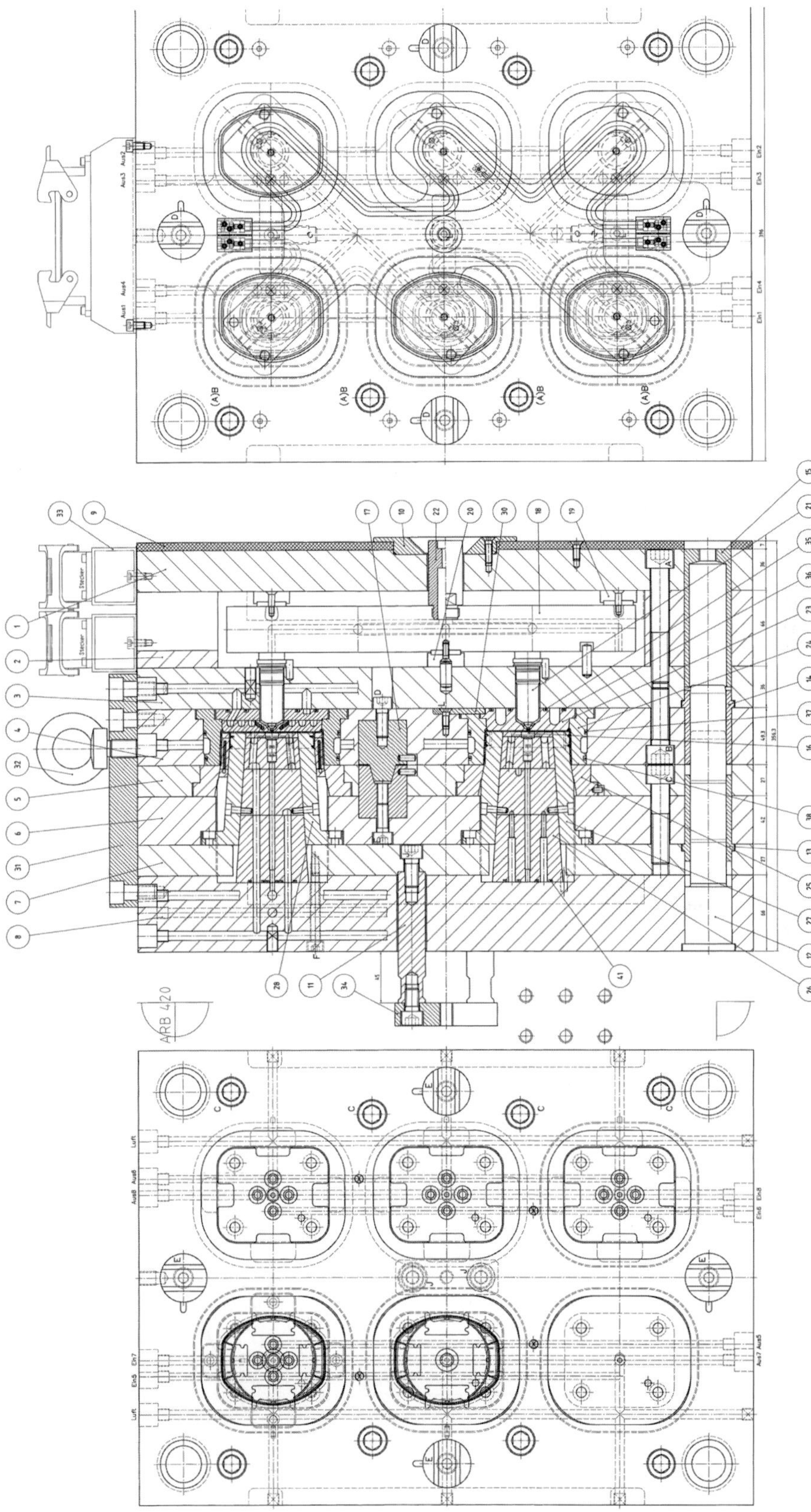

Figure 1 Six-cavity hot-runner mold for coffee cup covers made from propylene
1, 2, 3, 5, 8, 9, 12, 16, 18: hot-runner manifold, 21: heated sprue nozzle with tip, 22, sprue bush and filter, 23: head plate, 26: core, 27: external lifter, 28: internal lifter, 23, 26 to 28: System Contura
(Courtesy: Junghans, Hessisch Lichtenau, Germany)

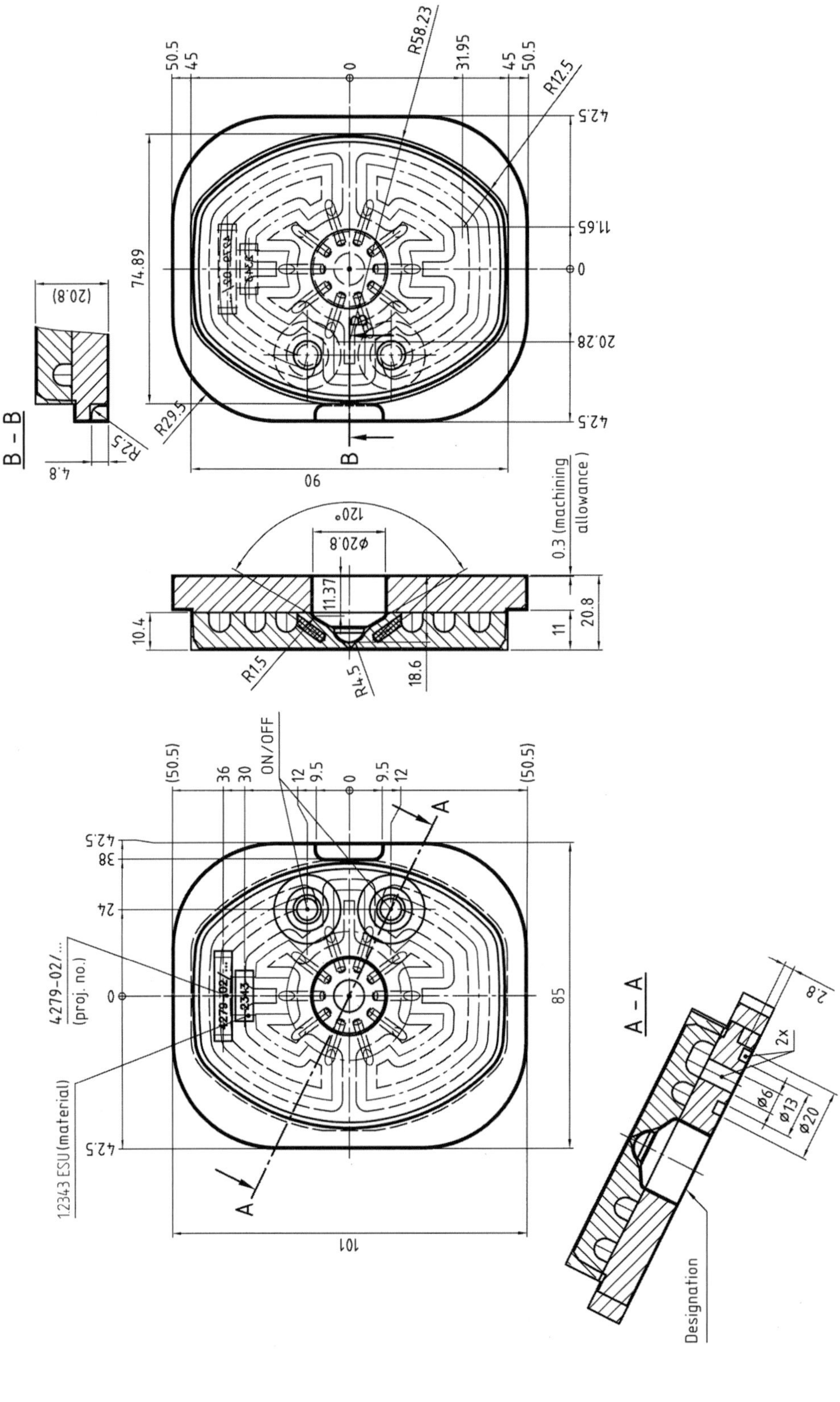

Figure 2 Insert 23 (from Fig. 1) with ten copper cores each for intensive thermal control of gate area, System Contura
(Courtesy: Contura, Menden, Germany)

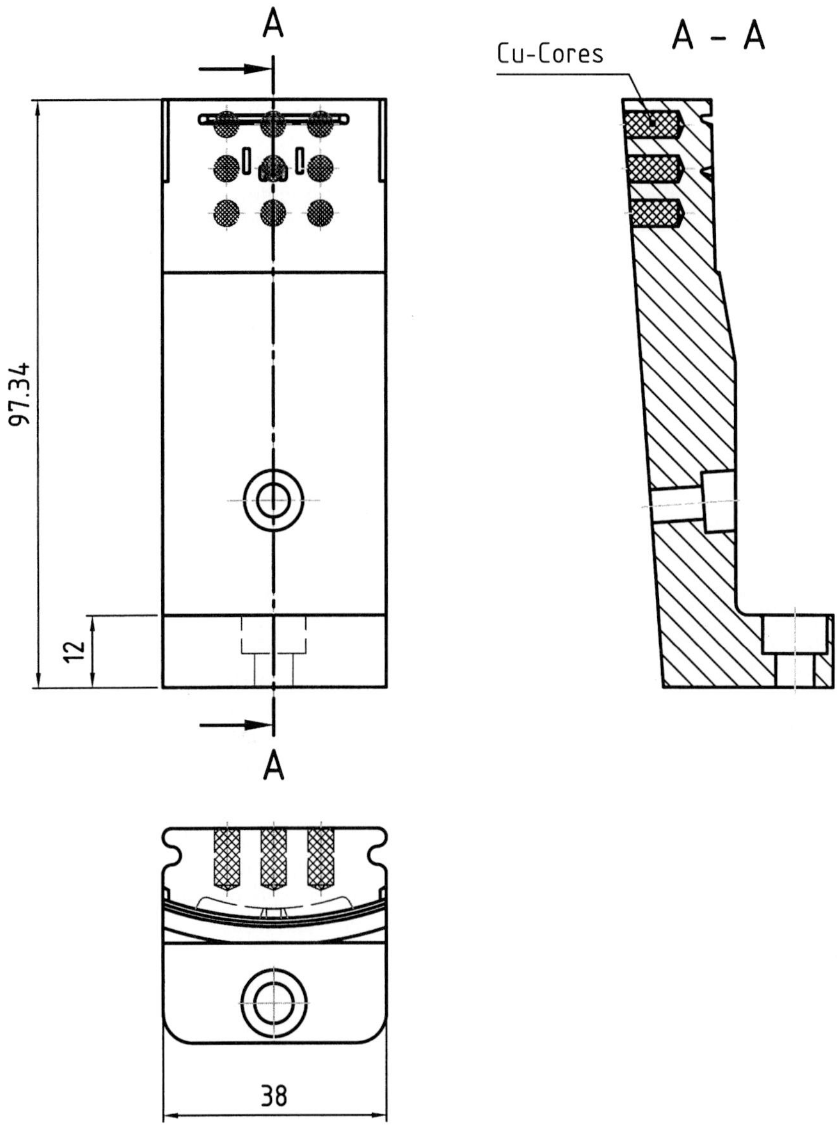

Figure 3 Internal slide 28 (from Fig. 1) with nine copper cores each for intensive temperature control of the thread segments, System Contura
(Courtesy: Contura, Menden, Germany)

Example 69, Two Injection Molds for Overmolding of Polyamide Tubing for Automobile Power Window Operators

Molded Part

In power window operators for automobiles, the operating force is transmitted from the drive mechanism to the lifting mechanism by means of a flexible gear rack that runs in plastic tubing. To hold the drive mechanism and mount it in the vehicle, two attachments of glass-fiber-reinforced polyamide are molded onto a piece of polyamide tubing (Fig. 1).

Mold

Two pieces of tubing (for the left and right-hand versions) are placed into one injection mold for the elbows (Figs. 2 to 4) and into another for the drive mounts (Figs. 5 to 7).

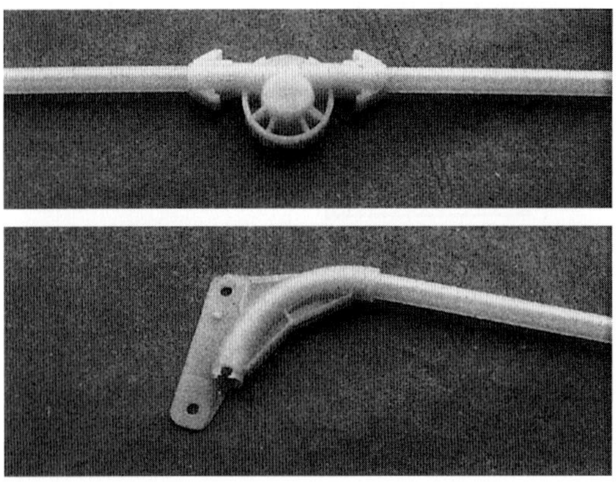

Figure 1 Guide tube with elbow and drive mount

Core inserts (1) are placed in the ends of the bent tubing and then in the elbow mold along with the tubing. Once this mold is closed, the core inserts are held in place by the heel blocks (2). Two sets of core inserts are available. The overhanging tubing is held in a bracket (9) attached to the side of the mold. The melt flows from a sprue through a runner system to fill each of the parts via two side gates.
For insert molding, the ejectors must be retracted to permit loading of the inserts. Accordingly, the elbow mold has a return spring (3) around the ejector rod (4). After ejection and prior to insert loading, the movable platen of the machine must be repositioned by an amount corresponding to the ejector stroke.
With the mold for the drive mounts, straight lengths of tubing are placed into the mold and overhang on either side. The overhanging tubing is held in spring-loaded retainers (1) at each end.
The parts are molded with the aid of a hot-runner system consisting of a hot-runner manifold (2) and six sprue nozzles (3) that feed six sprues with secondary runners. The sprue bushing (4) threaded into the hot-runner manifold has a decompression chamber to relieve pressure on the melt within the hot-runner system prior to opening of the mold.
The hot-runner manifold is heated by means of two cast-in heater coils (5) and is clad with insulating plates (6) to prevent heat loss.

Fig. 2

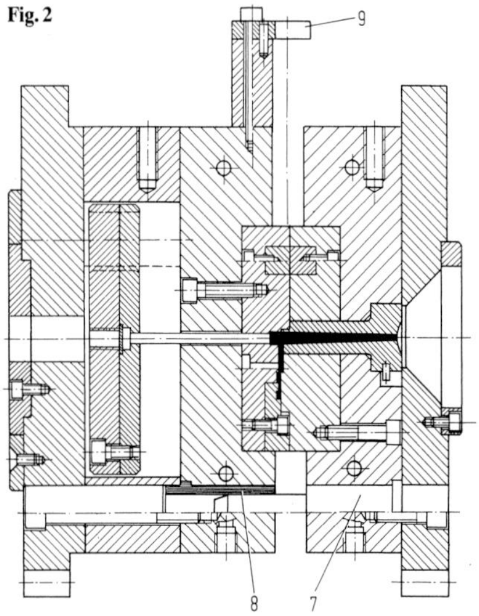

Fig. 3

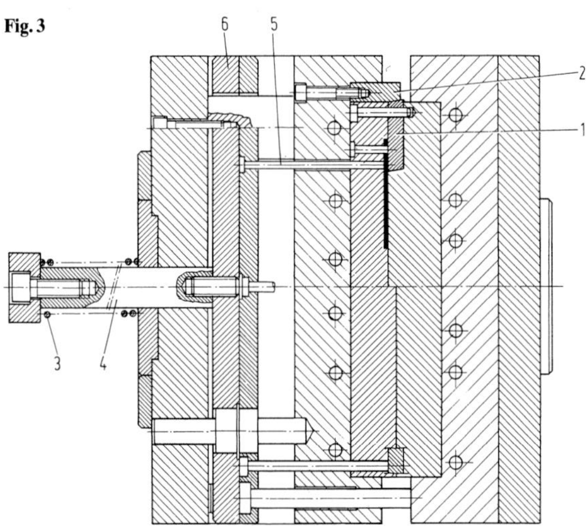

Figures 2 to 4 Injection mold for elbow
1: core insert; 2: heel block; 3: return spring; 4: ejector rod; 5: ejector; 6: ejector plate; 7: leader pin; 8: guide bushing; 9: bracket

With the gating selected, air may be entrapped at location (A) in the molded part. A date stamp (8) and vent insert (9) eliminate this danger.
In this mold as well, the ejectors must be retracted prior to loading the tubing. This is accomplished here by means of the hydraulic ejector in the machine, which is coupled to the ejector rod (10).

Fig. 5

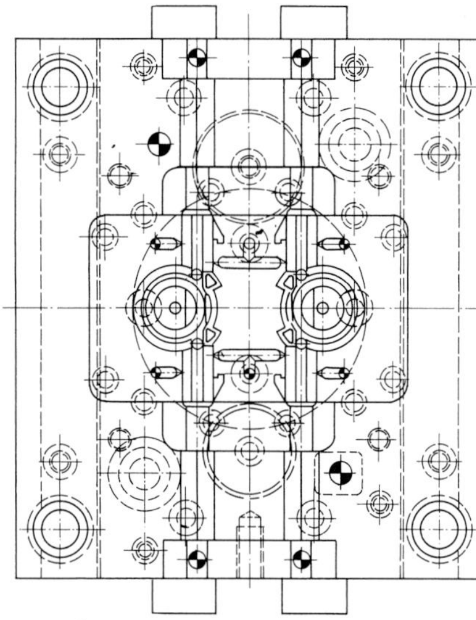

Figures 5 to 7 Injection mold for drive mounts
1: spring-loaded retainer; 2: hot-runner manifold; 3: hot-runner nozzles; 4: sprue bushing; 5: heater coils; 6: insulating plates; 7: insulating plate; 8: date stamp; 9: thread insert; 10: ejector rod

Fig. 6

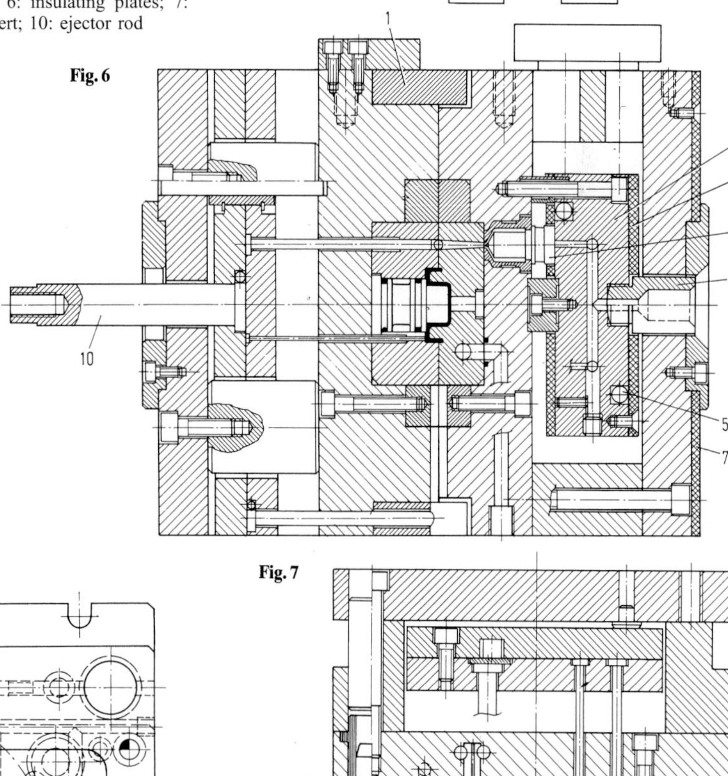

Fig. 4

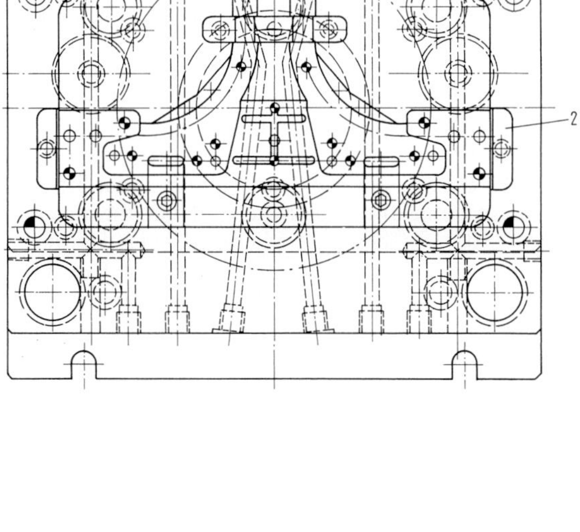

Fig. 7

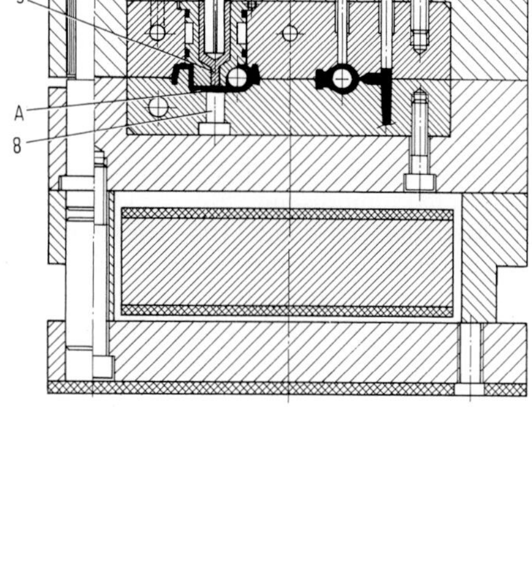

Example 70, Single-Cavity Injection Mold for a Housing Base Made from Polycarbonate

The housing (Fig. 1) has dimensions of 150 mm × 80 mm × 44 mm and has four threaded holes on its bottom and two each on the narrow ends. The narrow ends also have recesses between the threaded holes. The interior contains snap hooks, bosses and mounting eyes.

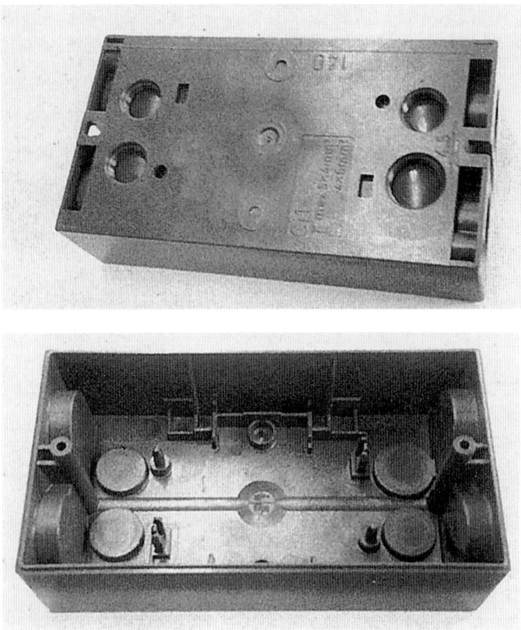

Figure 1 Housing base

Mold (Figs. 2 and 3)

The design of the part requires 8 unscrewing cores, two side cores and special measures for release of the snap hooks. Except for the part-forming components, the mold is constructed largely from standard mold components. The part is oriented in the mold with its bottom facing the injection nozzle. The four threaded cores (105) for the side holes are placed next to one another in pairs parallel to the mold parting line. These cores rotate and are supported at one end in bronze bushings (136) and at the other by means of journals (83) in ball bearings (82). They are further guided in threaded bushings (113) and are operated in pairs by a gear (111) attached to a pinion shaft (112). A rack (126) engages the pinion shaft.
The four threaded cores (106, 107) for the bottom holes are also guided in threaded bushings (114) and are also operated in pairs by means of gear racks (125). The two gear racks (125, 126) on each side of the mold are joined by a yoke (123) into which the piston rod of a hydraulic cylinder (138) is threaded.
The threaded cores (105) turn faster than the threaded cores (106, 107) because of the transmission ratio of the gearing (111, 112). This is required by the different thread lengths.
The four threaded cores (105) on the side each have a flange that is enclosed by the slide (88) and the retaining strips (89, 90).
When one of the two hydraulic cylinders moves, the four threaded cores driven by it unscrew from the molded part, carrying along the slide (88) and thus releasing the recess on the narrow side of the part. Since the drive mechanisms for the two sides of the housing are arranged in a mirror image with respect to the axis of the housing, the two hydraulic cylinders must operate in opposite directions (Fig. 3).
The part-forming inserts and the core are made of hardened steel (material no. 1.2083 (ESR)), the threaded cores are made of case-hardened steel (material no. 1.2764). The gear racks are made of inductively hardened C 45 K.

Runned System/Gating

The part is gated on its bottom and filled via a hot-runner nozzle (25) which is attached to a heated sprue bushing (108) which extends through the space required for the unscrewing mechanism in this half of the mold.

Mold Temperature Control

To the extent that space permits, cooling lines and bubblers with baffles (132) are provided in the core and stationary mold inserts.

Part Release/Ejection

Prior to mold opening, the piston rods of the two hydraulic cylinders (138) are moved in opposite directions so that the threaded cores unscrew from the threaded holes. The slides (88) retract from the recesses on the sides of the molded part and the mold can now open.
As the two-stage ejector (28) advances, the stripper plates (4), the ejector sleeves (30) and the mold cores (101, 102) jointly strip the molded part off the core. Mold core (103) remains stationary, thereby releasing the smooth back surface of the snap hook. After a distance of approximately 20 mm, the ejector plates (10, 12) stop and plates (9, 11) continue moving along with the stripper plate (4) and ejector sleeve (30). The snap hooks and the core pins for the mounting eyes are released and the part is now free to drop. Pin (37) and sleeve (99) also serve to vent the cavity for the bosses in the mold.

Fig. 2

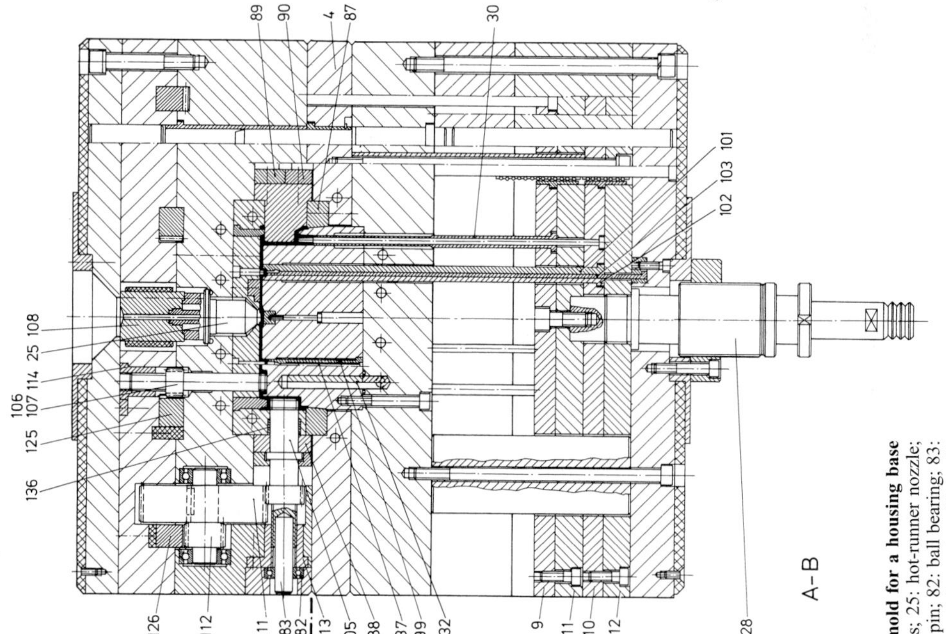

Fig. 3

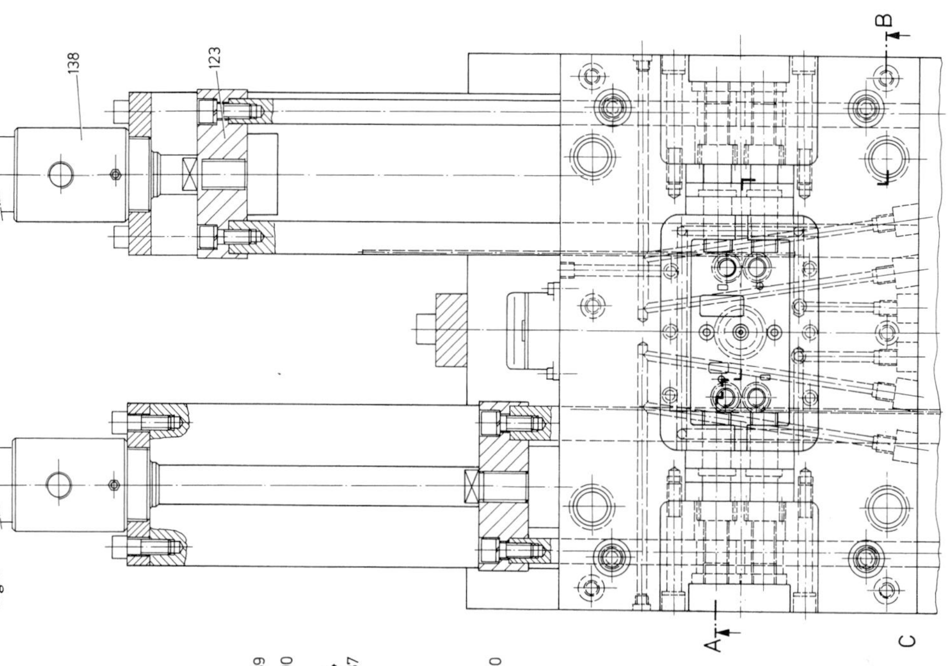

Figures 2 and 3 Single-cavity injection mold for a housing base
4: stripper plate; 9, 10, 11, 12: ejector plates; 25: hot-runner nozzle; 28: two-stage ejector; 30: ejector sleeve; 37: pin; 82: ball bearing; 83: journal; 87: stripper ring; 88: slide; 89, 90: retaining strips; 99: mold sleeve; 101, 102, 103: mold cores; 105, 106, 107: threaded cores; 108: sprue bushing; 111: gear; 112: pinion shaft; 113, 114: threaded bushings; 123: yoke; 125, 126: gear racks; 132: baffle; 136: bushing; 138: hydraulic cylinder

Example 71, Connector with Opposing Female Threads Made from Glass-Fiber-Reinforced Polyamide

The connector (Fig. 1) is 90 mm long in the direction of the through hole with the two opposing threads ($\frac{3}{4}$ in. NPT). Another hole intersects this first hole at a 90 degree angle and four additional holes pass through the connector parallel to it. The walls are 4 mm and the ribs 2 mm thick respectively.

Figure 1 Connector of glass-fiber-reinforced polyamide (PA)

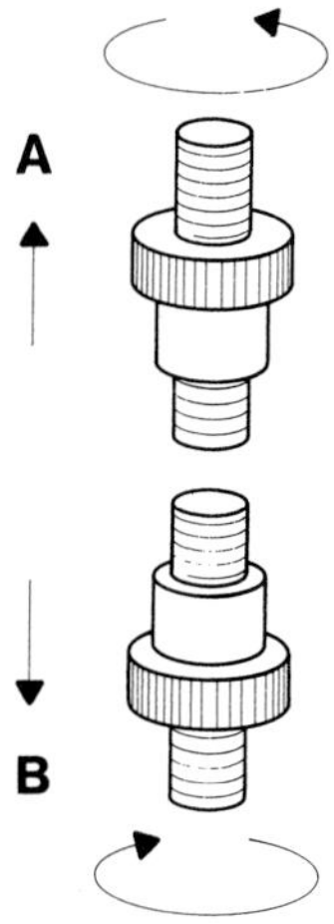

Figure 2 Threaded cores for releasing the two opposing threads

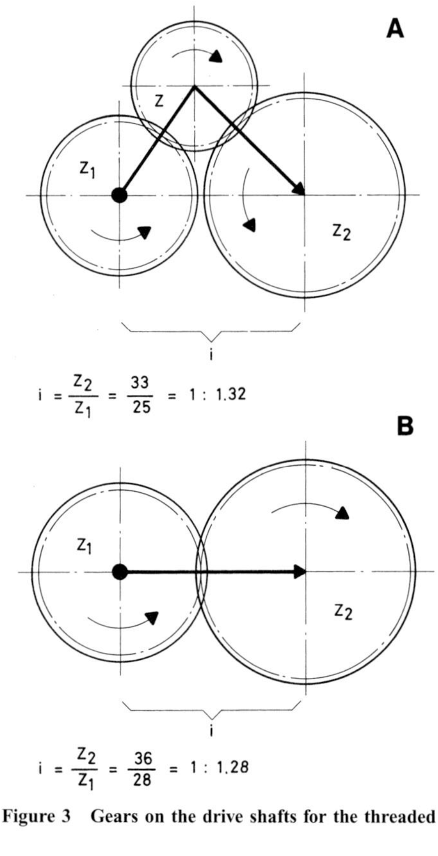

Figure 3 Gears on the drive shafts for the threaded cores

Mold

Unscrewing Mechanism

To release the two opposing threads and the hole between them requires two threaded cores, one at each end of the part. Dividing the length of the hole among the two threaded cores, each must move a distance of 45 mm or, with a pitch of 1.814 mm, make 25 revolutions.

A hydraulic unscrewing mechanism was selected to drive the threaded cores.

The cores rotate in opposite directions (Fig. 2) for which reason an intermediate gear Z is incorporated in the gear set A shown in Fig. 3. The gears Z 1 sit on the common drive shaft, while the gears Z 2 are located on the threaded cores. After 34 revolutions of the unscrewing mechanism, core A has moved 46.72 mm and core B 48.18 mm. The total displacement is thus approximately 94 mm, since the two cores engage one another slightly and thus center each other.

The unscrewing mechanism is attached to the stationary mold half (Fig. 4). Each of the two threaded cores (4) has a rectangular end (3) which slides in a mating hole in the gears (6). Threaded bushings (5) provide guidance for the threaded cores. The drive gears (8, 9) are mounted on the drive shaft (10). On side A, intermediate gear (7) is located between gear (8) and gear (6).

The core (16) for the internal shape, core (24) for the side hole and two cores for the holes passing through the connector are located on the moving mold half. Two additional cores (25) are attached to the stationary mold half, because there was not sufficient space to attach them next to core (24) on the moving mold half.

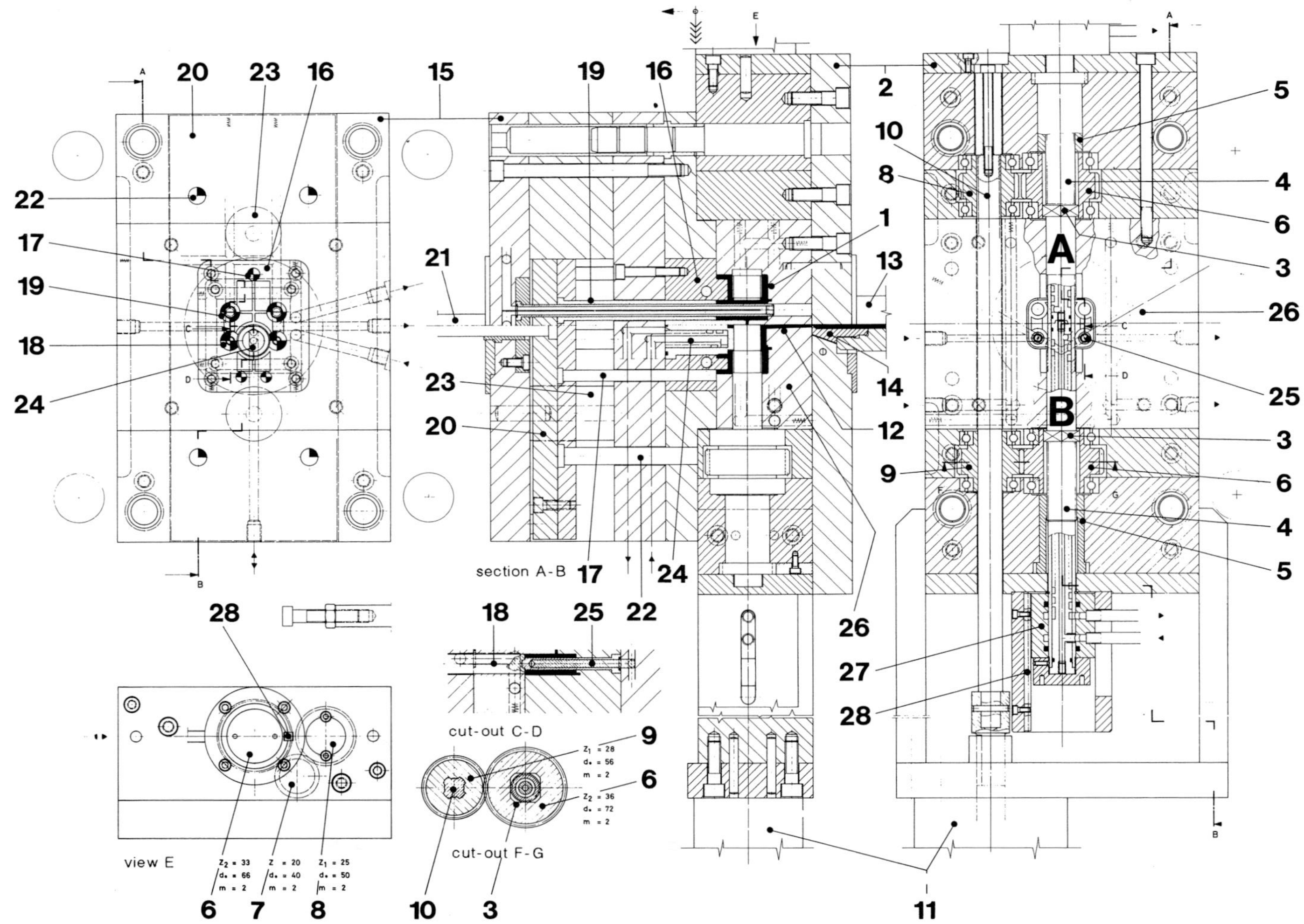

Figure 4 Injection mold with unscrewing mechanism for a connector
1: molded part; 2: stationary mold half; 3: rectangular end; 4: threaded core; 5: threaded bushing; 6: gear; 7: intermediate drive gears; 8, 9: drive gears; 10: drive shaft; 11: unscrewing mechanism; 12: sprue; 13: nozzle; 14: extension; 15: moving mold half; 16: mold core; 17, 18: ejector pins; 19: ejector sleeves; 20: ejector plate; 21: ejector rod; 22: pushback pin; 23: support pillar; 24: core; 25: core pin; 26: cavity; 27: manifold block; 28: key

Mold Temperature Control

The mold is operated at a temperature of 100°C (212°F). Channels are provided in the mold cavity (26) for mold temperature control. The two threaded cores (4) are hollow and each contain a single-flighted helical core. The temperature control fluid is supplied and returned through manifold blocks (27) with a rotating fit to the ends of the cores (4). Keys (28) prevent the manifold blocks (27) from rotating. Mold core (16) is provided with cooling channels. Core (24) is provided with helical cooling. The four core pins (25) are hollow and fitted with baffles to guide the temperature control fluid.

Part Release/Ejection

The ejector pins (17, 18) and ejector sleeves (19) serve to eject the molded part from the core. Ejector rod (21) advances the ejector plate (20) with the aid of the hydraulic ejector on the molding machine. Pushback pins (22) return the ejector plate to the original position. Pillars (23) support the core retainer plate.

Runner System/Gating

The part is filled through a sprue (12) with the aid of an extension (14) attached to the machine nozzle (13).
The cycle time is 98 s, and the cooling time is 57 s. Screwing the two cores (4) in and out each requires 10 s.

Example 72, Cylindrical Thermoplastic Container with Reduced-Diameter Opening – A Study in Part Release

Cylindrical containers such as paint cans have an inner rim and are sealed by means of a press-fit cover. Such containers can be packed in an especially space-saving arrangement. This study shows how such a container can be manufactured by means of injection molding and how it is released in the mold.

Mold

The mold (Figs. 1 and 2) consists of a cavity (5) and a central core (6) on which core slides (1) move along tapered surfaces. The core slides (1) can move additionally in a radial direction along a guide plate (7). When the mold is closed, the slides (3) between the core slides (1) seat against the central core (6) and together with the slides (1) form the cylindrical outer surface of the core.
The slides (3) can also move in a radial direction on guide plate (8). They are actuated by the cam pins (2) attached to the guide plate (7).
The rim of the container is formed by the stripper ring (4), which encloses the slides (1, 3) when the mold is closed.

Figure 2 shows the guides for the slides (1) and plate (7) at the top and the guides for the slides (3) and plate (8) at the bottom.
Not shown are the actuating mechanisms that control the part release sequence described in the following. The well-known actuators (hydraulic cylinders, latch arrangements, chains etc.) are suitable for this purpose.

Part Release/Ejection

1. The mold opens at parting line I; the core and molded part are withdrawn from the cavity.
2. Plate (7) and plate (8) separate from the central core (6) (parting line II). Because they move on guiding surfaces on the central core (6), the slides (1) move inward radially, while the slides (3) initially retain their original position. Tangential and radial spaces (Fig. 3) form between the slides (3).
3. Plate (7) comes to a stop and plate (8) continues to move (separation at parting line III). Now, the cam pins (2) pull the slides (3) inward; the molded part is held only by the stripper ring (4) (Fig. 4), which ejects it from the core (Fig. 5).

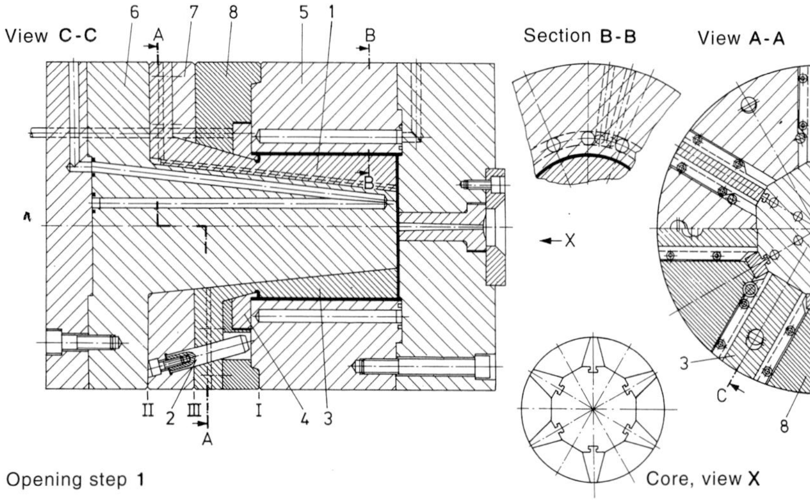

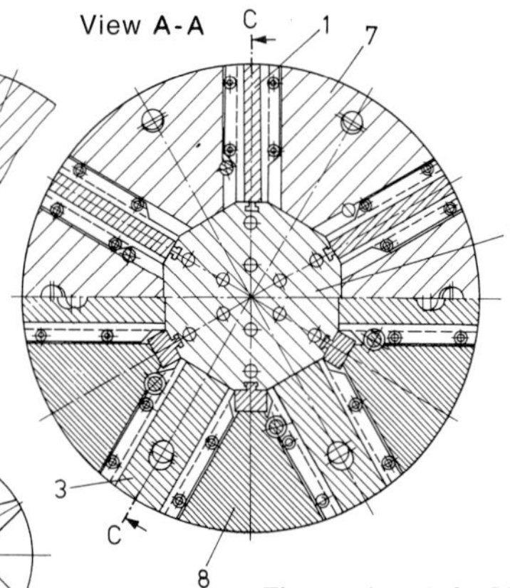

Figures 1 and 2 Mold for a cylindrical container with a reduced-diameter opening
1: core slide; 2: cam pin; 3: core slide; 4: stripper ring; 5: cavity; 6: central core; 7, 8: guide plate

Opening step 1

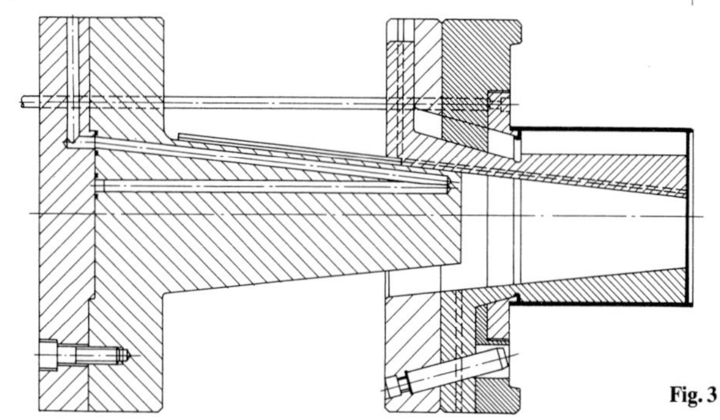

View X, Opening step 1

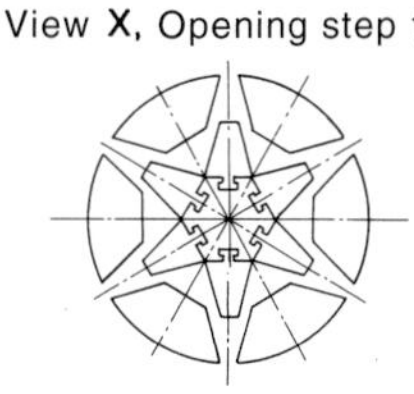

Fig. 3 Core slide (1) displaced inwards

Opening step 2

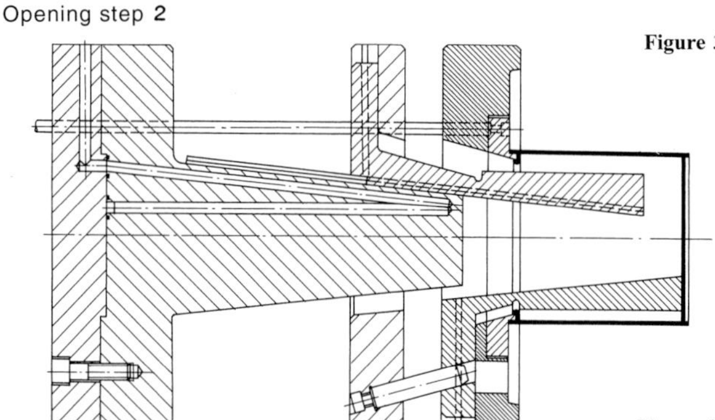

Figure 3 Core slide (1) displaced inward

View X, Opening step 2

Figure 4 Core slides (3) displaced inward

Opening step 3 and ejection

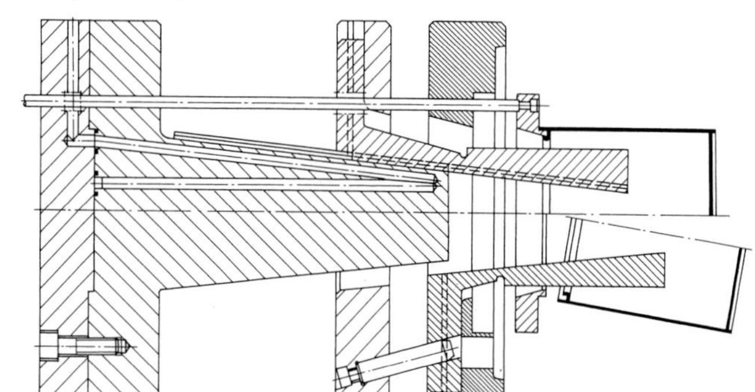

Figure 5 Ejection

Example 73, Single-Cavity Injection Mold for a Lighting Fixture Cover Made from Polymethylmethacrylate (PMMA)

The lighting fixture cover of PMMA with a diameter of 87 mm and a height of 76 mm (Fig. 1) is attached by means of three pockets into which hooks on the lighting fixture snap. Snapping the molded part off the core is not possible, therefore, three slides are necessary for part release..

Mold (Figs. 2 and 3)

A pneumatically actuated nozzle (37) is used to mold the part. Such nozzles are often used on single-cavity molds, because they leave a small, clean gate mark on the molded part and do not require any heaters or temperature controllers as does a heated sprue bushing, yet still permit fully automatic operation. They do, however, produce a small sprue with each shot.
Use of a granulator directly next to the injection molding machine and immediate reintroduction of the regrind into the hopper has proven successful. The operation of the nozzle (37) is described in Example 97.

Mold Construction

The mold is constructed of standard mold components. The cavity has been machined directly into the plate (2). The three cam pins (30) that actuate the slides (43) are also attached to plate (2).
The core is machined from plate (3) and is supported against the mold clamping plate (5) by plate (4), support ring (6) and support pillars (35). The core plate (3) contains three radial grooves which accommodate the slides (43) that form the pockets in the molded part.
When the mold is closed, the slides are held in position by the conical pins (31) as well as the taper pins (44).

Part Release/Ejection

As the mold opens, the following three sequences take place simultaneously:

1. The pushback pins (32) release plate (9), which can now displace the taper pins (44) with respect to the slides (43) by means of the compressed helical springs (40) such that a radial movement of the slides is now possible. Conical pins (31) release the slides (43) too.
2. With a slight delay resulting from the clearance “s”, the slides (43) are displaced inward by the cam pins (30) so that the pockets in the molded part are released.
3. The molded part is withdrawn from the cavity on the core.

Steps 1 and 2 are complete when plate (9) stops against plate (4).
Lastly, the molded part is stripped off the core by the ejector pins (33) which are actuated by the ejector plates (7, 8).
The ejectors (33) are hydraulically retracted prior to mold closing by means of the ejector plate (7, 8). In the final phase of closing, they also act as pushback pins. The taper pins (44) are pulled behind the slides (43) by the pushback pins (32) and plate (9) after the slides have been displaced outward by the cam pins (30).

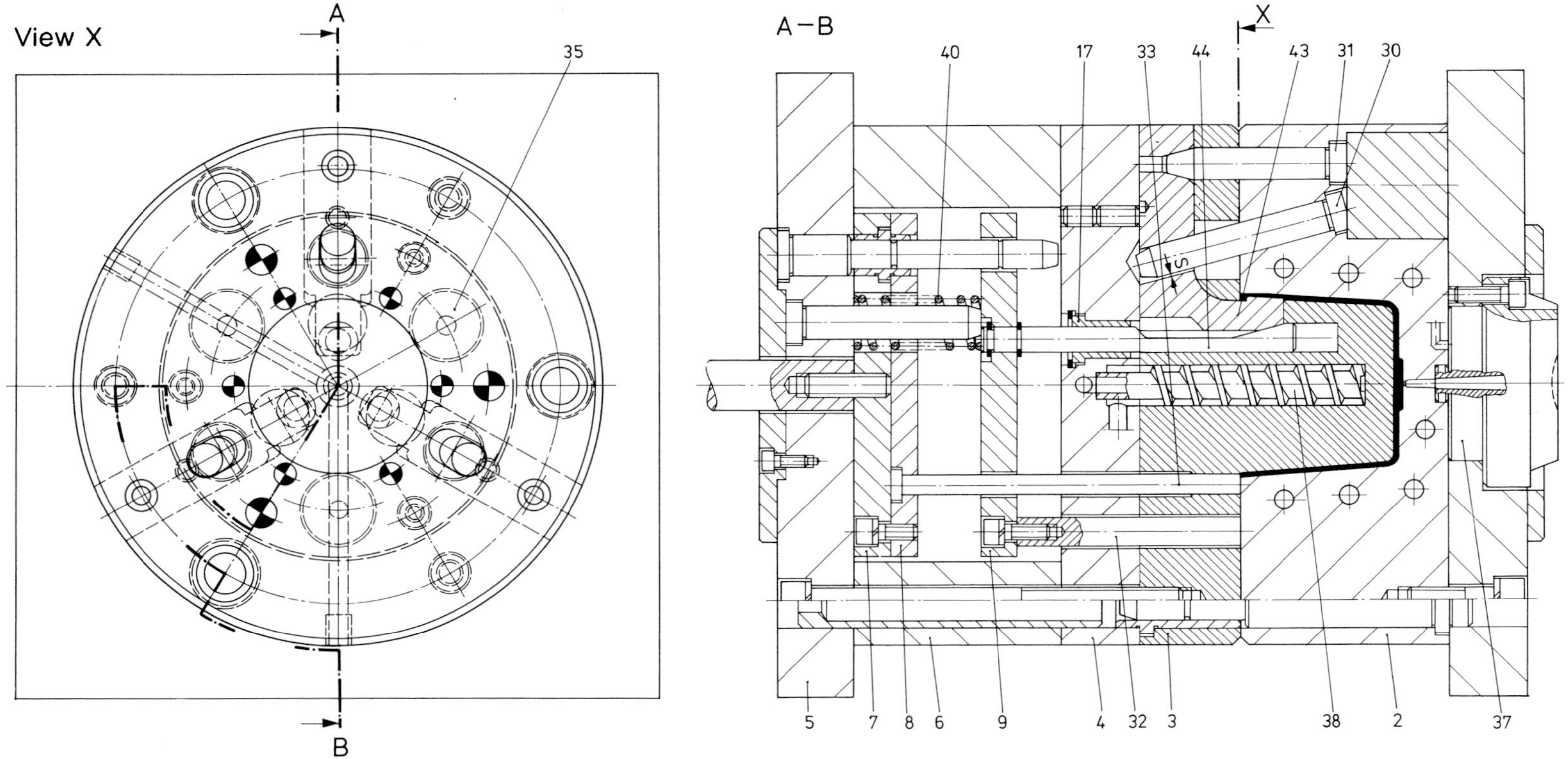

Figure 1 Lighting fixture cover of PMMA

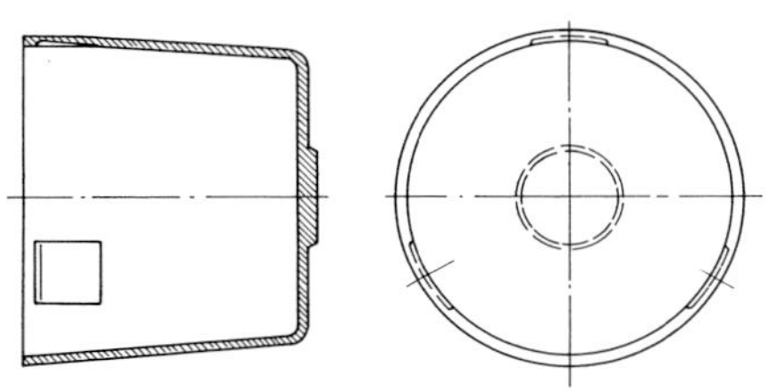

Figures 2 and 3 Single-cavity injection mold for a lighting fixture cover of PMMA
2: mold cavity plate; 3: core plate; 4: backing plate; 5: clamping plate; 6: support ring; 7, 8: ejector plates; 9: mold plate; 17: guide bushing; 30: cam pin; 31: conical pin; 32: pushback pin; 33: ejector pin; 35: support pillar; 37: pneumatically actuated nozzle; 38: helical cooling core; 40: compression spring; 43: slide; 44: taper pin

Example 74, Injection Mold for a Housing with a Thread Insert Made from Polycarbonate

The rectangular housing (Fig. 1) of polycarbonate has a neck with a thread at one end.

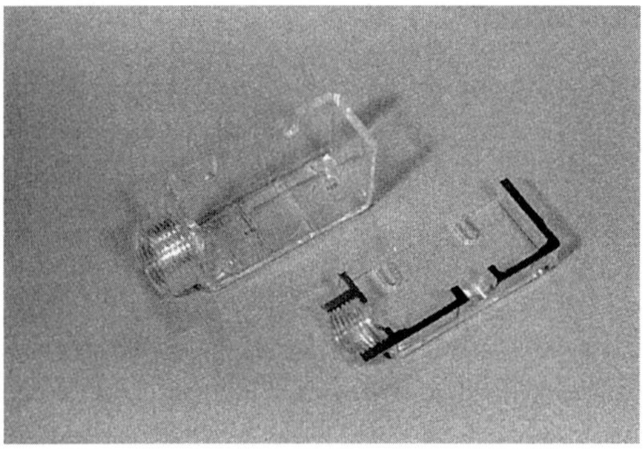

Figure 1 Housing with thread insert at one end

Unscrewing Mechanism

The unscrewing mechanism for the thread is actuated by the mold opening motion. Since the axis of thread is perpendicular to the direction of opening, a gear rack drive mechanism was selected.
The thread has a length of 10.7 mm with a pitch of 1.41 mm. An unscrewing stroke of 11.9 mm, requiring 8.5 revolutions of the threaded core, was selected.
The pitch diameter of the drive pinion on the threaded core must be at least 20 mm, so that a stroke of 20 mm $\times \pi \times 8.5 = 534$ mm would result with direct drive by means of the gear rack. The maximum possible opening stroke of the injection mold machine available, however, was only 200 mm. It is inadvisable, though, to use the entire 200 mm stroke available on a toggle machine, because the opening force of such machines is very low at the end of the opening stroke.
Accordingly, two pairs of gears with a transmission ratio to increase the speed were incorporated between the gear rack and threaded core (Fig. 2).

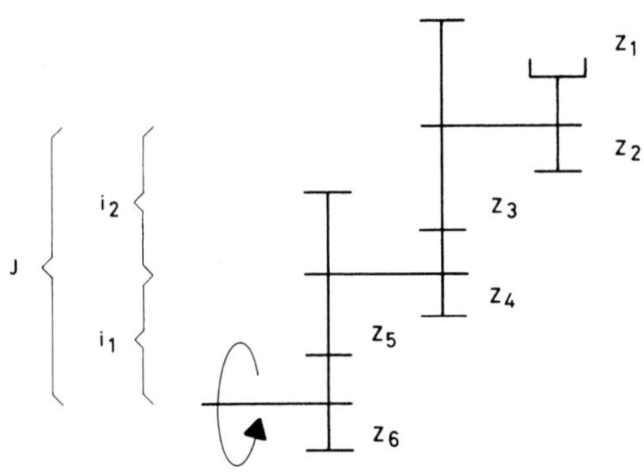

$z_1,\ z_2 = 20,\ z_3 = 44,\ z_4 = 18,\ z_5 = 34,\ z_6 = 20$

$i_1 = \frac{z_5}{z_6} = \frac{34}{20} = 1.7 : 1 \qquad i_2 = \frac{z_3}{z_4} = \frac{44}{18} = 2.44 : 1$

$J = i_1 \cdot i_2 = \frac{1.7 \cdot 2.44}{1} = 4.14 : 1$

Figure 2 Gearing arrangement

Runner System/Gating

The part is molded via a runner in the parting line which ends in a submarine gate on the end of the housing opposite the threads.

Molds Construction

The mold opens along the line a–b–c–d–e–f (Fig. 3). The gearing arrangement with the gears (6, 10, 11, 12, 13) and the threaded core (5) with the threaded bushing (8) are thus located in the moving mold half, while the gear rack (14) is attached to the stationary mold half.
With the overall transmission ratio selected (Fig. 2), the unscrewing stroke for the core is achieved with an opening stroke of 128 mm. An additional stroke of 3 mm is used for mold protection in the machine.
The stripper plate (16) is fitted around three sides of the mold core (15). It is actuated by means of rods (23) by the ejector plate (21), which is coupled to the machine hydraulic ejector by means of ejector rod (20). Pillars (19) support the core retainer plate against the mold clamping plate.

Mold Temperature Control

The mold temperature is supposed to be between 75 and 130°C (167 to 266°F). The cavity insert (17) is provided with cooling channels for the mold temperature control fluid. Bubblers with baffles (18) to direct the coolant are provided in the mold core (15).

Part Release/Ejection

Upon mold opening, the molded part (1) and mold core (15) are withdrawn from the cavity. During this motion, the threaded core is unscrewed from the molded part by means of the gear rack and gearing arrangement.
Simultaneously, the submarine gate shears off the molded part. Initially, the runner is retained on the cavity half of the mold, because the sprue puller (22) has been designed with a certain amount of axial play. This feature facilitates separation of the runner from the molded part. The runner is pulled out of the submarine gate only after the mold strokes this mount. Finally, the stripper plate (16) strips the molded part off the core and ejects the runner from the sprue puller. Prior to closing, the stripper plate is returned to its original position.

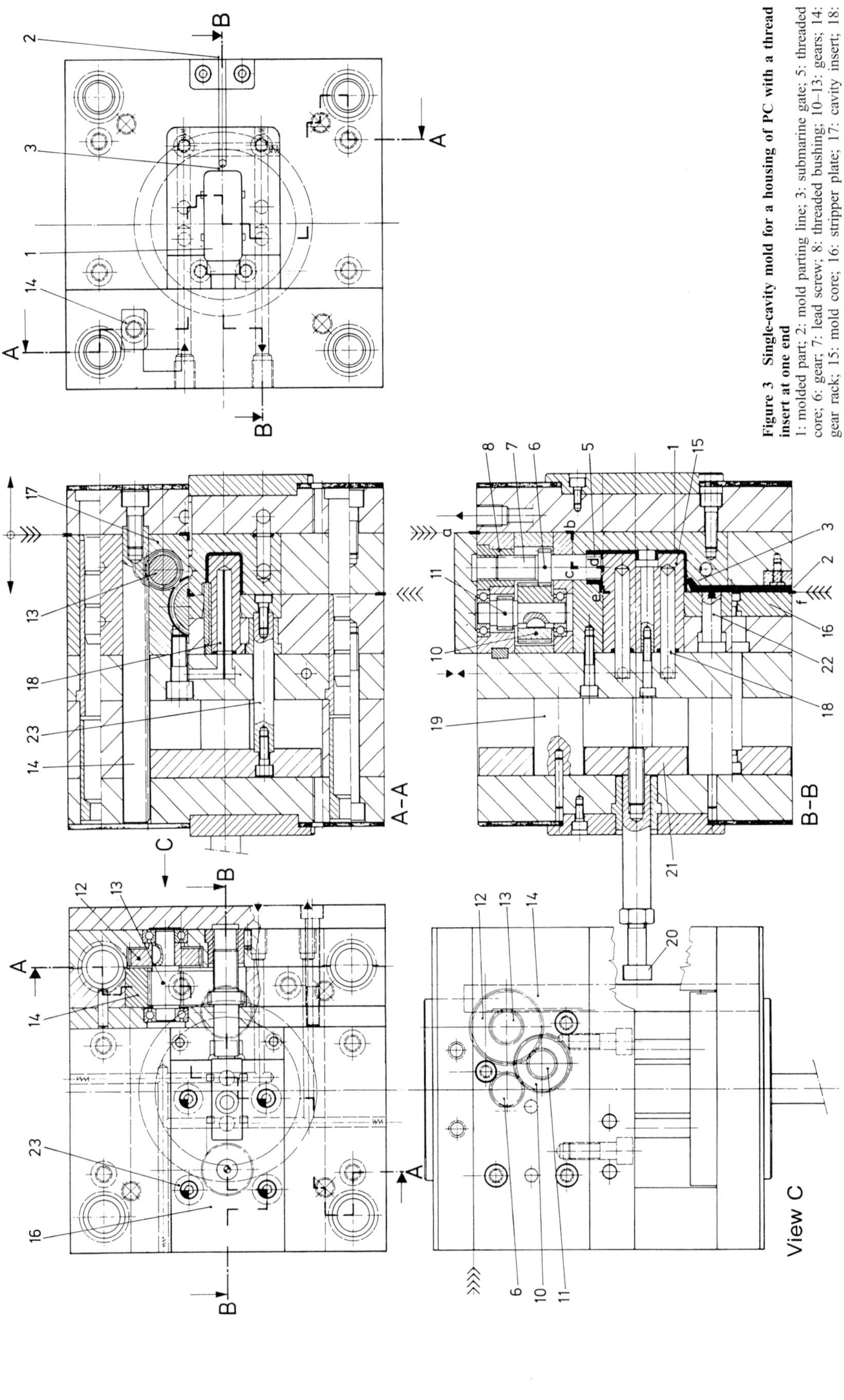

Figure 3 Single-cavity mold for a housing of PC with a thread insert at one end
1: molded part; 2: mold parting line; 3: submarine gate; 5: threaded core; 6: gear; 7: lead screw; 8: threaded bushing; 10–13: gears; 14: gear rack; 15: mold core; 16: stripper plate; 17: cavity insert; 18: baffle; 19: support pillars; 20: ejector rods; 21: ejector plate; 22: sprue puller; 23: ejector rod

Example 75, Mold for Long, Thin, Tubular Parts Made from Polystyrene

A large quantity of test-tube-like specimen tubes was to be molded in polystyrene. The tube was to have a flange at its larger end and a conical tip with axial opening at the smaller end. A four-cavity mold appeared both technically and economically reasonable. The length of the tubes, however, posed a problem. For the length of 170 mm, a mold daylight of 270 mm would have been required with a conventional mold design. Since only 40 cm^3 of melt would have been needed to simultaneously mold the four tubes, a small injection molding machine would have been adequate. As a rule, however, small machines do not have such a big mold opening. The clamping force required was slight, since the projected area amounted to only a few square centimeters, so that only a small machine would have been required from this standpoint as well. Because the outside surface of the tubes was not permitted to show any witness lines, it was not possible to place the cavities in the plane of the parting line.

Accordingly, a mold (Figs. 1 and 2) was designed that projected through an opening in the movable platen where the ejector mechanism normally would be mounted. A further slight modification of the machine was also required: the stripper plate (20) located in the stationary mold half leaves the normal guide pins (21) during the last portion of its stroke and thus needs an auxiliary means of guidance. Accordingly, two holes were drilled in the movable platen, through which two extended auxiliary guide pins (14) project. The stripper plate (20) runs in ball bearings (16) on these pins. To provide optimum filling, each cavity was provided with two submarine gates.

Bolts (18) and (19) are located between the cavity and auxiliary guide pin (14). Bolt (19) is mounted in the stationary-side clamping plate. Bolt (18) serves as the stripper bolt to actuate the stripper plate (20) located in the stationary mold half. This plate must not be actuated before the movable mold half has released the entire length of the molded parts. Accordingly, it is held in position by pins (22) that project into the runner system and become embedded in the solidifying melt. As soon as the movable mold half has released the molded parts, stripper bolt (18) actuates stripper sleeve (13), and the molded parts are stripped off the cores (6, 10) until the shoulder in sleeve (17) seats against the bolt (19) and stops further motion of stripper plate (20). The stripper plate is now supported on the auxiliary guide pins (14) and is returned to the molding position as the mold closes. This design proved successful both technically and economically.

Fig. 1

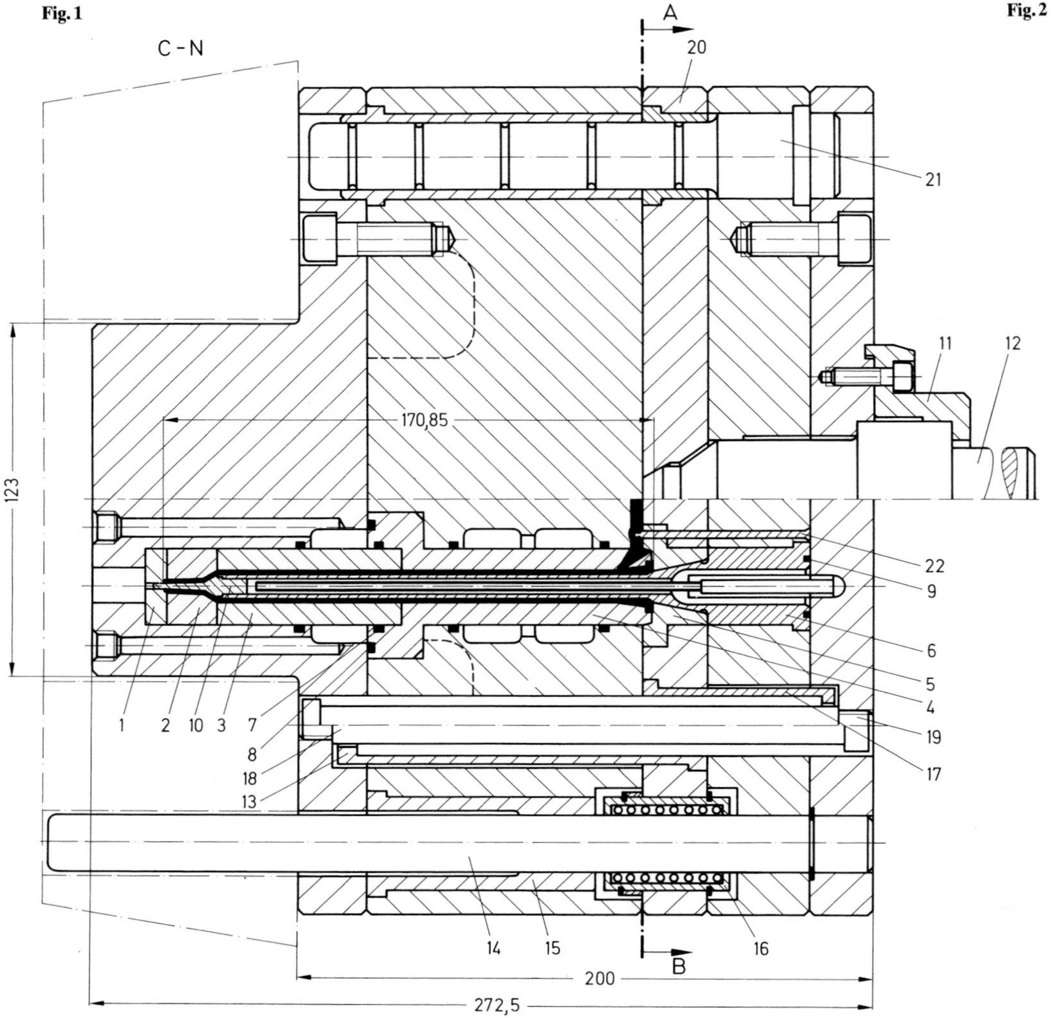

Fig. 2

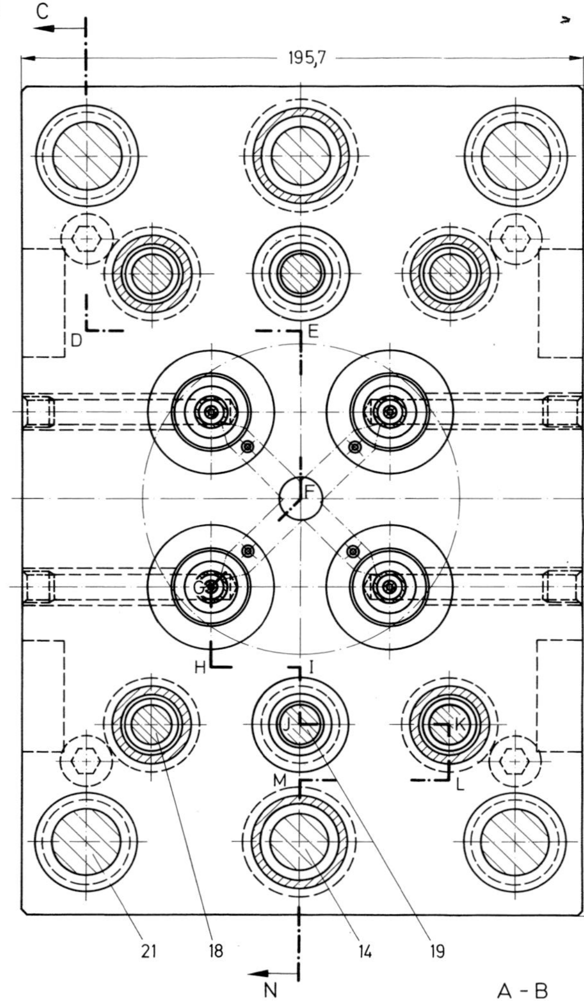

Figures 1 and 2 Four-cavity injection mold for test-tube-like polystyrene specimen tubes
1: core tip support; 2: cavity insert for tip; 3: cavity insert; 4: cavity insert for flange; 5: stripper ring; 6: core insert; 7, 8, 9: O-ring; 10: core insert tip; 11: locating ring; 12: hot sprue bushing; 13: stripper sleeve; 14: auxiliary guide pins; 15: guide bushings; 16: ball bearing; 17: stop sleeve; 18: stripper bolt; 19: stop bolt; 20: stripper plate; 21: guide pin; 22: sucker pin

Example 76, Insulated Runner Mold for Three Specimen Dishes Made from Polystyrene

In spite of the advanced state of hot runner system technology, there are still applications where insulated runner systems can be employed successfully. This is the case especially with easy-flowing resins for fast-cycling thin-walled parts, and for parts characterized by frequent material or color changes. The specimen dishes (dimensions: 40 mm × 40 mm × 26 mm; Fig. 1) are produced in a 3-cavity mold (Fig. 2). The star-shaped runner system is 20 mm in diameter and is thick enough that it does not solidify at cycle times of about 10 s. Cartridge heaters (14) in mold plate (1) provide "supplemental heating" in conjunction with a thermocouple and an insulating plate (6). The insulated-runner nozzles (7) are surrounded by air gaps on all sides to reduce heat transfer. The conical gate vestige is located in a dimple on the bottom of the dish.

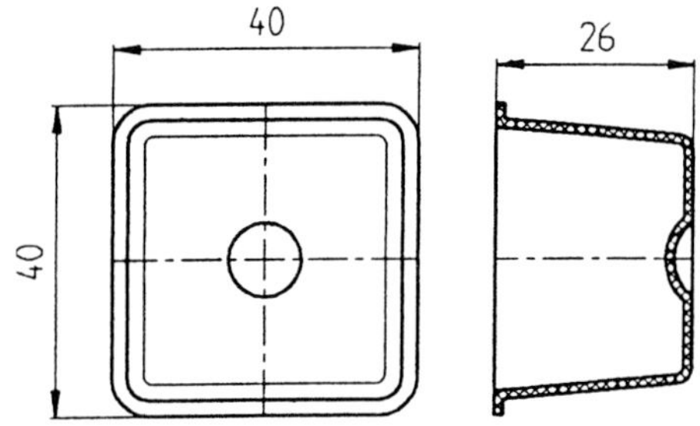

Figure 1 Specimen dish of polystyrene

When starting up the mold anew or when making a color change, the solidified runner must be removed. This is accomplished by loosening the bolt (13), reversing the strap (8), and opening the mold between plates (1) and (2).

Cooling

Mold cooling is provided by peripheral cooling grooves and channels in the cavity inserts (9) and by helical cooling channels (10) in the mold cores (11).

Part Release/Ejection

The parts are stripped off the cores by means of ejector pins located at the corners.

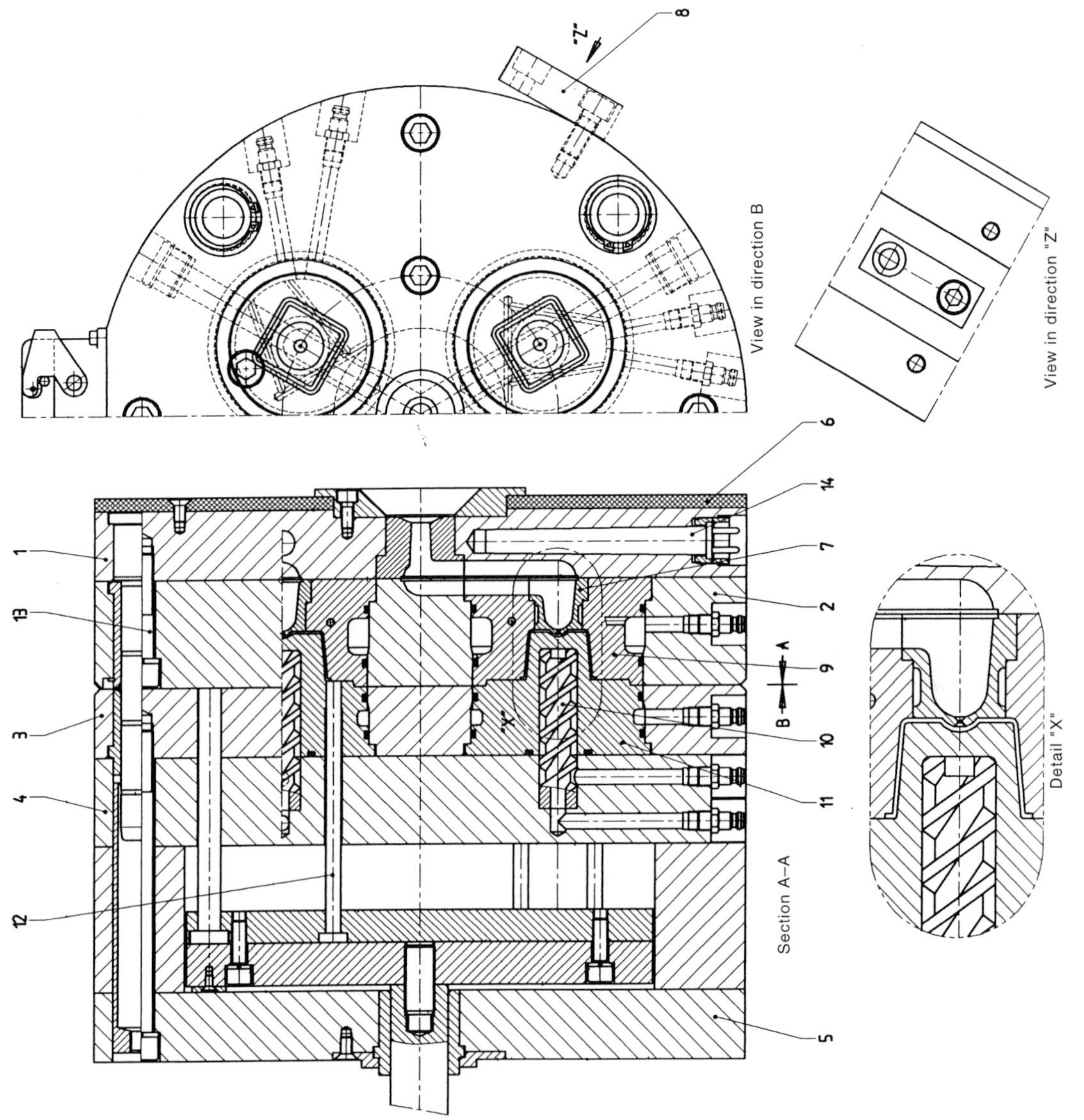

Figure 2 3-Cavity insulated-runner mold for polystyrene specimen dishes
1: clamping plate; 2: cavity plate; 3: core plate; 4: backup plate; 5: clamping plate; 6: insulating plate; 7: insulated-runner nozzle; 8: strap; 9: cavity insert; 10: helical cooling channel; 11: mold core; 12: ejector pin; 13: bolt; 14: cartridge heater
(Courtesy: Hasco)

Example 77, Single-Cavity Injection Mold for a Polypropylene Emergency Button

This emergency button (Fig. 1) is a hemispherical shell with a diameter of 100 mm. The attainable injection-molding cycle time depends on whether the cooling channels in the injection mold extend as far as the part-forming surfaces.

Figure 1 Emergency button

Figure 2 shows the original design of the injection mold which featured conventional cooling. The cycle time of 33 seconds is determined by the high mold temperature around the feed nozzle (75).
A significant increase in economics was achieved by redesigning the cooling, as shown in Fig. 3. The two mold cores (22, 61) and the mold cavity (60) are arranged according to the Contura® system so that the cooling channels closely follow the contours and the part-forming surfaces right up close to the gate. This ensures that heat dissipation is much more uniform and faster than before. The mold inserts each consist of two parts, whose mating surfaces contain the cooling channels. Full-face joining is effected with the aid of a special hard soldering or diffusion welding process, which makes homogeneous parts out of both. For this reason, these parts are each shown shaded in Fig. 3.
This cooling system allowed the mold temperature around the gate to be reduced from 90°C to 50°C (194°F to 122°F) and cut the cycle time by more than 20%.

Runner System/Gating

The molded part is gated via a pneumatically actuated nozzle with needle shutoff (75). The needle makes a slight indentation directly into the surface of the part (see Fig. 1) to ensure that there are no projecting sprue remnants that might cause injury.

Part Release/Ejection

The ejector plate (4) with the stripper ring (62) and stripper sleeve (62) are connected to ejector plates (8, 9). The snug fit between ejector ring (62) and core (61) is conical so as to avoid wear and flash formation on the molded part. When the ejector bolt (66) is actuated, the molded part is ejected via the ejector sleeve (21).

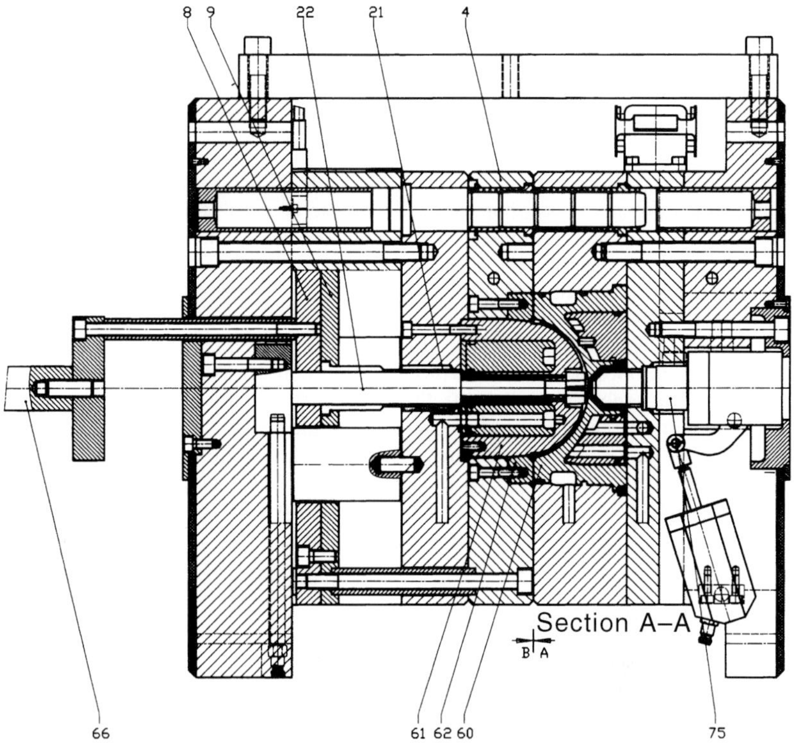

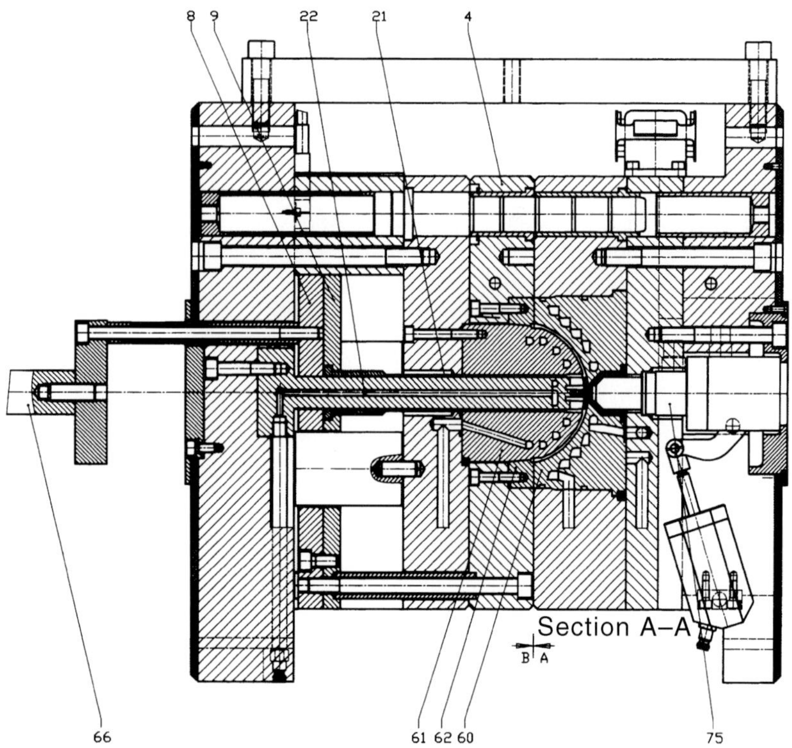

Figures 2 and 3 Single-cavity injection mold for polypropylene emergency button
4: stripper plate; 8, 9: ejector plates; 21: stripper sleeve; 22: core; 60: mold cavity; 61: core; 62: stripper ring; 66: ejector bolt; 75: heated sprue nozzle with needle shutoff
(Courtesy: Contura MTC GmbH, Menden, Germany)

Example 78, Eight-Cavity Injection Mold for Battery Caps with Undivided External Thread and Sealing Cone Made from Polypropylene

The external thread and the sealing cone of battery caps are usually molded in split-cavity tools, leaving a fine parting line seam.

If it has been specified that no parting line seam at all be visible, thread and sealing cone will have to be demolded by unscrewing. Figures 1 to 5 show such a tool designed as an eight-cavity mold. Its construction and operation are described below:

To avoid an externally visible gate, an easily severed pinpoint gate is situated inside the cap. The construction and operation of such molds have been described repeatedly. The external thread and the sealing cone of the caps are produced in the threaded sleeves (*a*). These are carried in bronze guide bushings (*b*), which have the same screw pitch as the battery caps. The threaded sleeves are provided with gear teeth on the other side, where they mesh with the central pinion (*c*). This pinion is driven by a central spindle (*d*), with a screw thread of low pitch, by the opening and closing movement of the mold. The mold drives the sleeves (*a*) counter-clockwise when opening and clockwise when closing. The caps grip the cores (*e*) sufficiently tightly through shrinkage during the unscrewing process, so that they remain stationary. The shape of the molded part also ensures that it does not turn.

As soon as the threads have been released, the ejector pins (*f*) push against the clamping cover (*g*) and the caps are ejected by the ejector pins (*h*).

Latches serve to open the sprue parting line and to eject the sprue.

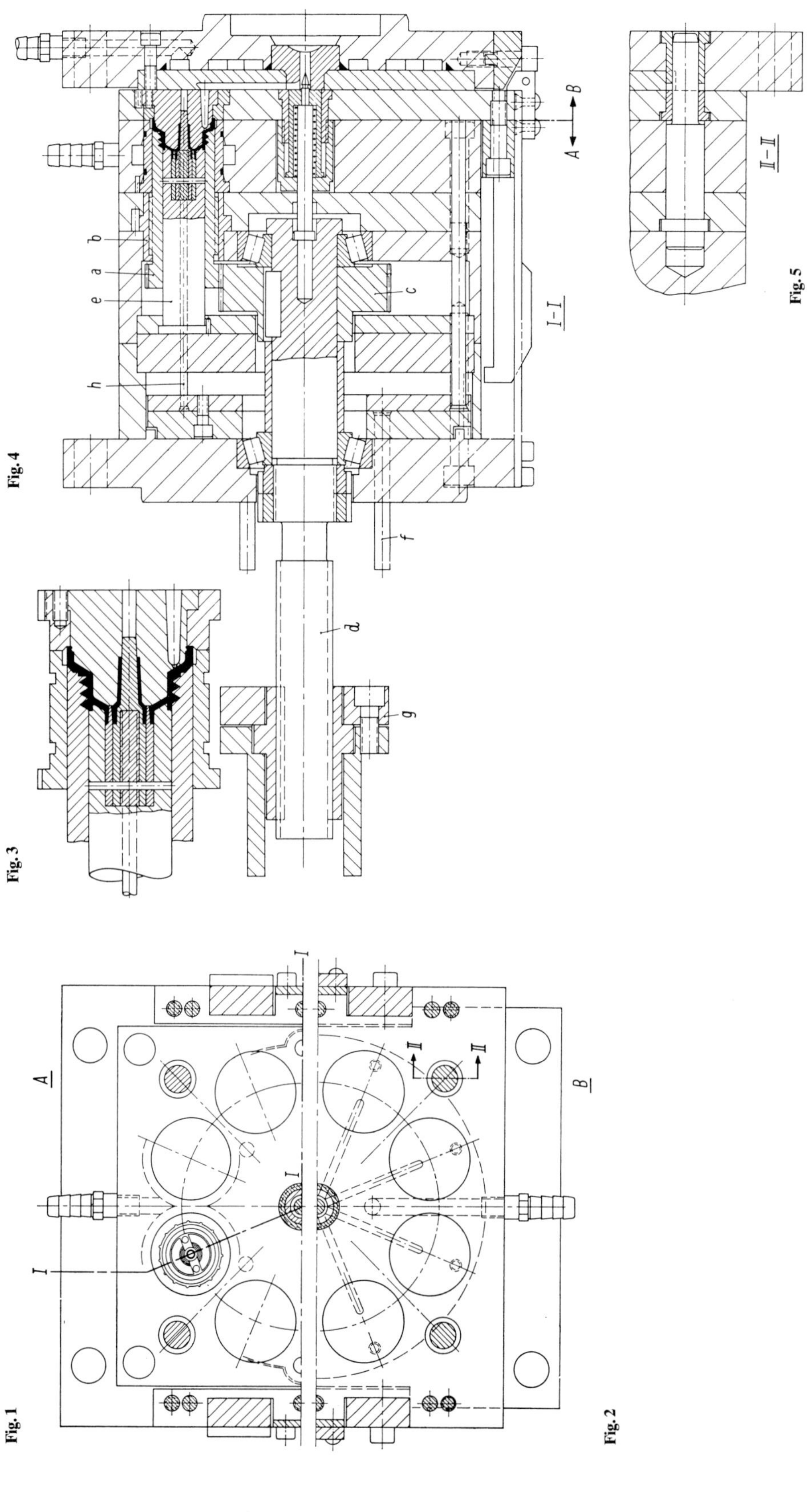

Figures 1 to 5 Injection mold for battery caps
a: thread sleeves; *b*: guide bushings; *c*: pinion; *d*: threaded spindle; *e*: cores; *f*, *h*: ejector pins; *g*: clamping cover

Example 79, Injection Mold for a Curved Pouring Spout Made from Polypropylene

Complicated part release processes result from curved, hollow molded parts, such as a pouring spout, for example. This particular one can be fitted with a retaining nut to be screwed onto the threaded mouth of a bottle for drinks. The production of this molded part calls for a curved core, requiring a stripper plate describing an arc that is matched to the contour of the molded parts.

The mold illustrated in Figs. 1 to 4 serves to describe the production of this type of pouring spout. For part release the tool opens in the parting-line plane. To start with, the molded part remains in the moving mold cavity of the mold plate (3), with the core carrier (9) being moved along the guide pin (24) by springs (26) until stopped by the discs (27) following the opening movement. On further opening, part release takes place from this mold cavity half as well and the freed molded part rests on core (13) between the mold halves. When the hinged latch (32) housed in the retainer (31) reaches the ejector rod (20), the latter through its downward movement rotates the stripper plate (10) against the force of spring (23) down around the fulcrum pin (11). The molded part is pulled off the core. Due to the conical shape of the core an ejector stroke of approximately 5 mm suffices to drop the molded part off the core.

With the mold closing movement the latch (32) moves into the retainer (31) via ejector rod (20) without shifting the ejector rod. When the injection unit moves back, it pulls the sprue out of the mold. The sprue must be removed from the machine nozzle after each shot in the version described here.

Fig. 1

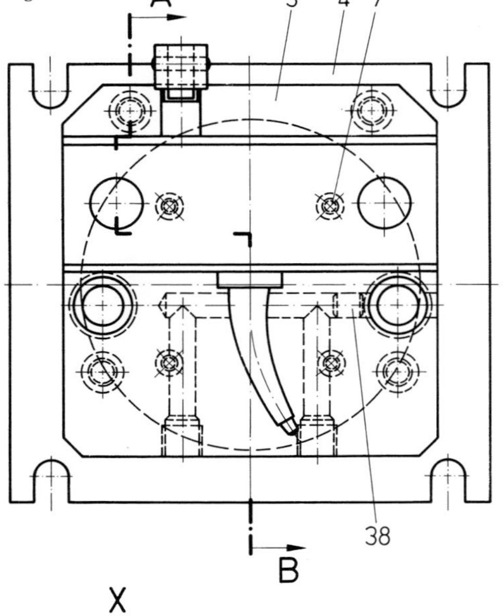

Fig. 2

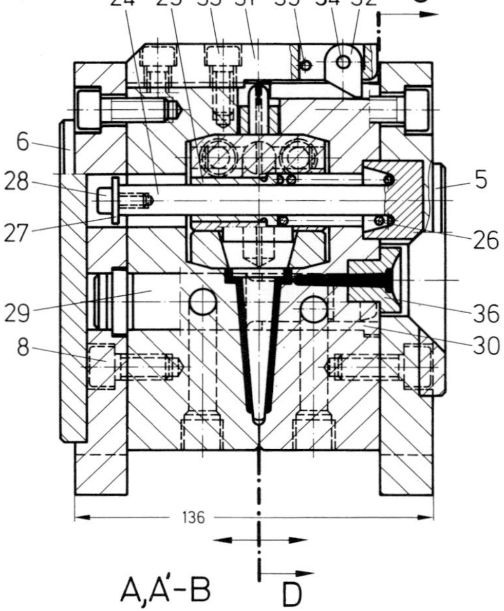

Fig. 3

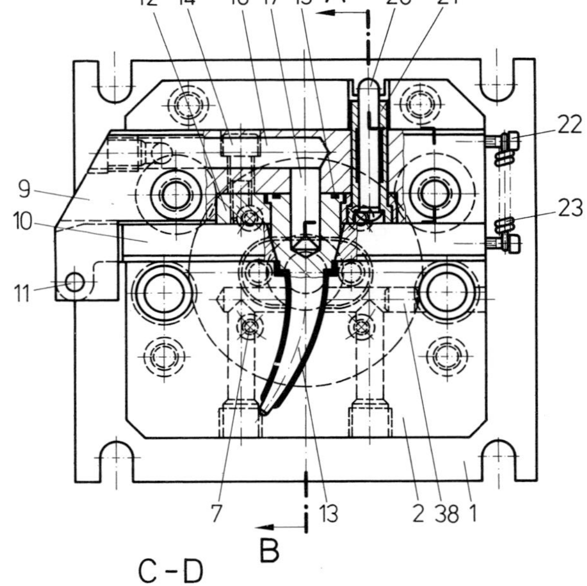

Fig. 4

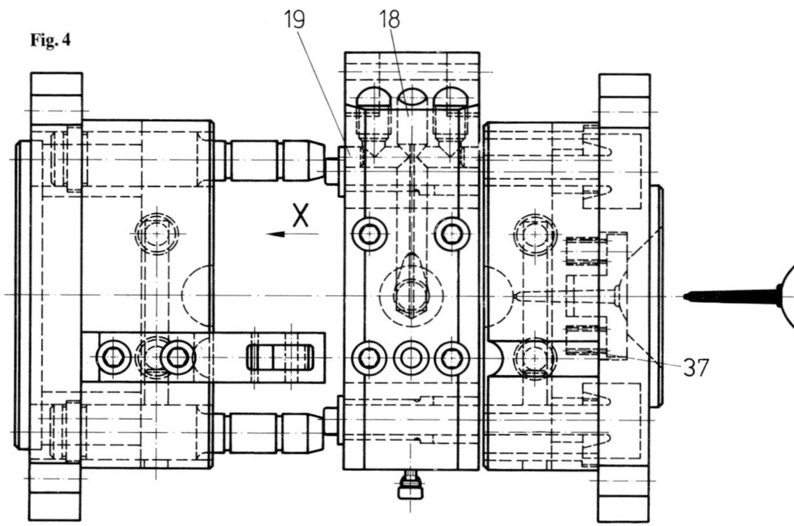

Figures 1 to 4 Injection mold for a curved pouring spout
1: fixed mold base plate; 2: mold plate, nozzle side, with mold cavity half; 3: mold plate, ejector side, with mold cavity half; 4: moving mold base plate; 5, 6: locating ring; 7, 8: Allen bolts; 9: core carrier; 10: stripper plate; 11: cylindrical pin; 12: retaining plate; 13: core; 14: Allen bolt; 15: O-ring; 16, 17: shim; 18, 19: plug; 20: ejector rod; 21: guide bushing; 22: Allen bolt; 23: tension spring; 24: guide pin; 25: guide bushing; 26: compression spring; 27: disc; 28: Allen bolt; 29: guide pin; 30: guide bushing; 31: retainer; 32: latch; 33: Allen bolt; 34, 35: cylindrical pin; 36: sprue bushing; 37: compression spring; 38: plug

Example 80, Injection Mold for an ABS Goggle Frame

The shape of a frame for protective goggles (Fig. 1) poses part-release problems for the mold designer.

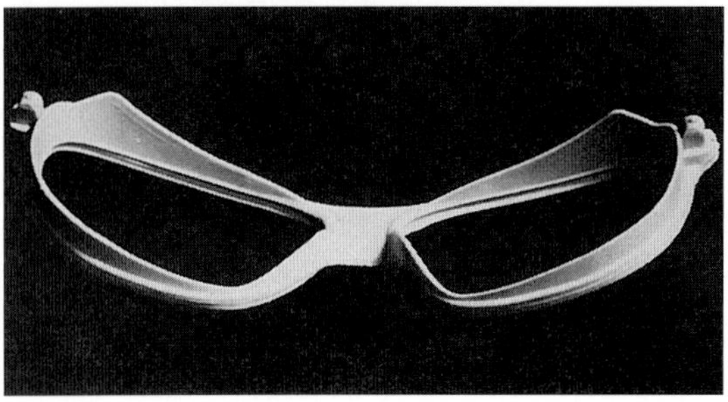

Figure 1 Goggle frame for protective goggles made of ABS

Undercuts on injection molded parts are usually formed in the mold by means of split cavities or slides. The production of the undercuts needed for subsequently fitting the lenses involves considerable design problems, especially since these slides would not lie at right angles to the direction of draw. The present design of an injection mold for making a spectacle frame (Figs. 2 and 3) shows how, by using collapsible cores, the problem can be simplified. These cores can be obtained as standard mold components.

The cavity for the frame is formed by plates 5 and 6 (Fig. 2). The lens surrounds (undercuts) are formed by the core segment sleeves (11) of the collapsible cores (9) (Fig. 4). When the mold opens (Fig. 3) the moving mold half (2) is separated from the fixed mold half (1) by the clamping plate (4), the frame being released from the cavity plate (5). During this operation, the pneumatic ram (15) of the cylinder (14) is subjected to compressed air via a control valve and the air inlet (L) of the cylinder, so that the parting line between the guide plate (13) and support plate (8) remains closed. After a certain opening stroke the air control valve shifts and the compressed air is passed via an air inlet (R) to the second piston surface of the double-acting pneumatic cylinder. This causes the plate (7) and the collapsible core elements (9) in it to move in the opposite direction together with the support plate (8). Hence the center cores (10) of the collapsible cores (9), which move outward, are held fast by the sliding blocks (16) of the guide plate (13) and so they slide out of the core segment sleeves (11).

Before the opening stroke is limited by the pneumatic piston, the stop bolts (18) in the guide plate (13) carry along the outer sleeve (12) of the collapsible core elements via the fastening flanges (17) after a certain distance has been covered. This causes the cores of the segment sleeves (11) to drop toward the inside. The undercuts (lens surrounds) of the goggle frame are thus released (Fig. 5).

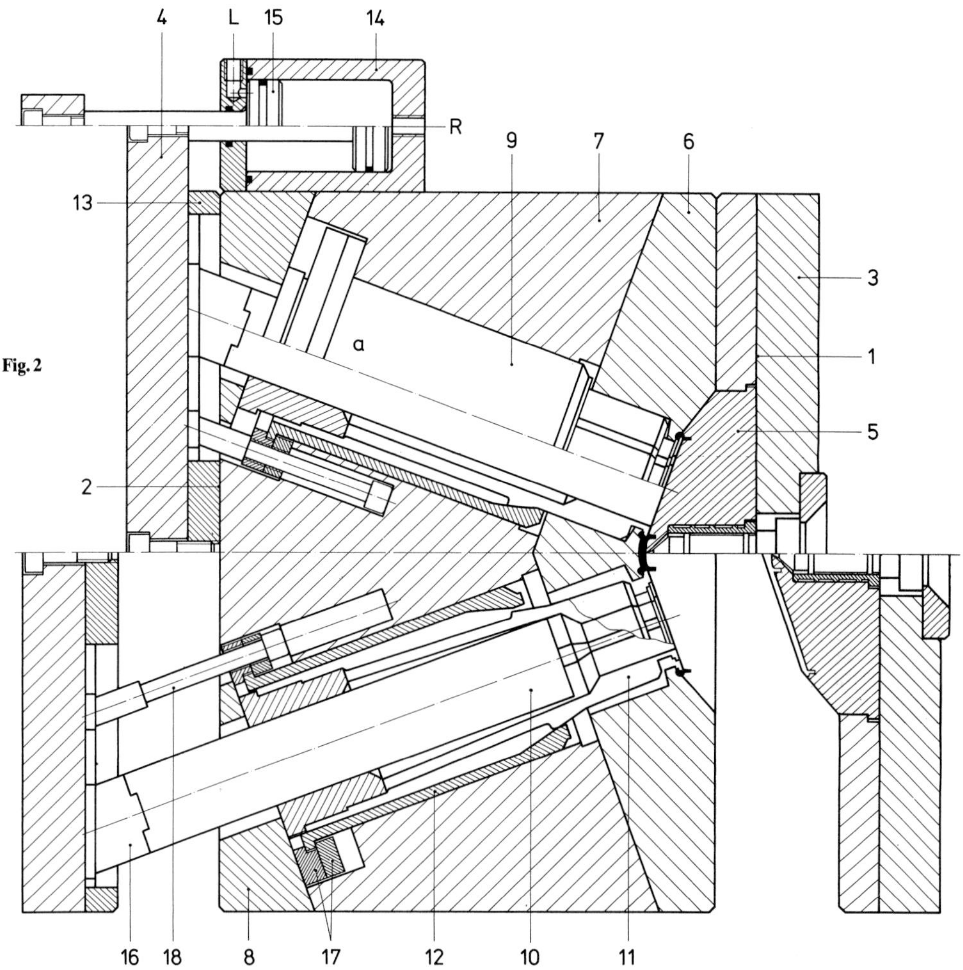

Fig. 2

Fig. 3

Figures 2 and 3 Injection mold for goggle frame
1: fixed mold half; 2: moving mold half; 3: fixed clamping plate; 5: cavity plate; 6: cavity plate; 7: plate; 8: support plate; 9: collapsible core element; 10: center core; 11: segment sleeve; 12: outer sleeve; 13: guide plate; 14: pneumatic cylinder; 15: double-acting pneumatic piston; 16: sliding block; 17: fastening flange; 18: stop bolt
View top: closed mold
View bottom: mold opened with collapsed core, part release is possible
a: collapsible core, L; R: air inlet holes

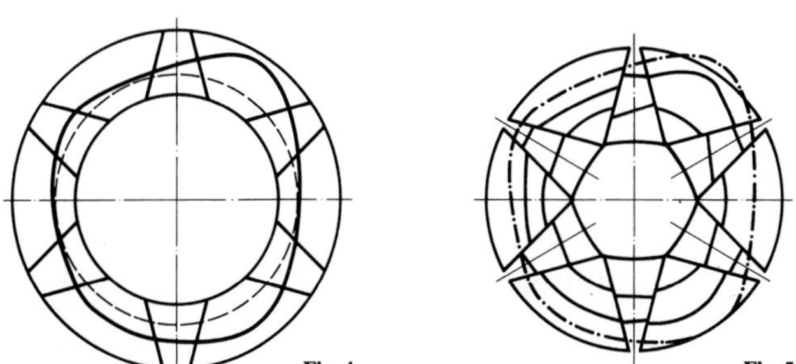

Fig. 4 Fig. 5

Figures 4 and 5 Contour of goggles near the lens surrounds on the core disc
Fig. 4 Contour on the core during molding
Fig. 5 Contour on the core during part release

Example 81, 4-Cavity Hot Runner Mold for an ABS-PC Front Ring

This axisymmetric molded part developed for use in sanitary facilities is mass produced by injection molding from ABS-PC polymer blend and electroplating (Fig. 1). The front ring has to satisfy high quality requirements. Outer diameter of the part is 30mm, its height is 7.5mm and its weight 1.4g. The collar-shaped inside of the part is tapered. The wall thickness of the "collar" thereby reduces from 0.8 to approx. 0.7mm. Four catches on the inner circumference serve to tighten the ring. Additional mold equipment is required to demold this segmented profile.

Mold

The chosen mold design (Fig. 2) has a hot-runner system and three parting lines, an unscrewing drive and cavity plate FS (1) connected with the ejector assembly (2) and serving as a stripper plate. The mold frame size is 296 × 296mm and is thus available as a standard modular construction. The hot side of the mold consists of the asymmetrically arranged clamping plate (3) made from 1.2312 tool steel, both intermediate plates (4) and (5), and cavity plate FS (6). The cavity plate material is 1.2764 high-alloy case-hardened steel; the remaining construction consists of 1.1730 non-alloy tool steel.

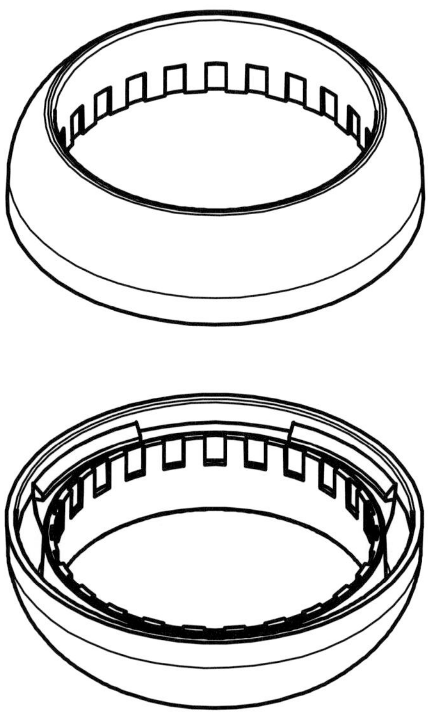

Figure 1 Front ring made from ABS-PC blend, diagram

The cavities are formed on the fixed side by round, cooled cavity inserts (7) and on the moveable side by cavity inserts (8) with the same outer dimensions. The four cavities are symmetrically arranged. The distance between cavities and mold center is determined by the semicircles of the intermeshing spur gear (10) and (34). The laterally flanged drive mechanism consists of a hydraulic cylinder (11) and toothed rack (12) for the mold core (35) as well as for the core bush (13) and the radially moveable core sleeves (14). The guide pillars are located on the moveable side in the platens (15) and (16). Guide pillar (17) length is only 155mm, but sufficient for the relatively short opening paths of parting levels I, II and III. The mold begins to open at parting level II. To ensure congruency at main parting level II, four tapered centering pieces (18) are incorporated.

Gating

The gating system with solidifying sub runner consists in part of a short sprue bar and tunnel gates on the internal diameter of the collar. This hot runner system consists of an H-shaped hot-runner manifold (20) and open sprue nozzles with tips (19). The head surfaces of the four externally heated sprue nozzles are non-positively joined to the hot-runner manifold by a sliding seal face. The hot-runner system is naturally balanced. The hot-runner manifold is centered by a dowel pin (21) and secured against rotation by a second pin. Adjustment of important mounting and functional dimensions ensures no-clearance installation and tension-free dimension compensation for thermal expansion, thus fulfilling manufacturer specifications. The support disks (22) made from titanium alloy reduce heat loss due to thermal conduction and support the weight of the hot-runner manifold. Aluminum reflector sheets (23) reduce radiation heat loss to the temperature-controlled mold. The indirectly heated (intermediate) sprue bushing (24) is equipped with a filter insert (25).

Installation was performed according to DIN 16765 version B. The electrical connections for power supply and thermal sensors are connected in the connection housing (26) according to VDE 0100.

Cooling

Close-contour water cooling is used in the cavity inserts (7) and (8) via cooling channels sealed by O-rings (27). The coolant supply and return circuits use deep holes bored in the cavity plates. The connections, easily accessible on one side, can

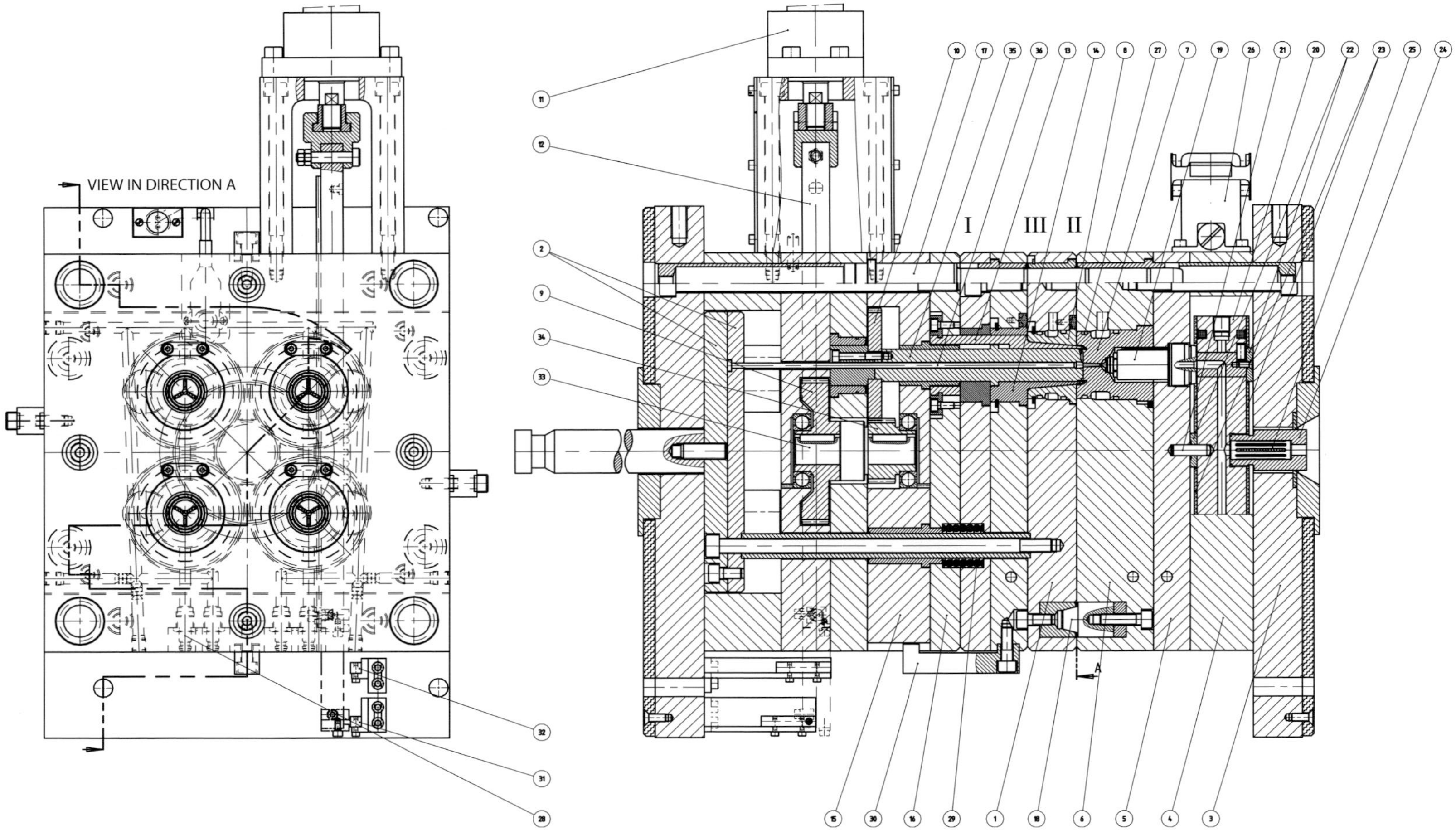

Figure 2 4-cavity hot-runner mold for front ring
1: cavity plate BS, 2: ejector assembly, 3: clamping plate FS, 4, 5: intermediate plate, 6: cavity plate FS, 7: cavity insert FS, 8: cavity insert BS, 9: central spur gear, 10: spur gear on mold core, 11: hydraulic cylinder, 12: gear rack, 13: core bushing, 14: core sleeve, 15, 16: platen, 17: guide pillar, 18: centering unit, 19: open sprue nozzle with tip, 20: hot-runner manifold, 21: dowel pin, 22: support disks, 23: reflector sheet, 24: sprue bush, 25: filter insert, 26: connection housing, 27: O-ring, 28: nipple, 29: pressure spring, 30: stop hook, 31: limit switch, 32: stops, 33: shaft, 34: spur gear, 35: mold core, 36: ejector
(Courtesy: Hasco, Lüdenscheid; Möller, Bad Ems)

be reached by quick couplings via the nipple (28). Diversions and plugging are executed according to a cooling diagram with standardized cooling system.

Demolding

The demolding sequence begins with the opening of parting level I. The entire plate assembly is held on the fixed side by the spring resistance of the pretensioned pressure springs (29). The traversing path of parting level I is limited by four lateral stop hooks (30). While the core bushes (13) remain on the moveable side, thus partially demolding the inner diameter of the article in segments, the cavity sleeves (14) with the undercuts still rest against the molded part. To demold the undercuts, the segments now have to be turned 45° in the space left open. This motion is performed by the machine-controlled hydraulic cylinder (11). Cylinder stroke is set with a limit switch (31) and secured by mechanical stops (32). The gear rack (12) engages with the central spur gear (9) which shares a common shaft (33) with a second spur gear (34). The mold cores (35) moved by a third spur gear (10) are then in extended position. The subrunner and gate retainer left in the mold core head (35) remain connected to the article. Molded part and sprue are not demolded until parting level II opens. When level III opens, the molded parts are stripped off by the cavity plate (1). Thereby, the gates are sheared off and the sprue is ejected by the ejectors (36).

Example 82, Two-Cavity Two-Component Injection Mold for a PC/ABS Bezel with a PMMA Window

The bezel (Fig. 1) is part of an audio storage system. The appearance surface has dimensions of 108 mm × 15.2 mm. On three sides it has a 6 mm high rim and a mean wall thickness of 1.3 mm. Inside there is a series of undercuts (*h*) along both long sides. A rectangular opening 20 mm × 10 mm in size is located at a distance of 3 mm from one end of the bezel, followed by a PMMA window 1.5 mm × 6 mm in size. This window, weighing 0.02 g, was originally attached to the bezel with the aid of an assembly fixture. Automatic assembly was associated with numerous difficulties, so that incorporation of the window as a second component during the molding process appeared desirable.

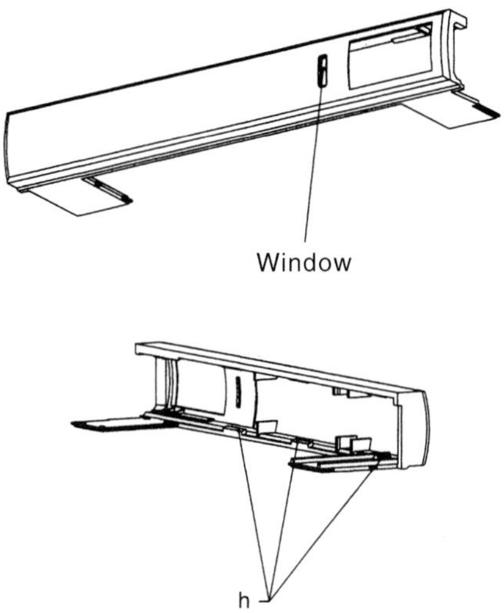

Figure 1 Bezel with window
h: undercut

As shown in Fig. 4, each bezel is tunnel gated at one corner. These gates are connected to a naturally balanced T-shaped runner system (*a*) fed by the centrally positioned sprue. The tunnel gates are located in the stationary-side mold insert (3).

Since the windows are not supposed to have any visible gate marks, a special gating approach was required. The melt for the windows is injected at the mold parting line into another naturally balanced T-shaped runner (*b*) by a vertically oriented injection unit. There it flows into a runner channel centered at each cavity and then, just before reaching the cavity, enters a tunnel gate (*c*) (Fig. 2). The tunnel gate is located in mold core (8). It passes beneath the 3 mm wide rim of the bezel and leads to an ejector bore (*d*). From here the runner channel continues along the 20 mm long opening in the bezel toward the window, where it once again enters a tunnel gate (*f*). This tunnel gate leads finally to an ancillary runner on the inside of the bezel that connects to the window.

The mold contains two retractable cores (16, Figs. 2 and 4) that seal off the ancillary runners and window openings while the bezel is being molded. These cores are connected via couplings (17) to the tapered surfaces of actuating rods (26, 34) such that they are moved in and out of the cavities by the motion of the hydraulic cylinder (39).

Once the bezel is molded and has cooled for a certain amount of time, the cores (16) are retracted. Immediately thereafter, the window is molded by the vertical injection unit.

Part Release/Ejection

As the mold opens, the sprue and runner (*a*) are pulled out of the sprue bushing (48) and tunnel gates, thus degating the molded parts. Following this, the two-stage ejector (46, Fig. 5) is actuated. Initially, ejector plates (22, 23, 24) advance simultaneously (distance *x* in Fig. 5).

- The lifters (10) to (14) release the undercuts (*h*) on the inside of the bezel.
- Ejectors (49) push the molded parts off the cores.
- The runner ejectors (51, 52) force the runner (*b*) for the window out of the tunnel gates (*c*) and (*f*), thus degating the window, break the runner into three pieces and eject them.
- Sprue (*a*) is ejected.

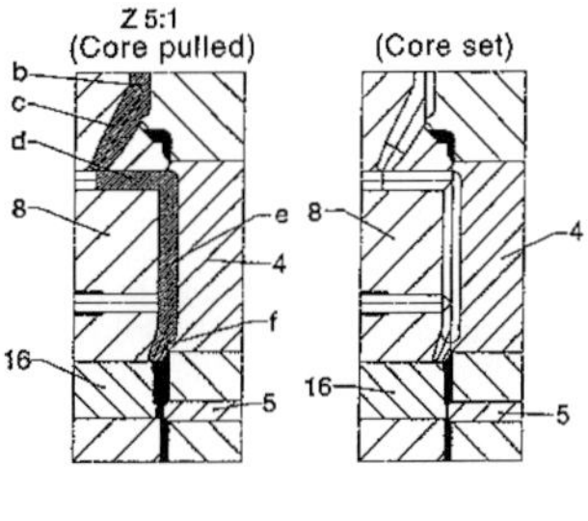

Figure 2 Section through the cavity with core pulled and core set

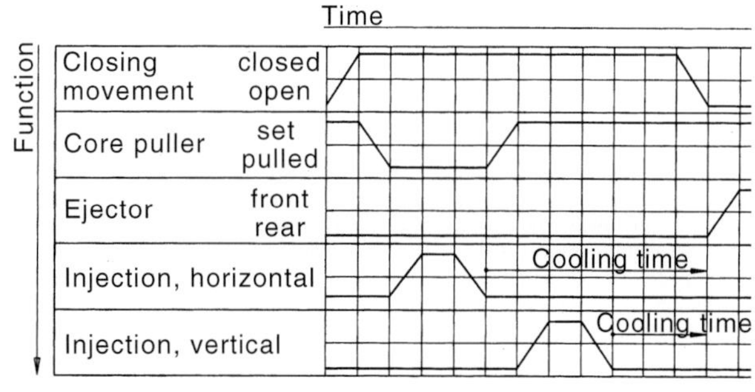

Figure 3 Sequence of operation

Plate (24) with the lifters then comes to a stop (distance *x*), and the ejectors push the parts, which are now free of the undercuts, further away from the lifters (distance *y–x*). Before the mold closes, all ejectors are retracted. Pushback pins (55) ensure that the ejector system returns to the proper end position prior to molding. The positions of the ejector plates and the core actuator are monitored by proximity switches (58). The sequence of operation is shown in Fig. 3.

Fig. 4

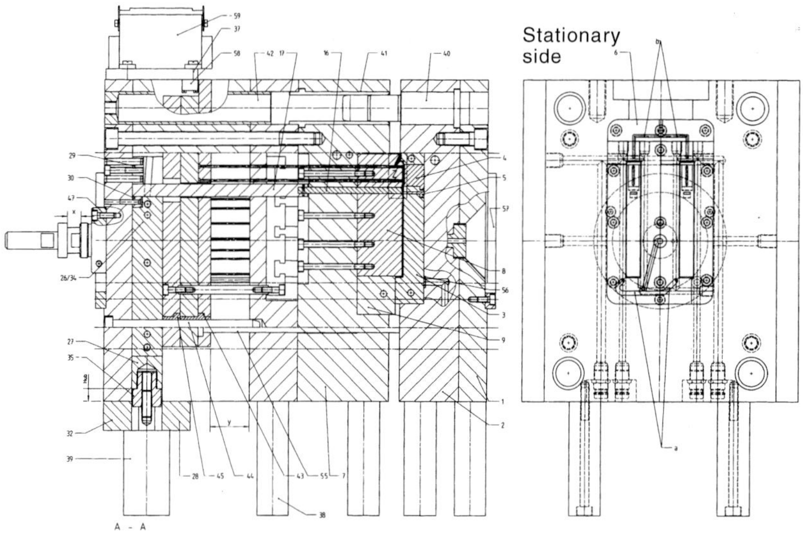

Fig. 5

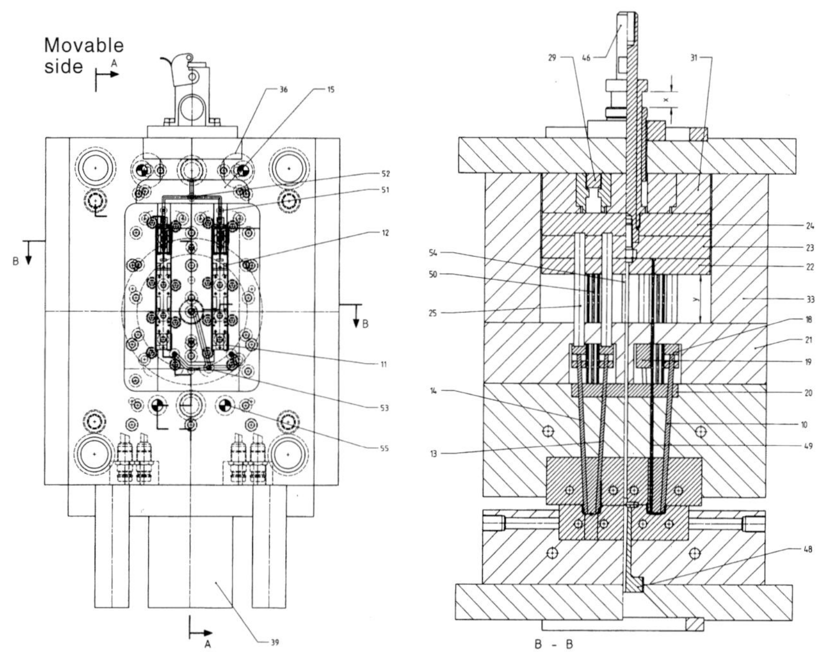

Figures 4 (top) and 5 (bottom). Two-cavity injection mold for a bezel
1: clamping plate (stationary side); 2: mold plate (stationary side); 3: mold insert (stationary side); 4: mold core A; 5: mold core B; 6: runner insert; 7: mold plate (movable side); 8: mold core C; 9: mold insert (movable side); 10: lifter A; 11: lifter B; 12: lifter C; 13: lifter D; 14: lifter E; 15: runner insert (movable side); 16: retractable core; 17: coupling for retractable core; 18: coupling A for lifter; 19: coupling B for lifter; 20: guide plate; 21: backing plate; 22: ejector retainer plate; 23: ejector plate; 24: thrust plate; 25: bolt; 26: retractable core actuating rod A; 27: connecting piece; 28: cylinder mount; 29: guide element A; 30: guide element B; 31: guide element C; 32: clamping (movable side); 33: spacer; 34: retractable core actuating rod B; 35: coupling for hydraulic cylinder; 36: stop; 37: mounting plate for connector; 38: support rod; 39: hydraulic cylinder; 40: leader pin A; 41: guide bushing A; 42: locating sleeve; 43: guide bushing B; 44: leader pin B; 45: guide bushing C; 46: two-stage ejector; 47: clamping ring; 48: sprue bushing; 49 to 54: ejectors; 55: pushback pin; 56: O-ring; 57: locating ring; 58: proximity switch; 59: electrical connector
Company illustrations: Fischer Automobile Systems, Tumlingen/
Waldachtal, Germany

Example 83, Two-Cavity Injection Mold for Runnerless Production of Polycarbonate Optical Lenses

The optical lens with a diameter of 30 mm shown in Fig. 1 was produced in polycarbonate via runnerless injection molding. One reason this material was selected was the necessary subsequent coating.

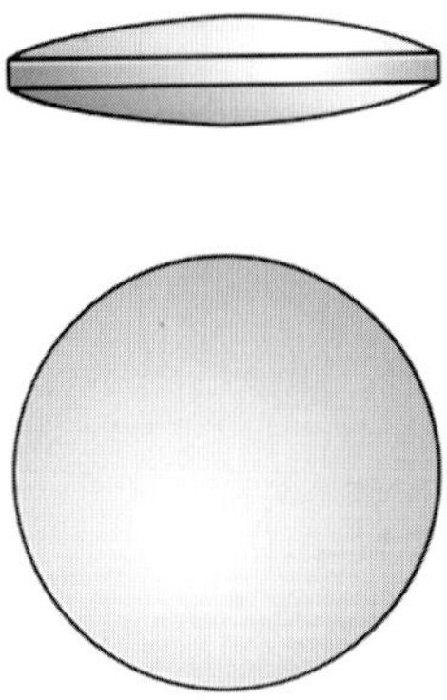

Figure 1 Polycarbonate optical lens

In contrast to conventional wisdom, according to which such optical components are to be gated via a thick sprue that very often exceeds the weight of the molded part and must, as a rule, be removed by machining, a standard split-cavity mold was employed in this case. The cavities for the two lenses are machined into the splits (24) and gated via a runnerless molding system.

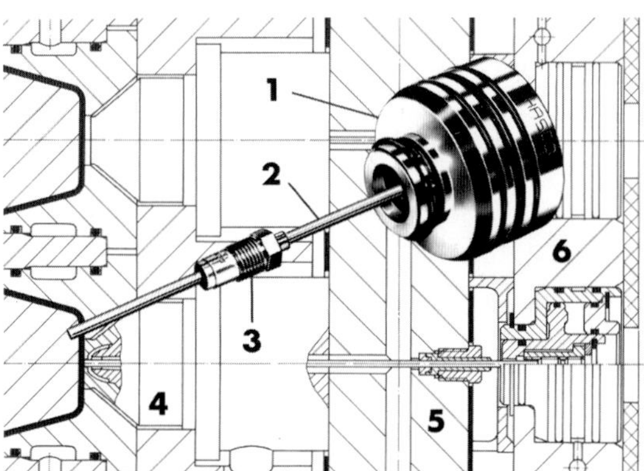

Figure 2 Individual components of the valve-gated hot-runner nozzles
1: standard pneumatically actuated valve gate; 2: shutoff pin (needle); 3: guide bushing; 4: hot-runner nozzle; 5: hot-runner manifold; 6: clamping plate

Thanks to this mold concept, it was possible to gate the parts on their outer edge, and then seal and smooth the gate using the shutoff pin, or needle.

Standard items were also used for melt-conveying components (Fig. 3). The melt flows from the machine nozzle through the inlet bushing (23) to the hot-runner manifold (10), and from there into the cavity via the hot-runner nozzles (17). The pneumatic actuating cylinders (15) for the valve gates are installed in the relatively cool clamping plate. The individual components of the valve-gated hot-runner nozzles are shown in Fig. 2.

To relieve pressure on the melt in the hot-runner system, the inlet bushing (23) is designed to accommodate a slip-fit machine nozzle, which executes a slight retract stroke, thus decompressing the melt in the hot-runner manifold. With a view to subsequent post-molding operations, the parts are removed by a part handling robot, which places them in the next processing station.

A slight undercut on the outer surface of the molded parts ensures that they always remain in the same split, from which they are removed by the part handling robot. The two rectangular projections (Fig. 3) on the right and left of each lens are provided for the grippers attached to the part handling robot.

The entire mold was constructed from standard mold components that are readily available from catalogs. The hot-runner manifold (10) is heated by heating elements located in slots machined into each side of the manifold. This ensures uniform heating of the manifold.

To reduce heat loss to the mold plates via radiation, reflector plates (27) are attached to the hot-runner manifold. These have been proven to save up to 30% of the energy required for heating. Likewise, the support pads (16) for the hot-runner system are fabricated from titanium, which because of its poor thermal conductivity means that as little heat as possible is conducted to the mold plates.

A thermal insulating plate (14) is installed on the stationary half of the mold to insulate it from the machine platen.

Prior to construction, the mold concept was simulated on a computer. It was possible in this manner to optimize the filling pattern specifically to such an extent that only minimal adjustments for fine tuning had to be performed on the machine. Particular attention was devoted to preventing jetting during filling of the cavities.

Production of these optical lenses via runnerless injection molding was a joint development with the IKV (Institut für Kunststoffverarbeitung), Aachen, Germany.

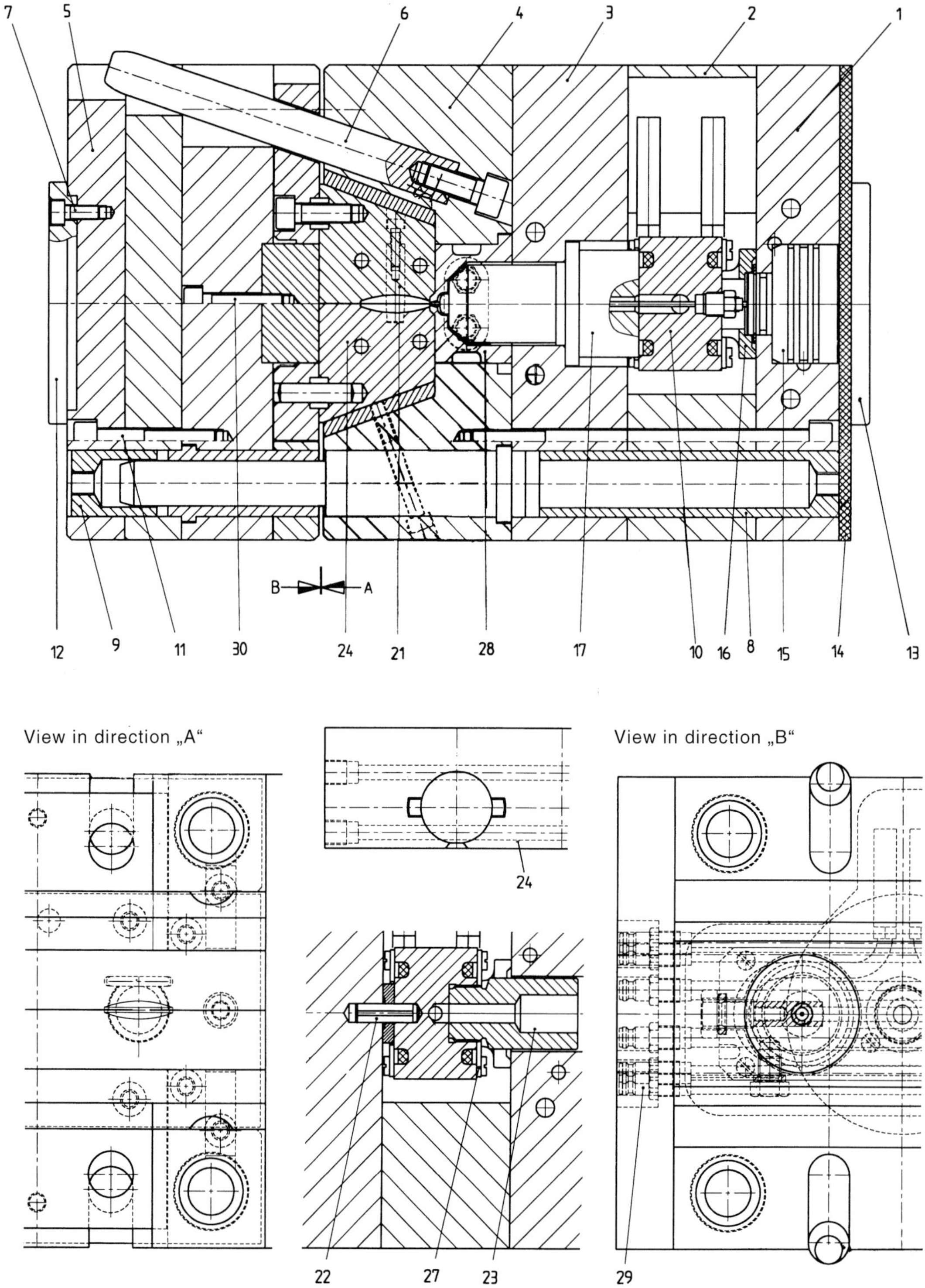

Figure 3 Injection mold for runnerless production of optical lenses
10: hot-runner manifold; 14: thermal insulating plate; 15: pneumatic actuating cylinder for valve gate; 16: support pads for hot-runner system; 17: hot-runner nozzle; 23: inlet bushing; 24: split; 27: reflector plate; 28: nozzle well insert
(Courtesy: Hasco, Lüdenscheid; IKV, Aachen, Germany)

Example 84, Injection Mold with Hydraulic Core Pull for a Cable Socket

Hydraulic core pulling equipment is usually required when long cores have to be pulled for which mechanical core pulling devices provide insufficient length of stroke. There are applications, however, which could be adequately served by mechanical core pulling, but where the employment of a hydraulically pulled core results in simplification of the mold design and therefore in economical advantages. This presupposes that the injection molding machine can be equipped with ancillary hydraulic equipment.

The two principles of design will be compared in the example of molds for cable sockets (Fig. 1). Figure 2 shows the mold construction incorporating mechanically operated slides, which has been chosen to give a better comparison with this example. The parting line of the mold runs vertically to the plane of the drawing along line (1) to the slide (3). As this molding is exposed to view and meant to carry the manufacturer's name on the back of elbow (10), it is injected through a submarine gate (2) on the peripheral rim in the parting line (1). However, to be able to demold the part, half of the lower part of the molding had to be machined into slide (3), although a cylindrical plug would have been quite adequate for the opening itself (4).

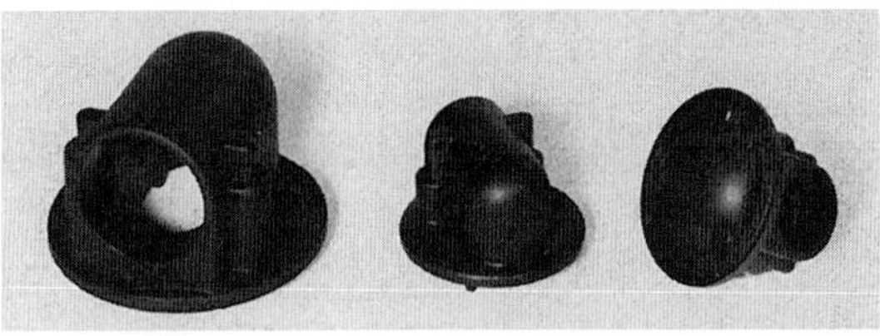

Figure 1 Cable sockets of varying sizes

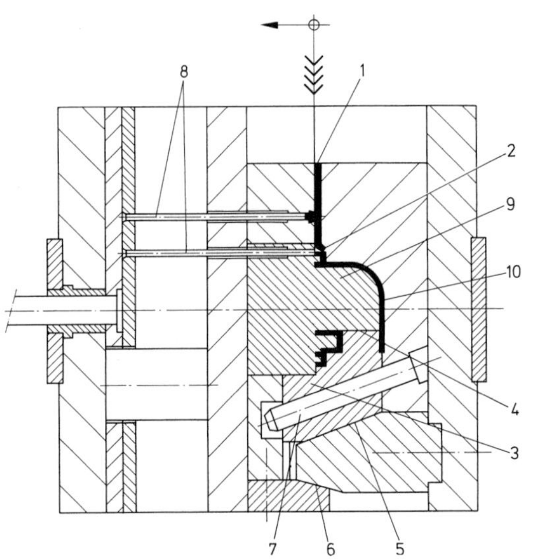

Figure 2 Cable socket injection mold with mechanical core pulling (slide)

1: parting line and point of injection; 2: tunnel gate; 3: slide; 4: opening; 5: wedge surface on the slide; 6: opposite wedge surface; 7: cam pin; 8: ejector pins; 9: core; 10: surface with manufacturer's lettering

When the mold opens, the angled surface (6) releases, so that the pressure on the slide (3) is released via the wedge area (5). The slide moves down, guided by cam pin (7) and sliding on the surface of wedge (5). This leaves the molding resting freely on the core (9), to be pushed off by the ejector pins (8).

Figures 3 to 8 show the application of hydraulic core pulling. The mold lies in the part line, which runs vertically to the plane of the drawing from top to bottom. Gating is also via tunnel gate (2), and injection takes place in the part line. Instead of the complicated slide arrangement, a cylindrical plug forms the opening in the mold. The plug is moved by a small hydraulic cylinder (4) that is mounted below the mold on an adaptor (5). The two electrical switches (8) and (9) are fitted to plate (6). They are actuated by an S-shaped switching rod (7), signaling the upper and lower position of cylinder (4) to the machine controls. As a guide for a slide is not required on the moving half of the mold, there is scope for surrounding core (10) by a stripper plate (11). This is linked with the ejector plates (13) by tie rods (12). Plate (11) acts simultaneously as push-back plate for the ejectors when the mold closes. The hollow space in the ejection system is supported by rolls supporting (14). Core (3) does not have to be mechanically interlocked, as the effective area for the injection pressure is not very large. This force can be absorbed solely by the hydraulic pressure of the cylinder (4), which is cushioned in both end positions.

The operating sequence starts with mold closing. Then the core (3) is moved in by the hydraulic cylinder (4). This is followed by the injection and cooling times. When they have elapsed, core (3) is withdrawn from the mold. The mold moves into the open position, and the hydraulic ejector – coupled with the tool through the ejector bar (15) – actuates the ejector plates (13) and pushes the mold part off the core (10) with the aid of the stripper plate (11) and the ejector pins (16). The mold is cooled in the mold cavity plate (17) as well as in the core (10).

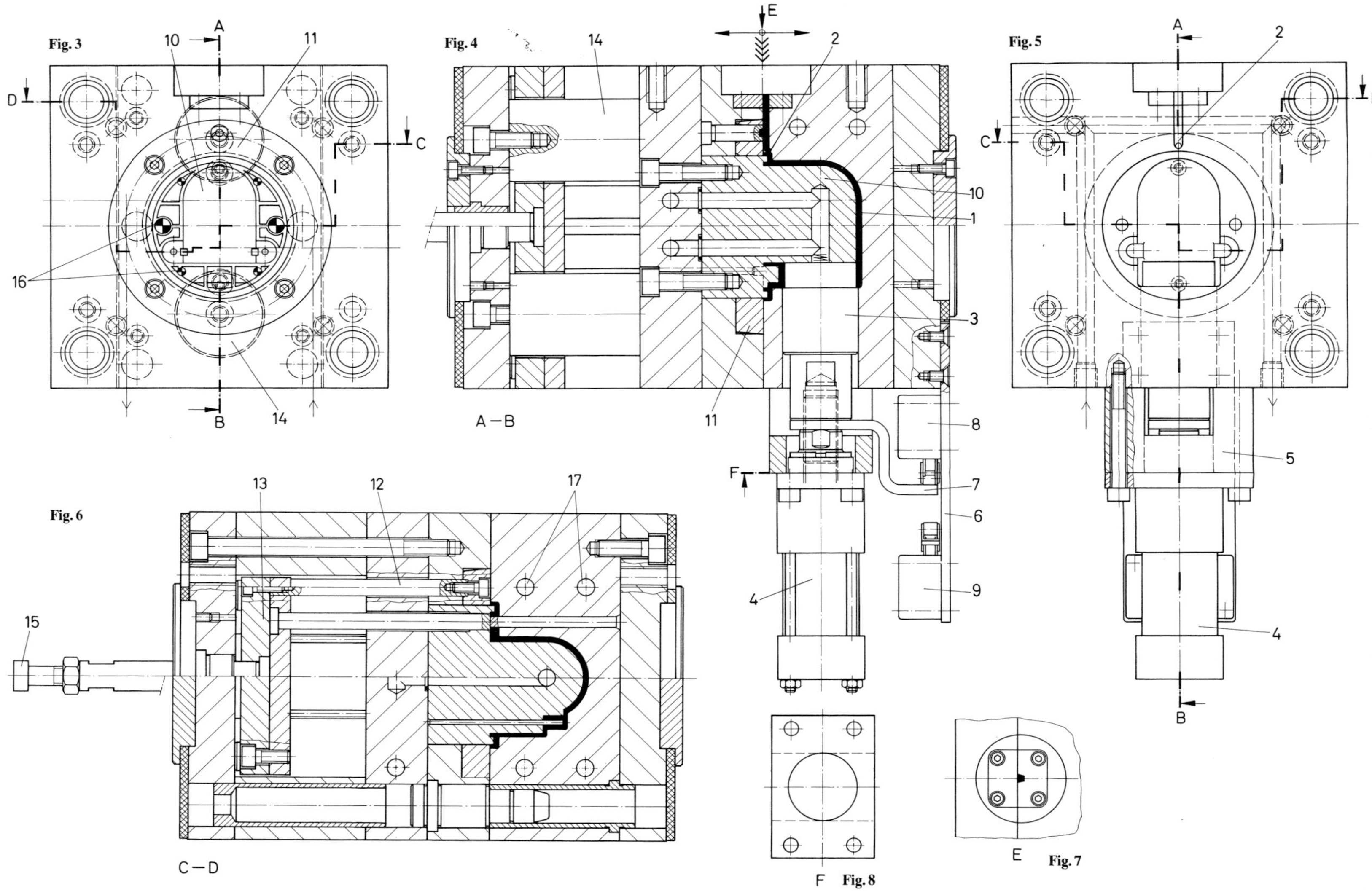

Figures 3 to 8 Cable socket injection mold with hydraulic core pull
1: molded part; 2: tunnel gate; 3: cylindrical plug for forming the opening in the molded part; 4: hydraulic cylinder; 5: adaptor; 6: plate; 7: switching rod; 8, 9: electrical switches; 10: core; 11: stripper plate; 12: tie rods; 13: ejector plates; 14: supporting rolls; 15: ejector bar; 16: ejector pins; 17: mold cavity plate

Example 85, Four-Cavity Injection Mold for Pipets Made from PMMA

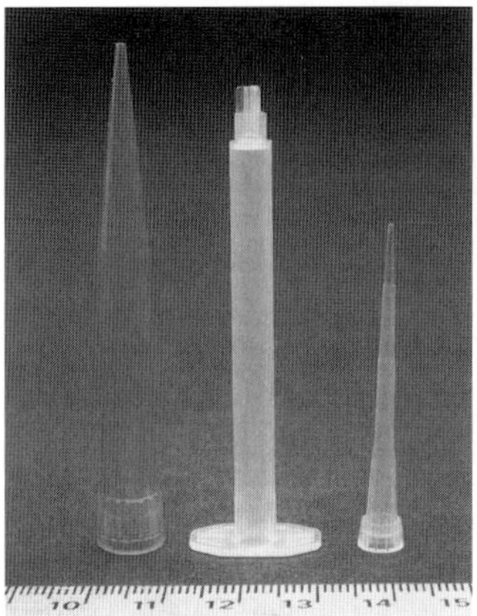

Figure 1 Injection molded pipets

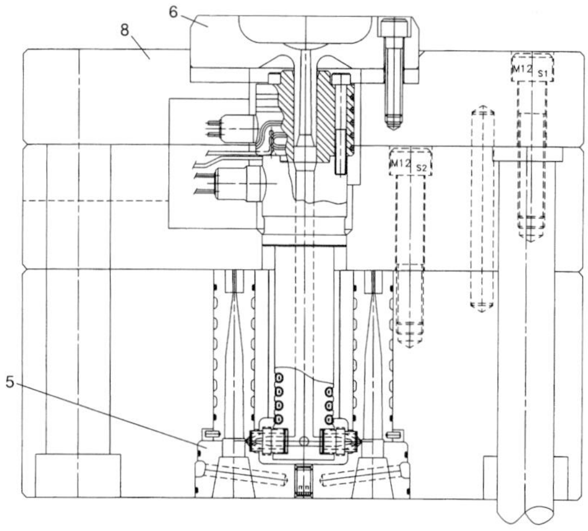

Figure 3 Nozzle with torpedoes, fully installed
5: retainer ring; 6: centering ring; 8: backing plate

Pipets are conical tubes, e.g. 70 mm long with an outside diameter which tapers from about 9 mm to about 1.5 mm at the tip (see mold cavity in Fig. 2). Injection systems consisting of a combination of hot runner nozzles and cold submanifolds and tunnel gates are not economical because of a relatively high shot weight and the occurrence of cold sprues. Therefore, the only possibility is direct gating using adequate hot runner nozzles.

As shown in Fig. 2 the torpedo 2 (heat conducting torpedo from a nickel-plated copper alloy or tungsten carbide, type: Horizontal Hot Tip) is set in the heated nozzle body (1) and screwed in tightly with a threaded bushing (3). The threaded bushing (3) is centered between the cavity insert (4) and the retainer ring (5) so that the tip of the torpedo is placed exactly in the center of the gate runner which is between the above parts (4, 5). Because of that it is possible to control the melt temperature in the gate runner all the way to the mold cavity to an optimum value.

The centering of the torpedo in the gate area has an additional consequence: the die body, because of its significant heat expansion ability, expands toward the nozzle of the machine. Therefore, it is necessary to apply at the block (6), where the machine nozzle is positioned, a device that compensates for these different expansions. Figure 3 shows the centering ring which also encloses the nozzle block and absorbs the changes in length of the nozzle body.

Installation of the Nozzle

Installation of the nozzle with its heating torpedoes into the mold is done in the following steps:

A The cavity block is divided along the gate runner (Fig. 2).

A1 The nozzle body without torpedoes is inserted into the cavity plate (7, Fig. 5).

A2 The torpedoes (2) and (3) are inserted and screwed into the nozzle body (Fig. 6).

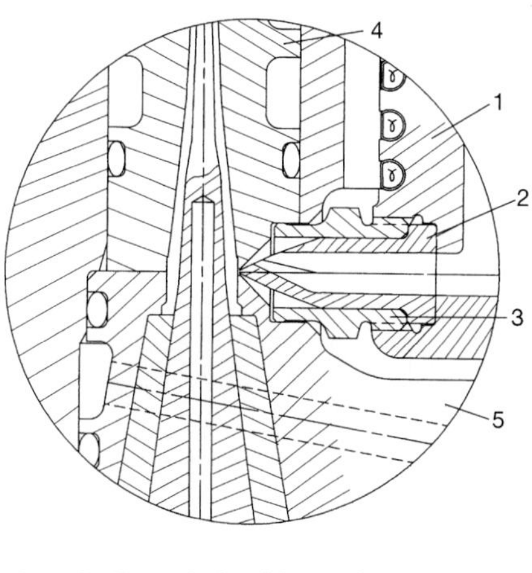

Figure 2 Nozzle body with torpedoes
1: nozzle body; 2: torpedo; 3: threaded bushing; 4: cavity insert; 5: retainer ring

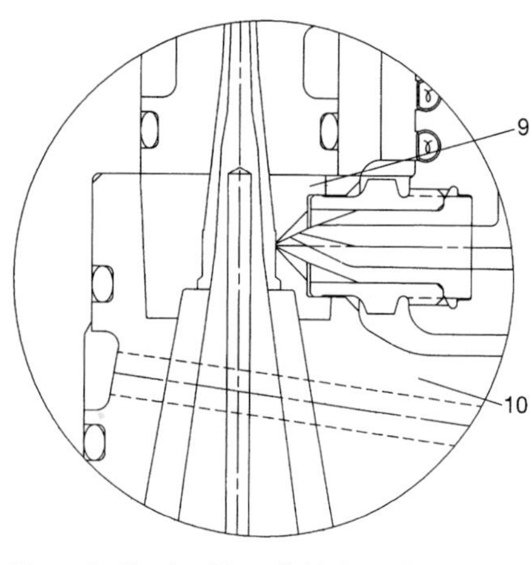

Figure 4 Nozzle with undivided gate insert
9: gate insert; 10: retainer ring

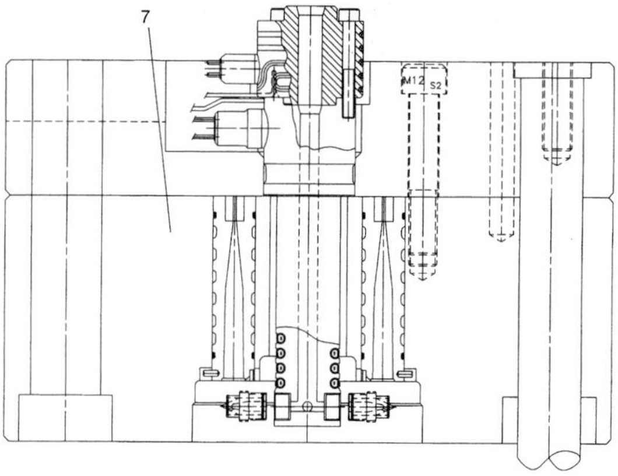

Figure 5 Nozzle installation, Step 1
7: cavity plate

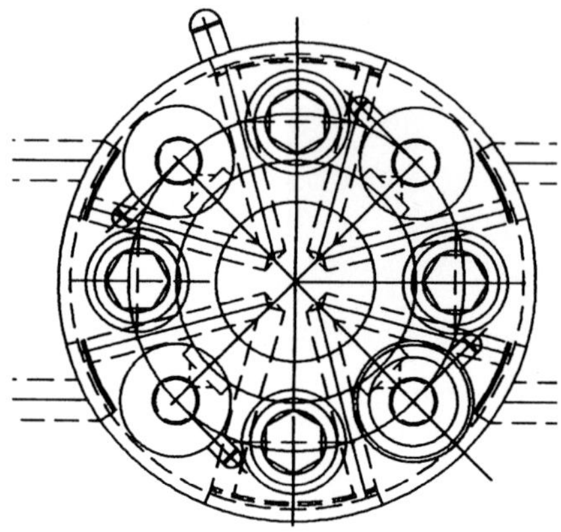

Figure 7 Arrangement of cooling channels in the retainer ring

A3 The nozzle with torpedoes is put into place, the retainer ring (5) is screwed on, the backing plate (8) and the centering ring (6) are attached (Fig. 3).

B The gate runner is not divided (gate insert 9, Fig. 4). During installation of this version the gate insert (9) is slid over the threaded bushing (3) before the retainer ring (10) is screwed on.

Temperature Control

Besides cavity inserts and mold cores, the retainer rings (5, 10) are also cooled intensively by star-shaped cooling channels (Fig. 7).
Particular advantages of this nozzle arrangement are:
- excellent thermal separation between the hot runners and the cooling system of the mold assuring a good homogeneity of melt flow,
- good monitoring and control of the nozzle temperature,
- short residence time of the melt because of the small cross-sections of the flow channels,
- favourable material behavior in general and during color changeovers in particular,
- small installed area, and thus better utilization of the mold area.

As depicted by Fig. 8, it is possible to accommodate 64 pipet mold cavities on the total mold area of 300 mm × 380 mm. Such a mold fits the majority of machines with 500 kN clamping force. This method has been developed by Mold Masters in cooperation with Cavaform Inc., St. Petersburg, Florida/USA. With this nozzle design it is possible to produce high quality thin, tubular injection molded parts, such as cartridge cases for ball point pens, pipets, hypodermic syringes and needle-cases dependably and effectively.
The arrangement described in this contribution is suitable for all semicrystalline plastics as well as for PPO, PMMA, PVC, CAB, TPU and all styrenic plastics.

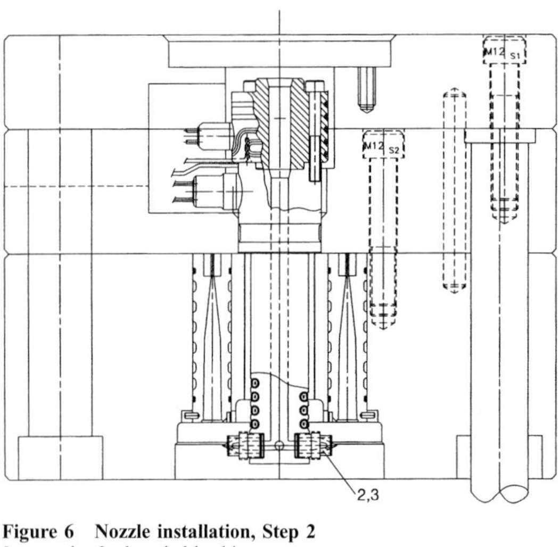

Figure 6 Nozzle installation, Step 2
2: torpedo; 3: threaded bushing

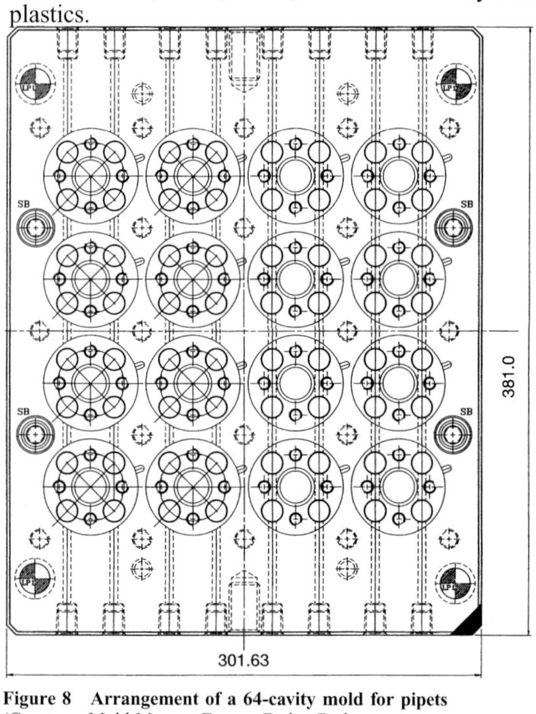

Figure 8 Arrangement of a 64-cavity mold for pipets
(Courtesy: Mold-Masters Europa Baden-Baden, Germany)

Example 86, Two-Cavity Mold for Water Tap Handles Made from PMMA

Decorative bathroom fittings are frequently made with transparent handles in which there is a second layer made from a non-transparent material. The mold illustrated in Figs. 1 to 5 is designed for the production of this type of part. Both of the differently colored materials are injected one after the other at two stations on the mold, thus enabling the part to be produced in one operation. It is also necessary to use an injection molding machine that has two injection units arranged at right angles to each other.

The main view (Fig. 2) shows the mold in its closed position. At the left mold station, the molding of the colored inner component of the handle is carried out by the injection unit on axis (*a*). At the same time, the outer transparent part of the handle is molded over this part using the unit on axis (*b*) through the sprue bushing (22). The wall thickness of the outer layer of the molding needs to be rapidly cooled and hence the mold cavity insert (15) is made with a narrowly machined helical cooling channel. On solidification of the molded part the mold is opened and the part is ejected. This occurs in position II, as shown in the right-hand side of the drawing (Fig. 4), by advancing the ejector bar (12) with the help of the pneumatic cylinder (20). Only after this first step can plate (4) be freed, and held in the appropriate position by the stop screw (30). The core retainer plates (5) and (6) with the cores (11), which carry the molded colored inner parts of the handle, move out of the plate (4) only as far as necessary to allow them to be turned through 180° under plate (4), so that on reclosing the mold the empty cores will reengage with the initial molding station and the cores containing the inner part will engage the final molding station. The turning movement is made by a four-cornered spindle (16), whose gear wheel (17) engages a pinion (18) moved by the pneumatic cylinder (19).

Figures 1 to 5 Injection mold for a transparent water tap handle containing a colored inner layer
1, 2: stationary-side plates; 3: mold cavity for preform; 4: mold cavity mounting plate; 5, 6: rotary plates; 7: guiding ring; 8: base plate; 9: spacer ring; 10: movable-side clamping plate; 11: collars; 12: ejector pins; 13: spring; 14: mold cavity retainer ring; 15: mold cavity insert; 16: spindle with four-sided head; 17: gear wheel; 18: pinion; 19: pneumatic cylinder for rotary plates 5, 6; 20: pneumatic cylinder for ejector of finished molded parts; 21: locating ring; 22: sprue bushing; 23: spring; 24: rod; 25: hook; 26: cross pin; 27: bearing housing; 28: spring; 29: cooling water connection; 30, 31: stop screws; 32: guide pin; 33: cross bolts; 34: fitting ring; 35: disc; 36: bushing; 37: nuts for 12; 38: seal ring; 39: O-ring; 40: telescoping sleeve for 30
▶

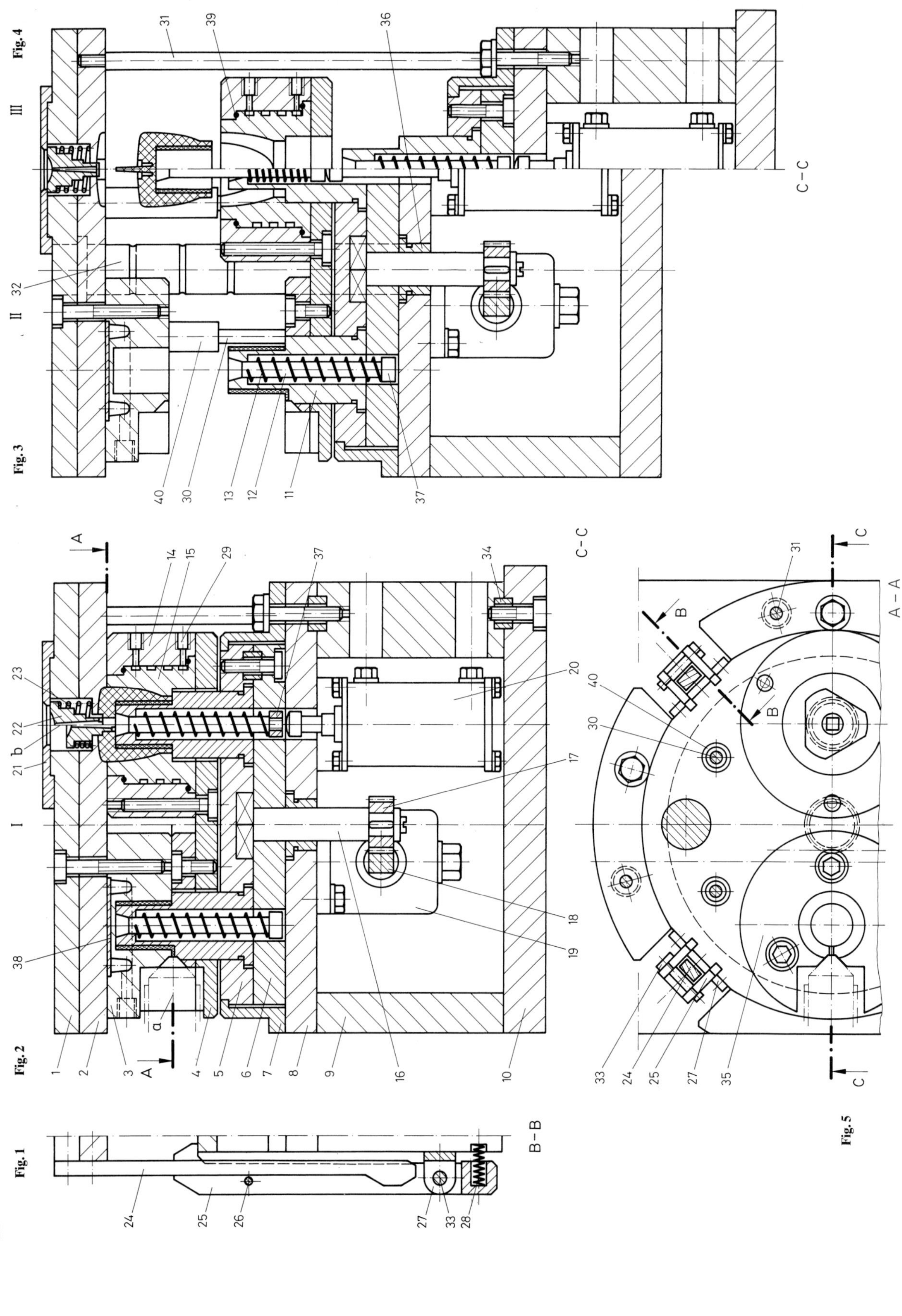
Fig. 1
Fig. 2
Fig. 3
Fig. 4
Fig. 5
A – A
B – B
C – C
I
II
III

Example 87, Two-Cavity Injection Mold for the Automatic Molding of Conveyor Plates onto a Wire Cable

Such granular materials as plastics pellets or grain can be transported by pipe conveyor systems. A conveying cable fitted with conveyor plates at fixed intervals runs through the pipe. These plates match the inside diameter of the pipe (Fig. 1).

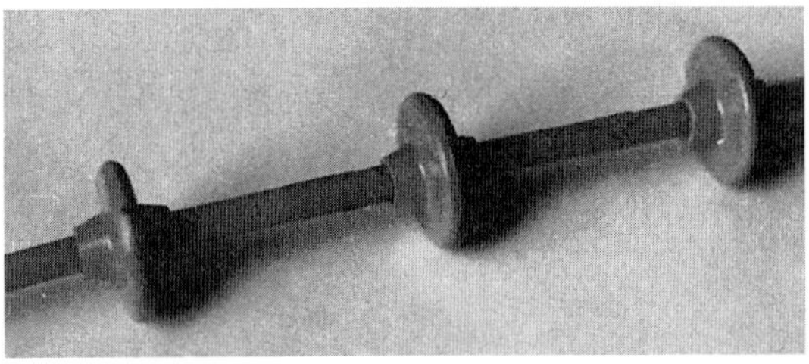

Figure 1 Conveying cable with plastic conveyor plates for mechanical pipe conveyor system

The mold shown in Figs. 2 to 4 was developed for the production of these conveying cables. Plates are molded simultaneously onto two parallel cables to increase productivity. There is no problem guiding the cables through the mold if the mold parting line is in the horizontal plane and the injection unit is mounted vertically on the machine.

To start production the two cables are pulled through the bores in part (8) and placed in the grooves in part (9). Automatic production can only commence once two plates each have been molded onto both cables. Up to that time the cables have to be advanced by hand. Thereafter the paddles (11) situated on the roll (10), which are rotated with each machine opening stroke, engage the molded plates and advance them by one division. To achieve this the cable (19) fixed to the bolt (30) lifts the double lever (13) against the resistance of spring (17) on bolt (16). The pawl (14) rotates the wheel (27) by engaging in its ratchet teeth, advancing the paddles (11) fitted to shaft (12) by 90°. The turning movement must only be allowed to start when the newly molded plates have been released from the lower cavity half (7) by lifting the mold components (8) and (9), followed with continued mold opening with release from the upper cavity half (6). Only then is the cable (19) put under tension. Therefore a total mold opening distance of at least 110 mm is required for part release and cable advancement. The length of cable (19) must therefore be matched to the opening movement of the injection molding machine.

Fully automatic operation of the mold necessitates interlocking with the injection molding machine controls. The functions of the mold and the presence of melt are supervised by switches K_1 to K_4. The siting of these switches is shown schematically in Figs. 2 and 3. Figure 5 shows the wiring diagram into which these switches have been integrated. With melt present, the switch K_3 is actuated during each cycle by the mold plates, closing relay J_1. Subsequently K_4 is also actuated (by the mold plates) as is K_1 (via the moving mold plate I), causing a relay in the injection molding machine control to indicate the end of the cycle so that a new sequence can be started. Should switch K_2 not be actuated due to a lack of melt, a subsequent machine cycle cannot take place. During mold closing the parts (8) and (9) are pushed back into the frame (3) again. Switch K_2 is thereby opened, which in turn opens relay J_1, so that the switching sequence for the next cycle is set up. Figure 6 shows the time sequence of these functions.

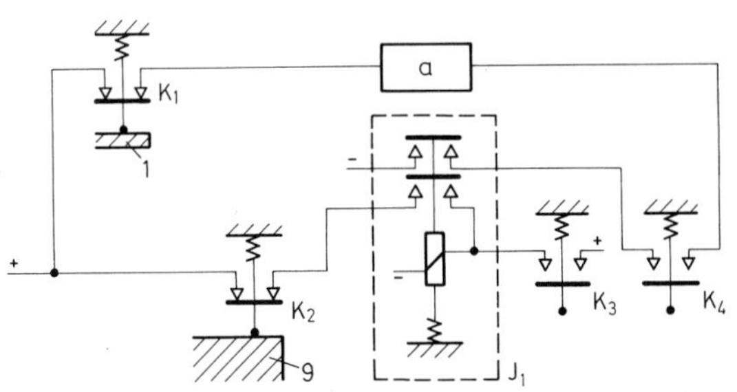

Figure 5 Wiring diagram
(K_1 to K_4) switches; (J_1) relay
(a) injection molding machine controls; (1, 9) mold components (refer to Figs. 2 to 4)

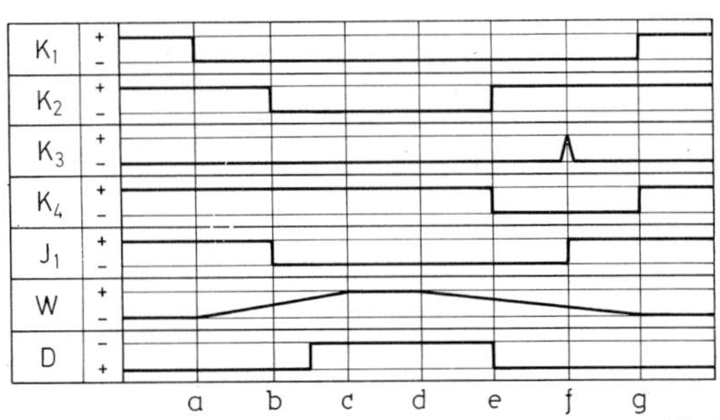

Fig. 6

Figure 6 Sequence diagram of the controls
a: mold fully opened, start of mold closing movement; *b*: mold halves are touching, so that K_2 and J_1 open; *c*: mold closed; *d*: start of mold opening; *e*: mold halves separate, cable advance starts; K_4: opens; K_2: closes; *f*: new plate temporarily closes; K_3: *g*: advanced plate closes; K_4: mold is open; *D*: wire cable; J_1: relay; K_1 to K_4: switches; *W*: mold
+: closed or tensioned; –: open or relaxed

Fig. 2

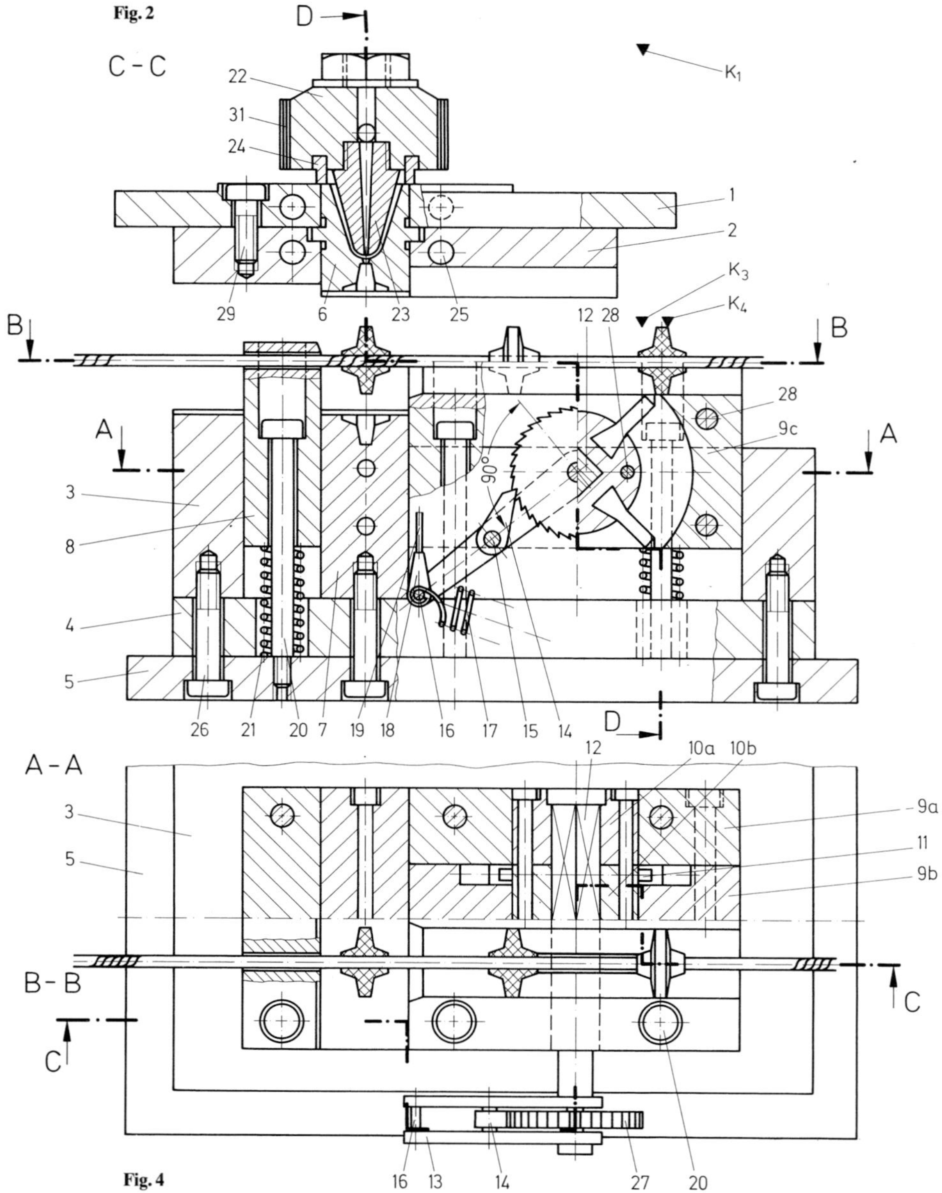

Fig. 4

Fig. 3

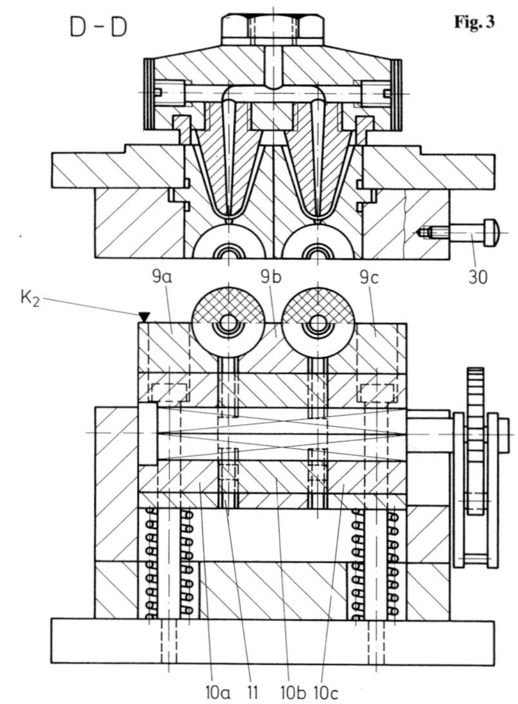

Figures 2 to 4 Injection mold for the automatic molding of conveyor plates onto conveying cables for mechanical pipe conveyors

1: upper mold clamping plate; 2: upper mold cavity retainer plate; 3: lower mold frame; 4: mold base plate; 5: lower mold clamping plate; 6: upper mold cavity half; 7: lower mold cavity half; 8: moving cable feed; 9: moving cable discharge; 10: rotation cylinder; 11: paddles; 12: square shaft; 13: double lever; 14: pawl; 15: pawl shaft; 16: cross pin; 17: spring; 18: eye on the draw cable; 19: draw cable; 20: screw; 21: spring; 22: injection head; 23: beryllium-copper nozzle; 24: pressure ring; 25: cooling water channels; 26: connecting bolt; 27: ratchet wheel with buttress teeth; 28, 29: connecting bolts; 30: bolts; 31: heating element

K_1 to K_4: switches

Example 88, 20-Cavity Hot-Runner Mold for Producing Curtain-Ring Rollers Made from Polyacetal Copolymer

Curtain-ring rollers (Fig. 1) are "penny articles." Nevertheless, their production requires considerable expenditure as far as the injection mold is concerned, which has to incorporate slides for forming the shafts carrying the small rollers and by requiring assembly of these rollers. These conditions are met with the present mold (Fig. 2) through the use of a hot-runner system that results in low manufacturing expenses and puts into practice a concept which already assembles the individual parts into finished ring rollers inside the mold itself (Fig. 3).

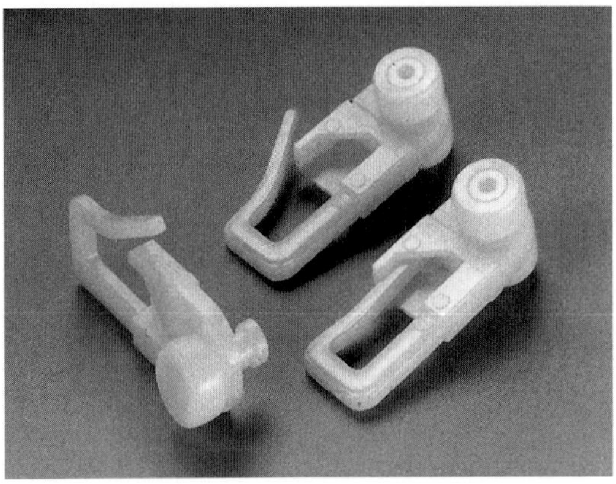

Figure 1 Curtain-ring rollers of polyacetal copolymer
left: curtain roller, shown with rear roller removed; center: curtain-ring roller with open hook; right: curtain-ring roller with closed hook

Mold Design

When calculating the mold, an optimum number n = 60 of cavities became established. Related to the number of complete curtain-ring rollers produced in the mold, this corresponds to a 20 cavity tool.
Gating between hot runner and molded parts is via small sub-runners with two submarine gates each of 0.8 mm diameter (Fig. 4). When dimensioning and designing the hot-runner system, reference was made to [1] (also refer to Fig. 36 there). The main dimensions arrived at for the torpedo were $d_T = 8$ mm for the torpedo diameter and $l_T = 52$ mm for the torpedo length. Six cavities each are fed by one torpedo (material specification 2.0060). The installed heating capacity amounts to 250 W/kg of hot-runner block, the latter being provided with two heating circuits each. In order to obtain intensive cooling of the cavities, copper cooling pins are employed. The mold is built up of standard components to material specification 1.1730, whereas 1.2162 with a surface hardness of HRC = 60 ± 1 has been chosen for the cavities and wearing parts.

Assembly of the Curtain-Ring Rollers Inside the Mold

The mold is technically interesting because of the fully automatic assembly of the curtain-ring rollers inside the tool, this being the subject of a patent [2]. In this case the rollers and the roller-carrier are injection molded spearately within the same tool. The shafts of the roller carrier have been provided with cylindrical clearances in the area of the undercut, so that there is as much elastic deformation as possible when the rollers are being fitted onto the shafts. The connection between roller and roller carrier is of the non-releasing cylindrical snap-fit type with a retaining angle of $\alpha_2 = 90°$ [3].
Once the cooling period has timed out, the roller carrier (*c*) (Fig. 3) is released by the mold-opening movement and in a subsequent step the rollers (*a*) and (*b*) are pushed home by the spring force acting on the ejector sleeves (*d*) and (*e*) (Fig. 3).
After assembly the finished article and the sheared-off runner are ejected, once the ejector sleeves and pins have been returned to their starting positions. Molded parts and runners are separated on the conveyor belt. The cycle time is 12 s.

Literature

1. Heißkanalsystem indirekt beheizter Wärmeleittorpedo, in: Berechnen, Gestalten, Anwenden (C.2.1), Schriftenreihe der Hoechst AG, 1982
2. DE-PS 2 528 903 (1979) F. & G. Hachtel
3. Berechnen von Schnappverbindungen mit Kunststoffteilen. In: Berechnen, Gestalten, Anwenden (B.3.1), Schriftenreihe der Hoechst AG, 1982

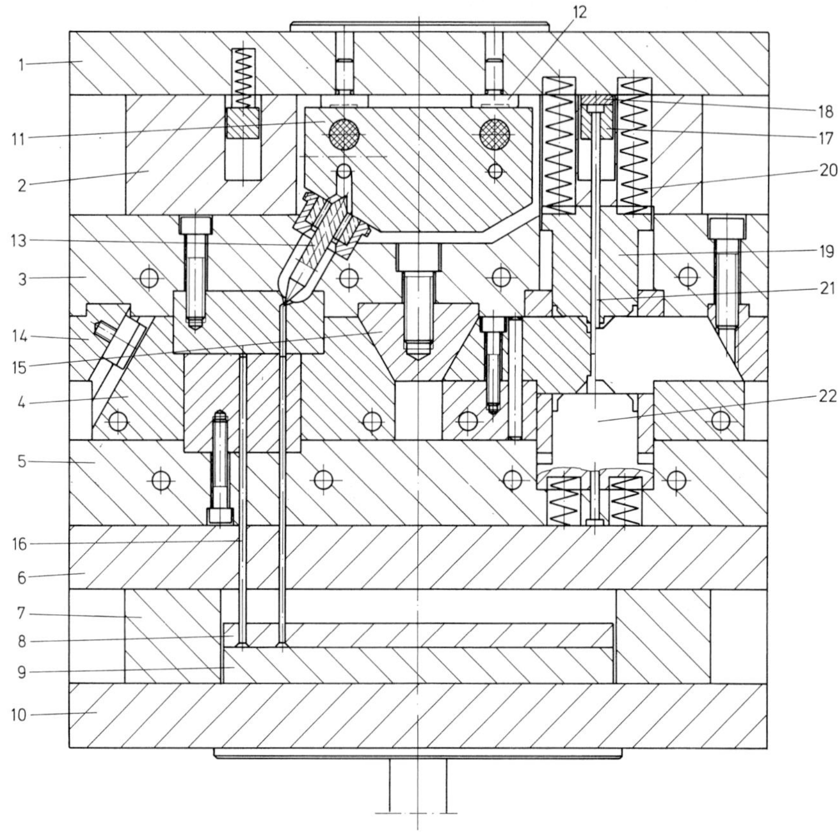

Figure 2 Section through the 20-cavity injection mold with hot-runner manifold and indirectly heated (thermally conductive) torpedo as well as an assembly facility for fitting the curtain-ring rollers together inside the mold (Courtesy: F&G Hachtel, Aalen, Germany)
1: mounting plate; 2: strip; 3: mold bolster; 4: slide; 5: mold plate; 6: plate; 7: strip; 8: ejector retainer plate; 9: ejector plate; 10: clamping plate; 11: hot-runner manifold; 12: support pad; 13: indirectly heated (thermally conductive) torpedo; 14, 15: heel block; 16: ejector pin; 17: insert; 18: strip; 19: stepped pressure piece; 20: compression spring; 21: ejector; 22: pressure slides

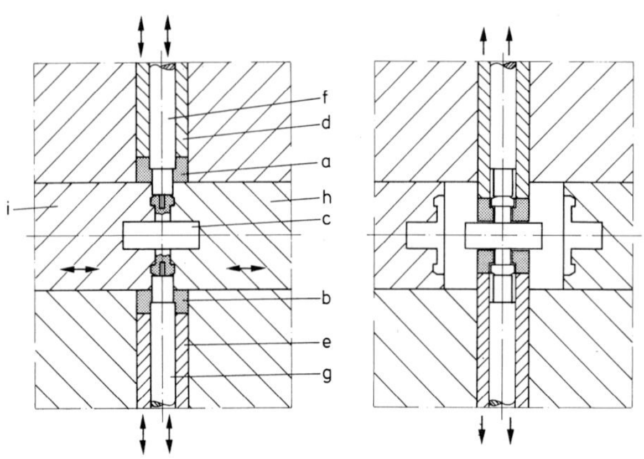

Figure 3 Assembling procedure for the curtain-ring rollers inside the mold
left: before assembly; right: after completed assembly
a, *b*: roller; *c*: roller carriers; *d*, *e*: ejector sleeves; *f*, *g*: ejector pins; *h*,

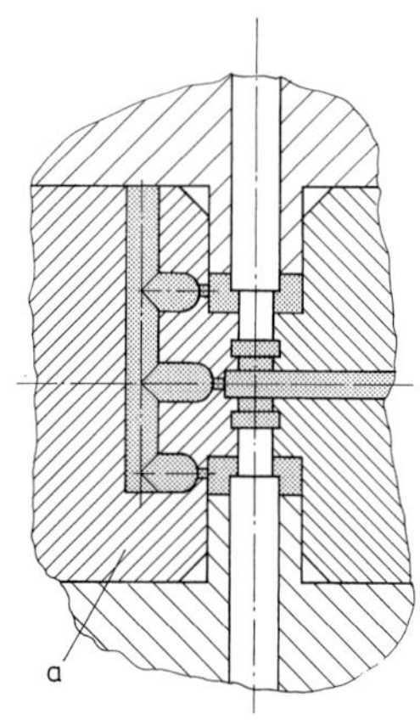

Figure 4 Gating to the molded parts in the mold
a: turned through 90° around the drawing plane

Example 89, Injection Mold with Attached Hydraulic Core Pull for Automatic Measuring Tubs Made from PC

A measuring tube for a liquid-distributing manifold was to be produced fully automatically. The molded part had to be comparatively thick walled, as operating pressures of up to 10 bar and operating temperatures of up to almost 100°C (212°F) occur. It proved expedient to inject from one face end to prevent unilateral stresses that would distort the tube to an unwelcome degree. In this case an injection molding machine capable of parting line injection is advisable. The smallest possible machine suitable can be employed without having to arrange the molded part eccentrically in the tool (Fig. 1), which would only result in long flow paths and unfavorable one-sided machine loading. Hydraulic core pulling is employed, as mechanical cores are unsuitable for such lengths of stroke. Insert cores would be unacceptable, because the requirement is for automatic production of the molded part.

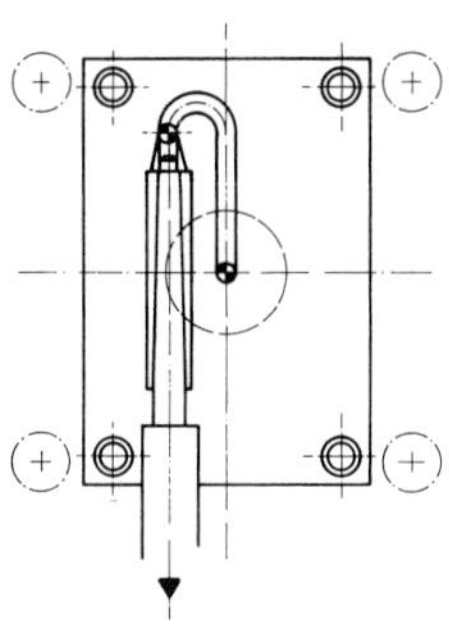

Figure 1 Required positioning of the molded part in the parting plane of the mold with the greatest possible utilization of the machine size and central injection

The mold design (Figs. 2 to 7) starts with central positioning of the measuring tube in the parting line. To obtain the clean scale graduation surface necessary for reading off the flow rate, these divisions have been machined into the fixed mold half. The core of the measuring tube is now located precisely in the center of the mold cavity inserts. It is centered at the end of the tube as well as at the entrance.

The sprue approaches the measuring tube via the end of the core in three adequately dimensioned sections. The melt flows around the core uniformly with this type of gating and is furthermore centered accurately. Below the mold on the moving half, core (3) is housed in a yoke (5), which is fixed in its direction precisely by guide rods (6). A cross plate (7), into which the hydraulic cylinder (8) has been screwed, is fitted to the end of the guide rods. The cylinder (8) has been additionally supported (9), to avoid any excessive vibrations from this long substructure during the travel movements of the mold. The piston rod (10) of the cylinder is coupled to the yoke (5). Heating/cooling channels (11) have been provided on the fixed as well as on the moving mold halves. Of great importance also is the possibility of core cooling. The core has been drilled for this purpose and divided into two chambers with a cascade by a separating baffle (12).

Hydraulic cylinders as well as connecting hoses to the hydraulic circuit of the machine are not part and parcel of the core pulling equipment, as is often assumed. The size of the cylinders has to be matched to the pressures occurring in the mold. This then also becomes the decisive factor in establishing whether the core to be pulled can be held just by the cylinder pressure or if it has to be mechanically interlocked as well. In the example presented interlocking is not necessary. It has proved advantageous for the cylinders to be equipped with cushioned end positions in both movement directions. A considerably gentler operation can be obtained in this way.

It is essential for the operating sequence of the controls to monitor the position of core (3) in its most forward and rearmost position electrically through limit switches (13, 14) and pass this information on to the machine control.

To describe the operating sequence, it is assumed that the mold is fully open and void of molded parts, i.e., in the starting position:

Core (3) is moved into the mold by hydraulic cylinder (8). The mold closes and the injection process starts. As soon as the injection, holding pressure and cooling times have elapsed the mold is opened for a few millimeters only. Due to the core (3) being mounted on the moving mold half, the measuring tube (1) with its scale (2) is released positively from the fixed mold half. Now core (3) is retracted completely from the measuring tube (1). The mold moves to the opened position and the hydraulic ejector of the machine moves forward. This is coupled with the ejector bar (15), which pushes the ejector plates with their built-in ejector pins (16) for the measuring tube and the sprue forward, ejecting the completed molded part from the mold. The core is moved in again and another cycle starts.

To make the mold more solid the hollow space required for the ejector plates contains support pillars (17). An essential feature of this mold is the quartz-crystal pressure transducer (18) in the vicinity of the gate for assessing the mold cavity pressure, which is then controlled in accordance with the data received to prevent sink marks and to reduce internal stresses in the molded part.

Fig. 2

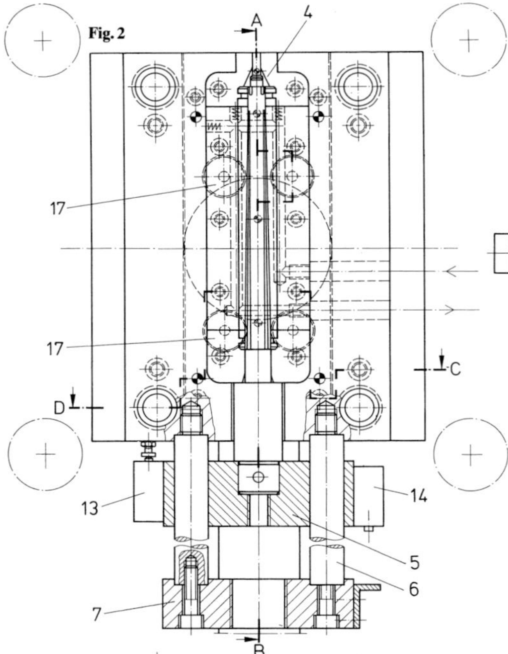

Fig. 3

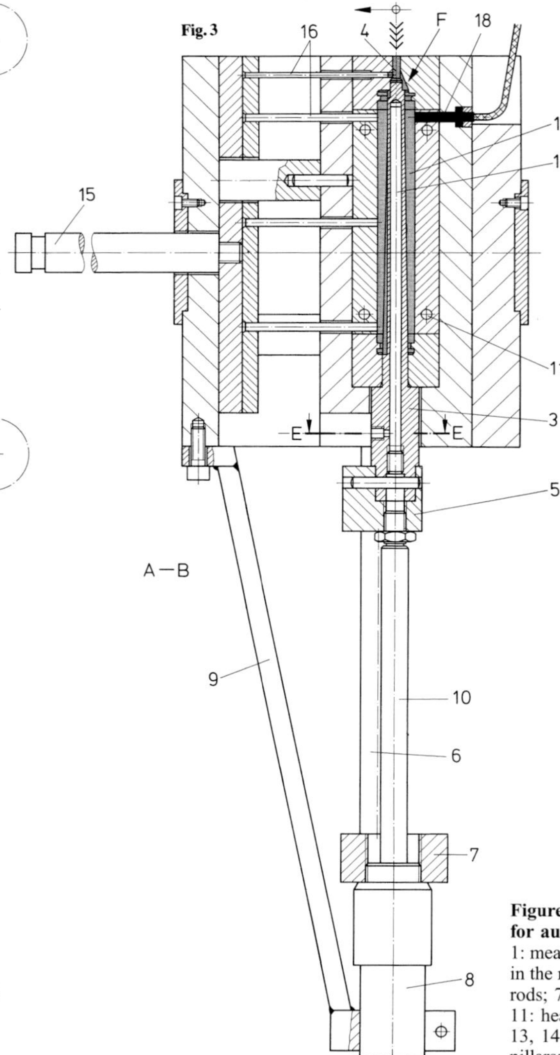

Fig. 4

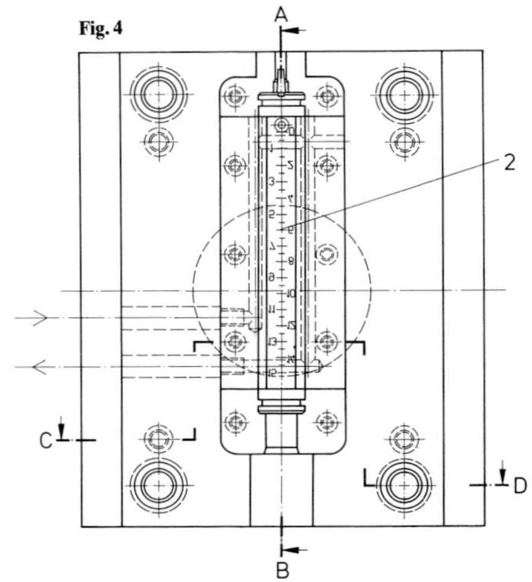

Fig. 5

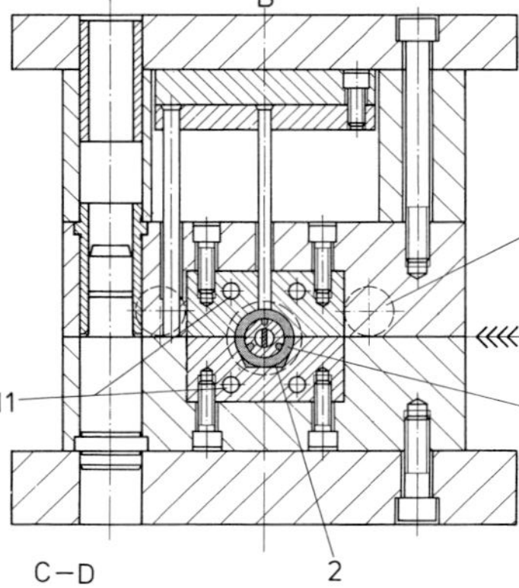

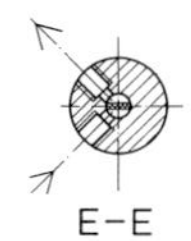

Fig. 6

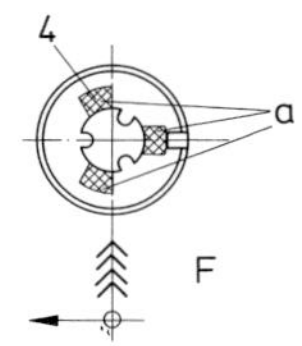

Fig. 7

Figures 2 to 7 Injection mold with attached hydraulic core pull for automatic measuring tube production
1: measuring tube; 2: graduation scale of the measuring tube engraved in the mold cavity of the fixed half; 3: core; 4: sprue; 5: yoke; 6: guide rods; 7: cross plate; 8: cylinder; 9: cylinder supports; 10: piston rod; 11: heating/cooling channels; 12: separating baffle in the core bore; 13, 14: limit switches; 15: ejector bar; 16: ejector pins; 17: support pillars; 18: quartz-crystal pressure transducer
a: Gate areas

Example 90, 48- and 64-Cavity Hot-Runner Molds for Coating Semi-finished Metal Composite with Liquid Crystalline LCP Polymer (Outsert Technology)

In this mold, two-piece electronic components are coated, and thereby encapsulated, with freely flowing, high-temperature LCP copolyester reinforced with 30% glass fiber content. Outsert technology is employed. The components joined together in a band are fed into the mold from coils and positioned therein in two rows of 24 cavities each. Subsequent to the encapsulating sequence, the next 24 are fed into the mold, etc. (Fig. 1). LCP was chosen in order to achieve the extremely thin wall thickness of approx. 0.2mm to protect the components (spools with ferrite cores) from mechanical damage. This requires a very flowable material. Since the components are soldered to circuit boards in an infrared oven (SMD technology), the polymer also has to have high shape stability. Additional properties, such as inherent flame resistance (UL 94 V-O) and a thermal expansion coefficient approximately corresponding to that of the metal material, LCP appears to be especially suited for applications in the electronics industry.

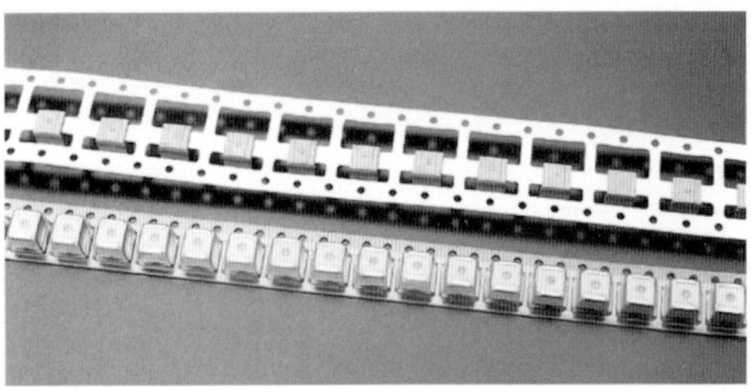

Figure 1 Metal rings coated by outsert technology.

However, this material has special characteristics that have to be considered during processing and when designing the mold. For one thing, *high* melt shear is *indispensable* for obtaining very low-viscosity. This can be achieved with very narrow channel diameters and high injection rates at high injection pressures. In this manner, long flow paths are feasible even for low wall thicknesses. However, the danger of jetting has to be considered.
Due to the abrasive effect of glass fibers combined with high flow rates, tool steels, such as 1.2721 and 1.2767, have proven insufficiently wear-resistant. Adequate service life can be achieved using P/M steels (see also Section 1.10.2.5).

Mold Design for 48 Cavities

Economic considerations led to the selection of a hot-runner system without subrunners. In order to eliminate irritating gate traces, among other things, special valve-gated nozzles with *conical* needle seats are used. The needles and annular pistons are moved independently from each other by a pneumatically controlled stroke plate. The specially developed hot-runner manifold with self-closing melt channels and the valve-gated nozzles with ring-shaped cross-sections of flow (3.5/2mm) are designed for high shear and shortest possible melt dwell times. The piston-type injection unit, Fig. 2, consists of the needle (2mm diameter) and a spring-loaded annular piston (3.5mm outer diameter).

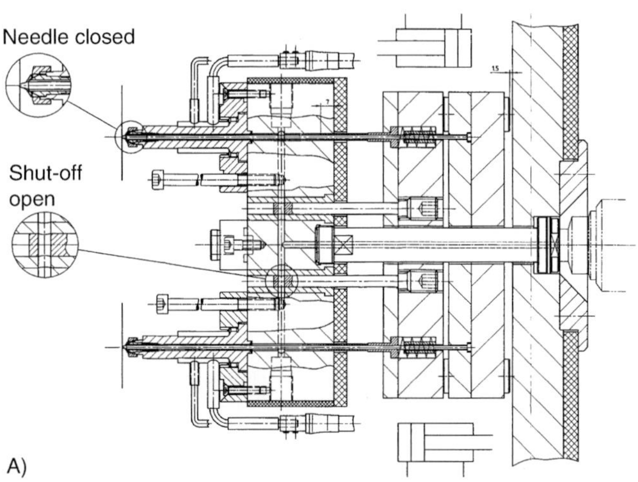

Figure 2 Construction and working principle of an injection mold whose cavity plates have two rows of 24 cavities each.
A: melt preparation, B: melt system closes, injection is prepared, C: injection and filling of 48 cavities, D: nozzles close, melt feed system traverses to start position

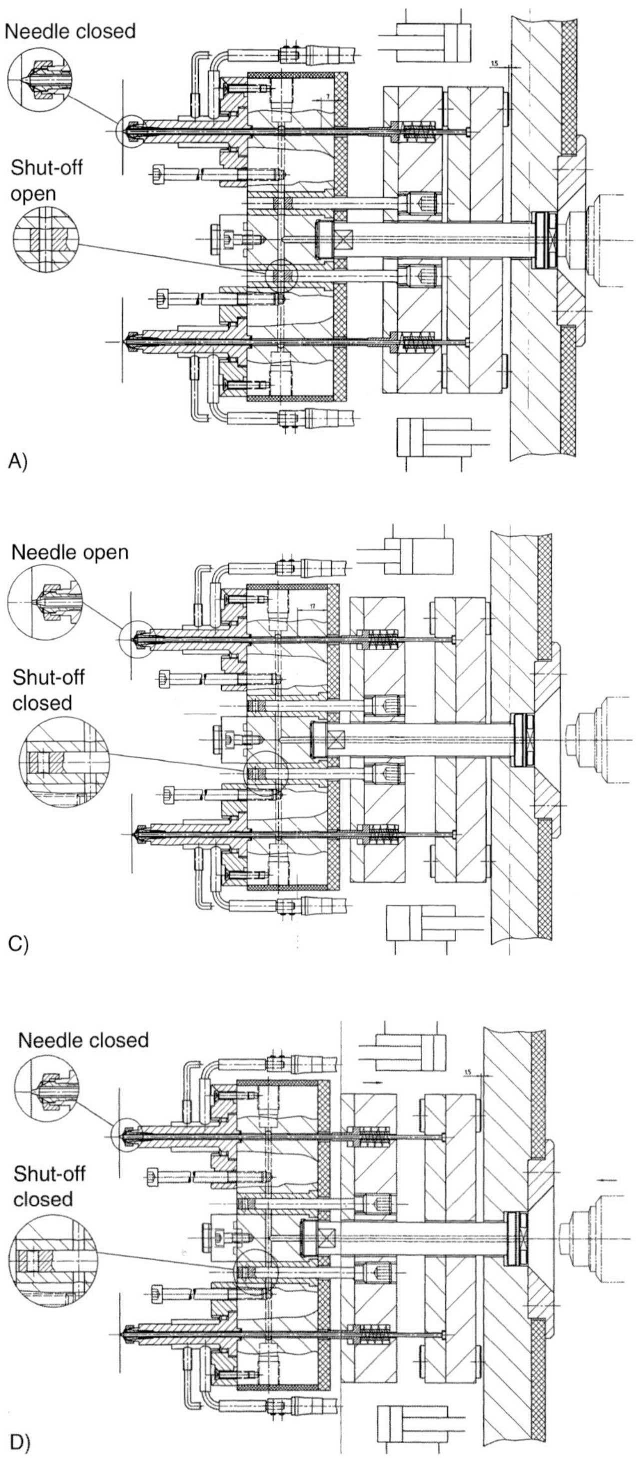

Figure 2 Continued

Figure 3 Metal bands coated by outsert technology

The injection molding sequence is illustrated in Figs. 2A to D. The melt is fed from a long, externally heated intermediate bush to the hot-runner manifold and to the valve-gated nozzles arranged in 2 rows of 24 each. Longitudinal center-to-center nozzle spacing is 12mm. The melt-feed channels in the hot-runner manifold can be directly opened or closed, depending on direction of movement, via the stroke plate (plate assembly) by two sliding frames, while each needle valve is activated by the stroke plate (plate assembly) for a 1.5mm stroke.

The stroke plate has two tasks to fulfill. In the first step, both sliders are actuated, thereby closing the melt feed channels of the hot-runner manifold. In a second step, the annular piston performs a stroke of 10mm to build up a maximum pressure of 2500 bar, injecting the melt into each cavity when the gate opens.

The hot-runner manifold is rheologically *imbalanced*, a situation which, due to the working principle of the piston-type injection unit, would have no advantage. Since they are jointly attached to one stroke plate, each piston-injection unit volumetrically filled with melt independently provides uniform molding conditions, such as uniform injection pressures and injection rates, and has strictly identical movement sequences. This system presumes extreme precision in the same-design valve-gating systems and piston-injection units. Special attention is demanded by the very narrow tolerances regarding perfect fit.

Following cooling time, both stroke plates are moved to their start position, thereby causing the needles to close the gates. Both slide plates open the melt feed channels of the hot-runner manifold, etc.

Despite the fact that *conical* valve-gating nozzles require frequent adjusting, this system was selected, since the conical ring gap can be set for optimum melt shear when the gate opens. If a cylindrical needle seat were used, this either would not, or only to a limited extent, be possible.

It should be noted that, as far as mold *and* processing technology are concerned, liquid crystalline

Figure 4 64-cavity hot-runner system with open flat sprue nozzles (Courtesy: Günther Heißkanaltechnik, Frankenberg)

LCP polymer does not differ much from other thermoplastics. However, it does require considerable experience.

64-Cavity Mold Design

This application is the same in principle; however, it uses externally heated, *open*, *flat* sprue nozzles with tips (Fig. 4, so-called twin-flat nozzles) using outsert technology. Thereby, molded parts (housings, Fig. 3) made from glass fiber reinforced LCP are injection molded onto a metallic punched strip. The flat nozzles are thermally separated from each other and consist of a pipe embedded in a brass nozzle body. They are heated by heater cartridges. Temperature at each gating point is separately regulated. To avoid excessive wear due to the glass fiber content, the nozzle seats are manufactured from a P/M hard metal, such as TZM. The hot-runner manifold is insulated on all sides to minimize radiation heat loss. Nozzle spacing is 9.35 mm × 1.5 mm. Molded part weight is 0.1 g. The entire hot side has 66 regulation points, one for each of the 64 gates and two for the hot-runner manifold. According to manufacturer specifications, minimum spacings of 7mm can be realized with the open flat nozzles described, whereas only 9mm is possible with similar valve-gating nozzles. By contrast to the forty-eight cavity mold, the choice of *open* sprue nozzles results in an otherwise negligible gate trace on the order of approx. 1/10mm. This system presumes, among others, intensive mold cooling, especially in the gate area, as well as minimal heat loss in the hot-runner system in order to realize the required thermal homogeneity.

Literature

1. *Gleisberg*, P. et al. "48 Einzelkavitäten in einer Platte "Kunststoffe 89 (1990) 5, p. 68-71
2. N.N. "Flüssigkristalline Polymere spritzgießen", Plastverarbeiter 47, Jahrgang Nr. 5, p. 92-94
3. DE-PS 19632315CI: Verfahren und Vorrichtung zur Herstellung von Kleinstspritzgießteilen (1996), Görlich, R.

Example 91, 24-Cavity Hot-Runner Injection Mold for Polyacetal Spool Cores

The spool cores serve to alternately wind the magnetic tape in an audio cassette. Since these cores must be absolutely free of any gate vestige, the injection mold (Figs. 1 to 3) was designed with a hot-runner system in which the gate orifices are closed by shutoff pins (valve gates).

Individually actuated shutoff pins in the hot-runner system would have resulted in a disproportionately large cavity spacing, given the 22 mm diameter of the molded parts. Instead, a three-tip valve-gated shutoff nozzle (Manufacturer: Primatechnik, Mannheim, Germany, patented) was employed. The three gate orifices in the nozzles are placed around a circle and are spaced 22 mm apart. Since the molded parts are gated off-center, a cavity spacing of 38 mm results.

The three shutoff pins (18, Fig. 2) are guided in bushings (19) and are precentered by a pilot taper in the nozzle body (20) prior to entering the cylindrical gate orifice (diameter: 1.1 mm). As the pins enter the pilot taper, the displaced melt in front of each pin flows backwards through specially designed relief channels. The shutoff pins are opened and closed by a pneumatic cylinder (21).

The nozzle body (20) contains a compression piece (22) that distributes to each of the three gates the melt coming from the diverters (23) in the hot-runner manifold (24). The inlet bushing (8, Fig. 1) of the hot-runner manifold is designed to accommodate a slip-fit extension on the machine nozzle. This permits decompression of the melt in the manifold prior to part ejection.

The distance between the manifold (24) and compression piece (22) is such that a preload of 0.03 mm results at the operating temperature. This preload ensures a tight seal at the interface, while at the same time permitting thermal expansion of the manifold with respect to the nozzles, which are mounted in the mold plate (5).

The hot-runner manifold is heated by tubular heaters (26) and fitted with insulating plates (29) to reduce radiative heat loss. The installed heating capacity in the manifold is 8000 W. The inlet bushing is fitted with a 500 W heater band, and each of the three-tip hot-runner nozzles has a 1000 W coil heater. A total of eleven control zones is provided.

Mold Temperature Control

The operating temperature of the piston rings in the pneumatic cylinders for the valve gates should not exceed 100°C (212°F). Accordingly, cooling channels are provided in mold plate (6) in close

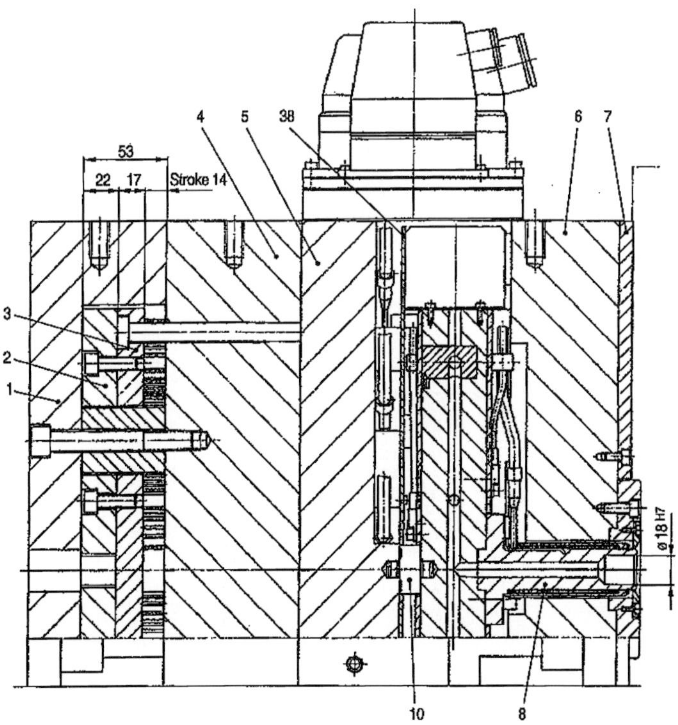

Figure 1 24-Cavity hot-runner injection mold for polyacetal spools
1: clamping plate; 2, 3: ejector plates; 4, 5, 6: mold plates; 7: insulating plate; 8: inlet bushing; 10: locator

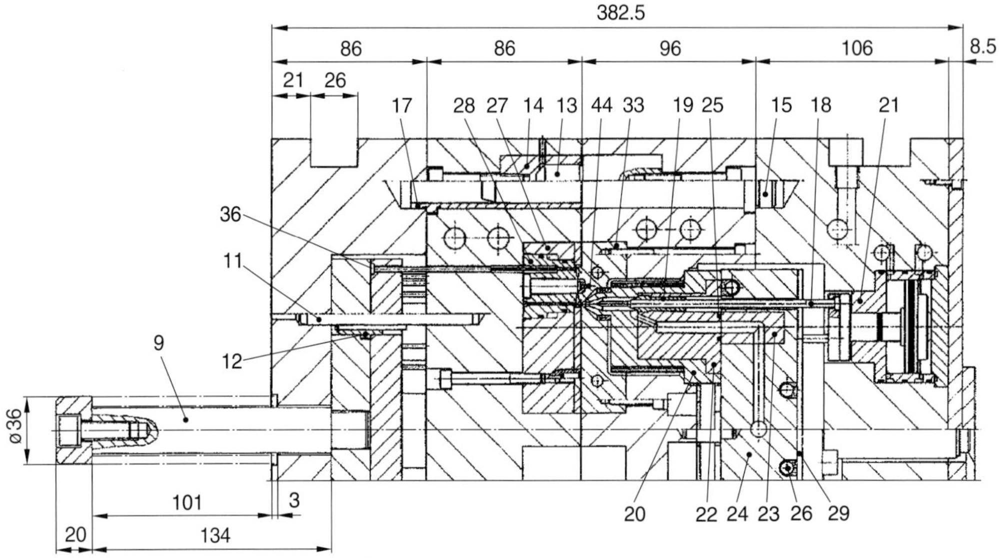

Figure 2 24-Cavity hot-runner injection mold for polyacetal spools
9: ejector rod; 11: guide pin; 12: ball bearing guide bushing; 13: locating pin; 14: locating bushing; 15: leader pin; 17: guide bushing; 18: shutoff pin; 19: guide bushing; 20: nozzle body; 21; pneumatic cylinder; 22: compression piece; 23: diverter; 24: hot-runner manifold; 25: O-ring; 26: tubular heater; 27, 28, 33: mold insert; 29: insulating plate; 36: knockout pin

proximity to the pneumatic cylinders and can be used to control the temperature of this plate prior to turning on the nozzle heaters. Each hot-runner nozzle is paired with a set of mold inserts (27, 33) containing three cavities. The mold inserts (28) on the movable side have cooling grooves around them, while the inserts (33) on the stationary side have drilled cooling channels. The cavity temperature is 60°C (140°F).

The mold inserts are made from steel grade 1.2344 and are hardened. The mold halves are guided by leader pins (15) and located with respect to each other by locating pins (13) and bushings (14). Part ejection is accomplished via knockout pins (36). The parts are molded at a cycle time 5.4 s. The mold dimensions are 214 mm × 430 mm × 382.5 mm (high).

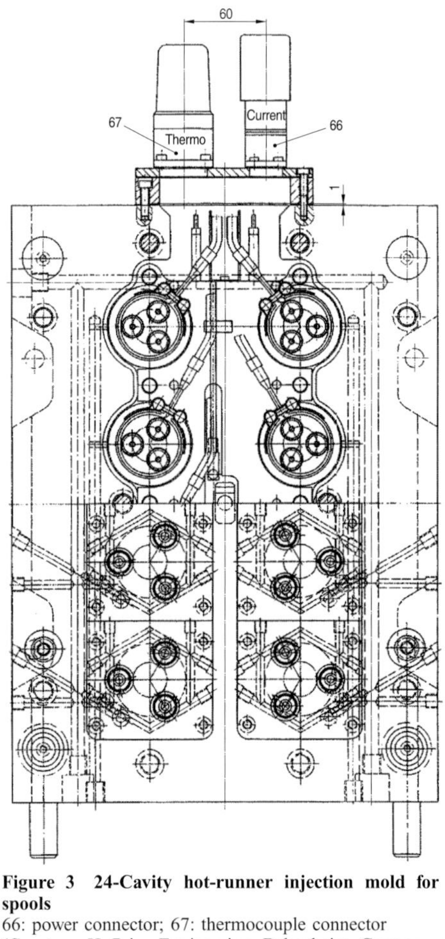

Figure 3 24-Cavity hot-runner injection mold for polyacetal spools
66: power connector; 67: thermocouple connector
(Courtesy: H. Prinz Engineering, Babenheim, Germany, now PSG)

Example 92, Two-Cavity Hot-Runner Mold for Loudspeaker Covers Made from Polyacetal

The grilles, which are some 150 mm × 180 mm in size (Fig. 1), are used as covers for loudspeakers in car interiors. For acoustic reasons, they are required to have a large number of perforations that will let through sound. The 8000 or so holes have a diameter of 1.1 mm and are positioned within a hexagonal honeycomb structure that has an inside clearance of some 13 mm, which was selected on both strength and flow engineering grounds. The structures are surrounded by a continuous wall around the outside. The wall thickness is approximately 1 mm for the most part. The outer contour of the grille is of a curved 3D design. The range on offer includes different grilles which are injection molded in pigmented compound. The material employed is an acetal copolymer with an MFR 190/2.16 of 28 g/10 min.

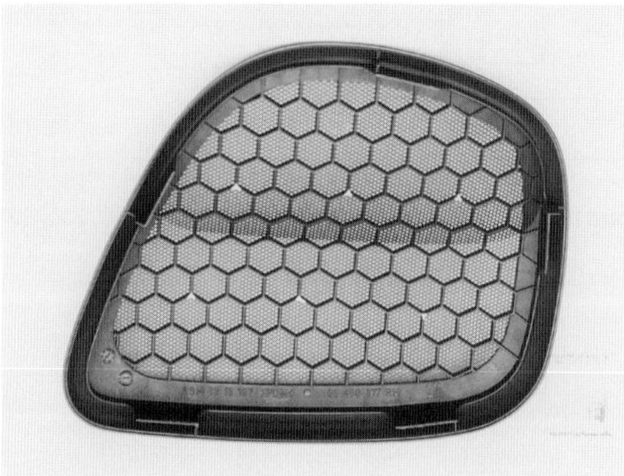

Figure 1 Loudspeaker grille

Mold

The mold (Figs. 2 and 3), which is 346 mm by 696 mm, and 403 mm high, is a two-cavity hot-runner mold. The melt passes through a connecting nozzle (1), into the hot runner plate (2), and from there into each of six cavity nozzles (3). The hot-runner system from Ewikon Heißkanalsysteme, Frankenberg, Germany, is fully controlled and is equipped with internal heating.

In order to improve on the color change behaviour of the hot-runner plate, flow-optimized packing boxes (4) were employed as deflector elements, with a polymeric coating offering a high heat resistance. The gate diameter is 0.8 mm. The 3D-curvature of the grille structure means that the cavity nozzles (3), have different shaft lengths of 69 and 76 mm respectively. The distance between two nozzles is minimal, at 42 mm. The torpedo ends protrude into the gate and thus form an annular gap. The tear-off height is ≤0.1 mm. The system only requires an installed heating capacity of 170 W overall. Mold wall temperatures are 80 to 90°C (176 to 194°F), and the melt temperature corresponds to the temperature in the hot-runner plate and the cavity nozzles, of approximately 210°C (410°F). Cycle times of between 35 and 40 s are attained. The heating and cooling of the mold is performed via 2 × 2 independent heating/cooling circuits. The mold halves are guided by four columns in the standard manner.

Demolding

Since the surface of the molded part has to fulfill stringent quality requirements on the visible side, the gates are positioned on the concave rear of the grille, which is then demolded on the gate side as well, with the ejectors acting solely on the surrounding contour. The ejector unit is guided by ball-bearing guides (6), and driven by hydraulic cylinders (7).

This concept means that the hot runner plate cannot be supported over its entire area but just by two support bolts (9). The molded parts are removed by a handling unit and deposited in defined position to cool to room temperature.

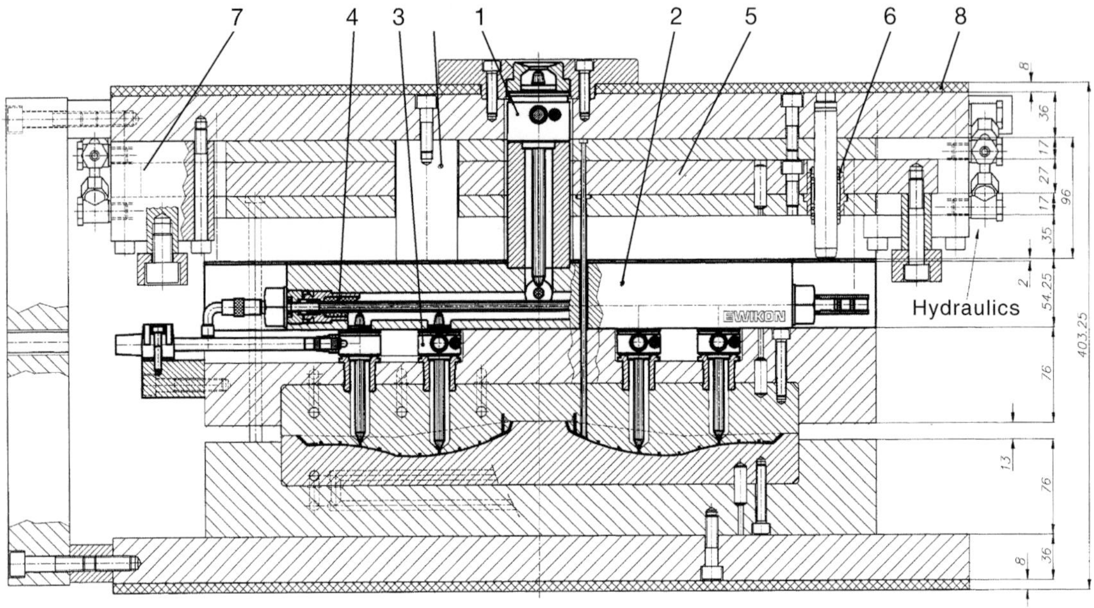

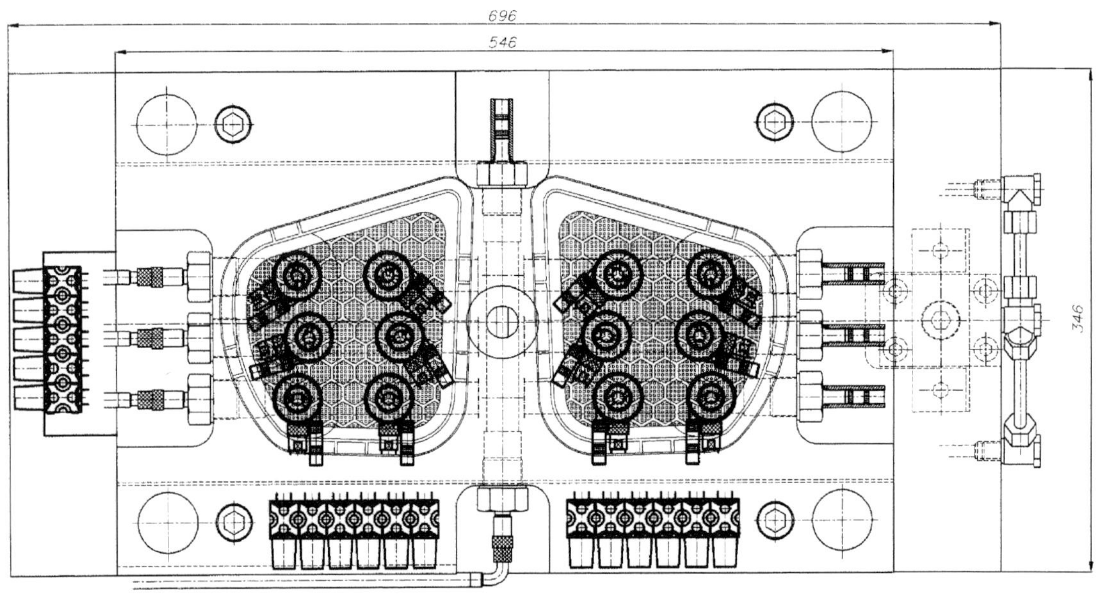

Figures 2 and 3 Two-cavity injection mold for loudspeaker grilles
1: connecting nozzle; 2: hot-runner plate; 3: cavity nozzle; 4: deflection element; 5: ejector plate; 6: ball-bearing guide; 7: hydraulic cylinder; 8: insulating plate; 9: support bolt
(Courtesy: Ewikon Hot Runner Systems, Frankenberg, Germany)

Example 93, Injection Mold with Air Ejection for Polypropylene Cups

In developing injection molding of thin-walled polypropylene packaging items, the greatest importance by far must be attributed to mold design. Thin-walled polypropylene cups cannot be produced economically and reliably in the types of molds used for polystyrene. Because of the greater enthalpy and lower thermal conductivity compared to those of polystyrene, the cooling must be more effective when processing polypropylene. Because of its reduced rigidity during ejection and greater tendency to shrink onto the core, ejection of thin-walled polypropylene cups by means of ejector rings creates problems. While air-assisted valve ejector systems can facilitate ejection, the less effective cooling possible with such a system does not permit extremely short cycle times. Roughening or draw polishing of the surface of the core is not a suitable solution for transparent cups because of the detrimental effects on the transparency. Air ejection by means of static or dynamic air valves is only of limited use in fast-cycling single-cavity molds, since at production rates of over 20 shots per minute rapid and exact closing of the valve is hindered by the air remaining in the valve stem. With side air ejection, stripper rings, which represent the most significant wearing part when producing cups by injection molding, can be eliminated. Figure 1 shows the design of such a core with side air ejection. With the mold illustrated in Fig. 2 acceptable cups were produced in a cycle time of $\leqslant 1.5$ s using side air ejection exclusively.

This ejection system can also be employed with multiple-cavity and stack molds (see examples 36 and 44). With polypropylene, air must also be introduced at the bottom from the cavity side. This serves not only to prevent the formation of a vacuum but also to sever the tough gate.

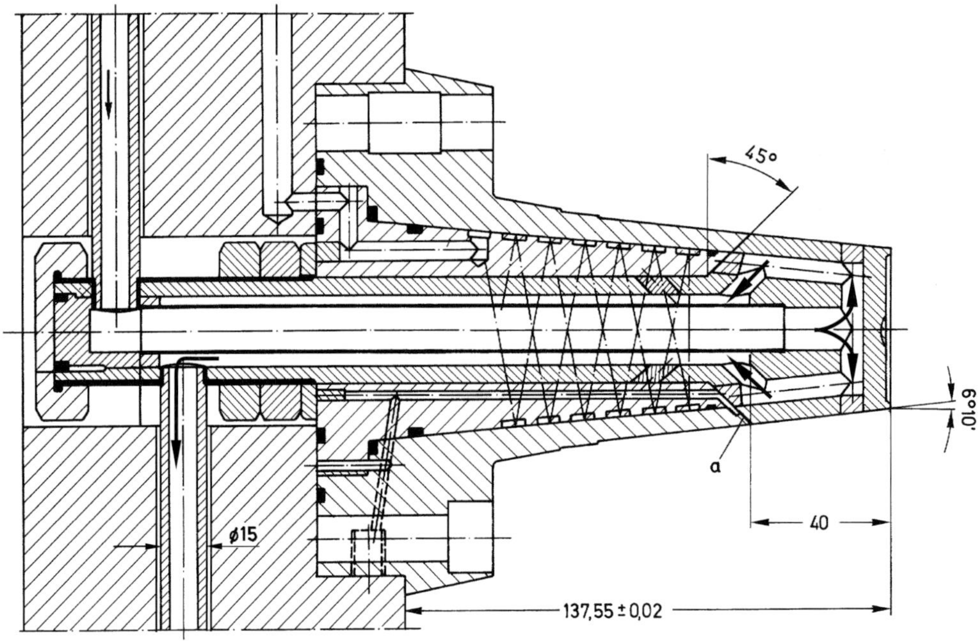

Figure 1 Core of a cup mold with side air ejection, dimensions in mm
a: annular gap with a width of 0.01 mm
(Courtesy: Hoechst AG)

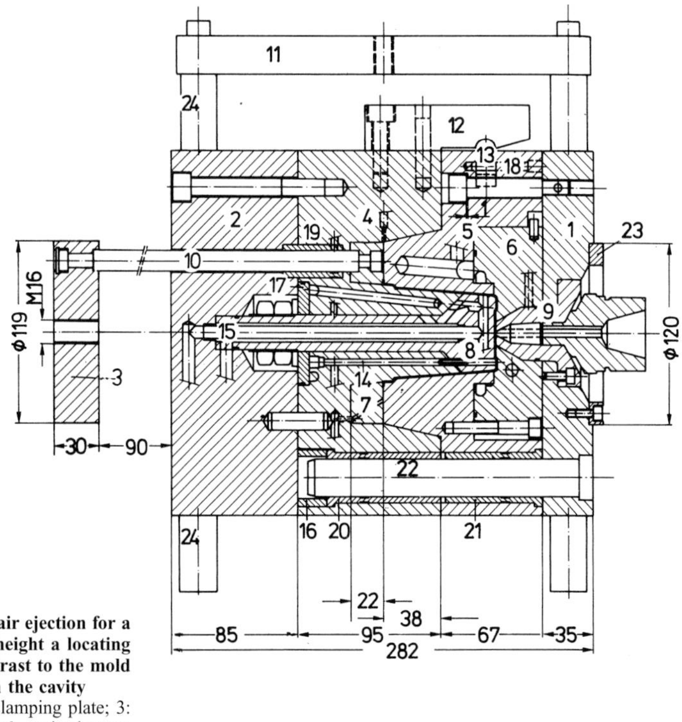

Figure 2 Single-cavity injection mold with side air ejection for a polypropylene dessert cup. To reduce the mold height a locating taper was provided parallel to the cavity, in contrast to the mold shown in Fig. 1, where the taper is in series with the cavity
1: stationary-side clamping plate; 2: movable-side clamping plate; 3: ejector plate; 4: core retainer plate; 5: cavity plate; 6: cavity bottom insert; 7: stripper ring; 8: core tip with cooling channels; 9: hot sprue bushing; 10: ejector rod; 11: strap for transporting the mold; 12: latch; 13: latch bolt; 14: static air valve; 15: central cooling water tube; 16: spacer sleeve; 17: sealing plate; 18: intended cavity take off; 19, 20, 21: guide bushing; 22: guide pin; 23: locating ring; 24: strap mounting bolts

Example 94, Molds for Manufacturing Optical Lenses Made from PC

When designing molds for optical plastic parts, standard mold components and column-guided mold frames are used. This guarantees that the individual parts are interchangeable and reduces maintenance and repair times. Additional advantages include stocking of spare parts and reusability of mold components after the completion of a product.
A distinction is made between injection molds with mechanically and those with hydraulically operated compression.

Injection Mold for Mechanical Compression

Figure 1 shows a two-cavity injection compression mold for objective lenses (meniscus lenses) that is set up for mechanical compression. The compression step in this case is carried out indirectly by the clamping unit of the injection molding machine.

The Compression Sequence

The mold is closed at low pressure before the filling step, so that there is a gap of 0.2 mm between the mold mounting plate (2) and the floating plate (8). The gap is maintained by means of spring washers (19), which exert pressure through the tapered slide (14) and pin (18) on the mounting plate (2). During injection, the spring washers (19) must act against the injection pressure and keep the main parting line between the two mold plates (6 and 7) closed by their spring force, preventing the injection mold from opening. When the filling process is complete, the injection molding machine switches to full clamping pressure, which acts as compression pressure through the lens stamper after the force of the spring washers has been overcome and closes the initial gap of 0.2 mm.

Design Details

The sprue with attached runner conveys the melt to the pinpoint gate on the edge of each cavity. Ejector pins are positioned around the edge of each lens. The mold is vented through the parting line; the vent gap should not exceed 0.02 mm. The internal cavity pressure is measured piezoelectrically at a runner and stored as a measurement and control parameter. A prerequisite for perfect lens elements is non-porous surfaces and optically perfect stamper depressions in compliance with DIN 3140. The lens thickness can be corrected by adjusting the position of the stamper via the tapered slide (14).
The mold locating means is separate for each cavity. The inserts on the stationary half are firmly fitted into the mold plate (7), while the movable-side inserts have a certain radial play to permit alignment with the stationary-side inserts.

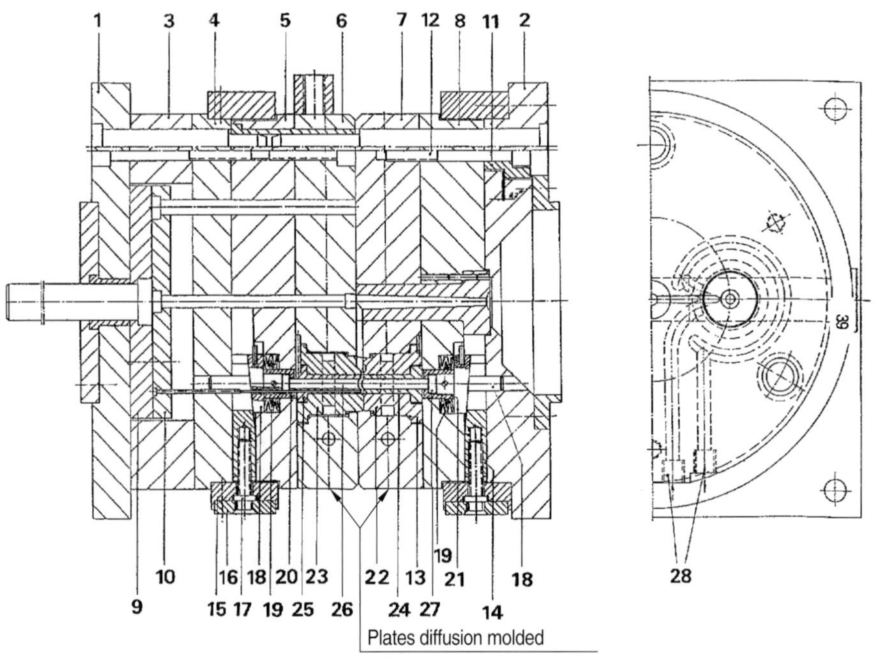

Figure 1 Two-cavity injection mold with mechanical compression for objective lenses
1: clamping plate; 2: clamping plate; 3: spacer ring; 4: backing plate; 5: backing plate; 6: mold plate; 7: mold plate; 8: backing plate; 9: ejector plate; 10: ejector retainer plate; 11: spacer; 12: screw; 13: O-ring; 14: tapered slide; 15: retaining ring; 16: plate; 17: adjusting screw; 18: wedge pin; 19: spring washers; 20: cap; 21: cap; 22: mold insert; 23: mold insert; 24: mold sleeve; 25: mold sleeve; 26: lens stamper; 27: lens stamper; 28: connection for mold cooling

The mold can be heated with heaterbands. A fluid circulating temperature control system with PID controls is provided for each mold cavity. The channels (28) are machined in the mold plate halves, hard chrome plated and joined to a single system by diffusion welding. O-rings (27) are used to seal the inserts (22 and 23).
All movable functional parts such as stampers and ejectors have been provided with appropriate play at their sliding surfaces so that they slip at the operating temperature of the mold and do not seize.

Injection Mold for Hydraulic Compression

An injection mold for hydraulic compression is shown in Fig. 2. In this case, the injection molding machine needs a pressure cushion, i.e. a separate hydraulic cylinder for the compression step. The compression step is initiated after the holding pressure stage.

The Compression Sequence

After the filling step and while the injection pressure is still active (0.3 to 1.0 s), the transfer from injection pressure to holding pressure is made as a function of the filling pressure. The compression step is initiated independently of holding pressure by a pressure cushion, i.e. an additional hydraulic cylinder, with approx. 0.1 s delay.

Design Details

Here, too, the cavity is filled via a pinpoint gate. Ejector pins and ejector sleeves are used (18), which permit stress-free ejection of the lenses. The mold is also vented via the parting line.
The lens stampers (16, 17) are made of ESR steel (material no. 1.2842, with a hardness of 63 Rockwell C or as a combination of a steel holder with a ceramic insert. Repositioning of the lens stamper, which must be carried out after final polishing or to adjust the lens thickness, is accomplished by turning the threaded spindle (12) and worm gear (13). The worm gear (13) changes the axial position of the stamper holder (14, 15) via the adjusting thread. The adjusting thread must be dimensioned to withstand the mold buoyant force at an injection pressure of 1000 bar.
The mold halves are located by means of three conical locating units (10). For stringent requirements with regard to the concentricity (e.g. 0.010 to 0.015 mm) and surface quality of the plastic lenses, each mold cavity has its own locating unit with short guide and very tight tolerances. With these elaborate measures, very high-precision lens radii are obtained and any lateral movement of the mold cavities that might be caused by the play between the machine tie bars and guide bushings is eliminated.

Stamper Inserts

The quality of the injection molded parts depends largely on the precision of the stamper as regards surface, centering and life expectancy during operation. A precision of 0.5 to 2 (Newton) rings (2 rings = wave length of light λ) is required at diameters of 5 to 10 mm. The stamper surface must be prepared with similar accuracy.

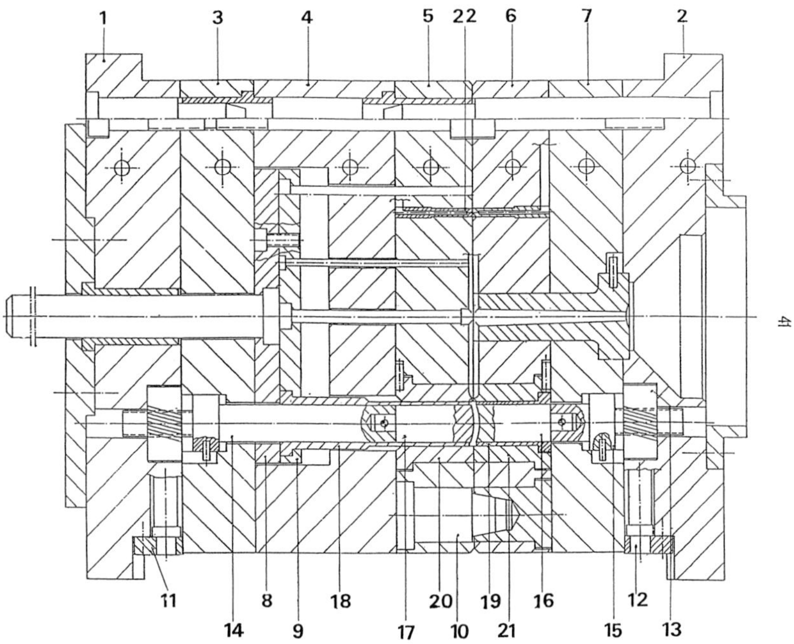

Figure 2 Six-cavity injection mold for meniscus lenses
1: clamping plate; 2: clamping plate; 3: backing plate; 4: backing plate; 5: mold plate; 6: mold plate; 7: backing plate; 8: ejector plate; 9: ejector retainer plate; 10: locating unit; 11: retainer plate; 12: threaded spindle; 13: worm gear; 14: stamper holder; 15: stamper holder; 16: lens stamper; 17: lens stamper; 18: sleeve ejector; 19: sleeve; 20: mold insert; 21: mold insert; 22: thermocouple

Example 95, Two-Cavity Injection Mold for a Polycarbonate Steam Iron Reservoir Insert

The insert for the reservoir of a steam iron (Fig. 1) is of a complicated shape due to the functions it has to fulfill. The insert serves as closure of the opening on the face of the reservoir, for instance. A spray nozzle is screwed onto the retaining thread (Fig. 2). The associated spray pump is mounted on a supporting strip. A connecting tube runs between spray pump and nozzle. This tube is pushed onto the connection stud A at the rear of the nozzle-retaining thread.

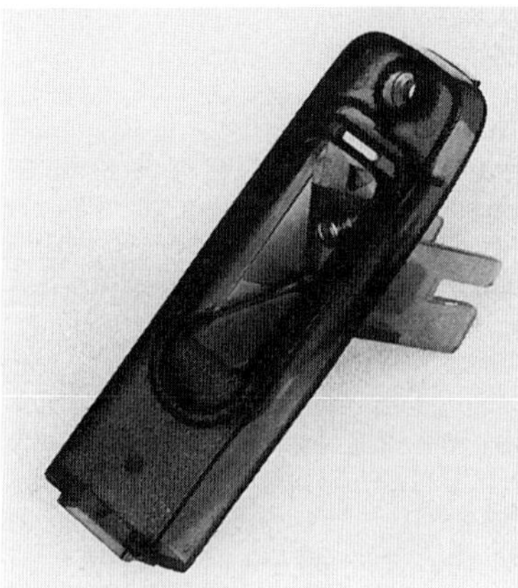

Figure 1 Reservoir insert of hydrolysis-resistant polycarbonate, color: transparent blue

Mold

The mold has been constructed to incorporate two cavities and a conventional runner (Fig. 3). Due to the angle of the spray nozzle in relation to the plane in which the reservoir is employed, the mold is equipped with an unscrewing device for both cavities and angled lever-operated demolding mechanisms for the undercuts formed by the connecting studs and their bores. Both cavities are filled through submarine gates on the lower insert rib in a nonvisible area (arrows in Fig. 2). The gates are severed with the simultaneous ejection of the molded parts and the runner. The cavities proper have been machined into cavity inserts (40–43).
Cooling channels have been arranged in the mold plates (9) and (11) outside the cavity inserts. Only the supporting strips of the two molded parts are served by a cooling pin (52), which penetrates through the ejector plate into the cooling sleeve (53) situated in the clamping plate (4), where it is surrounded by cooling water. The cavity inserts and the threaded cores as well as the contour cores are made of hardened steel (material. no. 1.2343).

Mold Operation

Once the mold cavities have been filled and the cooling time has elapsed, the unscrewing cores (38) are rotated and withdrawn with the aid of the guide thread in the guide nut (37) by displacing the rack (49) via the hydraulic cylinder (50) and the pinions (61) and (62) before the two-plate tool is opened. Simultaneously with the thread-forming sleeve of

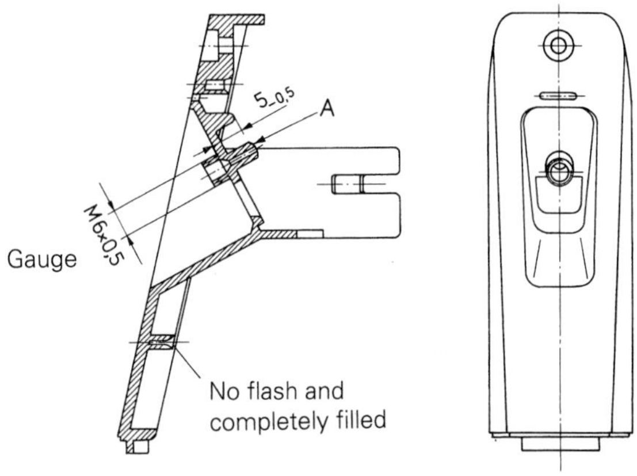

Figure 2 Inspection diagram for the reservoir insert; bore holes and openings without flashes, no sink marks, voids, scratches or flow marks on visual faces; part must fit into the frontal reservoir opening

the unscrewing cores (38) the contour pins (35) arranged centrically inside them are demolded. These contour pins also locate the core pins (63) at the tip. At the conclusion of the unscrewing sequence and actuation of the limit switches (51), the mold opening movement is initiated. The

Figure 3 Two-cavity injection mold with unscrewing mechanism for the reservoir insert shown in Fig. 1
1, 2: spacer rails; 3: movable-side locating ring; 4: movable-side clamping plate; 5: ejector plate; 6: ejector retainer plate; 7, 8: shoulder bolts; 9: movable-side mold plate; 10: sprue puller bushing; 11: stationary-side mold plate; 12: stationary-side clamping plate; 13: sprue bushing; 14: stationary-side locating ring; 15: limit switch, unscrewing strip; 16: limit switch; base plate; 17: limit switch, spacer plate; 18: cylinder mounting plate; 19: limit switch support; 20: washer; 21: cylinder spacer strip; 22: cylinder unscrewing strip; 23: shim for sprue puller bushing; 24: guide bushing for ejector rod; 25: ejector rod; 26: ejector plate; 27, 28: articulated bushing; 29: rocker retainer strip; 30: rocker; 31: guide bushing; 32: spring guide bushing; 33: movable-side contour core; 34: actuating strip for rocker; 35: contour pin; 36: stationary-side contour core; 37: guide bushing; 38: unscrewing core with lead thread; 39: bearing bushing; 40, 41: movable-side contour inserts; 42: large movable-side contour insert; 43: large stationary-side contour insert; 44: bearing bushing for rocker; 45: shaft; 46: spacer bushing; 47: rack guide; 48: bearing bushing; 49: gear rack; 50: hydraulic cylinder; 51: limit switch; 52: cooling pin; 53: cooling sleeve; 54: cooling sleeve plug; 55, 56: runner ejector; 57 to 60: ejector pins; 61: pinion; 62: gear; 63: core pin; 64, 65: springs ▶

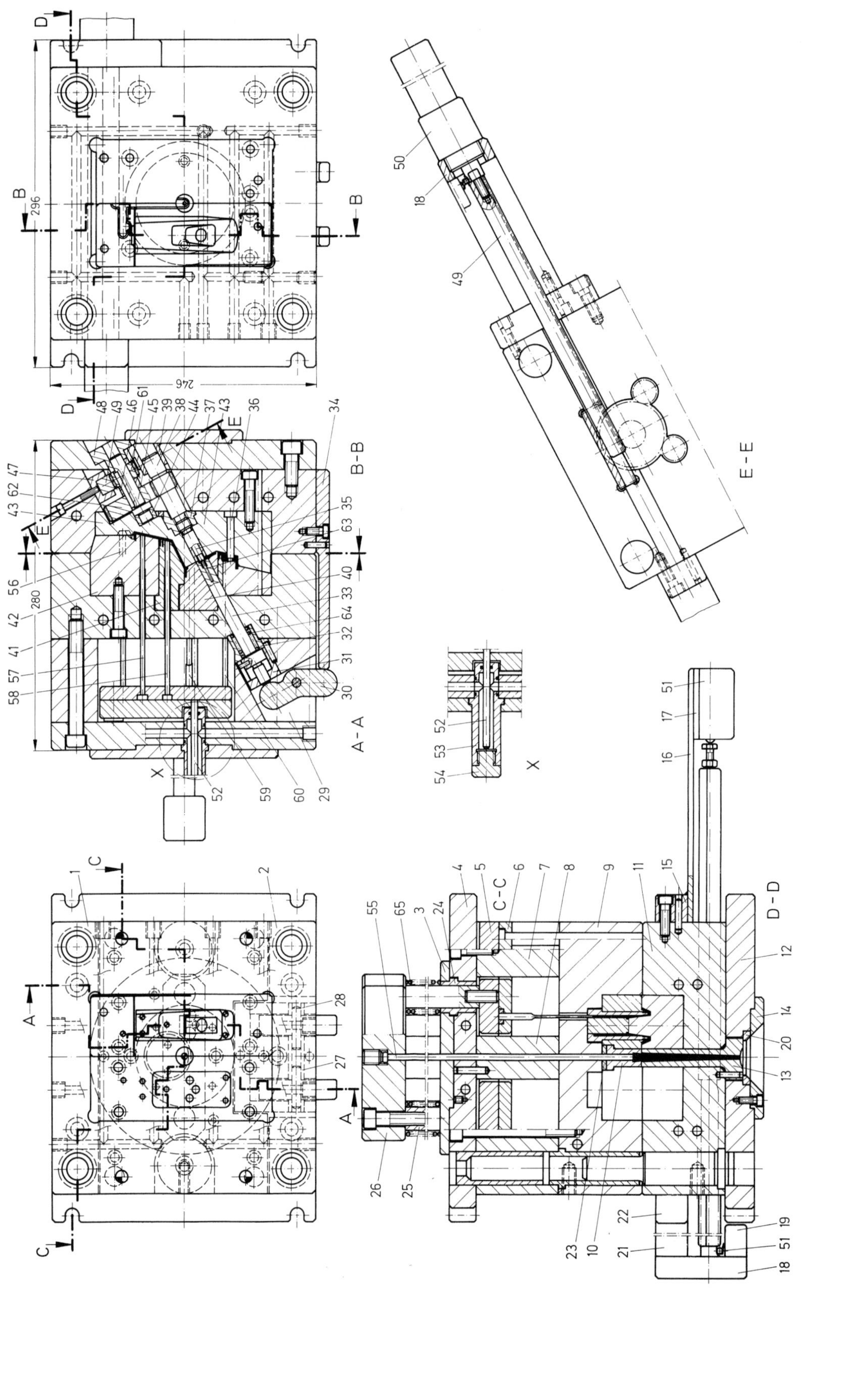
A-A
B-B
C-C
D-D
E-E
X

actuating strip (34) releases the rocker (30), enabling the compression spring (64) to push the contour core (33) away from the molding, thereby allowing the tube connection on the reservoir insert to be demolded internally and externally (core pin 63). Thus the obstruction to demolding the article has been removed. Only after this release must the plate (26) at the end of the ejector bars (25) be allowed to contact the fixed machine ejector during further opening movement of the mold, thereby pushing the ejector pins forward to demold the articles as well as the runner.

Prior to mold closing, the ejectors must be retracted. During the closing motion, the contour core (33) is returned to the molding position by means of rocker (30) and actuating strip (34). The unscrewing cores (38) are advanced after the mold has closed.

Example 96, Injection Mold with Pneumatic Sprue Bushing for a Headlight Housing Made from Polypropylene

The simpler the design and operation of an injection mold the more economical it is for volume production. Housings for car headlights which can be retrofitted as an optional extra are parts that fall into this category. The following description will deal with a mold for these lamp housings (Fig. 1), which are produced in flame-retardant polypropylene reinforced with 15 wt.% glass fiber. The dimensions of the headlight housing are 80 mm × 170 mm × 60 mm. The wall thickness is 2 mm; part weight is 84 g. The cycle time is 12 s.
The mold was constructed with standard mold components. Using the selection tables and catalogues from standard-component manufacturers, it is possible to determine the appropriate gating system [1,2]. The decision to produce a single-cavity mold was made due to cost considerations arising from the planned production quantities. The pneumatic sprue bushing was selected in order to have a smooth running mold without the need for additional control equipment required for a hot-runner.

Mold Design

Figure 2 shows the design of the mold, which has been assembled mostly using standard mold components. The lamp housing is gated via the pneumatic sprue bushing (25), which is available ready for installation. The part is stripped off using ejector pins. The ejector sleeves (21) are provided for the bores in the brackets which are connected with a film hinge. The internal bosses are released and the core (34) pulled via the lifters (37), which are mounted and actuated by the ejector plates. In order to be able to accommodate the support pillars (19) as well as the ball guides (12) within the ejector plates (7, 8) of the relatively small mold, an enlarged ejector plate version has been selected. The support rails are not positioned in the usual manner, but only as corner pieces so as to allow a larger working area. For precise pressure monitoring, a pressure transducer (15) is located behind the ejector pin (16) for pressure-dependent switching from injection to holding pressure. The ideal pressure characteristic is recorded and each mold set-up will be done in accordance with this curve [3].

Figure 1 Lamp housing of polypropylene, reinforced with 15 wt.% glassfiber, flame-retardant

The quick disconnect couplings (29) with suitable nipples allow the heating/cooling and air lines to be connected both quickly and reproducibly. This has a favorable effect on the set-up times. The helical core (26) ensures effective temperature control of the mold core.
The cavity plates (2, 3) are made of steel grade 1.2767. This through-hardening steel is very advantageous if the contours are to be hardened after rough machining and then finished via EDM; this prevents any distortion caused by subsequent hardening. For the same reason, both plates have a grinding allowance in the guide bores. The lifters (37) are produced from precision ground flat steel, also of steel grade 1.2767. This steel, machined precisely on all sides, is available in a wide range of dimensions and is particularly suitable for manufacturing mold components of these and similar types.
The adjustable date insert (32) complies with the requirement of the automobile industry for injection molded parts to be clearly marked with the manufacturing date. These new standardized date inserts can be set from the contour side of the mold using a screwdriver. They show the month and year of production in raised characters on the injection molded part.

Operation of the Mold

The cavity is filled via the pneumatic sprue bushing (25) shown on the right in Fig. 3. In most cases, the front portion of the sprue is machined directly into the cavity plate; with very abrasive resins, a nozzle insert (Fig. 3, left) can also be used as a wear part.

Figure 3 Pneumatic sprue bushing (right) and interchangeable nozzle insert (left) for thermoplastics processing

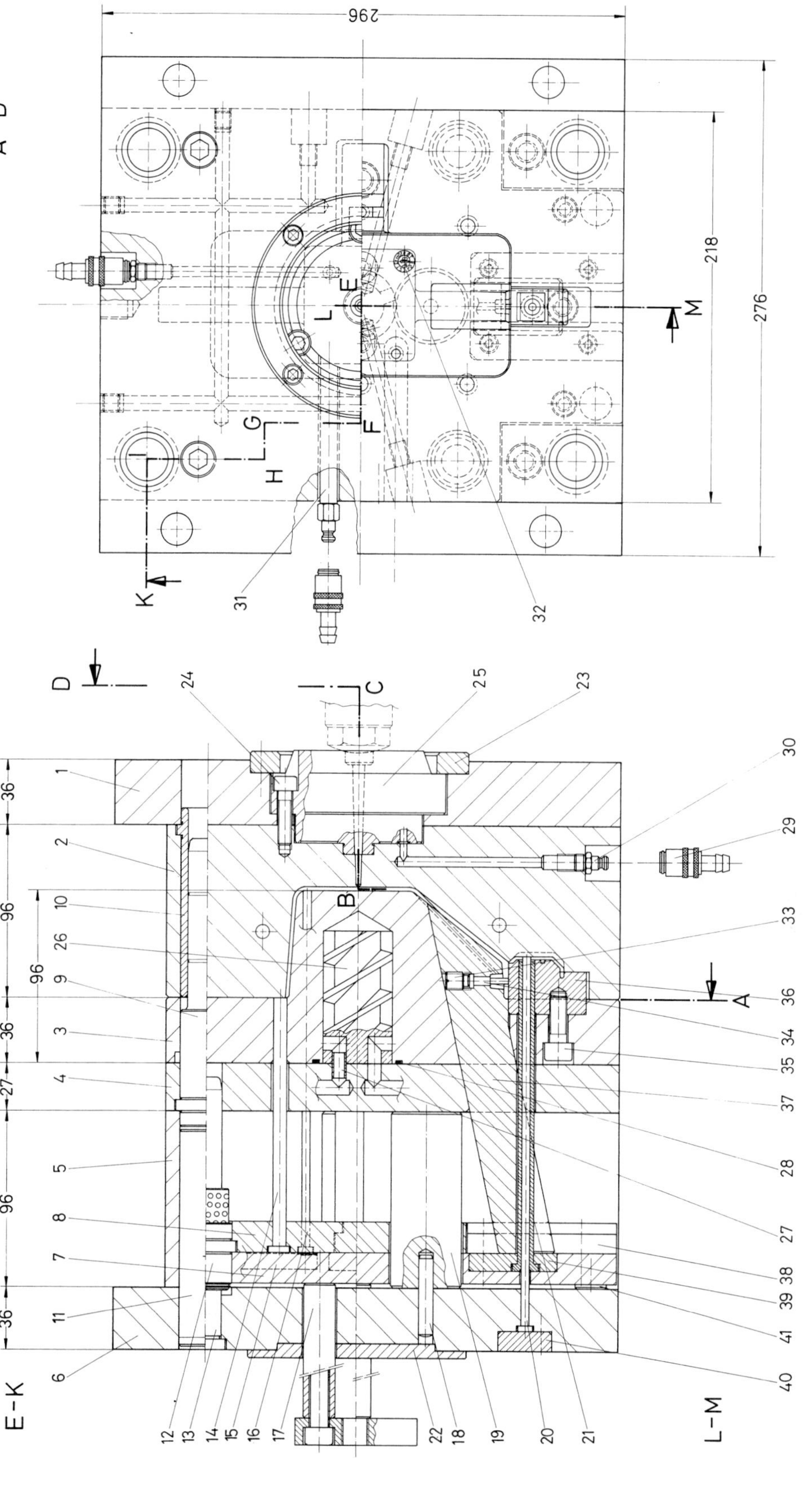

Figure 2 Injection mold with pneumatic sprue bushing for lamp housing
1: clamping plate; 2: cavity plate; 3: cavity plate; 4: backing plate; 5: support rails; 6: clamping plate; 7, 8: ejector plate; 9: guide pins; 10: guide bushing; 11: centering sleeve; 12: ball guide; 13: guide pin; 14: ejector pin; 15: pressure transducer; 16: ejector pin; 17: guide sleeve; 18: dowel pin; 19: support pillar; 20: ejector pin; 21: ejector sleeve; 22, 23: locating ring; 24: socket head cap screw; 25: pneumatic sprue bushing; 26: helical core; 27: brass tube; 28: O-ring; 29: quick disconnect coupling; 30: connection nipple; 31: extension nipple; 32: date insert; 33: hexagon socket set screw; 34: core pin; 35: socket head cap screw; 36, 37, 38, 39, 40: ground flat steel; 41: stop disk

The pneumatic sprue bushing is an alternative to the three-plate mold and to the hot-runner and shares the advantages of the conventional sprue. Existing tools can also be converted with this sprue bushing.
Figure 4 shows the function of the pneumatic sprue bushing, which is screwed directly to the cavity plate (2) in Fig. 2. After filling the mold and ending of the holding pressure time, the machine nozzle *f* retracts. Compressed air is introduced through the connection (30) in Fig. 2 and the bore *h* into the hollow piston *c* via a pilot valve. This pulls the sprue *e* from the part and releases air for the piston *d* which, aided by an air stream, ejects the sprue *e*. Before the next injection cycle starts, the machine nozzle forces the pistons *c*, *d* of the pneumatic sprue bushing back into their initial positions. The bore *g* allows additional temperature control for the injection area.
The ejector plates are connected to the hydraulic ejector of the machine via guide sleeves (17). When the ejector plates advances, the lifters (37) automatically move inward and release the inner contour. The ejector plates are guided precisely via the ball guides (12). The ejectors and lifters are pulled back hydraulically before the mold closes. The lateral ejector pins (14) act as return pins in the final mold closing phase. They push the ejector plates into home position.

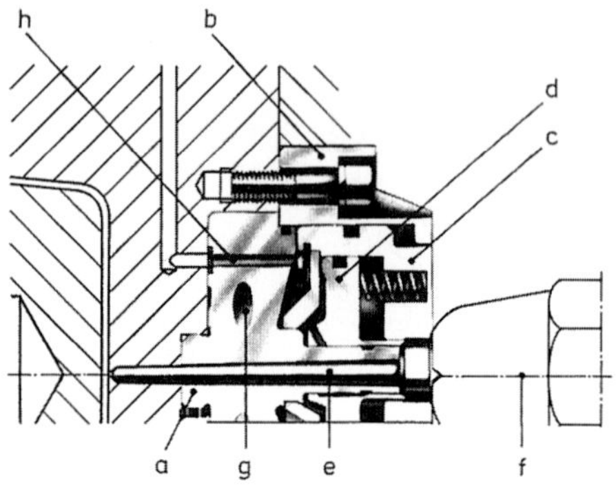

Figure 4 Section through the pneumatic sprue bushing (for explanation, refer to text)

References

1. Heuel, O. Kunststoffe 18(1984)a, p. 24–26
2. Heuel, O. Kunststoffe 71(1981) p. 866–869
3. Heuel, O. Plastverarbeiter 32(1981) p. 1496–1498

Example 97, Injection Mold for a Mounting Plate (Outsert Technology)

By means of the so-called outsert technique, one or more functional POM parts can be molded through openings onto both sides of a substrate, usually a metal plate, in a single step. Usually, assemblies produced in this manner are functional without any secondary finishing operations. Production of individual components and subsequent assembly are thus eliminated.

The outsert technique utilizes the specific properties of both the substrate material – the metal plate – and the plastic employed. The pronounced stiffness of the metal plate and its relatively low coefficient of thermal expansion are combined with the properties of the plastic, such as:

- good frictional behavior,
- chemical resistance,
- good vibration-damping characteristics, etc.

A decisive aspect is the economical production of high-quality multi-material assemblies. This technique has been employed successfully for years in the precision manufacturing sector.

In the present case, a mounting plate with more than 60 individual parts was produced for the Mini 14 cassette drive in an audio cassette radio (Blaupunkt, Hildesheim, Germany) through use of the outsert technique (Fig. 1).

Figure 1 Mounting plate

Galvanized sheet steel 1 mm thick conforming to DIN 1544 was selected for the metal plate (dimensions: 110 × 140 mm). Following the rough and final stamping operations, the metal plates were formed, straightened and then decreased.

The requirements to be met by the individual components that were to be injection molded – e.g. friction bearings, springs, mounting bosses, guides – resulted in selection of a polyacetal with an MFI of 190/2.16 = 13 g/10 min, which represented the best compromise from a technical standpoint.

Moreover, the relatively high shrinkage of this material – about 2.3% in this application – proved advantageous in that it promoted firm attachment of the components to the metal plate.

Mold

The three-plate mold has dimensions of 280 mm × 296 mm × 344 mm shut height (Fig. 2) and is fitted with an externally heated hot sprue bushing (33). After the metal plate is loaded into the mold, the melt is injected into the cavities, through the hot sprue bushing and runner system, via 20 gates – gate orifice 0.8 mm – either indirectly or with the aid of subrunners. Because of the severe spatial constraints, the cavities can be cooled only indirectly by two cooling channels in the mold plates (18, 22).

Part Release/Ejection

The parts molded onto the metal plate as well as the subrunners remaining on the plate must be released from both the stationary and movable mold halves. At the end of the cycle, the mold opens first at parting line I; this severs the 20 pinpoint gates. During this motion, parting line II is opened by latches (not shown); this releases the runner system. Before parting line II has opened completely, the stripper bolts (19) actuate the stripper plate (25), which pulls the runner off the sucker pins (58). Stripper bolt (27) limits the stroke of plate (25). The runner is removed from the mold from above by a part handling device, regranulated directly at the molding machine and subsequently processed in less demanding applications.

After the mold has opened completely, the ejector plates (10, 12) open parting line III, the stroke of which is limited by stripper bolt (41), through the action of the two-stage ejector (44, 45). This motion loosens and/or strips off the cores the molded parts located on the moving mold half. Further motion of the ejector plates (10, 12) completes part release and ejection. The mounting plate is held in position by locating pins (49) and can thus also be removed from the mold from above by the part handling device.

The mold plates are guided by leader pins (35, 47). Exact positioning of parting line I is accomplished with the aid of conical locating elements (not shown). The cavity wall temperature measure about 80°C (176°F); the melt has a temperature of 210°C (410°F).

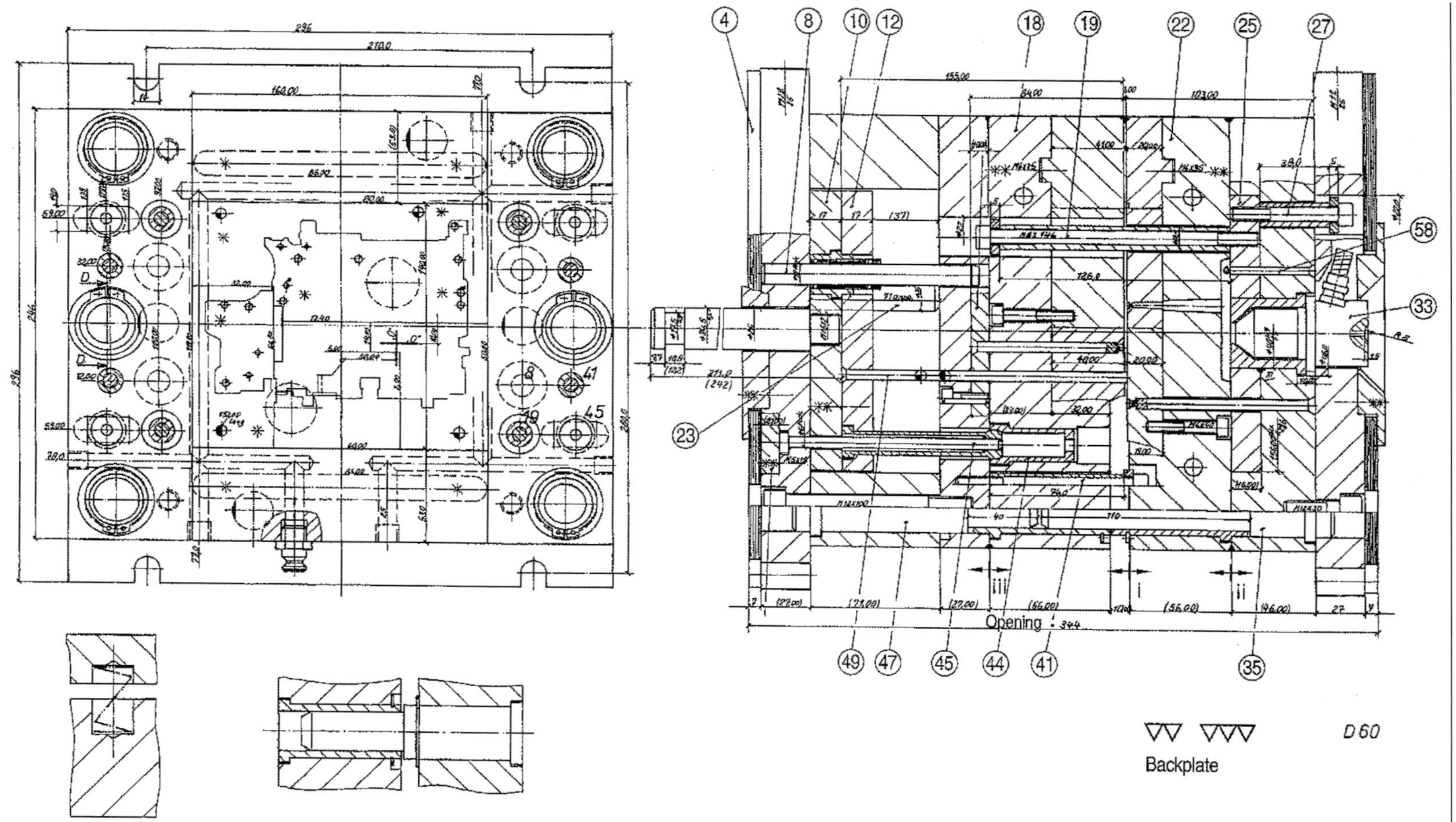

Figure 2 Injection mold for a mounting plate
4: thermal insulating plate; 8: guide pin; 10, 12: ejector plates; 18, 22: mold plates; 19, 27, 41: stripper bolts; 23: support pillar; 25: stripper plate; 33: hot sprue bushing; 35, 47: leader pins; 44, 45: two-stage ejector; 49: locating pin; 58: sucker pin
(Courtesy: Blaupunkt, Hildesheim, Germany)

Example 98, Twelve-Cavity Hot-Runner Mold for a Polyphthalamide (PPA) Microhousing

Microhousings with metal contacts (Fig. 1) were to be made by the outsert technique. The partly competing demands of

- economic production and
- low thermal damage to the polyphthalamide (PPA) through short dwell time in the runner system

were met by using a hot runner mold in which the molded parts were direct-gated via double nozzles from Günther Heißkanaltechnik, Frankenberg/Germany.

The thermoplastic material to be injection molded is a semicrystalline polyphthalamide containing 33% glassfibers that is made by outsert molding onto a perforated strip (1) of tin-plated bronze. The strip is unrolled mechanically, positioned in the mold by index pins, encapsulated by injection and moved further by an external step motor twelve times the distance from the mold cavity. The encapsulated strip with the finished microhousings is then rolled up again and processed further.

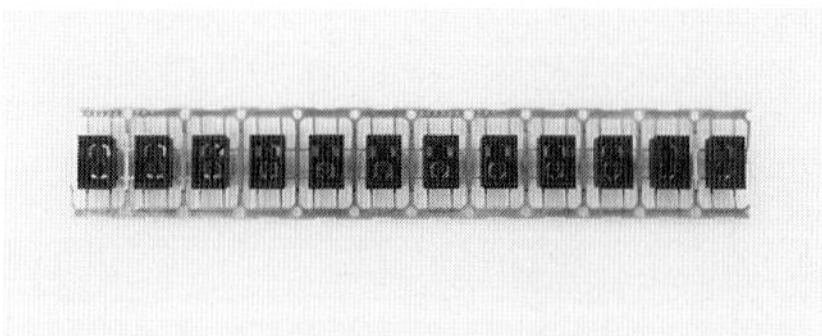

Figure 1 Molded parts with punching lattice (see from stationary mold half)

Without metal insert, the molded part weight is 0.28 g, the walls are between 0.15 and 2.7 mm thick, and the molded part measures 8 mm × 11 mm × 6 mm.

Mold

The mold (Figs. 2 to 4) is a twelve-cavity hot runner mold with external heaters for both the hot runner manifold (2, 23 V) and the six openly heated nozzles (24 V), each of which is controlled.

The gate diameter is 0.75 mm.

The pivot on the molded part, where gating occurs, has a diameter of 0.8 mm. The mold cavities have inside caliper dimensions (center-to-center spacing) of 12 mm, so that with six double nozzles with a mean distance of 24 mm, twelve molded parts can be gated at once (Fig. 5).

The six double nozzles, which when heated press directly against the hot runner, are all located in a housing (4) measuring 160 mm × 40 mm × 43 mm. Air pockets ensure minimal energy loss via heat conduction. The heating capacity of each nozzle is 200 W and that of the hot runner block is 2 × 650 W. Because the molded part weight was low at 0.28 g (without metal insert 3.36 g for 12 parts), no rheological balancing of the hot runner block was provided, but this did not affect quality. The theoretical dwell time of the melt in the hot runner system is around 30 s. The nozzles and hot runner temperatures are 340°C (644°F), while the mold wall temperatures range between 80°C and 160°C (176°F and 320°F). The mold has four different cooling circuits.

Machine

The working method requires an injection molding machine with a vertical injection and clamping unit. The melt is injected into the mold at a pressure of around 1100 bar. The injection unit is not retracted after injection. To rule out drooling from the open nozzles, the screw has to be vented. The height of the gate remnant is less than 0.3 mm.

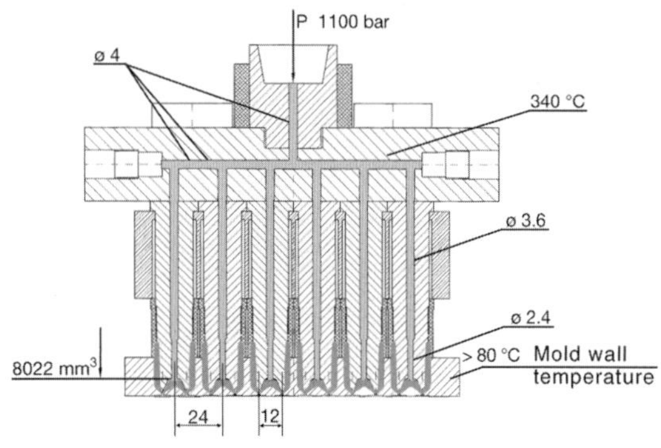

Figure 5 Hot runner layout

Throughput: 3.3 g/shot, 8 shots/min; dwell time in system at 8022 mm^3 = 28 s

Fig. 2

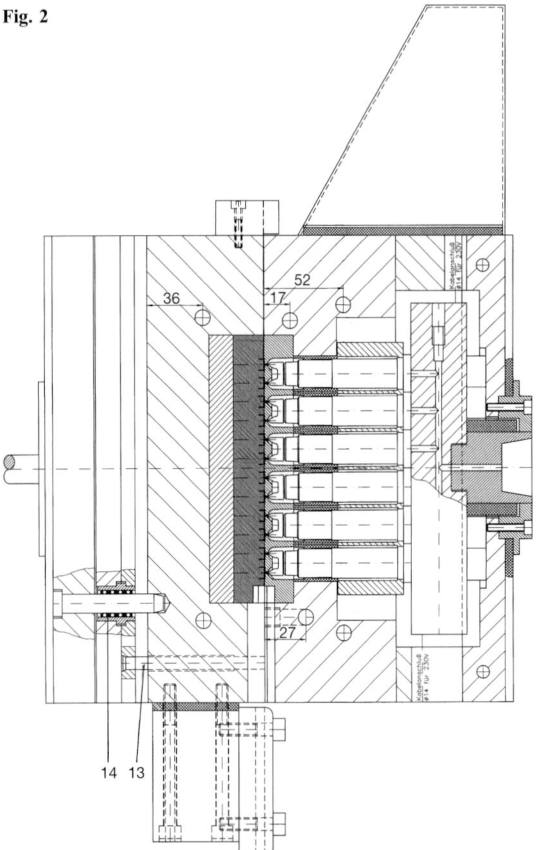

Fig. 3

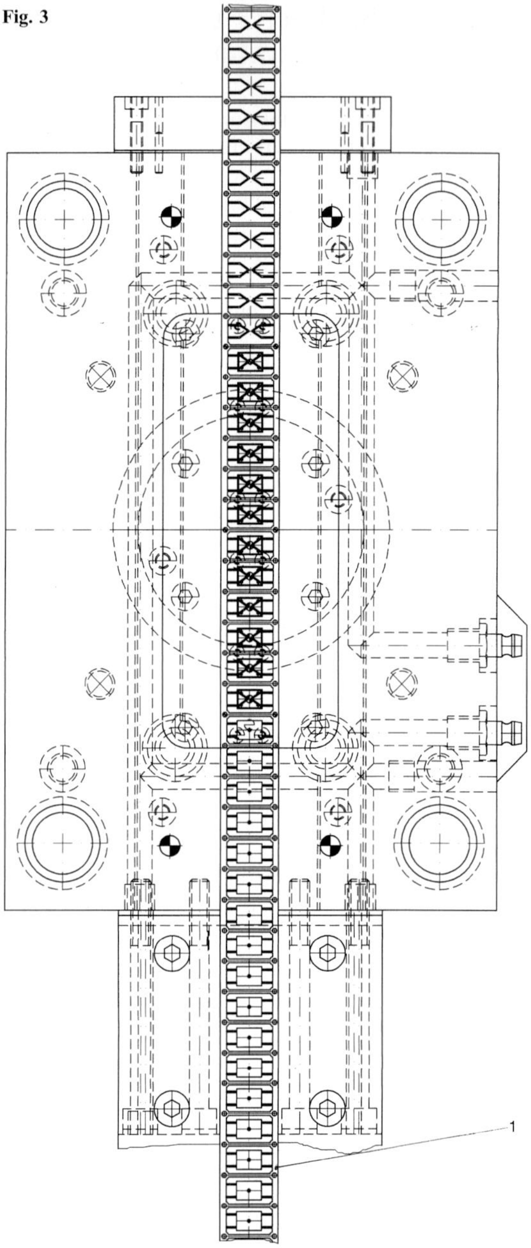

Fig. 4

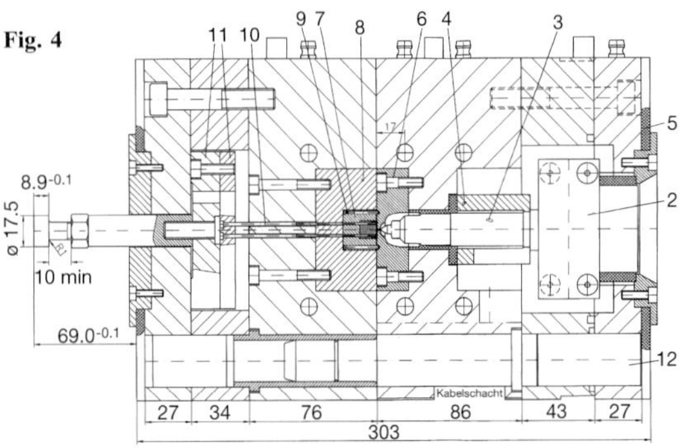

Figures 2 to 4 Twelve-cavity hot runner mold for PPA microhousing
1: punching lattice; 2: hot runner manifold block; 3: heated nozzle; 4: housing; 5: thermal insulation plate; 6 to 8: mold inserts; 9: index pins; 10: ejector pin; 11: ejector plates; 12: guide pillar; 13: return pin; 14: ball guide bushings
(Courtesy: Günther Heißkanaltechnik, Frankenberg; Reiter Präzisions-Spritzguß + Formenbau GmbH, Hilpoltstein, Germany)

Example 99, Two-Cavity Injection Mold for Handle Covers Made from Glass-Fiber-Reinforced Polyacetal

This mold (dimensions: 246 mm × 396 mm × 328 mm shut height) differs in that the major components were produced from a high-strength AlZnMgCu alloy (brand name: Fortel 7075, Almetamb, Stuttgart, Germany); designation as per DIN EN: AlZnMgCu 1,5; material no. 3.4365.

The material was machined in a stress-relieved condition without heat treatment and employed in the as-machined state. Compared to tool steels, this aluminum alloy is characterized by the following differences:

- low specific gravity (2.8 g/cm^3),
- lower modulus of elasticity (70 000 N/mm^2),
- very good thermal conductivity (about 160 W/m·K),
- very good machinability,
- very high removal rates during electrical discharge machining (EDM).

As a result of the approx. one-thirds lower modulus of elasticity, mold plates, for instance, exhibit three times the deflection of a steel mold with identical dimensions when subjected to a mechanical load. Since the deflection f is inversely proportional to the product of the modulus of elasticity E and the moment of inertia I, i.e. $f \sim (E{\cdot}I)^{-1}$, the stiffness of steel can be achieved by increasing the plate thickness by about 44%. Even in this case, the weight of an aluminum mold is about only half that of steel. With regard to abrasive wear, unprotected aluminum is clearly inferior to steel when exposed for the same length of time, but this can be corrected to a great extent with a suitable surface treatment, for instance, electroless nickel plating. Note that hard surface layers may break on a relatively soft substrate, such as aluminum alloy, which would render them more or less ineffective.

The mold was used to produce a limited quantity of covers (< 100 000) in 30% glass-fiber-filled polyacetal. Cores and cavities were EDM'd using a sinker-type machine, and polished, but not given any subsequent surface treatment.

Since a pairing of Al with Al can result in galling under sliding conditions (if the surfaces are not treated), dissimilar materials were paired as necessary. The decision in favor of an aluminum mold for the quantity of parts required resulted largely from the lower manufacturing costs versus a steel mold.

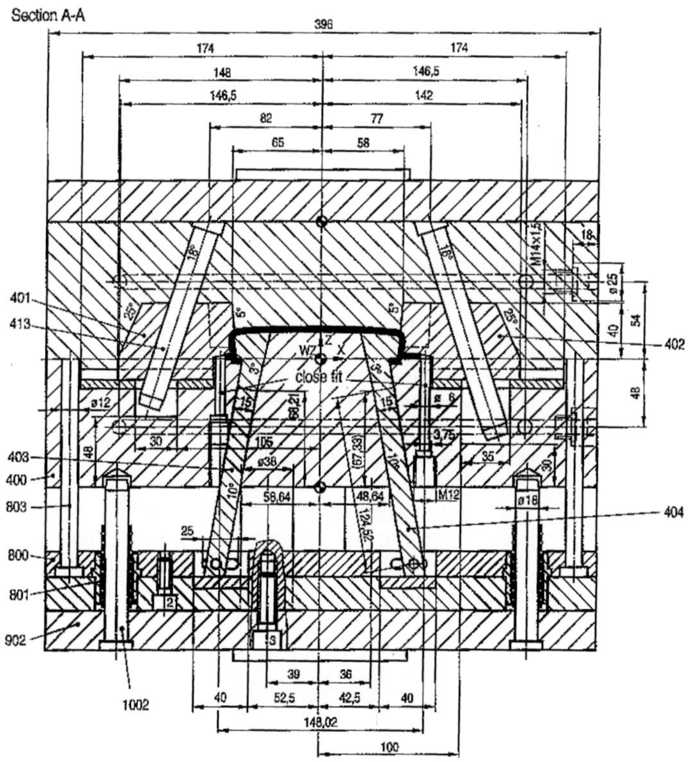

Figure 1 Two-cavity injection mold for handle covers of glass-fiber-reinforced polyacetal
400: mold plate; 401, 402: slides; 403 404: lifters; 413: cam pin; 800: ejector plates; 801: ball guides; 803: pushback pin; 902: wear plate; 1002: guide pinc

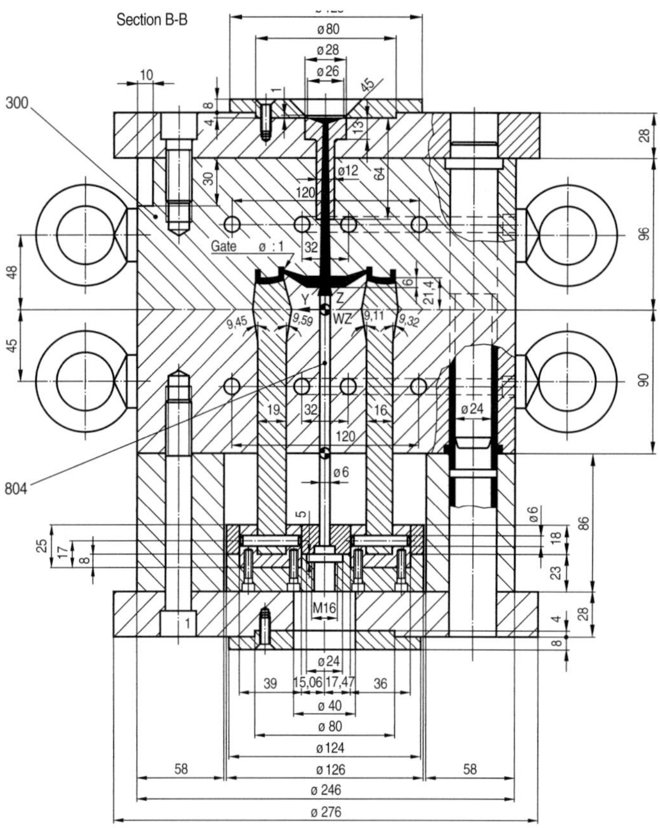

Figure 2 Two-cavity injection mold for handle covers made from glass-fiber-reinforced polyacetal
300: mold plate; 804: ejector

Because of the good thermal conductivity, shorter cycle times are achievable than with conventional steel molds.

Mold

The molded part (Figs. 1 to 4) exhibits both internal and external undercuts, which must be released via slides.

The slides (401, 402, Fig. 1) that release the external undercuts are actuated during the opening motion by four steel cam pins (413). In the open position, these slides are held by ball detents (408, Fig. 4). The slides are of bronze, the wear plates (407) and guides (405) of hardened steel, the stationary-side mold plate (300, Fig. 2) of aluminum. Support plates, e.g. of steel, between the stationary-side mold plate and slides were dispensed with, that is, the injection pressure is absorbed directly by the angled contact surfaces.

The bronze lifters (403, 404, Fig. 1) needed to release the internal undercuts run in the aluminum mold plate (400). These lifters are supported by the aluminum ejector plates (800). The wear plates (902) are also of hardened steel here. The ejector plates (800) move on steel guide pins (1002) in conjunction with ball guides (801).

The sprue ejector (804, Fig. 2) is made of bronze. The ejector mechanism is returned to the molding position by push-back pins (803, Fig. 1, diameter: 12 mm) as the mold closes. Each mold half is provided with a separate temperature control circuit.

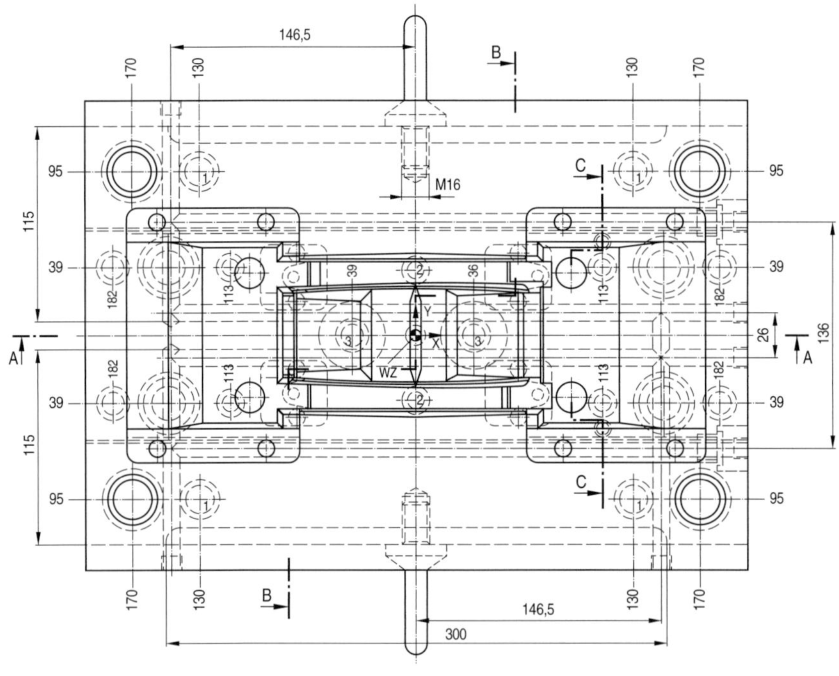

Figure 3 Two-cavity injection mold for handle covers made from glass-fiber-reinforced polyacetal

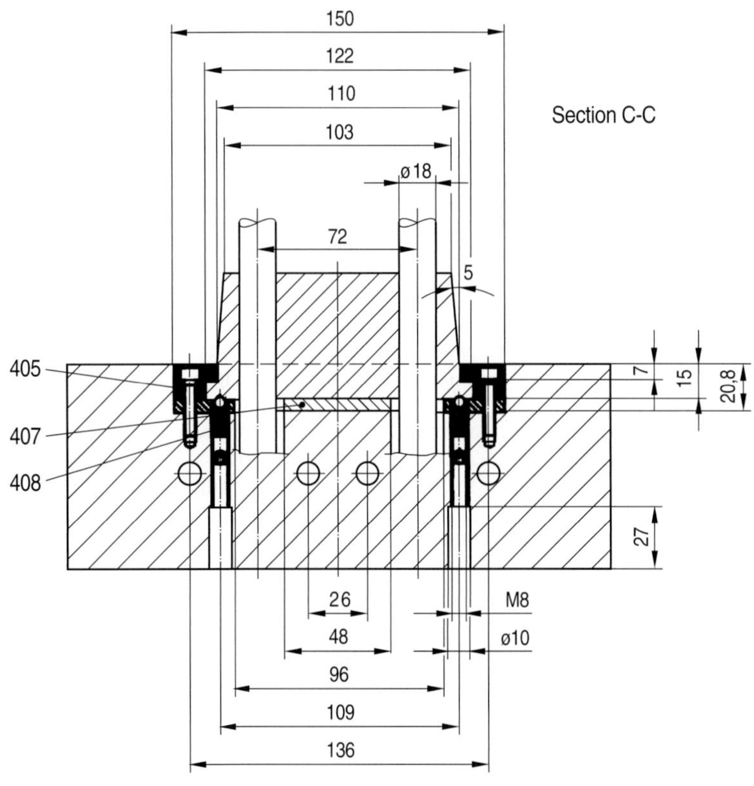

Figure 4 Two-cavity injection mold for handle covers made from glass-fiber-reinforced polyacetal
405: guide for slides; 407: wear plate; 408: ball detent
Company illustrations: Almet amb GmbH, Düsseldorf, Germany

Example 100, Four-Cavity Injection Mold for Thin-Walled Sleeves Made from Polyester

A four-cavity mold with parting line injection was needed for a thin-walled sleeve having a wall thickness of only 0.5 mm for a length of 26 mm (Fig. 1). Parting line injection was necessary, because an extremely long hydraulic ejector was needed for the mold. The material to be molded was a polyester (polyethylene terephthalate) with good flow properties that is especially suited for thin-walled parts with a high flow length/wall thickness ratio.

compressed air introduced through openings (6). As the release bar (7) disengages the latch (4), parting line (2) is opened by means of bolt (8). Parting line (3) is held closed by means of latch (9). Undercuts (10) retain the runner system and in this manner shear off the submarine gates (3). Opening at parting line (2) continues until the runner system can drop out properly. Release bar (11) then disengages latch (9) as plate (12) is held by stop (13), so that parting line (3) now opens. As the mold reaches the full-open position, the hydraulic ejector (14) is actuated, thereby ejecting the sleeve from the cooled cavity insert (16). Simultaneously, plate (17) actuates plate (18). The ejector pin (19) mounted in plate (18) is located behind the retaining undercut (10) for the runner system, which is now ejected. It does not drop out of the mold, however, until ejector pin (15) is retracted by the hydraulic ejector.

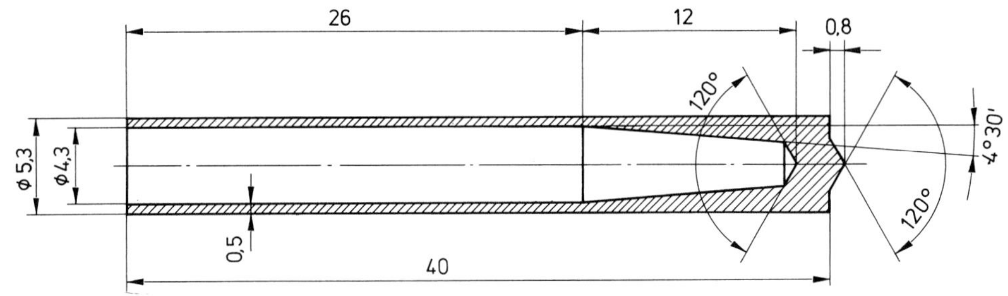

Figure 1 Polyester sleeve

To permit fully automatic operation, the sleeves were to be ejected separately from the sprue and runner system. Furthermore, the outer surface of the sleeves was not permitted to have any witness line. The closed end with conical tip had to be smooth and clean. The best solution thus appeared to be to gate the sleeve at its thick-walled end by means of two submarine gates on opposite sides (Fig. 2).

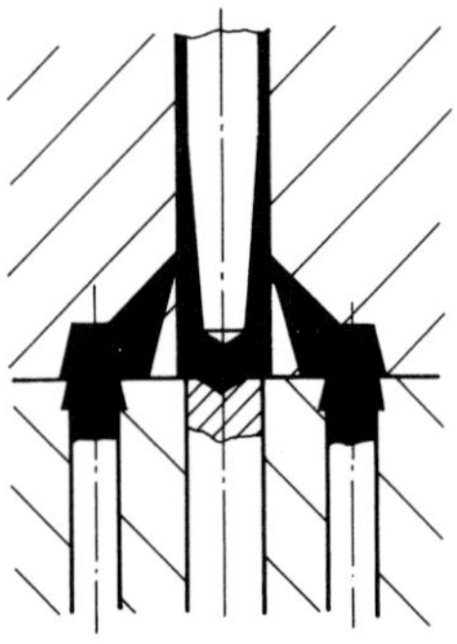

Figure 2 Gating of the sleeve by means of two opposite submarine gates

Ejection without damaging the thin walls of the molded part takes place by first withdrawing the core (5) from the sleeve (1) while it is still completely contained in the cavity. The mold (Figs. 3 to 12) first opens at parting line (1). Parting lines (2) and (3) are held closed by latch (4). During the opening stroke, the cores (5) are cooled by means of

The position of ejector plate (17) is sensed by two roller switches, which are actuated by switch rods (20) and (21), and determine the machine sequencing. Ejector plate (18) is returned to the molding position by pushback pin (22) as the mold closes.

The closed end of the sleeve exhibits the same 120° tip as does the inner core to ensure that this inner core cannot be deflected toward one side as the sleeve is filled through the two gates (Fig. 2). In addition, the ejector pin (15) is spring-loaded (23). When the mold is closed, the end of ejector pin (15) seats against the inner core (5) and centers it in the corresponding recess. As the melt enters the cavity, the core is held centered until the cavity pressure overcomes the force of the spring located behind ejector pin (15) and forces it to its retracted position. By this time, the core (5) is surrounded by melt to such an extent that it can no longer be deflected. This precautionary measure in the mold design was found to be absolutely necessary on test molding with the completed mold.

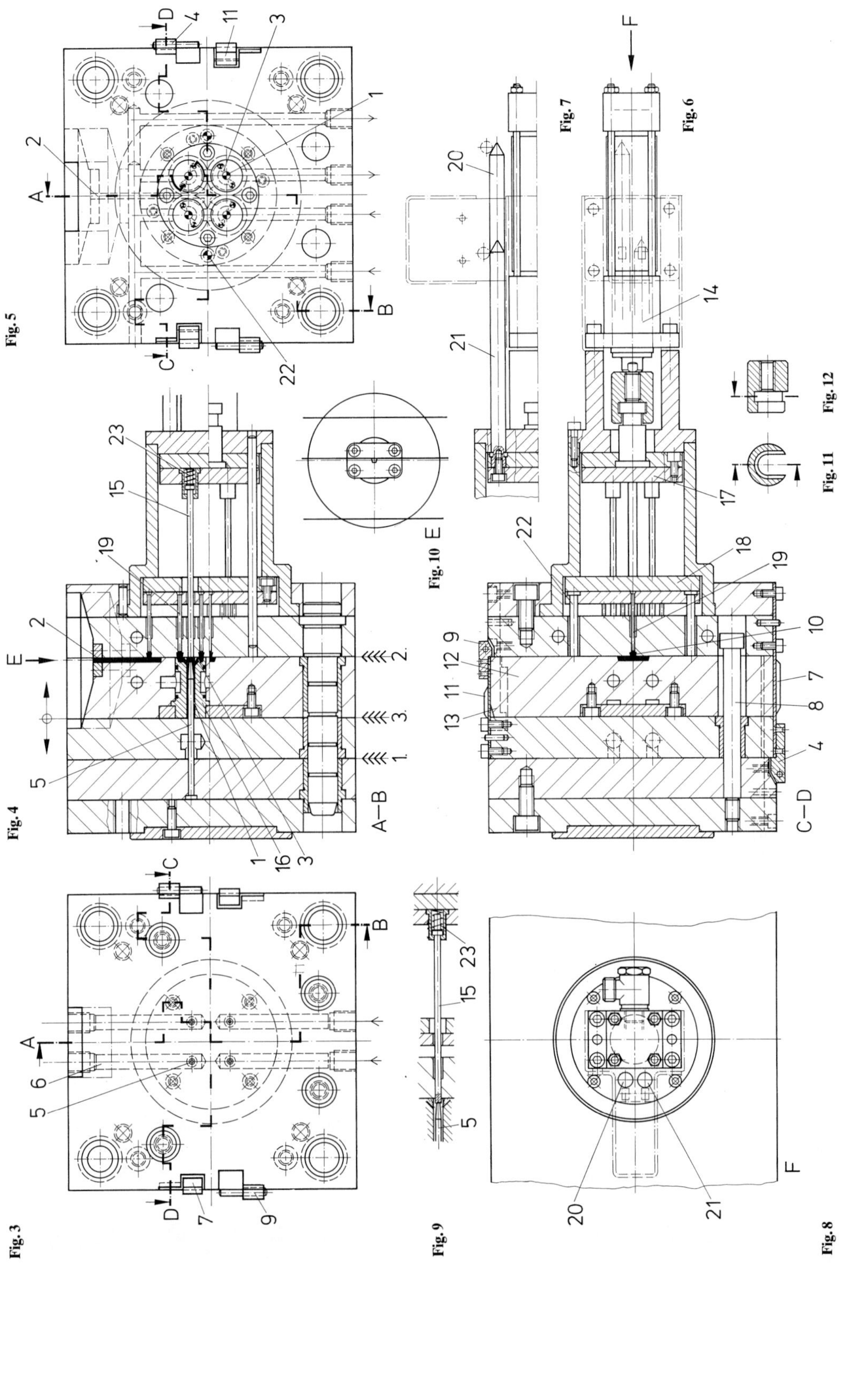

Figures 3 to 12 Four-cavity injection mold for automatic molding of thin-walled polyester sleeves
1: sleeves; 2: sprue bushing; 3: submarine gates; 4: latch; 5: core; 6: opening for cooling air; 7: release bar; 8: bolt; 9: latch; 10: undercut; 11: release bar; 12: cavity retainer plate; 13: stop; 14: hydraulic ejector; 15: ejector pin; 16: cavity insert; 17: ejector actuating plate; 18: ejector plate; 19: ejector pin; 20, 21: switch rods; 22: pushback pin; 23: spring

Example 101, Injection Mold for a Microstructure Made from POM

Components with microstructures are being employed increasingly for mechanical and optical elements in sensors and actuators for the automotive industry, minimally invasive surgery, aeronautics, and even consumer items. The best-known application is found in ink jet printers.

To produce the mold cavities, traditional methods such as micromachining, micro-EDM and laser machining can be employed, along with the LIGA method (German acronym for lithography, electroforming, mold making). Structures produced using the LIGA method have extremely smooth side walls, so that even thermoplastic parts with minimal draft can be released and ejected. Three-dimensional parts, however, require considerably more work to produce than do parts lying in a plane (two-and-a-half dimensional parts).

Figure 1 shows a microstructure injection molded in polyacetal; Fig. 2 shows a section of the part, together with the corresponding section of the mold, which was produced using the LIGA method.

The molded part (dimensions: 4 mm × 4 mm) represents the IKV logo and consists of 1100 individual columns. The columns, which are hexagonal in shape, have a diameter of 80 μm and a height of 200 μm. The distance between the columns, i.e. the wall thickness of the partitions in the mold, is 8 μm. On the one hand, the mold must satisfy requirements for a great deal of flexibility of the cavities employed in order to permit practical experiments involving a variety of different possible applications. On the other, it must take into account the special nature of micro-injection molding. The required flexibility is achieved by constructing the mold from a standard mold base (Fig. 3) that can accommodate a variety of interchangeable mold inserts for the tests of interest.

In order to permit problem-free filling and prevent damage to the fragile cavities, it is essential that the mold be heated in the region of the cavities to almost the melt temperature of the plastic being processed. Following this, the part-forming components must be cooled to the ejection temperature of the particular plastic employed.

The processing method described above is often called the "Variotherm process". It is important, above all, that the cycle time for processing be kept within acceptable limits. This is achieved by maintaining the mold at a constant temperature of 130°C (266°F). This is also the ejection temperature of the plastic employed. Only the region around the cavity and runner is heated briefly and locally to approximately the melt temperature.

To facilitate filling of the cavity, the process is performed under vacuum. The use of vacuum also counteracts any potential "dieseling" during filling. Using the process sequence described here, cycle times of 1.5 min are achieved.

Total shot weights of <0.3 g have already been achieved with a very high degree of reproducibility with an injection molding machine that was developed specifically for this process.

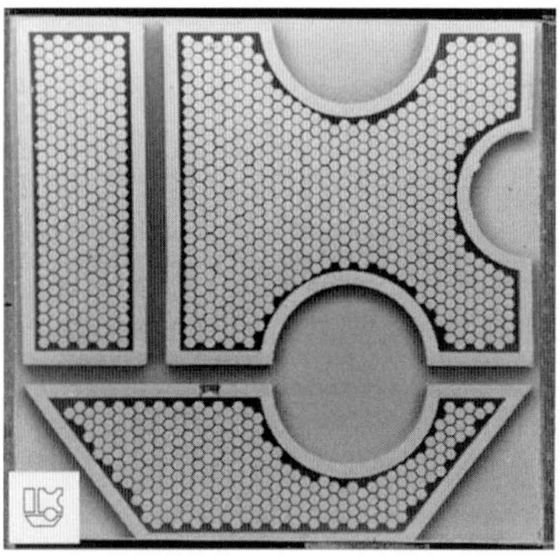

Figure 1 Injection molded POM microstructure

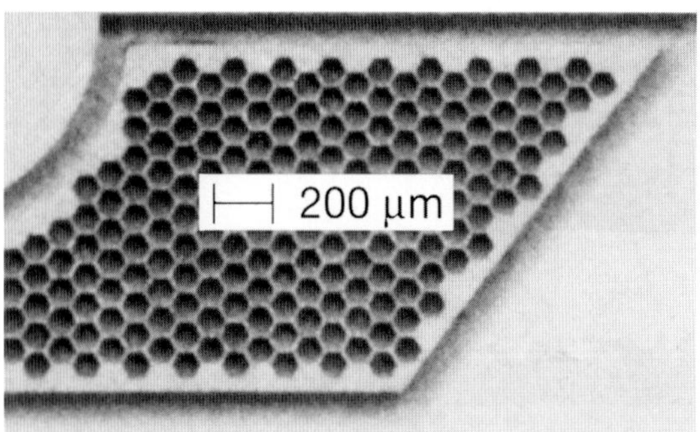

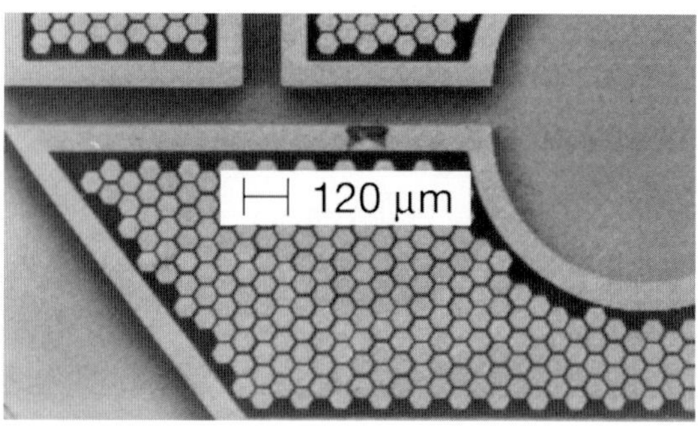

Figure 2 Magnified views: left = cavity, right = molded part

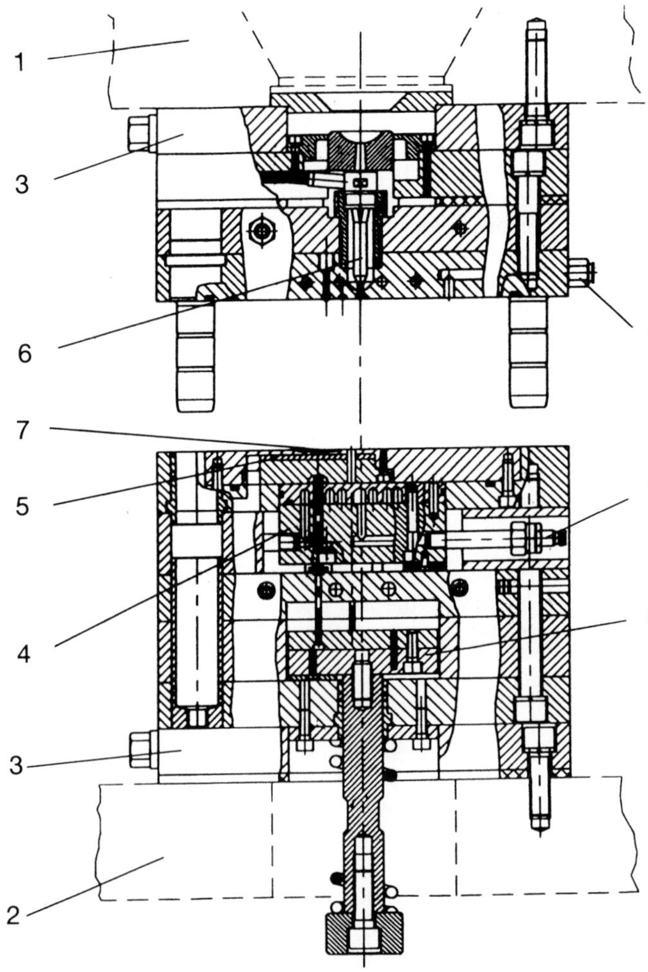

Figure 3 Injection mold (sectional view) for thermoplastic microstructures
1, 2: machine platens; 3: mold quick change system; 4: Variotherm unit; 5: electric heating plate; 6: sprue nozzle (system spear, seiki); 7; LIGA cavity; 8: ejector system; 9: vacuum connection; 10: connection for fluid temperature control
(Courtesy: Hasco, Lüdenscheid; IKV, Aachen, Germany)

Example 102, Injection Mold for Production of Adjustable Climate Control Vents via 3-Shot Molding

With the objective of improving part quality and reducing manufacturing costs, a production mold in which adjustable climate control vents consisting of 7 individual parts were molded in a ready-to-install condition was shown in Düsseldorf for the first time in 1992 (Fickenscher, Selb, Germany).

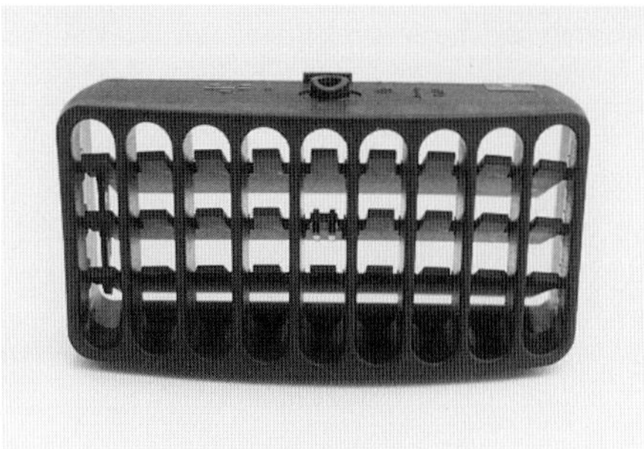

Figure 1 Climate control vent

The climate control vent consists of a housing with louvers that can be rotated around their axis lengthwise. In addition, a linkage is connected to the louvers in a manner that permits smooth, simultaneous adjustment of all louvers and thus directional control of the air flow. The vent was "assembled" from different thermoplastic materials in 3 mold stations with the aid of the multi-shot/multi-component injection molding technique.

The plastics are selected largely on an empirical basis and, in addition to meeting a large number of specific requirements (gloss, scratch resistance, combustion characteristics), must exhibit an "incompatibility" in order to ensure movement of the individual components. In other words, bonding (welding) of the components to one another must be prevented under all circumstances. A design feature can assist in this regard: the smaller the surface areas in contact, the less likely is the risk of any "initial melting."

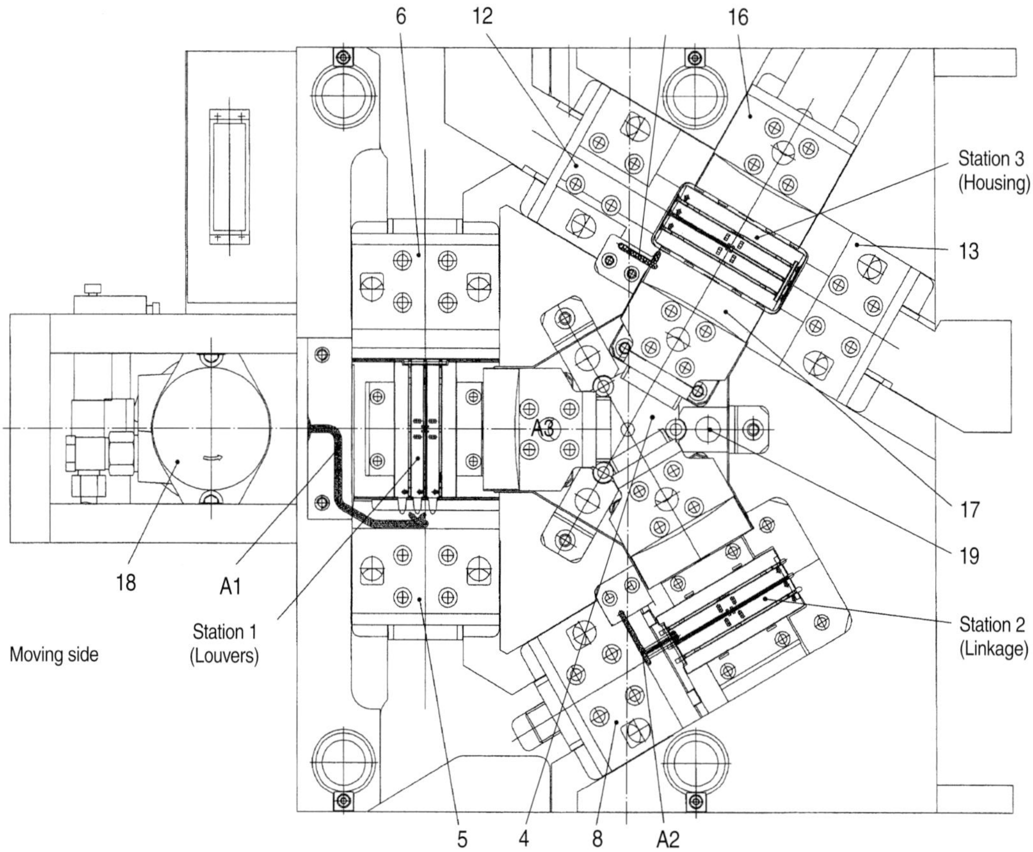

Figures 2 to 5 Injection mold for climate control vents

Figure 2a View of parting line (moving side)
A1, A2, A3: runners; 4; indexing mechanism; 5, 6, 8: slides; 12, 13, 16, 17: split cavities; 18: hydraulic motor: 19: locator

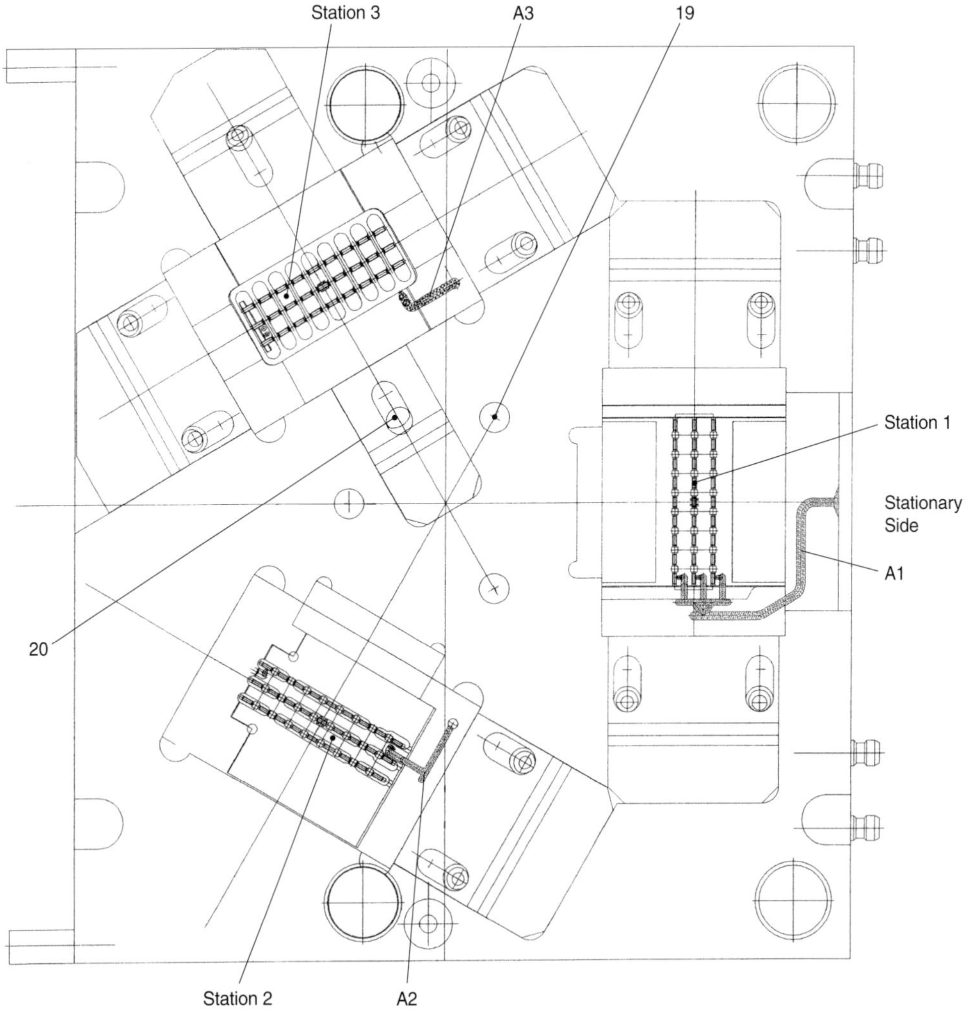

Figure 2b View of parting line (stationary side)
A1, A2, A3: runners; 19: locator; 20: cam pin

The mold described here serves to produce the climate control vents for a mass-produced automobile. This climate control vent (Fig. 1) consists of 5 individual components:

- three louvers of 20% glass-microsphere-filled PBT,
- one linkage of 20% talc-filled PP,
- one housing of modified PPO.

The measurements are roughly 110 mm × 150 mm × 35 mm. The principle underlying the mold is covered by two world patents. An injection molding machine with three plasticating units, i.e. with two units on top of each other and an injection unit at right angles to them is needed.

Mold – Principle of Operation

The five individual components are produced in one mold with three stations (Figs. 2a and 2b). The three louvers are molded at Station 1; the linkage at Station 2 and the housing surrounding the louvers and interconnecting linkage at Station 3. The linkage operates the three louvers, each of which rotates around two projecting studs (which function as journals) and is guided by eight support surfaces in the housing. The three plastics selected for the respective components are injected simultaneously at all three stations. At the end of each cycle, three runners are ejected and the completed climate control vent is removed and deposited at a specified location by a part handling device.

The connecting element between the three mold sections (stations) is the "indexing plate" (4) that is integrated into the moving mold half. This plate advances (moving the mold sections out of the moving mold half after each cycle and after the finished part is ejected at station 3), rotates by 120° and then retracts. This motion transfers the molded louvers from Station 1 to Station 2, and the louvers with overmolded linkage from Station 2 to Station 3.

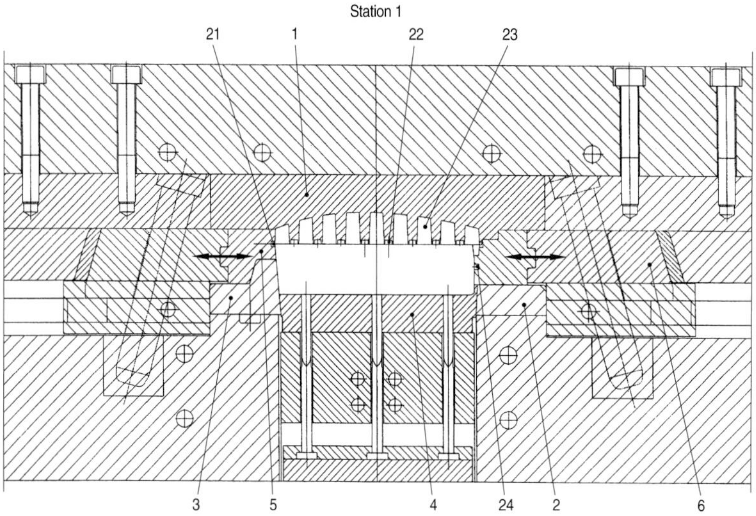

Figure 3 Station 1: Louvers
1, 2, 3: mold inserts; 4: indexing plate; 5, 6: slides; 21: support journal; 22: guide surface; 23: tongue; 24: linkage journal

When the mold starts up, the louvers are molded at Station 1 during the first cycle. During the second cycle, louvers are molded at Station 1, while the linkage is molded at Station 2. Injection takes place at all three stations only as of the third cycle.

Station 1 – Mold Section for the Louvers (Fig. 3)

The three louvers are formed by the stationary-side mold insert (1), the moving-side mold segments (2, 3), the mold core on the indexing plate (4) and the two slides (5, 6) (Fig. 3). Slide (5) serves two functions: on the one hand, to form the linkage journal without a parting line; on the other, to provide space for ejection of the louver runner system as the mold opens. Located between the tongues (23) of the louvers are guide surfaces (22) in the shape of half-cylinders that permit smooth, defined movement of the louvers within the subsequently overmolded housing.
Slide (6) forms both the support journal (21) and the linkage journal (24) without a parting lines. In addition, this slide forms a portion of the adjacent louver edge. The reason for this is that space must be provided for subsequent forming of the linkage. The three louvers are filled via a runner (A1, Fig. 2) with submarine gates. Only the runner is ejected at Station 1. The louvers remain in the mold core on the indexing plate (4) and are ejected at Station 3 only after two more cycles.

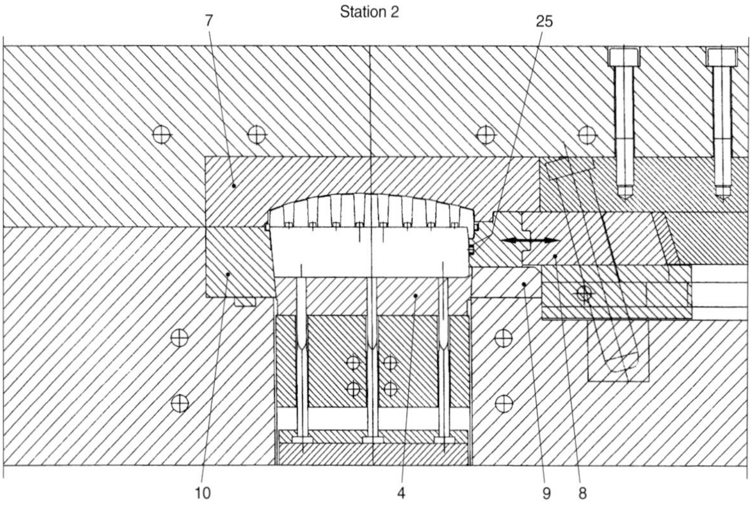

Figure 4 Station 2: Linkage
4: indexing plate; 7: mold insert; 8: slide; 9, 10: mold inserts; 25: linkage

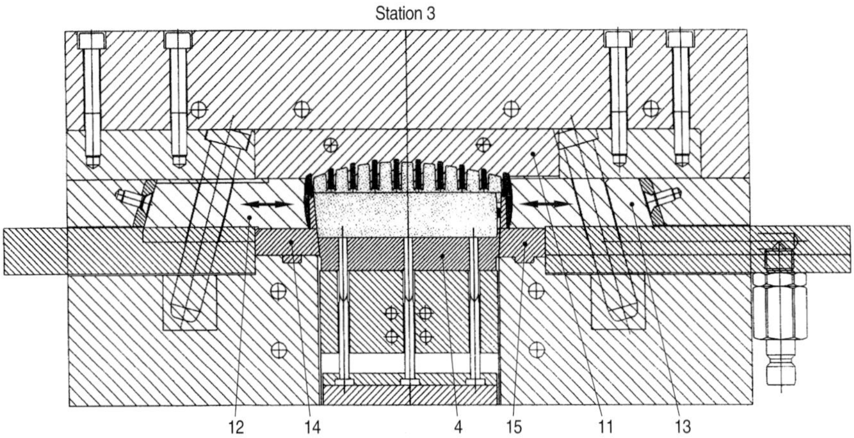

Figure 5 Station 3: Housing
4: indexing plate; 11, 14, 15: mold inserts; 12, 13: splits
(Courtesy illustrations: Fickenscher + Co. GmbH, Selb, Germany, Now TRW)

Station 2 – Mold Section for the Linkage (Fig. 4)

The linkage (25) is formed by the slide (8), the mold core on the indexing plate (4), and the three louvers from Station 1. It is also filled via a runner (A2, Fig. 2) with submarine gate. Only the runner is ejected in this station as well.

Station 3 – Mold Section for the Housing (Fig. 5)

The function of the "housing" mold section is more extensive than the function of the mold sections at Stations 1 and 2. The four outside surfaces are formed by the splits (12, 13, 16, 17; Figs. 2 and 5). The inside surfaces of the housing are formed by mold segments (14, 15) and the mold core on the indexing plate (4). The stationary-side mold insert (11) forms the visible surface of the climate control vent. The louvers molded at Station 1 serve to form the support area between the housing and louvers. As at Stations 1 and 2, filling takes place via a runner (A3) with submarine gate. The "completely assembled" climate control vent and runner are ejected by a two-stage ejector mechanism. The necessity for this is twofold: on the one hand, the ejector pins behind the louvers may be moved only at this station; on the other, the stroke of the louver ejector pins (for reasons of space and stability) is shorter than the stroke of the ejector pins for the runner and housing.
The ejector mechanism is actuated by two hydraulic cylinders integrated into the mold.

Indexing Plate – Connecting Element between the Stations

The indexing plate (4) participates at all three stations in the formation of the individual molded parts, which differ from one another in their geometry. Each of the three "arms" forms a part of the mold section at a station. The mold cores on the indexing plate shut off against other part-forming inserts or slides at all three stations. Because of this, it is necessary that the mold be manufactured with a high degree of precision.
Of the three slides (17) attached to the center of the indexing plate, only one (that at Station 3) is in the molding position and forms an outside surface of the housing (compare cam pin (20) in Fig. 2b). The machine ejector advances and retracts the plate; the 120° indexing motion is performed by a hydraulic motor (18) and gear belt. In order to achieve the most accurate possible positioning when stopping, two different speeds are employed during rotation. The initial, relatively high speed is intended, above all, to minimize the cycle time. The second, lower speed, which starts after a rotation of about 100°, is intended to stop the indexing plate as accurately as possible at the 120° position. Final positioning is accomplished with the aid of three locating pins (19) in the mold.

Mold Temperature Control

The mold incorporates three independent cooling circuits. Extreme temperature differences within the mold result in binding as the result of differential thermal expansion. The temperature of the splits (12, 13, 16, 17) at Station 3 is selected to provide a good surface on the housing without visible flow lines.
A change in cavity wall temperature in certain areas of the mold can have a negative impact on louver motion, and thus the functionality of the climate control vent.

Example 103, Two-Cavity Hot-Runner Injection Mold for an ABS Cover

Molded Part

The cover (dimensions: 125 mm × 110 mm × 28.8 mm) is in the shape of a rectangular box and has a number of internal and external undercuts. Four ribs with snap fits, together with reinforcing and sealing ribs, are located on the underside of the cover. Inside, there are two pockets for hinge pins along the narrow side and, opposite these, four detent grooves.

Mold

The mold is constructed largely of standard mold components. For instance, a standard mold base with dimensions of 246 mm × 346 mm is employed. The mold shut height is 396 mm. Plate thicknesses are also standard dimensions. An exception to this is the plate forming the frame around the hot-runner system. It is sized to accommodate the manifold and hot-runner nozzles.
The two cavities and their mating mold cores on the moving mold half are identical in shape. However, for reasons related to the mechanical design and part ejection, they are oriented symmetrically around the vertical axis of the mold (Fig. 3). Steel grade 1.2767, through-hardened, was employed for the mold cores. For the stationary-side mold plate, steel grade 1.2764, case-hardened, was employed. The slides and associated actuating components as well as the ejector blades are also fabricated from steel grade 1.2764.
The slide actuating components, which are subjected to lateral forces during ejection, are guided by bronze wear strips. The two ejector plate assemblies (AW1, AW2) are guided by four guide pins (16). These ejector plate assemblies are actuated by a two-stage ejector (1) that divides the stroke of the machine ejector into two successive partial strokes. All guiding and wear components have been treated with Lamcoat, a soft, self lubricating coating based on tungsten disulfide. By using this coating, friction is reduced up to 70% and the dry-running properties are improved significantly. Treating part-forming components with Lamcoat has, in certain cases. Resulted in a cycle time reduction of up to 25%.

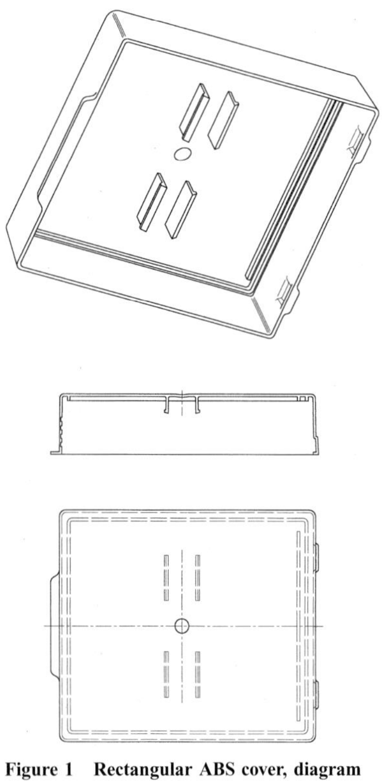

Figure 1 Rectangular ABS cover, diagram

Runner System

The parts are filled via a hot-runner system for runnerless molding. Standard mold components are used exclusively. The melt enters at the inlet bushing (13) and proceeds through the straight-bar manifold (14) to the two heated hot-runner nozzles (15), which lead directly to the cavities. Optimal thermal separation between the hot-runner system and cavity permits use of open hot-runner nozzles with a cone point. The gate vestige remaining on the molded parts is minimal. Through the use of appropriate controls for the hot-runner system, extremely tight temperature control is possible in conjunction with the other mold and machine parameters.

Part Release/Ejection

Step 1:

The mold opens.

Step 2a:

The machine ejector actuates the two-stage ejector (1), which advances the front ejector plate assembly (AW1) by the amount H1. This motion actuates the following ejector components:
The rectangular gear rack cores (2) between the snap-fit ribs are retracted by the rack (3) and pinion (4) mechanism, thus creating space to permit the ribs to be deformed for ejection.
The slide actuators (5) withdraw the slides (6) from the internal undercuts in the molded parts via a helical gear mechanism. The space required for this type of actuation is less than that for lifters. However, the small surface area of the gears can withstand only relatively low injection pressures. Accordingly, such a design can be employed only for undercuts with a small surface area. On the other

hand, because of the long slide strokes that are possible, very deep undercuts can be released.
The slide actuators (7) withdraw the slides (8) from the internal undercuts in the molded parts via a tapered surface with dovetail guides. This is also a space-saving design. Because the injection pressure is spread over a large support surface for the slides (8), this design can also be employed for undercuts with a large surface area. However, because of the limited slide strokes that are possible, only shallow undercuts can be released.
At the end of stroke H1, the ejector plate assembly (AW1) and the retracted slides (2, 6, 7, 8) are held by the two-stage ejector.

Step 2b:

The rear ejector plate assembly (AW2) advances by the amount H2. The parts are stripped off the mold cores (10) by the stripper bars (11). Simultaneously, ejector pins (12) aid in release and ejection of the snap-fit ribs. These ejector pins (12) are located between the snap-fit ribs on the molded part and run in guide bores in the rectangular gear rack cores (2).

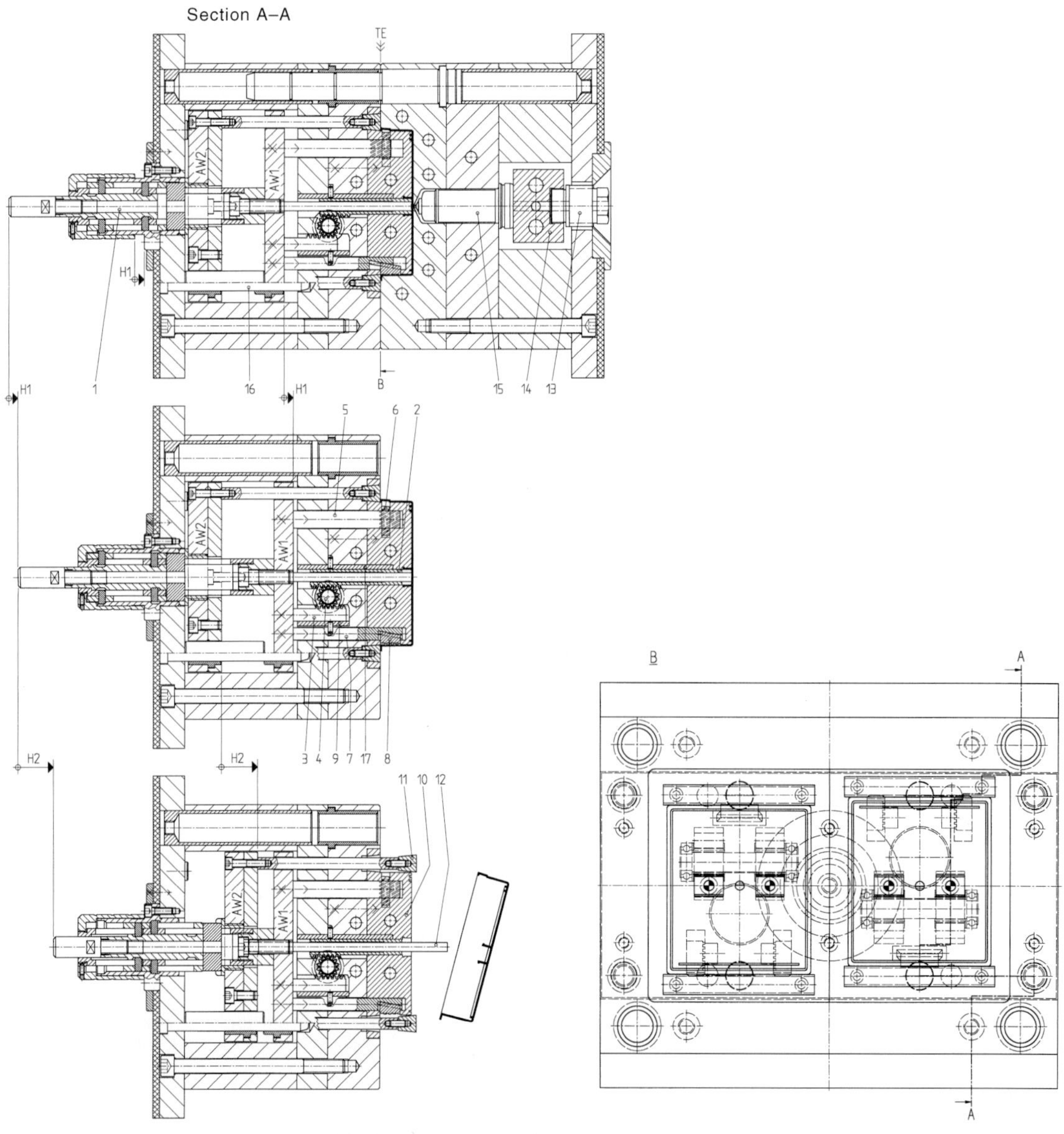

Figures 2 and 3 Two-cavity hot-runner injection mold for an ABS cover
AW1, AW2: ejector plate assemblies; 1: two-stage ejector; 2: gear rack core; 3; gear rack; 4: pinion; 5, 7: slide actuator; 6, 8: slides; 9: wear strip; 10: mold core; 11: stripper bar; 12: ejector pin; 13: inlet bushing; 14: hot-runner manifold; 15: heated hot-runner nozzle; 16: guide pin; 17: wear strip
(Courtesy: EOC Normalien, Lüdenscheid, Germany, now DME)

Example 104, Six-Cavity Injection Mold for Retaining Nuts Made from Polyamide with Metal Inserts

Retaining nuts on electrical instruments are provided with a threaded copper insert part to ensure good contact. To prevent the thread from becoming contaminated with the plastics material injected into the mold cavity these inserts are usually screwed onto a core. This process requires a considerable amount of time during loading as well as demolding of the finished part. Although it would be expedient to employ a rotary table injection molding machine, the loading and demolding time of a six-cavity mold determines the cycle.

A slight alteration of the insert can change this. If the insert is provided with a collar on the bearing area that rests against the locating mandrel, this can prevent plastics melt from entering the threads from below. In this case a smooth pin can be used as a holder at the upper end of the ejector (13), so that loading and demolding are greatly facilitated. The pressure bolt (11) together with the disc springs (12) compensates for the uncalibrated length of the insert, the pressure bolt sealing off the thread from the top.

Fig. 1

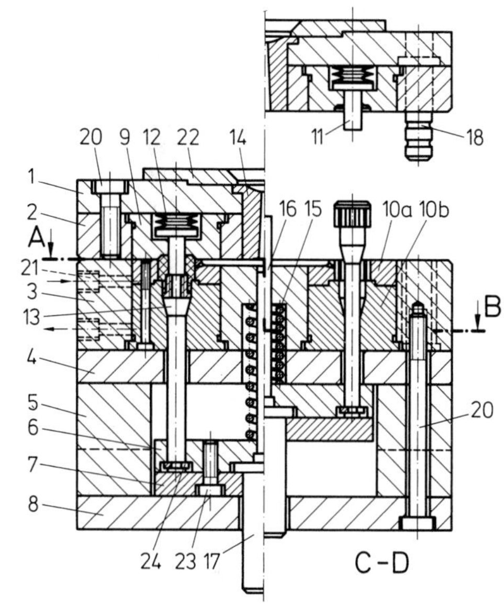

Fig. 2

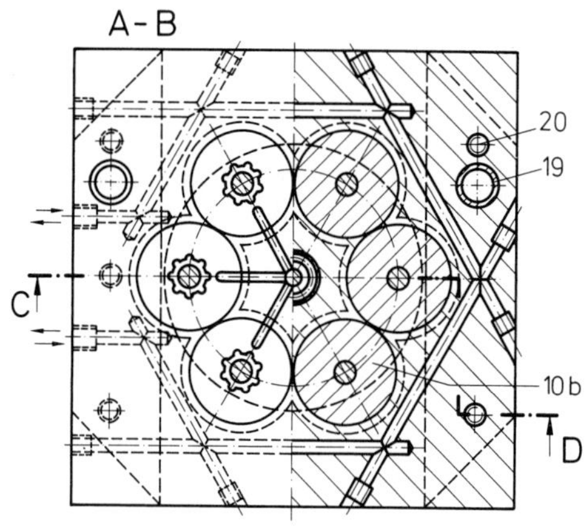

Fig. 3

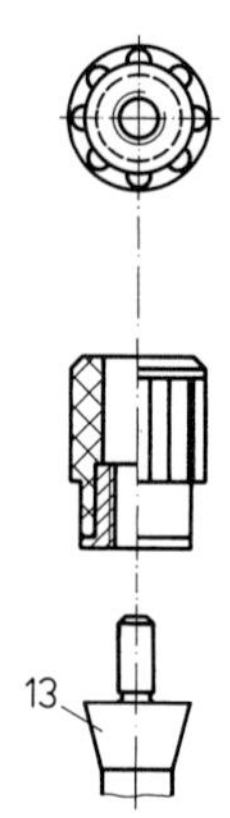

Fig. 4

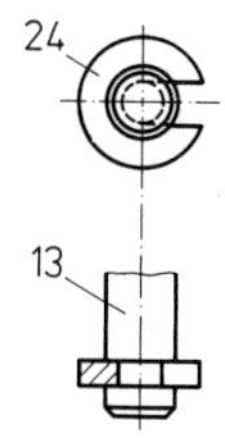

Figures 1 to 4 Six-cavity injection mold for retaining nuts with metal inserts
1: upper clamping plate; 2: insert retainer plate; 3: mold plate; 4: base plate; 5: spacer plate; 6, 7: ejector plates; 8: lower clamping plate; 9: upper inserts; 10 a, b: cavity inserts; 11: pressure bolt; 12: spring disc; 13: ejector rod; 14: sprue bushing; 15: springs; 16: sprue ejector; 17: ejector bar; 18: guide pin; 19: guide bushing; 20: bolt; 21: cooling water connection; 22: locating ring; 23: connecting bolt; 24: slotted washer

Example 105, Single-Cavity Injection Mold for a Switch Housing Made from Polyacetal

Luxury cars have, among other things, a level control system that receives its information as to the vehicle orientation by means of mercury switches. The inclination in three planes is sensed, for which reason the switch housing (Fig. 1) has three obliquely positioned holes. Each of these three holes has two small openings for cable connections at its end.

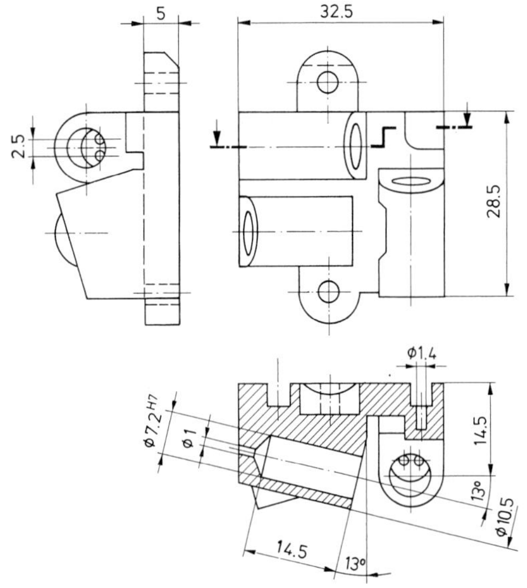

Mold (Figs. 2 to 5)

The orientation and position of the holes require three angled side cores (16, 17, 53) running in different directions. The cores used to form the holes visible in the sectional drawing of Fig. 1 cross one another, i.e. one penetrates the other. The slides (17, 53) for these two cores must be sequenced such that during mold opening and part release the penetrating core (17) is pulled first and the penetrated core (53) is not pulled until the first has been withdrawn.
Sequencing of the slide motions is accomplished with the aid of a cam plate (18). The shape of the cam tracks is shown with dotted lines in Fig. 3. A cam plate is shown in the "slides withdrawn" position. Guide pins (2) attached to the slides run in the cam tracks. The slides run on obliquely positioned guide strips (10, 32, 36). When the mold is closed, they are held in the molding position by heel blocks (4). The cam plate (18) is pinned to a gear (21); both are mounted in needle bearings (20) so as to permit rotation. A gear rack (6) operated by a hydraulic cylinder (45) engages the gear (21) to rotate the cam plate. The end positions of the hydraulic cylinder and of the ejector plate (25) are monitored by limit switches (40, Figs. 2 to 5).

Runner System/Gating

The molded part is filled via a sprue with pinpoint gate. The sprue is held by an undercut on the machine nozzle when the injection unit retracts and is sheared off the molded part. It is subsequently removed from the machine nozzle by a special mechanism.

Mold Temperature Control

The mold is operated at a temperature of 95°C (203°F). Channels for mold temperature control have been provided in the mold plates (1, 5). Insulating plates (13) prevent heating of the machine platens.

Materials

Mold inserts, slides, cam plate and sprue bushing are made of hardened steel, material no. 1.2718.

Part Release/Ejection

The mold opens in the plane of gating; the slides are now free to move. The gear rack is now advanced into the mold so that the cam plate (18) finally reaches the position drawn in Fig. 3. From the shape of the three cam tracks, it can be seen that the slide (17) moves first and that slide (53) does not move until slide (17) has been withdrawn. Slide (16) also remains stationary for a while before it is withdrawn. After the side cores have been pulled, the molded part is ejected with the aid of the ejector pins (24) and ejector sleeves (42).

Fig. 2

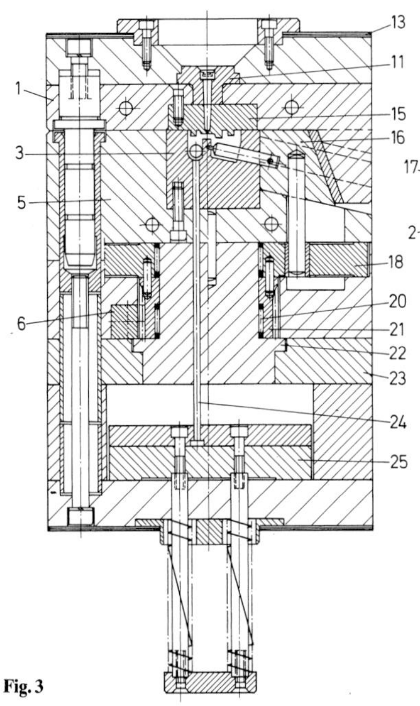

Fig. 4

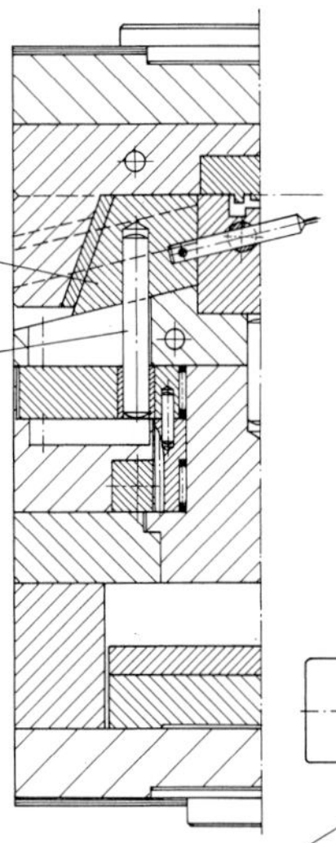

Fig. 5

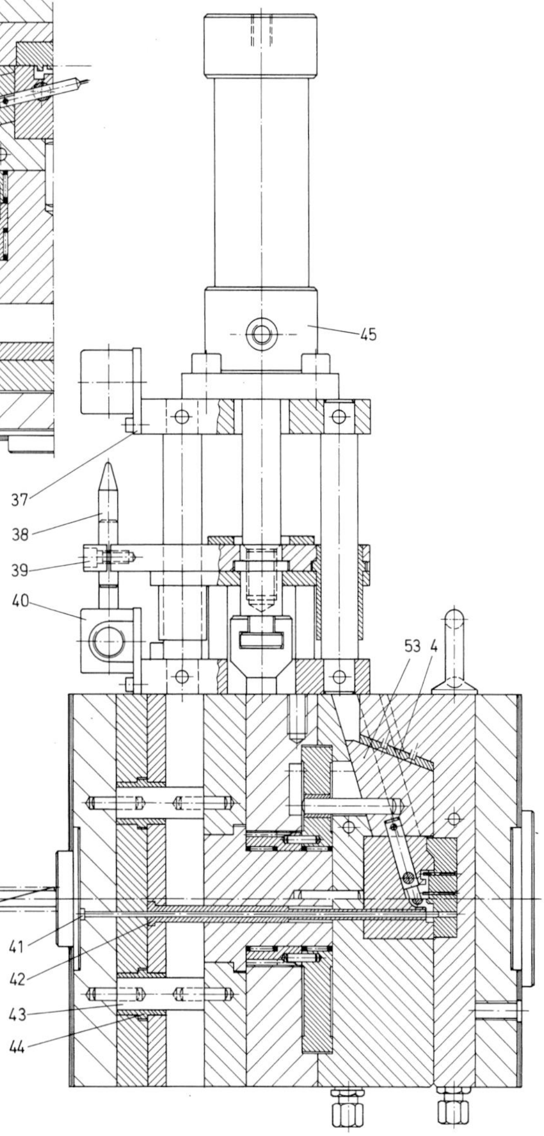

Fig. 3

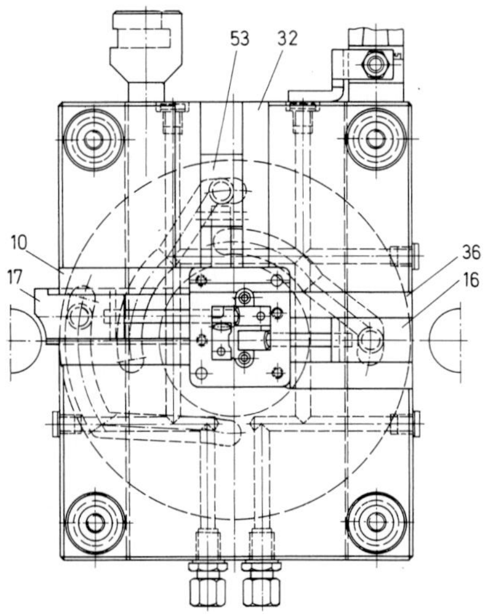

Figures 2 to 5 Single-cavity injection mold for a switch housing
1: mold plate; 2: guide pin; 3: mold insert; 4: heel block; 5: mold plate; 6: gear rack; 10: guide strip; 11: sprue bushing; 13: insulating plate; 15: mold insert; 16, 17: slides; 18: cam plate; 20: needle bearing; 21: gear; 22: backing block; 23: plate; 24: ejector pin; 25: ejector plate; 32: guide strip; 36: guide strip; 37: switch holder; 38: actuating pin; 39: clamping screw; 40: limit switch; 41: core pin; 42: ejector sleeve; 43: support pillar; 44: guide bushing; 45: hydraulic cylinder; 53: slide

Example 106, Single-Cavity Injection Mold for a Snap Ring Made from Polyacetal

The snap ring (Fig. 1) is attached to metal parts by being snapped on. Originally, the two undercuts on the ring were forcibly released, but this did not provide a satisfactory snap fit. Machining of the two undercuts was too expensive, so that a suitable split-cavity mold was designed.

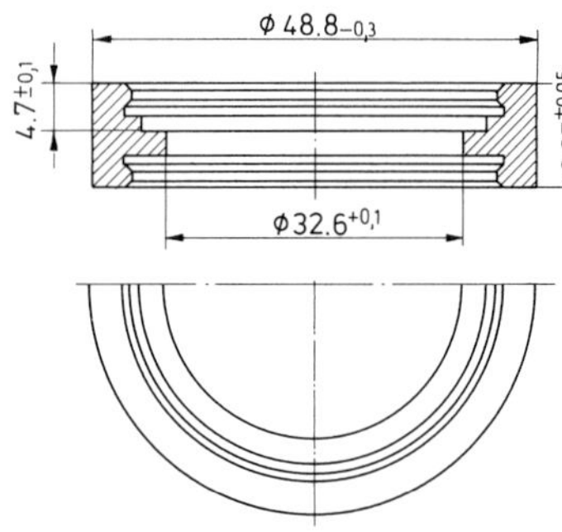

Figure 1 Snap ring of polyacetal

Mold

The mold (Figs. 2 to 5) has been designed with a single-cavity. A system of slides (11, 14) that form the undercuts in the ring is located on each side of the mold cavity. Four (narrow) slides (11) each are guided on a cone (9, 17). Four (wide) slides (14) acted upon by springs (15) are located between the narrow slides. In the molding position, all eight slides on a mold half are seated on their inner surface on the guide surfaces of the cones (9, 17), while the outer surface is seated in a conical opening in the mold plate (1) and stripper plate (12). The mold opens in parting lines (I, II, III) to actuate the slides and to release the molded part. Latches (10, 31) and ball detents (20, 22) are provided for this purpose.

Runner System/Gating

With the slide arrangement chosen, injection into the parting line (II) is most favorable. The part is filled via a single submarine gate on its circumference.

Mold Temperature Control

The mold must be operated at a temperature of 120°C (248°F). The two cones (9, 17) are hollow and are fitted with "helical cores" (37). A cooling circuit is provided in mold plate (1).
Two insulating plates (2) inhibit heat transfer to the machine platens.

Mold Steel

The two cones (9, 17), the slides and the mold ring (13) are made of hardened steel, material no. 1.2718.

Part Release/Ejection

The mold opens at parting line (I), because the latch (31) initially holds the mold plates to the left together. During this motion, the cone (17) is axially displaced with respect to the slides around it. The (narrow) slides (11) shift inward and make it possible for the (wide) slides (14) between them to also move inward through the action of the springs (15). The right-hand undercut on the molded part is now free.
After the latch (31) is released the opening motion of parting line (I) is limited by the shoulder bolt (34).
Parting line (III) is held closed by the ball detents (20, 22) (Fig. 3), so that parting line (II) now opens. The molded part, held by the slides on the left, is withdrawn from the mold ring (13). The submarine gate shears off the molded part, while the sprue is held by the sprue puller (16).
As soon as the hook of latch (10) reaches plate (3), the ball detents (20, 22) are released and the mold opens at parting line (III). Cone (9) is now also axially displaced with respect to the slides around it, and the left-hand undercut on the molded part is now free. The shoulder bolt (27) limits this opening motion.
Finally, the molded part, now held loosely by the slides on the left, is ejected by the stripper plate (12). The sprue is withdrawn from the sprue puller and can drop. The increased expense for such a mold with two collapsible cores is amortized after approximately 100,000 parts as a result of the elimination of machining.

Fig. 2

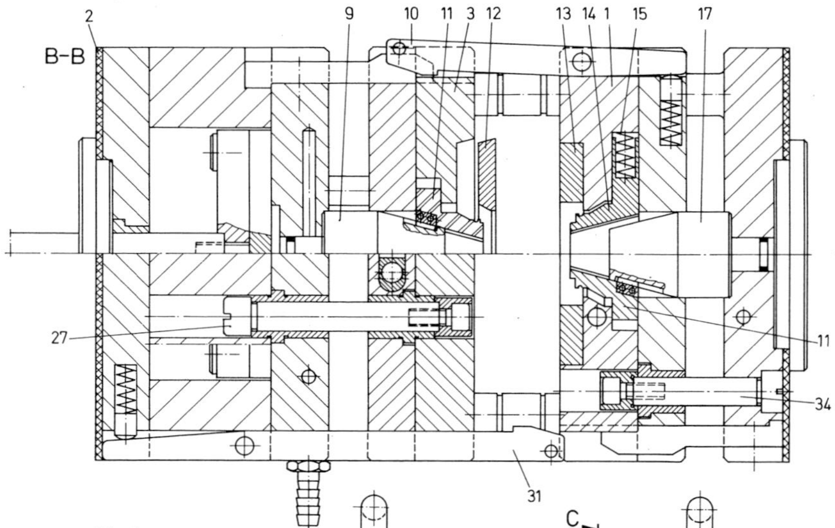

Fig. 3

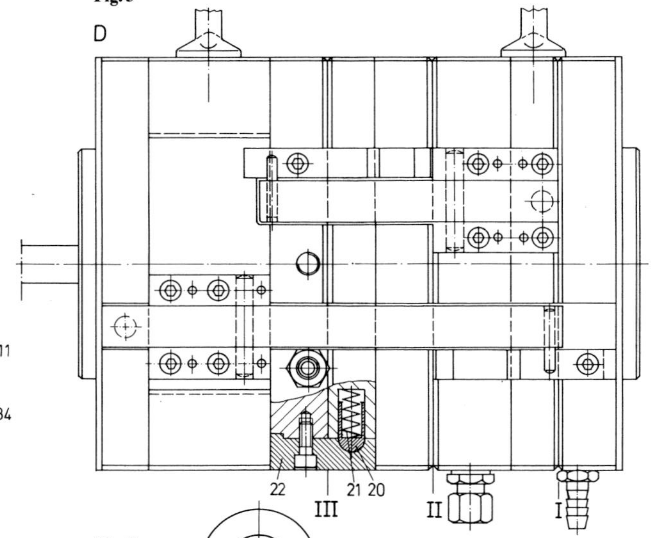

Fig. 4

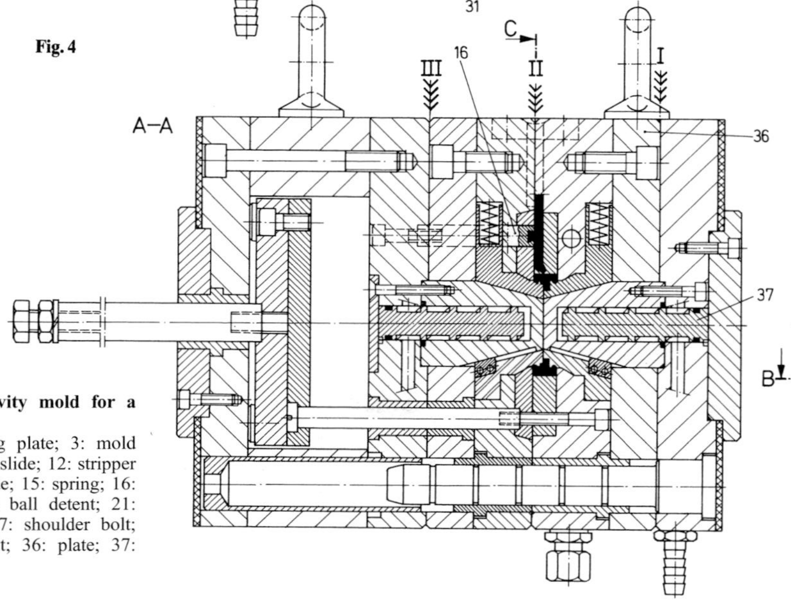

Fig. 5

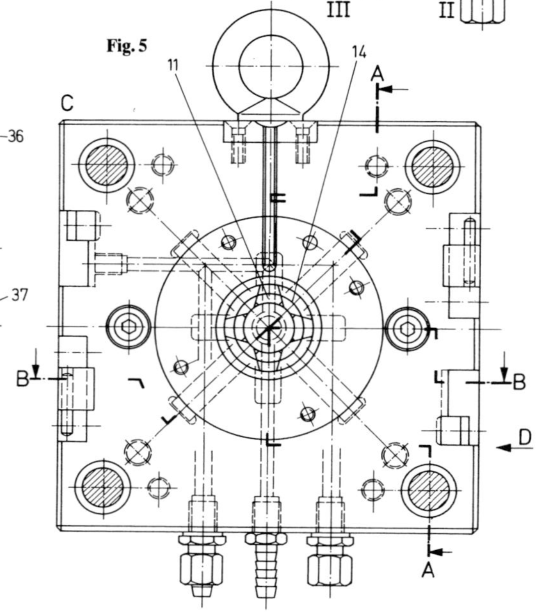

Figures 2 to 5 Single-cavity mold for a snap ring
1: mold plate; 2: insulating plate; 3: mold plate; 9: cone; 10: latch; 11: slide; 12: stripper plate; 13: mold ring; 14: slide; 15: spring; 16: sprue puller; 17: cone; 20: ball detent; 21: spring; 22: detent block; 27: shoulder bolt; 31: latch; 34: shoulder bolt; 36: plate; 37: helical core

Example 107, Single-Cavity Hot-Runner Injection Mold for High-Density Polyethylene (PE-HD) Trash Can Lids

The lid belongs to a large trash container of 1100 liter capacity. It is 1280 mm long and 1030 mm wide. The upper side is curved like a cylinder. Both broad sides have two internally ribbed arms, each of which has hinges at its lower end for opening the lid. These lateral arms give the lid an overall height of 770 mm.

Mold

The mold (Figs. 1 to 3) measures 1950 mm × 1500 mm × 1806 mm (height) and weighs around 40 metric tons.
The cavity (1) forms the upper side of the lid and the two external surfaces that bound its broad sides in the vicinity of the curve.
The mold core consists of a core plate (2), which carries a core shoulder (3). This shoulder forms most of the internal contour in the region between the two side arms (Fig. 3).
As a result of their hinge bores and ribs, the two side arms have external and internal undercuts, which are ejected with external splits (4) and internal splits (5) (Figs. 1 and 3). To accommodate this pair of splits, the core plate (2) has deep pockets in which the slide pairs move in angled guides (6, 7).
Within the guideway (7) for the internal splits, there is a link drive (8) for guiding the retainers (9). These retainers press the hinge sleeves of the side arms of the mold from the internal splits when these move inward during ejection.
External and internal splits are actuated by ejector rods (10, 11), which are attached firmly in the ejector plate (12) and in the sides of the slides such that they can be moved. The ejector plate is moved by hydraulic cylinders.
Vents (14) in the cavity and the core prevent a vacuum from forming on mold opening and ejection of the part from the core.
Aside from guide columns, wear plates (15) ensure exact centering of the core and cavity.

Gate

The lid is sprue-gated at three points on its top side via a hot runner manifold block (16). The injection pressure is controlled via pressure sensors.

Temperature Control

The top and bottom of the lid are cooled by cooling channels that run parallel to the contour in the mold plate and core. The cooling channels in the two pairs of splits are supplied with water via angled cooling tubes (17, 18). These cooling tubes are inserted into the splits and screwed into guide pieces (19), which are guided in the ejector plate (12) so that they can move sideways.
Cooling coils (20) are screwed into the sprue bushings by cutting and are connected via bores (21) to the cooling water.

Ejection

On mold opening, compressed air is fed into the stationary-mold-half vents (14); the molded part remains on the core.
When the ejector plate (12) is pushed forward, the ejector pins and the two split pairs demold the molded part from the core. The slides move away from the lid arms, releasing them from the sides. This process is supported by air blown into the core-side vents.

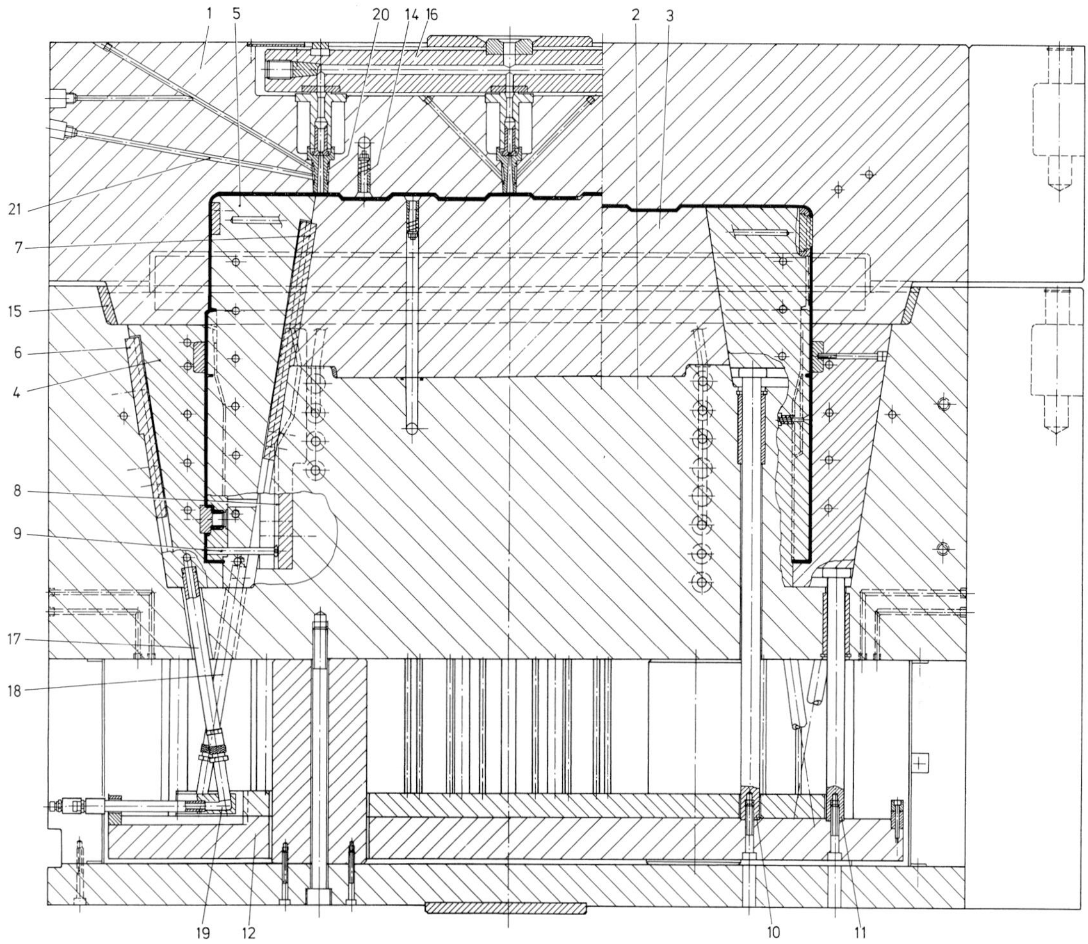

Figure 1 Injection mold for PE-HD trash can lid
1: cavity plate; 2: core plate; 3: external split; 5: internal split; 6, 7: angle guide; 8: link drive; 9: retainer; 10, 11: ejector rod; 12: ejector plate; 14: vent; 15: wear plate; 16: hot runner distributor block; 17, 18: cooling pipe; 19: guide piece; 20: cooling coil; 21: cooling water bore

Fig. 2

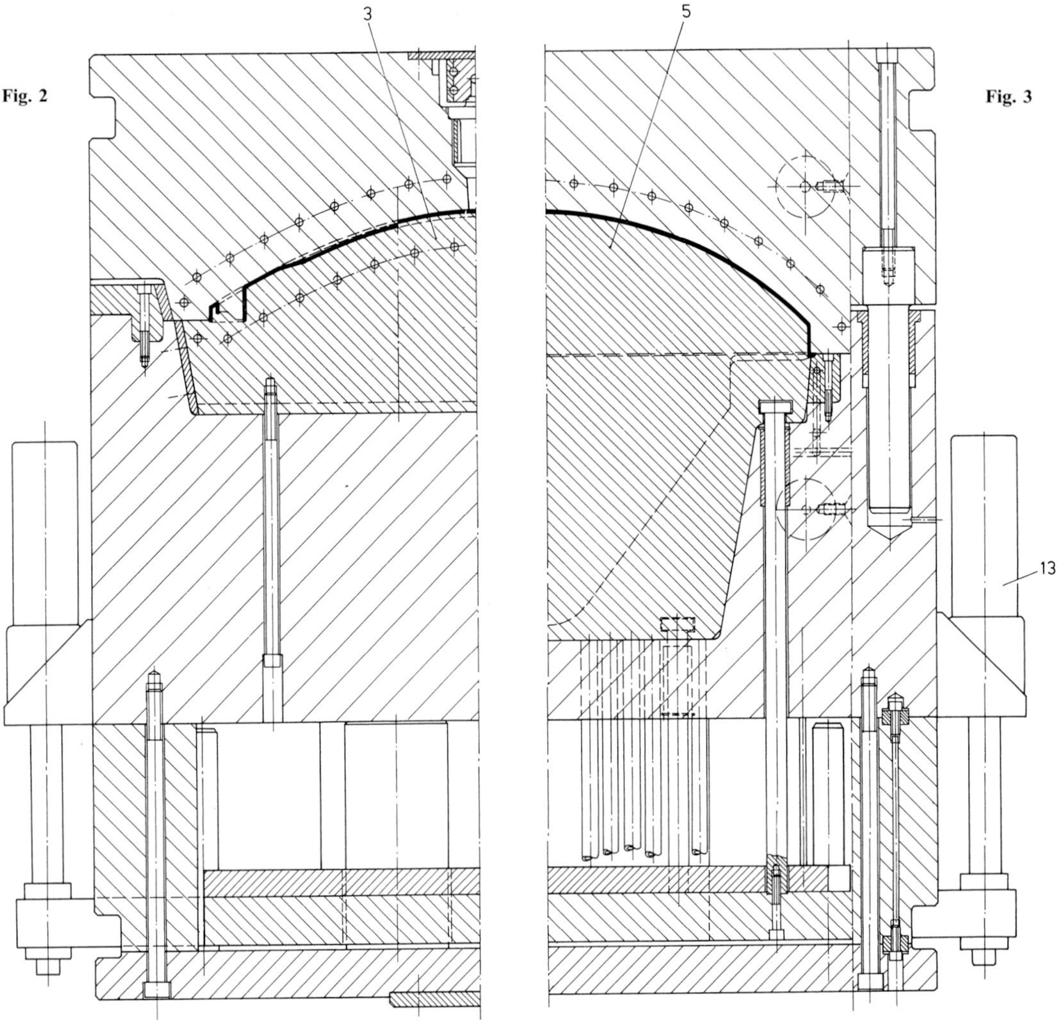

Fig. 3

Figures 2 and 3 Injection mold for PE-HD trash can lid
3: core shoulder; 5: internal split; 13: hydraulic cylinder

Example 108, Single-Cavity Hot-Runner Injection Mold for an Air Vent Housing Made from Acrylonitrile Butadiene Styrene (ABS)

The frame-shaped air vent housing is part of the ventilation system for the passenger compartment of an automobile. It has dimensions of approx. 100 mm × 70 mm and must hold five adjustable vent flaps (vanes) with which the air flow is regulated. In addition, mounting points are located on the outside along the two long sides of the housing. The shafts for the air vents fit into holes also located along the two long sides. Each pair of opposite holes must be in exact alignment. The specified hole diameters do not permit the pairs of aligning holes to be formed by single long cores. Accordingly, the design shown in Figs. 1 and 2 was chosen.

Mold

The mold is constructed largely from standard mold components. The molded part is located between cavity insert (11) and core insert (12). The cavity insert (11) and the surrounding mold plate (13) have an opening on one side for a slide (14) that runs in guide (15). Two rows of core pins (16, 17) that form the holes from both the outside of the housing (core pin 16) as well as from the inside (core pin 17) are attached to the slide (14). To accommodate core pins (17), the slide has a hook-shaped end that protrudes into a recess in the cavity insert.

The slide (14) is operated by a latching cylinder (18). The piston of this cylinder is displaced by means of oil pressure and is held in the mounting position (forward) mechanically (Fig. 3).

The operation of the cylinder is illustrated schematically in Figs. 4 and 5. Figure 4 = unlatched; Fig. 5 = latched. When the piston moves forward, the segments (2) are forced into the annular groove in the piston rod (1) by the latching sleeve (3), thereby holding the piston in position.

This mechanical latching is ensured even in the absence of hydraulic pressure and is several times stronger than the hydraulic force.

Prior to retraction, the latching sleeve (3) is shifted by the hydraulic fluid, thereby unlatching the cylinder. The sensor pin (4) provides an exact indication of the position of the latching sleeve (3).

The cylinder (18) is threaded into the flange (19) and locked in position by the slotted nut (20).

Two slides (21, 22) actuated by cam pins (23, 24) attached to the cavity half are provided to release the mounting points on the long sides of the air vent housing.

The part-forming inserts of the mold are made of case-hardening steel, material no. 1.2764.

Runner System/Gating

The part is filled from the outside via three pinpoint gates located on one of the long sides. Since the molded part is centered in the mold, the melt flows from the sprue bushing (25) via an externally heated hot-runner manifold (26) and hot-runner nozzle (27) to a conventional runner. This runner is machined into the upper surface of the slide (21) along with the three submarine gates leading to the mold cavity. Pockets (28) and sprue puller (29) hold the runner in the slide (21) as the mold opens. A pressure sensor (32) monitors the internal cavity pressure.

Mold Temperature Control

Channels for mold temperature control have been machined into the part-forming inserts as well as the two slides. Thermocouples (33) monitor the mold temperature.

Part Release/Ejection

Prior to mold opening, the slide (14) is pulled outward by the cylinder (18), thereby withdrawing the core pins (16, 17) from the molded part.

The molded part is retained on the core half as the mold opens. During this motion, the two slides (21, 22) move outward, releasing the external mounting points on the air vent housing. Slide (21) also pulls the runner away from the molded part, shearing off the gates. The molded part is subsequently stripped off the core by ejectors.

Since the conventional runner is located in the moving slide (21), a special mechanism is needed to eject it. A runner ejector (30) that is pulled back by a spring (31) is located behind the undercut (29) that holds the runner. Once the slide (21) is in the fully opened position, the machine ejector is actuated and the ejector pin (34) strikes the end of the runner ejector (30) located in slide (21). The runner is now ejected.

The machine ejector must be retracted prior to mold closing in order to avoid damage to the ejector pin (34) by the inward moving slide (25). The position of the ejector plate is monitored by the limit switch (35).

Fig. 1

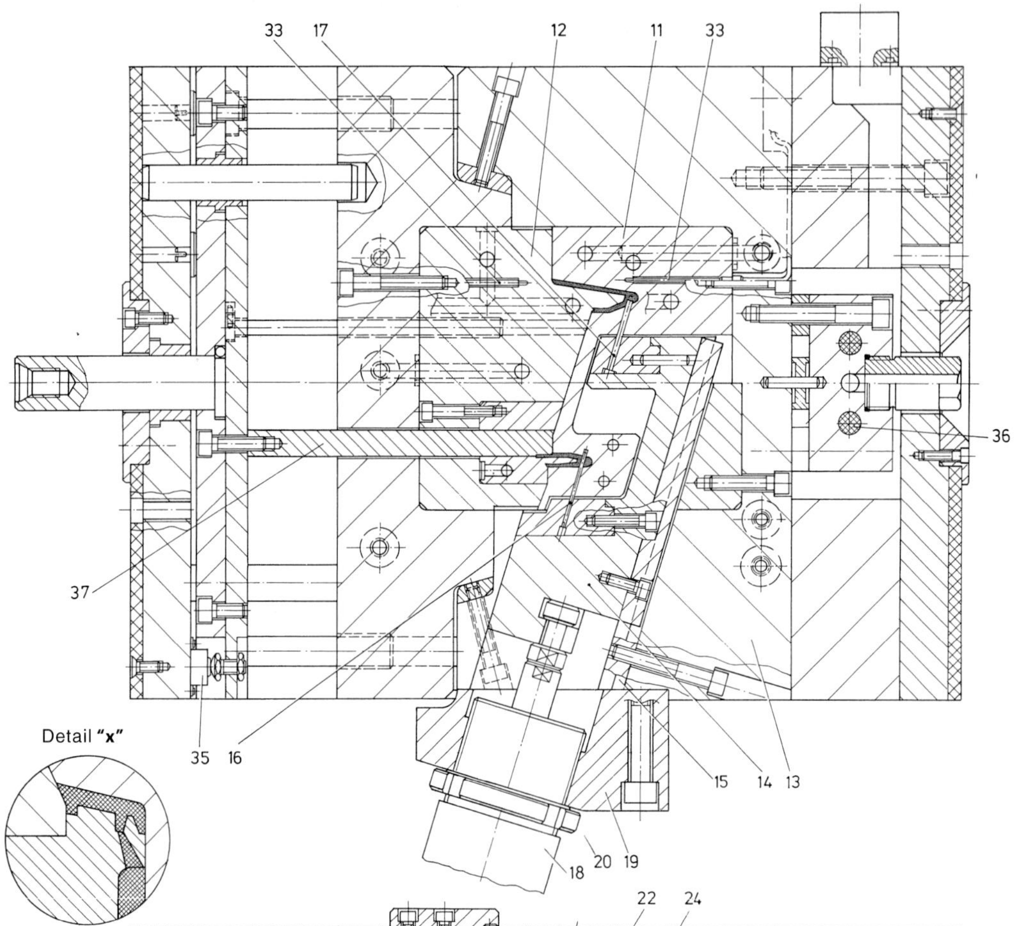
33
17
12
11
33
36
37
Detail "x"
35
16
15
14
13
20
19
18

Fig. 2

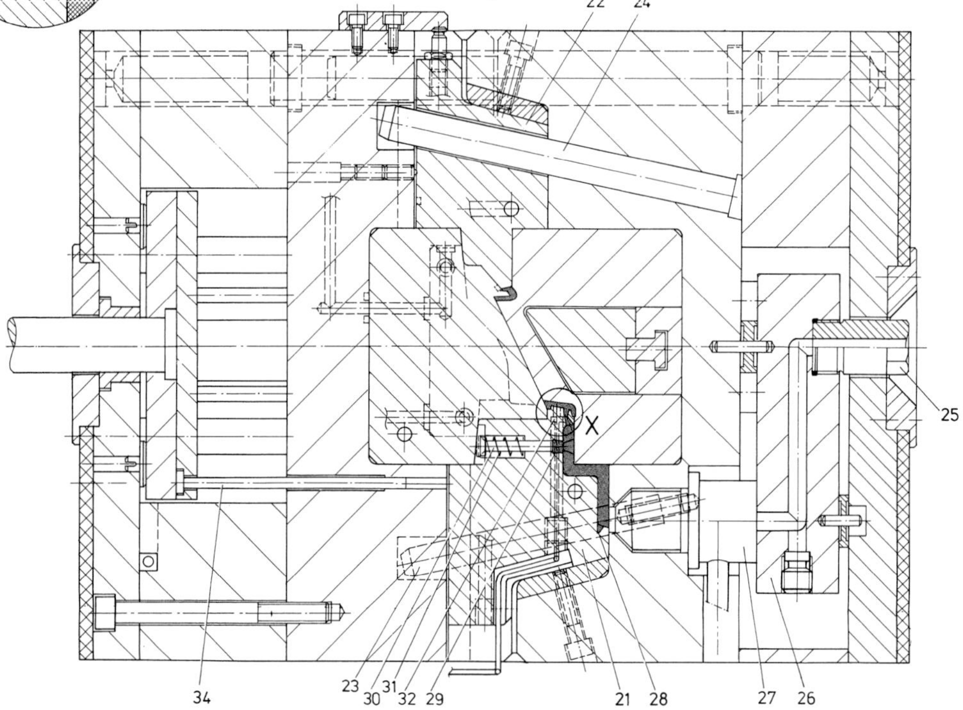
22
24
X
25
34
23
30
31
32
29
21
28
27
26

Figures 3 to 5 Latching cylinder
1: piston rod; 2: segments; 3: latching sleeve; 4: sensor pin; 5: slotted nut
(Courtesy: Hasco)

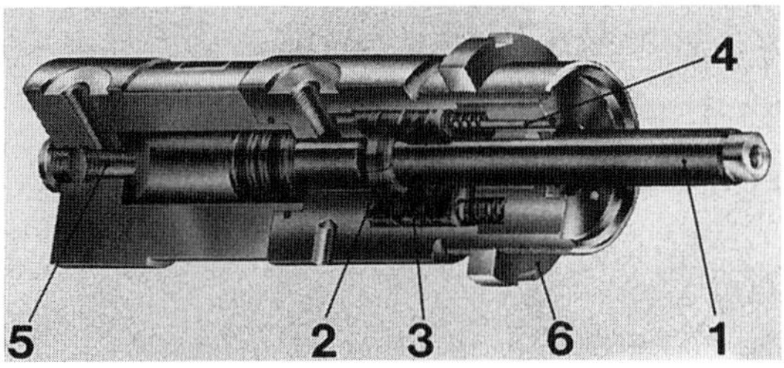

Fig. 3

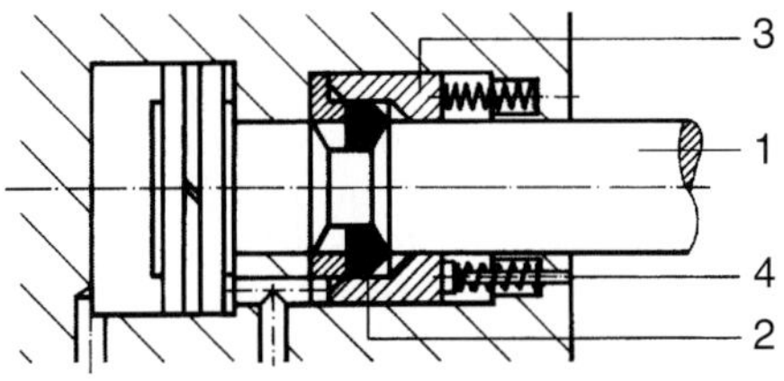

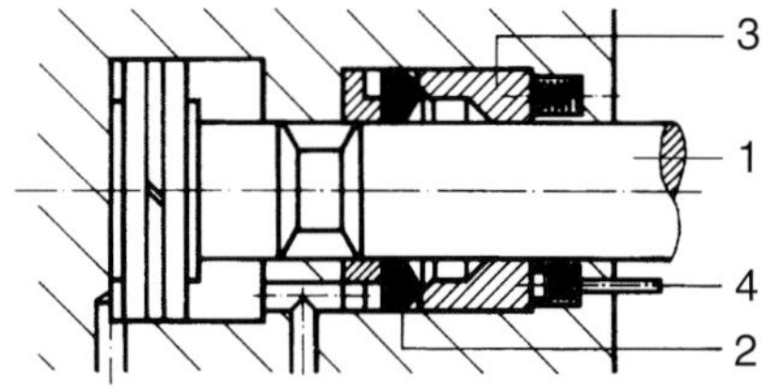

Fig. 4

Fig. 5

Figures 1 and 2 Single-cavity hot-runner injection mold for an air vent housing of ABS
11: cavity insert; 12: core insert; 13: mold plate; 14: slide; 15: guide; 16: core pin; 17: core pin; 18: latching cylinder; 19: flange; 20: slotted nut; 21, 22: slides; 23, 24: cam pins; 25: sprue bushing; 26: hot-runner manifold; 27: hot-runner nozzle; 28: pocket to hold runner; 29: undercut to pull runner; 30: runner ejector; 31: spring; 32: pressure sensor; 33: thermocouple; 34: ejector pin; 35: limit switch; 36: cartridge heater; 37: ejector core

Example 109, Single-Cavity Hot-Runner Injection Mold for an ABS Housing

Molded Part

The part is a rectangular cover with dimensions of 111.4 mm × 192.6 mm × 35.6 mm. On the inside, there are two pairs of snap-fit ribs which hold circuit boards that are inserted between three ribs located in the interior on a long side of the part. The two round openings accommodate rotary switches; the four stand-offs are used to securely attached the cover.

Mold

The mold is constructed largely of standard mold components. For instance, a standard mold base with dimensions of 246 mm × 246 mm is employed. The mold shut height is 377 mm. Plate thicknesses are also standard dimensions. An exception to this is the mold plate on the stationary half. It is sized to accommodate the cavity and hot sprue bushing.
Steel grade 1.2767, through-hardened, was employed for the mold core on the moving half. For the stationary-side mold plate, steel grade 1.2764, case-hardened, was employed. The lifters (2, 15) and the slide blocks (4) are also fabricated from steel grade 1.2764. The lifter rods (3) are mounted in extended support blocks (5) fastened to the ejector plate assembly (AW2), while the slide blocks (4) are guided in corresponding grooves. The extended support pads (5) are space-saving, permit the lifter assemblies to be shortened and can be incorporated into the mold base. The lifter rods (3) are standard guide pins and run in guide bushings (6) to ensure reliable operation. The design based on lifters permits release of deep undercuts through the use of long lifter strokes. The two ejector plate assemblies (AW1, AW2) are guided by four guide pins (13). These ejector plate assemblies are actuated by a two-stage ejector (1) with composite stroke that divides the single, continuous stroke of the machine ejector into two combined partial strokes.
All guiding and wear components have been treated with Lamcoat, a soft, self lubricating coating based on tungsten disulfide (see also Example 104).

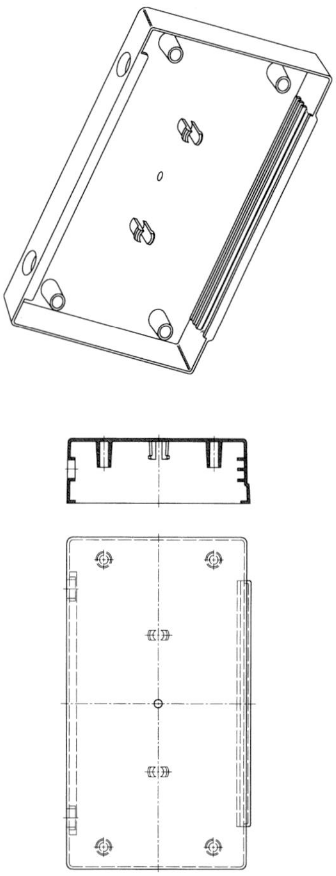

Figure 1 ABS housing, diagram

Runner System

The cavity is direct-gated via a hot sprue bushing with cone point (14) for runnerless molding.

Part Release/Ejection

Step 1:

The mold opens.

Step 2a:

The machine ejector actuates the two-stage ejector (1), which advances the ejector plate assemblies (AW1, AW2) by the amount H1. This motion actuates the following ejector components:
All ejector components advance and lift the molded part off the mold core (12) by the amount H1.
The core pins (9) for the external snap fits on the molded part do not move. They are attached to the clamping plate of the moving mold half.
During this motion, the molded part is guided by the core pins (7) for the mounting stand-offs as well as the rib-forming sleeves (10) for the snap-fit ribs. The lifter assemblies (2, 3, 4, 15) release the internal undercuts.
At the end of stroke H1, the ejector plate assembly (AW2) with the core pins (7), sleeves (10) and lifter assemblies (2, 3, 4, 15) are held by the two-stage ejector and stop moving.

Step 2b:

The front ejector plate assembly (AW1) advances by the amount H2. The ejector sleeves (8) strip the molded part off the core pins (7) for the mounting stand-offs.

Simultaneously, additional ejector sleeves (11) aid in release and ejection of the snap-fit ribs from the rib-forming sleeves (10). The ribs are deformed inward during this step.

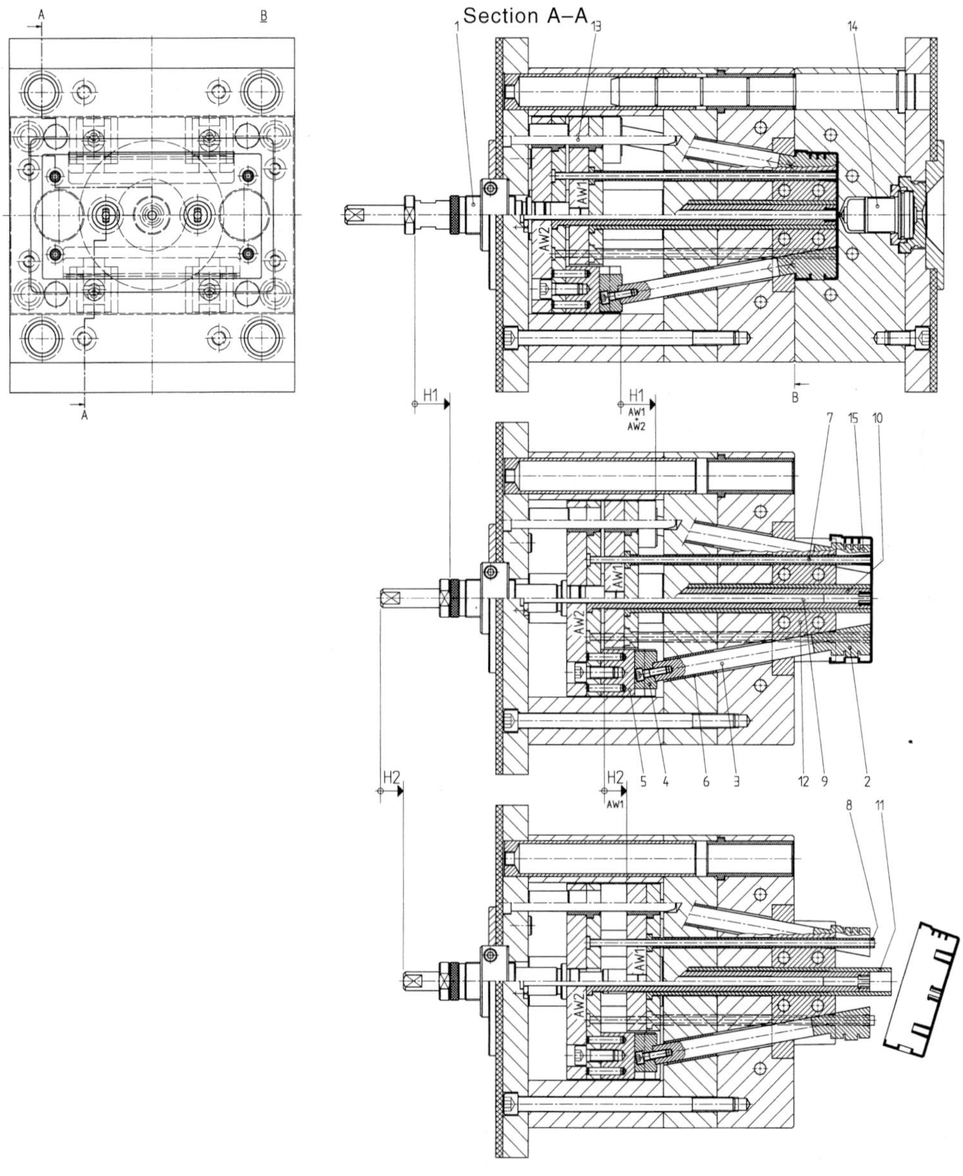

Figure 2 Single-cavity hot-runner injection mold for an ABS housing
AW1, AW2: ejector plate assemblies; 1: two-stage ejector; 2: lifter; 3: lifter rod; 4: slide block; 5: support pad; 6: guide bushing; 7: core pin; 8, 11: ejector sleeves; 9: core pin; 10: rib-forming sleeve; 12: mold core; 13: guide pin; 14: hot sprue bushing; 15: stripper bar
(Courtesy: EOC Normalien, Lüdenscheid, Germany, now DME)

Example 110, Single-Cavity Runnerless Injection Mold for a Polystyrene Junction Box

Molded Part

The junction box (dimensions: 136.8 mm × 117.8 mm × 28.8 mm) has four round openings on two opposite sides close to the back wall. No witness marks from side cores or slides are permitted on the outer surface of the box between the upper edge and these openings. On the other hand, sink marks are allowed near the openings on both the side walls and the bottom of the box, since these areas are not visible once the box is installed.

Mold (Fig. 2)

The core (4) of the mold, a helically grooved inner core and four small side cores (11) are of conventional design. All remaining components are standard or off-the-shelf items.
For instance, a standard mold base with dimensions of 246 mm × 246 mm is employed. The mold shut height is 271 mm. Plate thicknesses are also based on standard dimensions, except for the mold plate (3) on the stationary half (FS). It is sized to accommodate the cavity and sprue bushing.

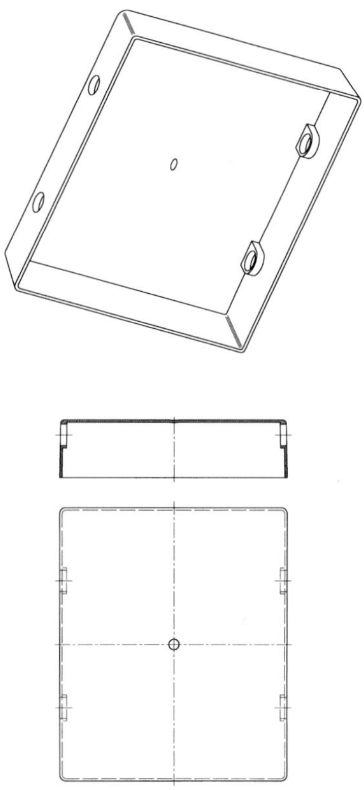

Figure 1 Polystyrene junction box diagram

Steel grade 1.2767, through-hardened, was employed for the mold core on the moving half (BS). An inner core with helical grooves provides cooling. For the stationary-side mold plate (3), steel grade 1.2764, case-hardened, was employed. The side cores (11) are also fabricated from steel grade 1.2767 and are treated with chromium nitride (CrN) and Lamcoat®.
Given the part geometry and the requirement that there be no witness lines on the outside surface, the side cores for the round openings must be placed on the stationary-side mold plate FS (3). Latch locks (1) located inside the mold actuate these side cores (11) as an auxiliary parting line (TEF) opens. The four latch-lock actuating rods (7) also serve to guide the ejector plate assembly (5).
The ejector plate assembly is actuated by the ejector rod (6), which is connected to the machine ejector. All guiding and wear components have been treated with Lamcoat® (see also Example 104).

Gating/Runner System

For runnerless molding, the cavity is direct-gated via a hot sprue bushing with cone point (13) housed in a bushing well insert (14). A slight gate vestige that does not disturb functionally or visually remains in a dimple on the back wall of the part.

Part Release/Ejection

Step 1:

Stroke H1:

The mold opens. The stationary-side mold plate FS (3) remains connected to the moving mold half (BS) by the actuating rods (7) for the latch locks (1). The auxiliary part line (TEF) opens by the amount H1. As a result of this motion, the cam pins (10) attached to the stationary-side clamping plate (2) withdraw the side cores (11) in mold plate (3) from the openings (undercuts) in the molded part.
Once the undercuts have been released at the end of stroke H1, mold plate (3) is locked in this position by disengaging the latch-lock actuating rods (7) while simultaneously engaging the detent segments (15) in the latch-lock housing (9).
The central sprue bushing (13) and bushing well insert (14) are attached to the stationary-side clamping plate FS (2) and do not move as the auxiliary parting line (TEF) opens. A conical slip fit between the stationary-side mold plate (3) and bushing well insert (14) serves to minimize wear. When the mold is closed, this insert shuts off against the part-forming surface on the front of the mold.

Stroke H2:

At the start of stroke H2, the actuating rods (7) for the latch locks (1) disengage as the detent segments (15) engage. The moving mold half (BS) continues to move and opens the primary parting line (TEE) to permit part ejection (the stationary-side mold plate FS (3) is held at its position and no longer connected).

Step 2:

The machine ejector advances, actuating the ejector rod (6), which, in turn, advances the ejector plate assembly (5). The ejector pins (12) strip the molded part off the mold core (4).

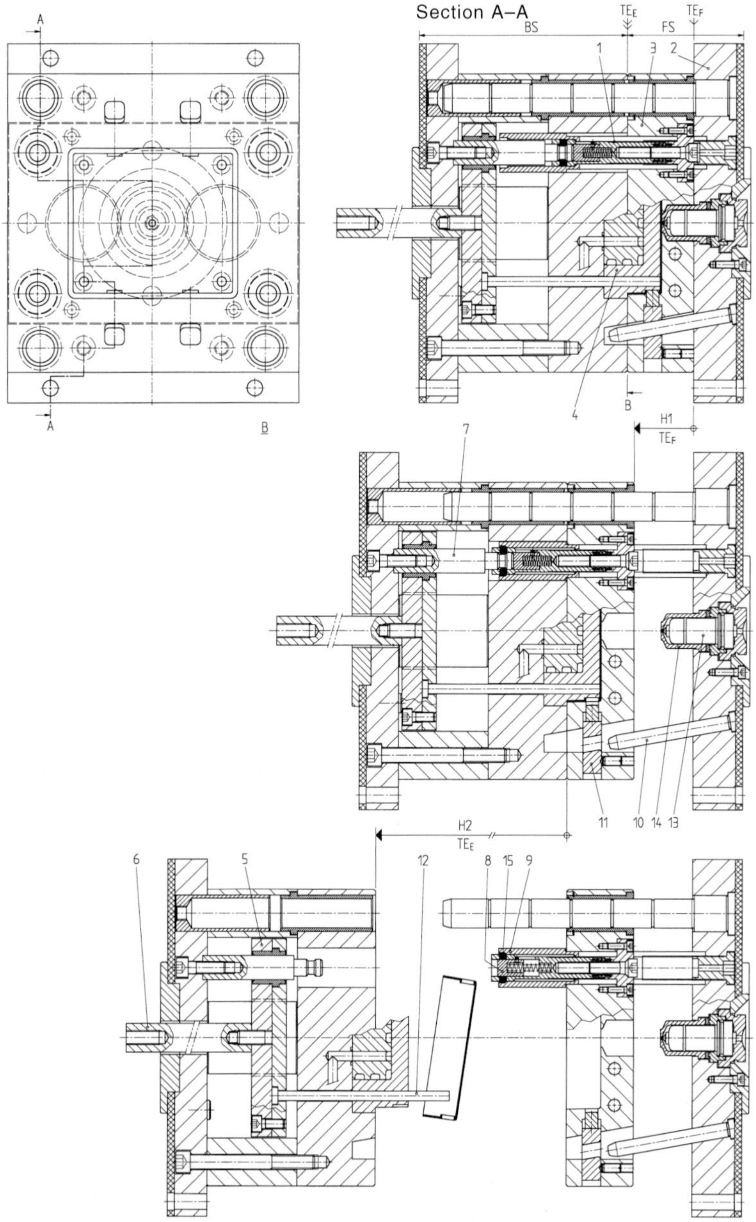

Figure 2 Single-cavity runnerless injection mold for a polystyrene junction box
BS: Moving mold half; FS: stationary mold half; TEF: auxiliary parting line; TEE: primary parting line; 1: latch lock; 2: stationary-side clamping plate FS; 3: stationary-side mold plate FS; 4: mold core; 5: ejector plate assembly; 6: ejector rod; 7: latch-lock actuating rod; 8: retaining bushing; 9: latch-lock housing; 10: cam pin; 11: side core; 12: ejector pin; 13: hot sprue bushing; 14: bushing well insert; 15: detent segment (Courtesy: EOC Normalien, Lüdenscheid, Germany, now DME)

Example 111, Four-Cavity Hot-Runner Injection Mold for a Polyamide 6,6 Joining Plate

Molded Part

The joining plate (dimensions: 173.8 mm × 160.4 mm × 10 mm) is essentially rectangular in shape. At each corner, there is a cylindrical boss with an M5 female thread. An additional four symmetrically positioned recesses serve as counterbores to accept screw heads. There are also reinforcing ribs between the bosses.

Mold

The mold is constructed largely of standard mold components. For instance, a standard mold base with dimensions of 246 mm × 246 mm and two independently actuated ejector plates is employed. The mold shut height is 314 mm. Plate thicknesses are also based on standard dimensions, except for the mold plate (2) on the stationary half (FS) and the rails (5) in the ejector housing. These are sized, respectively, to accommodate the cavity and sprue bushing as well as the components of the ejector assembly.
Steel grade 1.2764, case-hardened, was employed for the mold plate (2) on the stationary half (FS) and the stripper plate (3). The core pins (8) and the threaded cores (9) are of steel grade 1.2767 and are coated with chromium nitride (CrN). The core retainer plate (8a) is of steel grade 1.2312.
The two ejector plate assemblies (AW1, AW2) are guided by four guide pins (6). These ejector plate assemblies are actuated by a two-stage ejector (1) that divides the single stroke of the machine ejector into two successive partial strokes H1 and H2.
All guiding and wear components as well as the gear drive elements have been treated with Lamcoat®, a soft, self-lubricating coating based on tungsten disulfide (see also Example 104).

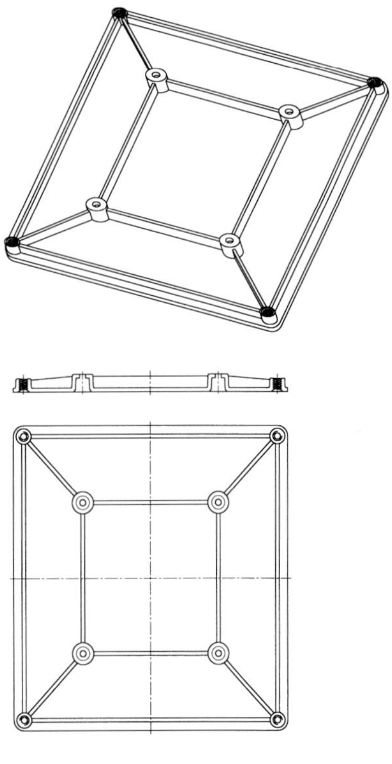

Figure 1 Nylon 6,6 joining plate diagram

Gating/Runner System

The cavity is direct-gated via a hot sprue bushing with cone point (17) for runnerless molding.

Part Release/Ejection

Step 1:

The mold opens.

Step 2a:

The machine ejector actuates the two-stage ejector (1), which advances the front ejector plate assembly (AW1) by the amount H1. As a result of this motion, the threaded drive nuts (11) located in the ejector plate assembly (AW1) cause the threaded drive shafts (10) to rotate. Pinion gears (12) mounted on the threaded shafts rotate the threaded cores (9), which run in threaded guide bushings (14), and unscrew them from the female threads in the part. During stroke H1, the molded part remains on the core pins (8) and is not ejected by the stripper plate (3).
To simplify assembly of the mold and facilitate adjustment of the threaded cores (9), the threaded drive nuts (11) contain elongated holes for the mounting bolts (15). Provision for a spanner wrench (16) permit the threaded drive nuts to be rotated. Rotating the drive nuts causes the drive shafts (10) and the attached pinion gears (12, 13) to rotate, which results in axial displacement of the threaded cores (9) by the amount required to seal off the face of the core against mold plate (2) on the stationary mold half (FS).
At the end of stroke H1, the ejector plate assembly (AW1) and the unscrewed threaded cores (9) come to a stop and are locked in position by the two-stage ejector.

Step 2b:

As the machine ejector continues its motion, the rear ejector plate assembly (AW2) advances by the amount H2. The attached ejector rods (7) actuate the stripper plate (3), which strips the molded part off the core pins (8).

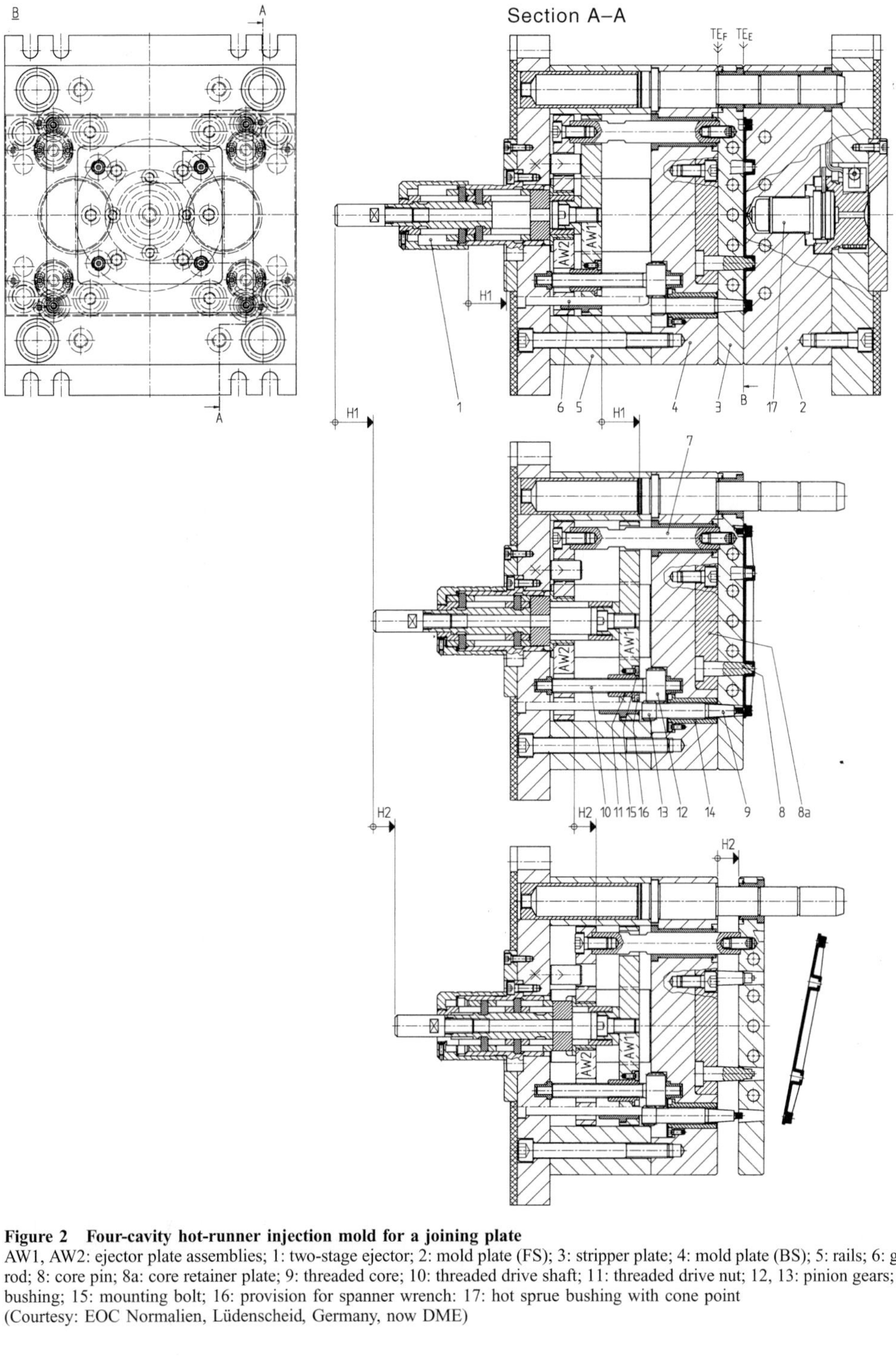

Figure 2 Four-cavity hot-runner injection mold for a joining plate
AW1, AW2: ejector plate assemblies; 1: two-stage ejector; 2: mold plate (FS); 3: stripper plate; 4: mold plate (BS); 5: rails; 6: guide pin; 7: ejector rod; 8: core pin; 8a: core retainer plate; 9: threaded core; 10: threaded drive shaft; 11: threaded drive nut; 12, 13: pinion gears; 14: threaded guide bushing; 15: mounting bolt; 16: provision for spanner wrench: 17: hot sprue bushing with cone point
(Courtesy: EOC Normalien, Lüdenscheid, Germany, now DME)

Example 112, 2 × 4-Cavity Hot-Runner Stack Mold for Hinged Covers

Molded Part

The hinged cover (dimensions: 89.2 mm × 89.2 mm × 13.6 mm) is shaped like a square box. Two external undercuts accept hinge pins. The internal undercut serves as a catch to hold the cover securely closed.

Mold

This stack mold is constructed largely of standard mold components. For instance, a standard mold base with dimensions of 546 mm × 446 mm is employed. The mold shut height is 582 mm. Plate thicknesses are also based on standard dimensions, except for the four mold plates (14, 15), which are sized to accommodate the part shape and standard hot-runner nozzles. The 2 × 4 mold inserts (16, 17) are identically shaped and oriented symmetrically around the center of the mold. Steel grade 1.2767, hardened, was employed for the mold inserts.
Steel grade 1.2764, case-hardened, was employed for the lifters (9). The lifter rods (10) have a circular cross-section, are guided in inserts with angled bores (11), and are fastened to slide blocks (12). The slide blocks ride on guide pins (13) held in the ejector plate assemblies (AW).

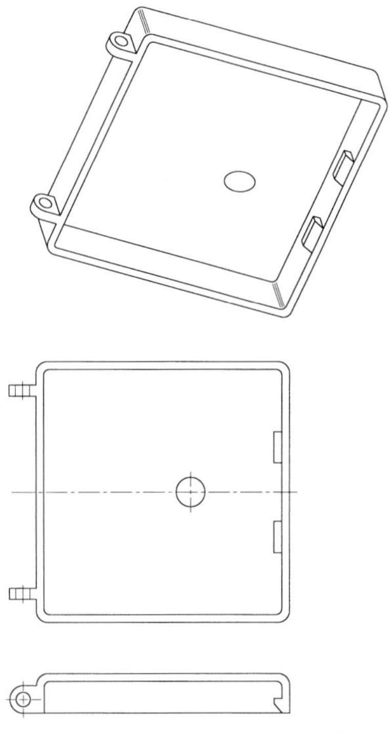

Figure 1 Polypropylene hinged cover diagram

The 16 slides (6) that release the hinges are of steel grade 1.2764, case-hardened, and are actuated by cam pins (5). The slides (6) move in T-slots (detail "D", Fig. 3).
The threaded double shafts (1, 2) are mounted in the center block in a manner that permits rotation. The threaded drive nuts (4) in the ejector plate assemblies (AW) are made of steel and, to minimize wear, have a threaded insert made from a special nylon. The threaded drive nuts (3) in the mold blocks are made of brass grade 2.0550.
All guiding and wear components have been treated with Lamcoat® (see also Example 104).

Gating/Runner System

The parts are filled via a hot-runner system. From the inlet bushing (18), the melt flows through the heated bushing extension (19) and H-shaped hot-runner manifold (20) to the eight heated hot-runner nozzles (21). The entire hot-runner system is mounted in the movable center plate. Optimal thermal separation between the hot-runner system and cavity permits the use of open sprue nozzles with a cone point. The gate vestige remaining on the molded parts is minimal. Through the use of appropriate controls for the hot-runner system, extremely tight temperature control is possible in conjunction with the other mold and machine parameters.

Part Release/Ejection

Step 1:

The mold opens by the amount H. The twin threaded shafts (1) ensure uniform opening of both parting lines TEE. These are mounted in the center plate in a manner that permits rotation but not axial displacement. Rotation of the twin threaded shafts (1) is brought about by the threaded drive nuts (3), which are mounted securely in both the movable mold half (BS) and stationary mold half (FS).
While the movable mold half (BS) moves the distance H, the center plate moves the distance H/2. Cam pins (5) on the center plate pull the 16 side cores (6) out of the hinges as the parting lines TEE are opening (detail "D", Fig. 3).
The ejector plate assembly (AW) on the movable side (BS) is connected to the machine ejector (MAW) during mold opening and moves together with it. Since the twin threaded shafts (2) that actuate the ejector plate assembly (AW) on the stationary side (FS) are mounted in the center plate along with the twin threaded shafts (1), the ejector

plate assembly on the stationary side (FS) remains in its initial position.

Step 2:

The machine ejector (MAW), advances the ejector plate assembly (AW) on the movable side (BS). The threaded drive nuts (4) in the ejector plate assembly (AW) on the movable side (BS) cause the twin threaded shafts (2) mounted in the center plate to rotate. The oppositely threaded other halves of the twin shafts (2) pull the ejector plate assembly (AW) on the stationary side (FS) toward the center of the mold with the aid of the threaded drive nuts (4).

The ejector plate assemblies (AW) in the stationary (FS) and movable (BS) mold halves actuate lifters (9 to 13) that release the internal undercuts, while ejector pins (7, 8) knock the molded parts off the mold cores (16). To prevent the parts in the upper half of the mold from catching on the lifters (9), the ejector pins (7) close to the lifters are designed as profiled blade ejectors. These extend out to the edge of the mold cores (16) and prevent the molded parts from "hanging up" on the lifters (9).

Mold Closing

The mold closes in the reverse sequence.

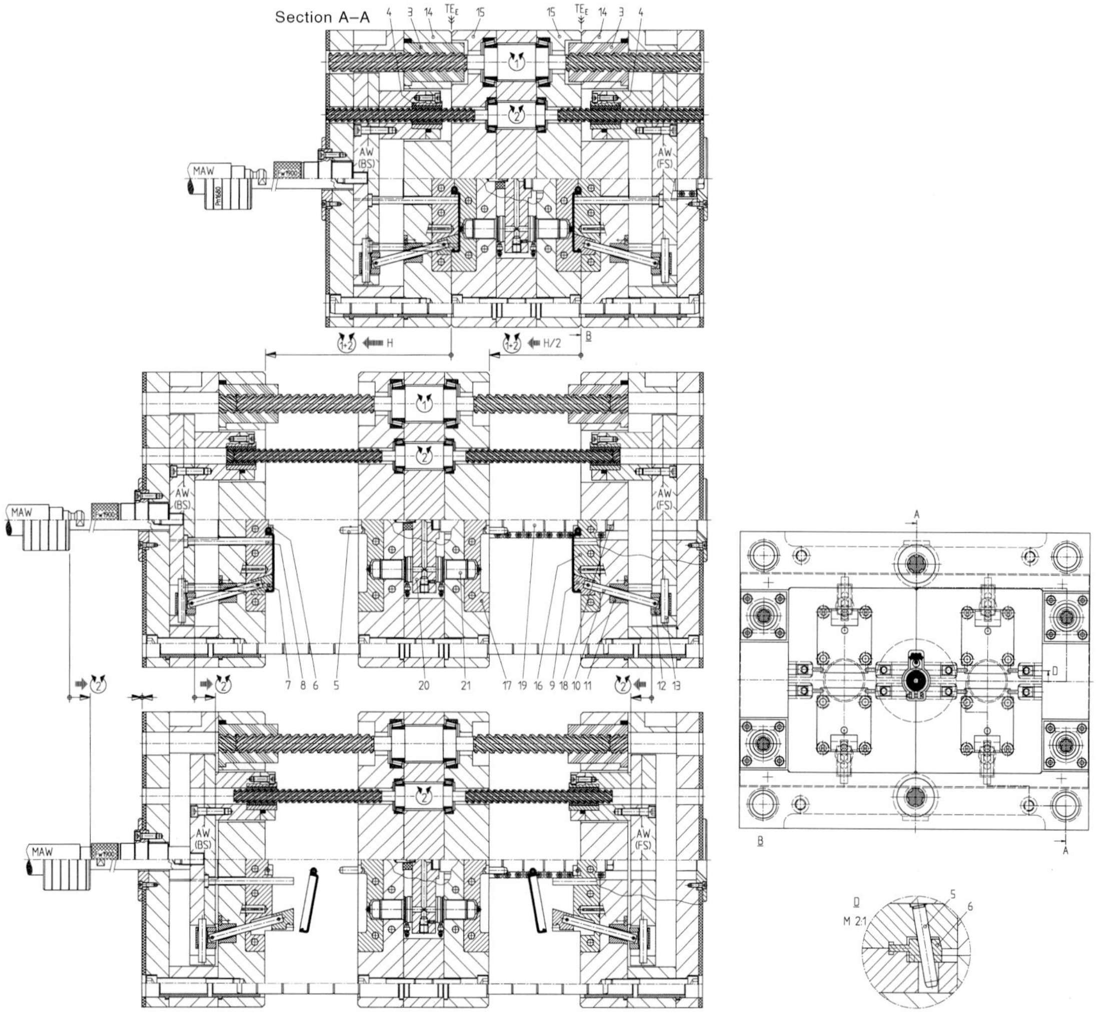

Figures 2 and 3 2 × 4-Cavity hot-runner stack mold for hinged polypropylene covers
AW (FS): ejector plate assembly (stationary mold half); AW (BS): ejector plate assembly (movable mold half); 1, 2: Twin threaded shafts; 3, 4: threaded drive nuts; 5: cam pins; 6: side cores; 7: ejector blade; 8: ejector pin; 9: lifter; 10: lifter rod; 11: insert with angled guide bore; 12: slide block; 13: pin; 14, 15: mold plates; 16, 17: mold inserts; 18: inlet bushing; 19: bushing extension; 20: hot-runner manifold; 21: hot-runner nozzles
(Courtesy: EOC Normalien, Lüdenscheid, Germany, now DME)

Example 113, 16-Cavity Mold with Cold-Runner System for Liquid Silicone Rubber (LSR) Caps

Molded Part

The cap (dimensions 28 mm × 21.6 mm; Fig. 1) protects the power door lock switch in an automobile from moisture and dirt. At the open end, two snap rings for fastening purposes are located inside, while a seal lip runs around the lower edge on the outside. The closed end forms a bellows and has a mean wall thickness of 0.8 mm.

Mold

The split-cavity injection mold (Figs. 2 to 4) consists of the part-forming section KE and the cold-runner system KKS, and is constructed of standard components to the greatest possible degree, starting with a standard mold base having dimensions of 296 mm × 496 mm. The shut height of the mold, including cold-runner system, is 384.2 mm. Mold plates are also of standard thickness. Part-specific mold components include the split-cavity halves (2) and the support blocks (3, 4) as well as the mold inserts (5) in mold plate (1) on the stationary mold half FLS.
Steel grade 1.2767 (hardened) is used for the mold inserts (5) in the stationary mold half FS, while

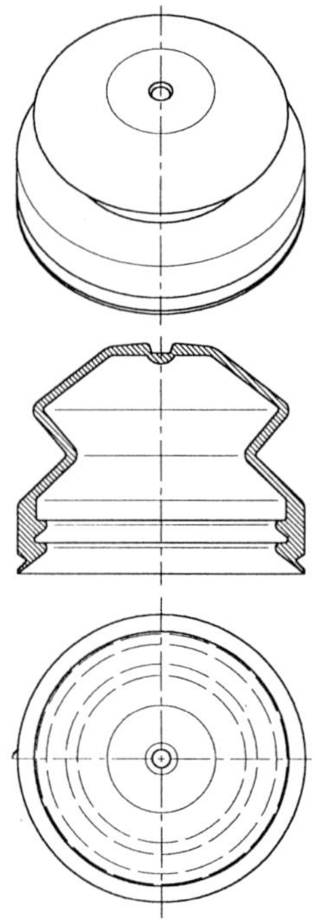

Figure 1 Protective cap of LSR diagram

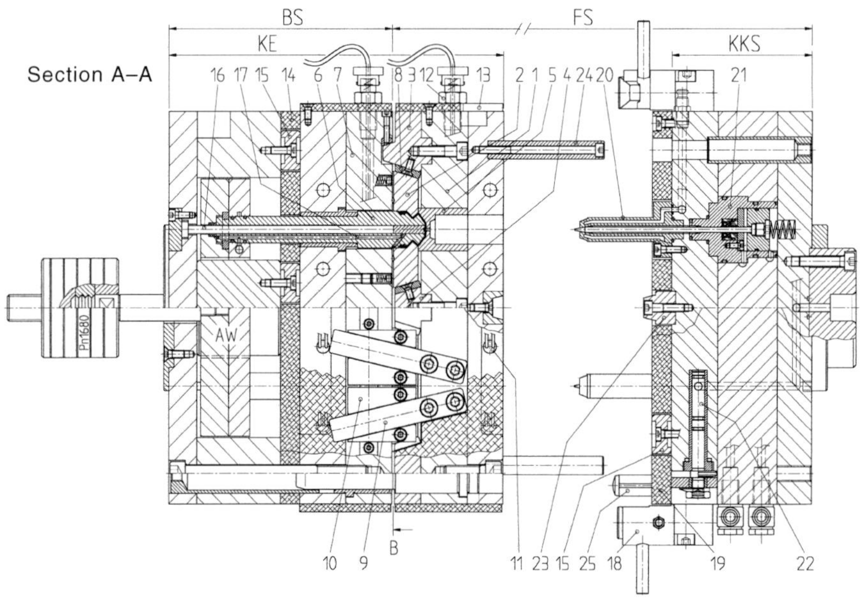

Figure 2 16-Cavity injection mold with cold-runner system for LSR caps
KKS: cold-runner system; KE: part-forming section of mold; FS: stationary mold half; BS: moving mold half; AW: ejector mechanism; 1: mold plate on stationary side 2: split-cavity half; 3: outer support block; 4: center support block; 5: mold insert on stationary side; 6: mold core on moving side; 7: mold plate on moving side; 8: support rail; 9: cam rail; 10: wear plate; 11: cartridge heater; 12: thermocouple; 13: 6 mm insulating plate; 14, 19: 15 mm insulating plate; 15: support block; 16: pin; 17: compressed-air channel; 18: quick-clamping element; 20: nozzle; 21: needle shutoff system; 22: flow control; 23: center locating pin; 24: pilot locating pin; 25: dowel pin

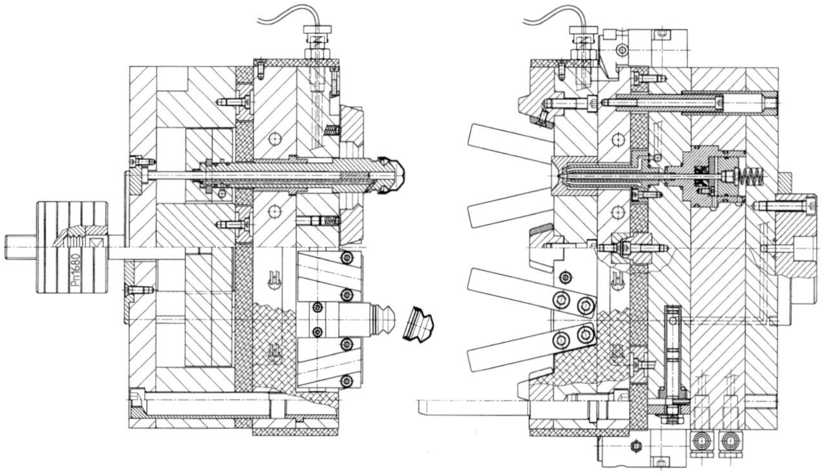

Figure 3 Part release and ejection sequence

grade 1.2312 is used for the mold plate FS (1). Steel grade 1.2764 (case-hardened) is employed for the split-cavity halves (2) and the mold cores (6) in the movable mold half Bp. Grade 1.2312 is used for the support blocks (3, 4) while grade 1.2842 (hardened) is employed for the support rails (8). Wear plates (10) and cam rails (9) are of steel grade 1.7131 (hardened).

The split-cavity construction with inserted support blocks (3, 4) permits mold plate (1) on the stationary mold half to be machined from a standard mold plate (7). In addition, with this approach the support blocks can be more readily adjusted to the design.

When processing synthetic rubbers that crosslink at elevated temperatures, the part-forming section of the mold must be heated, in the case of liquid silicone rubber (LSR) to about 180°C (356°F). To achieve this, the mold plates incorporate electric cartridge heaters (11) that are controlled by thermocouples (12) placed as close to the cavities as possible. Uncontrolled heat loss due to thermal radiation is prevented through the use of 6 mm thick insulating plates (13) fastened to the outside of the mold plates and backup plates. A 15 mm thick insulating plate (14) is sandwiched between the backup plate on the moving side and the ejector housing. Special support blocks (15) absorb the clamping and injection forces.

The cavities are vented via existing parting surfaces as the mold fills. LSR has an extremely low

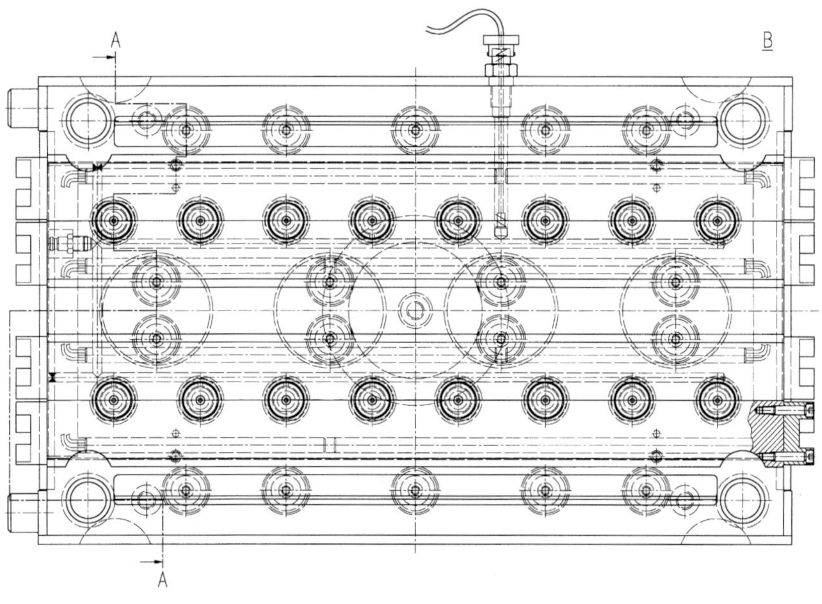

Figure 4 View of the moving-side parting line
(Courtesy: EOC Normalien, Lüdenscheid, Germany, now DME)

viscosity and penetrates gaps smaller than 0.01 mm, so that care must be taken when machining part-forming components in order to prevent the formation of flash. In conjunction with optimum process parameters, it is possible to mold flash-free parts.

Runner System

The standardized cold-runner system KKS can be installed and removed as a separate entity independent of the mold, and can be considered part of the mold or part of the machine. With the aid of quick-clamping elements (18), it can be separated very quickly from the part-forming section of the mold KE (and then subsequently reconnected) to prevent crosslinking of the LSR in the cooled runner system in the event of an interruption in production. The 15 mm thick insulating plate (19) sandwiched between the mating surfaces and nozzles (20) (which are cooled down to their tips) ensures good thermal separation between the runner system and part-forming section of the mold.
Each nozzle (20) is equipped with a pneumatically actuated needle shutoff system (21), resulting in only a minimal gate mark. "Cold slugs" and "stringing" are prevented. Integrated flow controls (22) at each nozzle provide for optimum balancing of material flow into each cavity. A central locating pin (23) guarantees proper alignment of the cold-runner system KKS with the part-forming section of the mold. Pilot locating pins (24) protect the nozzles from damage during assembly to or disassembly from the mold. The dowel pin (25) prevents rotation of the cold-runner system with respect to the mold.

Part Release/Ejection

The elasticity of the LSR material greatly simplifies part release and ejection. The internal undercuts can be forcibly ejected. As the mold opens (Fig. 3), the split-cavity halves (2) are forced apart by the cam rails (9). This releases the external undercuts on the molded parts.
The mold cores (6) are fastened to the ejector mechanism AW. At the center of each mold core is a pin (16) that is held in the moving-side clamping plate. As the ejector mechanism and the attached mold cores (6) advance, the pins (16) remain stationary. The motion of the mold cores (6) relative to these pins opens a channel (17) inside each of the cores. These channels are pressurized with compressed air via a connection on the ejector housing. The molded parts, which are sitting air-tight on the cores, are expanded by the build-up of air pressure between the parts and cores, and are ejected abruptly.

Example 114, Two-Cavity Injection Mold for a Styrene-Acrylonitrile Safety Closure

The safety closure shown in Fig. 1 (dimensions: 48 mm dia. × 31 mm) has three openings on its circumference as well as a central opening and three smaller holes in its face.

Mold

The mold (Fig. 2) is constructed from standard mold plates with dimensions of 246 mm × 346 mm and has a shut height of 286 mm.

Gating/Runner System

The melt travels from the sprue bushing (1) through two naturally balanced runners to the cavities, at which point the runners divide into two submarine gates in the walls of the two cavity inserts (2).

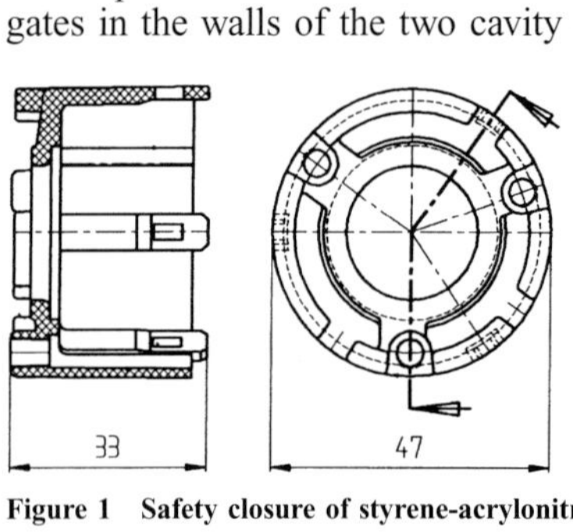

Figure 1 Safety closure of styrene-acrylonitrile

Mold Temperature Control

The cavity inserts (2) are cooled by circumferential cooling channels, while the mold cores (3) are provided with helical cooling inserts.

Part Release/Ejection

To release the openings on the circumference of the parts, each cavity has three slides (4) that move radially along gibs (8) attached to mold plate (5). Pins (6) locks the slides in position when the mold is closed. Under the action of the coil springs (9), the mold opens first at parting line I (partial section B-B). The pins locking the slides in position disengage and the cams (7) cause the slides to move radially. The opening motion at parting line I stops after a distance "H", and the mold opens at the primary parting line II. The parts are pulled from the cavities; the submarine gates shear off and snap out. During this motion, the runners are held on the sucker pins (10) and sprue puller (11). To permit ejection of the molded parts, the mold finally opens at parting line III. The two ejector plates (12, 13) and the attached ejector sleeves strip the parts off the cores and the runners of the sucker pins.

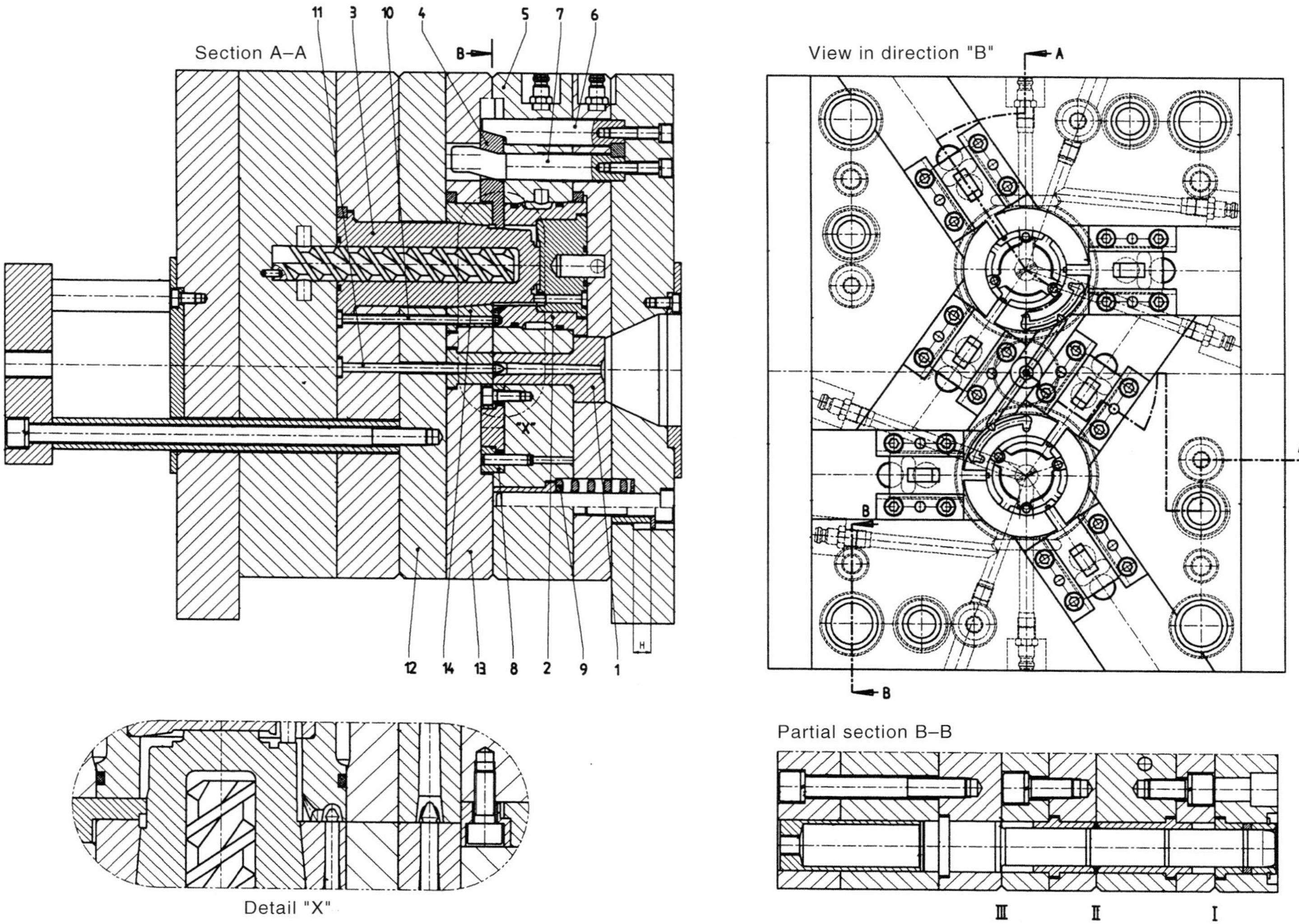

Figure 2 Two-cavity injection mold for a safety closure of styrene-acrylonitrile
1: Sprue bushing; 2: cavity insert; 3: mold core; 4: slide; 5: mold plate; 6: locking pins; 7: cam; 8: gib: 9: coil spring; 10, 11: sucker pins, sprue puller; 12, 13: ejector plates; 14: ejector sleeve (Courtesy: Hasco, Lüdenscheid, Germany)

Example 115, Four-Cavity Unscrewing Mold for Threaded Polypropylene Closures

The closures are 50.7 mm in diameter, 28.6 mm high and have, in addition to an internal thread, a circular sealing lip 8 mm high on the inside. When designing the mold cores, care had to be taken to ensure that, during mold filling, defects would not occur in the lip as the result of entrapped air.

Mold

The four-cavity mold (dimensions: 196 mm × 196 mm × 240 mm) is constructed from standard mold plates. Each of the four mold cores comprises an unscrewing outer core (1), a stationary inner core (2), and bearing bushings (3) in between. The threaded core forms the threads in the molded part and is supported radially in the mold by two needle bearings (4) and axially by one cylinder roller bearing (5). Gearing on each core engages the drive pinion (6), which is mounted on a motor-driven shaft (7). The parts are filled via a sprue and runner system with submarine gates.

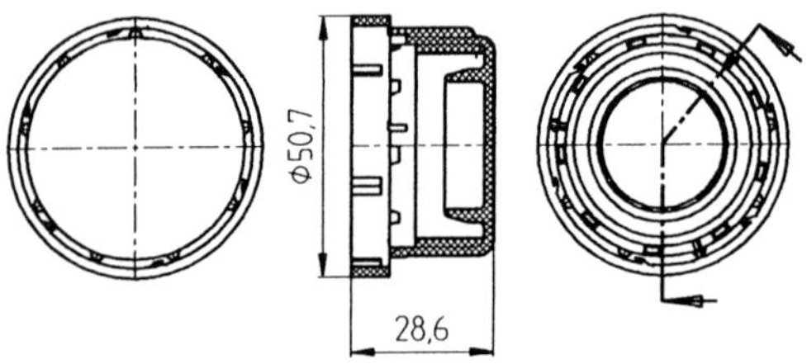

Figure 1 Polypropylene closure

Part Release/Ejection

The mold opens at A-B. The molded parts are pulled out of the cavity inserts (8), shearing off the gates. The runner system is held by the sprue puller (9). The outer cores (1) now begin to unscrew. The parts do not rotate, because they are held on the sleeves (10) by their internal ribs. These sleeves are mounted between plates (11, 12), which, under the action of the compression springs (13), shift the closures axially with respect to the unscrewing cores (1) by the amount that the cores (1) unscrew from the parts. During this step, the force applied by the springs (13) is transmitted by the threads in the molded parts into the unscrewing cores (1) and from them into the axial support bearings (5). After a distance "H", unscrewing is complete and the molded parts sit on the sleeves (10). They are subsequently stripped off by the ejector sleeves (14) as plate (15) comes to a stop against the shoulder of stripper bolt (16) at the end of the opening stroke.

In contrast to designs based on the commonly employed unscrewing cores with lead threads, the mold can close sooner with these motor-driven cores, because the cores do not need to be rethreaded to their starting positions. It must be noted, however, that, as a consequence, the starting point of the threads with respect to the other features will shift around the circumference of the part from shot to shot.

Venting

During filling, the air trapped in the portion of the cavity that forms the circular sealing lip is vented through the gap between the outer unscrewing core (1) and the stationary inner core (2).

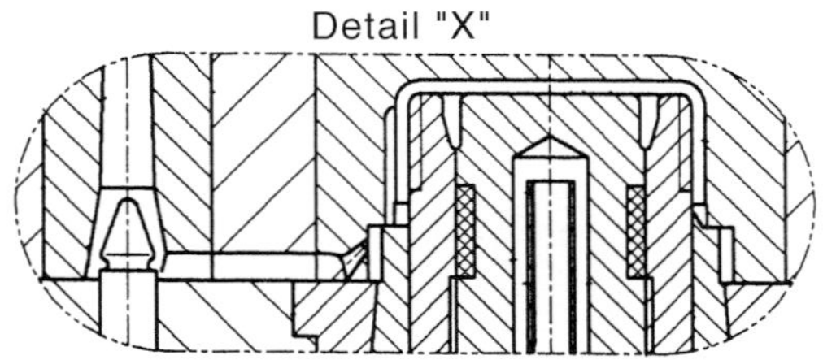

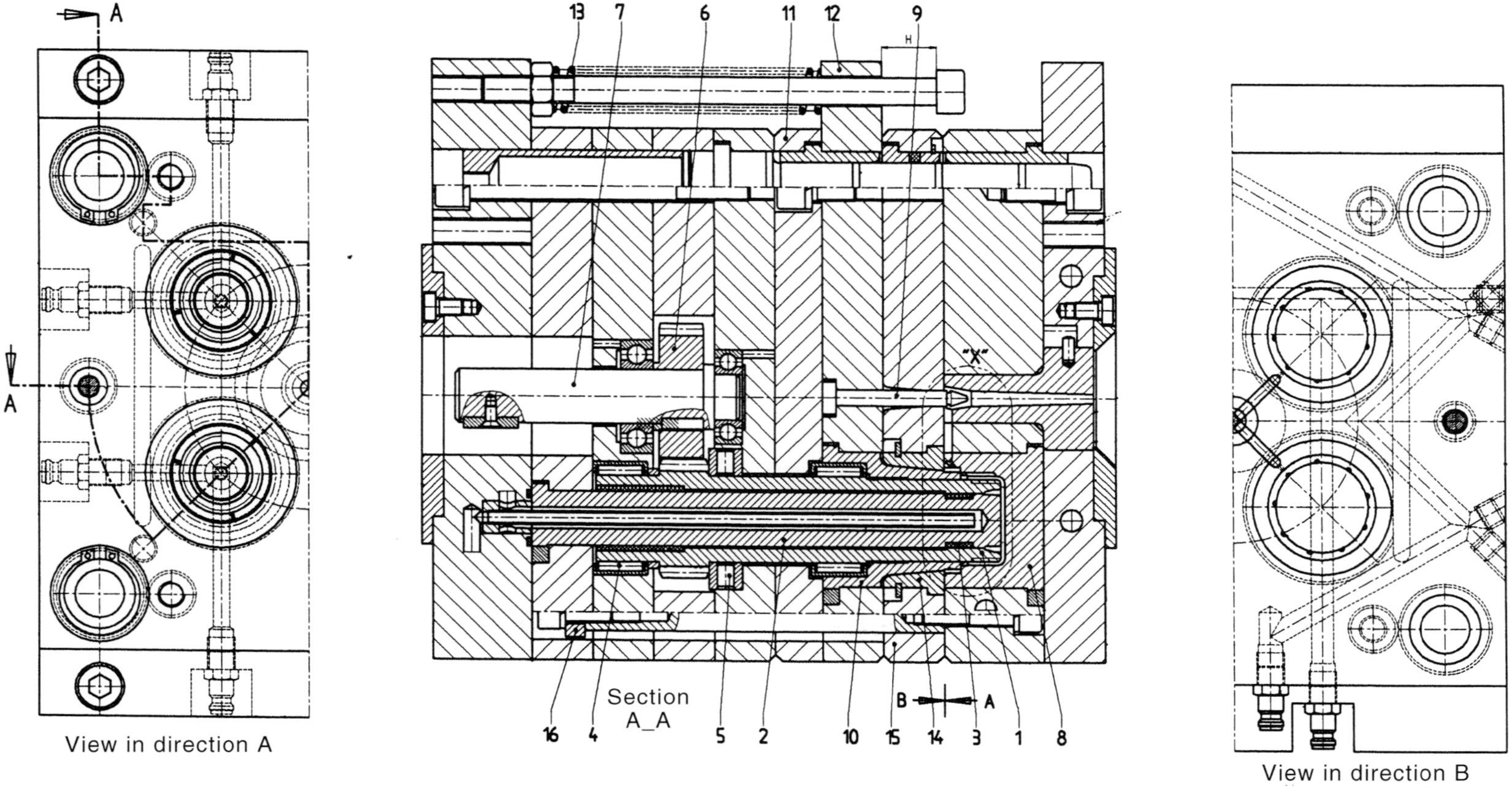

Figure 2 Four-cavity unscrewing mold for polypropylene closures
1: outer unscrewing core; 2: stationary inner core; 3: bearing bushing; 4: needle bearing; 5: cylinder roller bearing; 6: drive pinion; 7: drive shaft; 8: cavity insert; 9: sprue puller; 10: sleeve; 11, 12: mold plates; 13: compression spring; 14: ejector sleeve; 15: mold plate; 16: stripper bolt
(Courtesy: Hasco, Lüdenscheid, Germany)

Example 116, Four-Cavity Injection Mold for Polyester Dispenser Heads

The dispenser head shown in Fig. 1 (dimensions: 60 mm × 30 mm × 25 mm) fits on a container and serves to dispense a pasty liquid. Its characteristic feature is a discharge spout with a rectangular channel 7.5 mm × 2 mm and a length of 40 mm. This channel connects with a central bore inside the dispenser head. The channel in the spout forms an angle of 87° with the vertical axis of the dispenser head.

Mold

The mold has dimensions of 346 mm × 246 mm × 224 mm. The four cavities are arranged symmetrically with respect to one another in the mold (Fig. 2). Two slides (1), each of which holds two rectangular cores (2) for the channels in adjacent parts (see Detail "X"), serve to release the rectangular channels. The slides (1) are guided on the angle plate (3) and held in position by the heel blocks (4) when the mold is closed. Cam pins (5) actuate the slides. At a cam pin angle of 20°, a relative motion between cam pin and slide of at least 110 mm in the mold opening direction is needed to release the rectangular channels. As the mold opens, the cam pins withdraw from the guide bores in the slides. Ball detents (6) hold the slides in the open position.

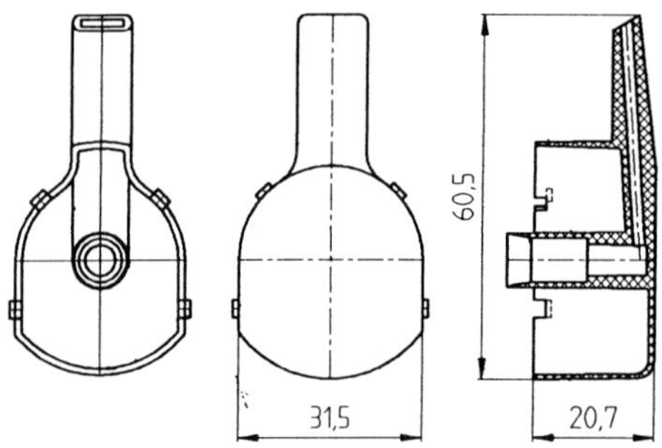

Figure 1 Polyester dispenser head

Gating/Runner System

The mold is filled via a sprue bushing (7) that connects to a transverse runner channel cut into the mold parting line. Submarine gates at each end of the runner channel lead to the cavities in the cavity inserts (8).

Temperature Control

The cavity and core inserts are cooled by cooling channels. In addition, the cores contain helical cooling inserts (9).

Venting

The central pipe inside the molded part is vented via the inserted core pin (10).

Part Release/Ejection

As the mold opens at the parting line A-B, the parts are pulled out of the cavity inserts. Actuated by the cam pins, the slides withdraw the rectangular cores from the molded parts. The sprue is pulled out of the sprue bushing, while the submarine gates shear off and snap out. Plate (11) moves to the right to strip the parts off the cores. Stripper rings (12), which partially enclose the core inserts, eject the dispenser heads and runner.

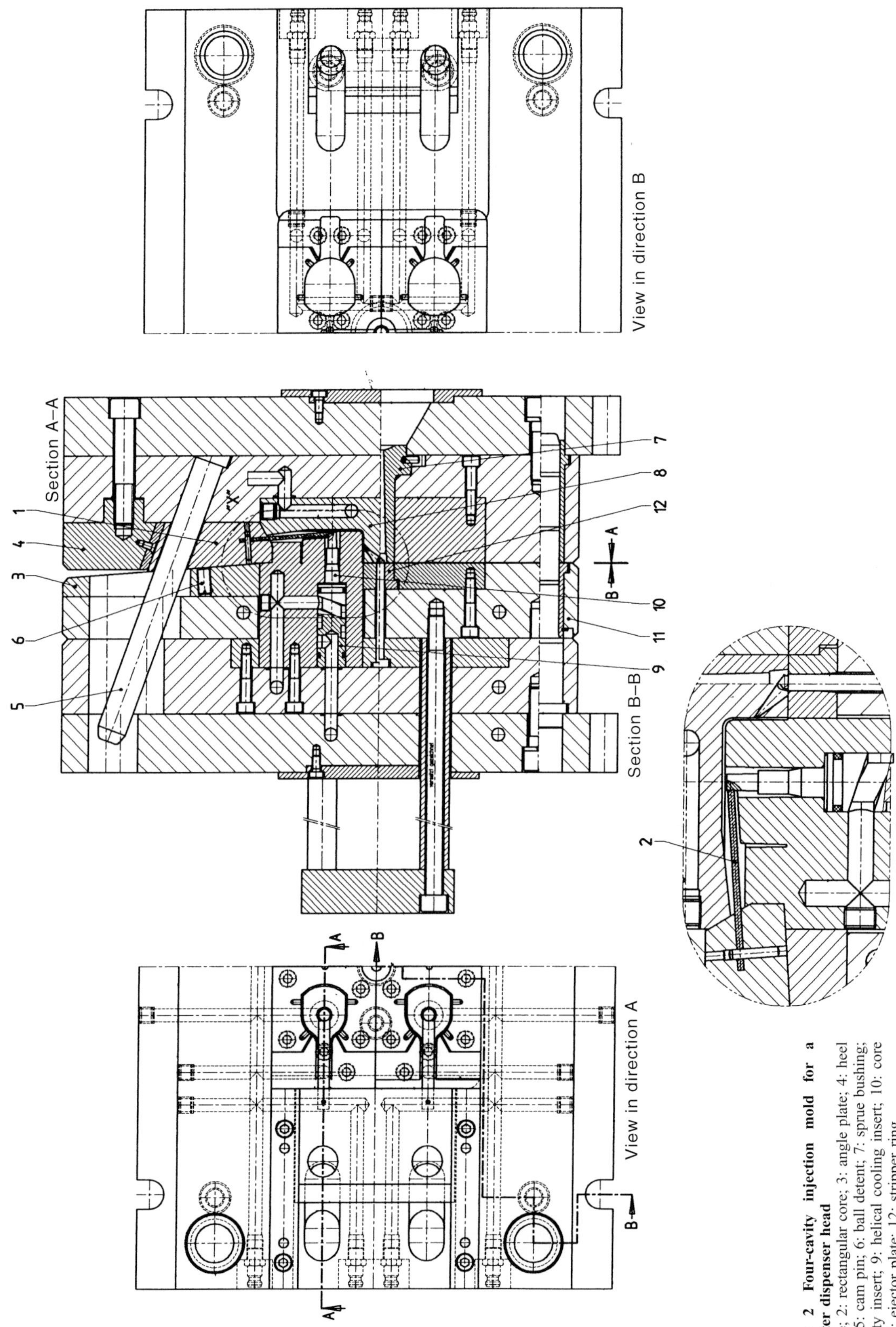

Figure 2 Four-cavity injection mold for a polyester dispenser head
1: slide; 2: rectangular core; 3: angle plate; 4: heel block; 5: cam pin; 6: ball detent; 7: sprue bushing; 8: cavity insert; 9: helical cooling insert; 10: core pin; 11: ejector plate; 12: stripper ring
(Courtesy: Hasco, Lüdenscheid, Germany)

Example 117, Two-Cavity Injection Mold for PMMA Lighting Fixture Cover

The lighting fixture cover shown in Fig. 1 (dimensions: 95 mm × 34 mm × 27.5 mm) exhibits on the inside two mounting ribs with narrow slits. At one of the narrow ends, there is a tapered surface extending inward from the rim at an angled of 45°. The adjacent rib has a similarly angled surface.

Mold

The mold dimensions are 246 mm × 296 mm × 280 mm. The cavities are oriented in such a manner that a single slide (1), to which two detail inserts (7) are attached, can release the tapered surfaces in both parts. The slide moves at an angle of 45° and is actuated by two cam pins (2).

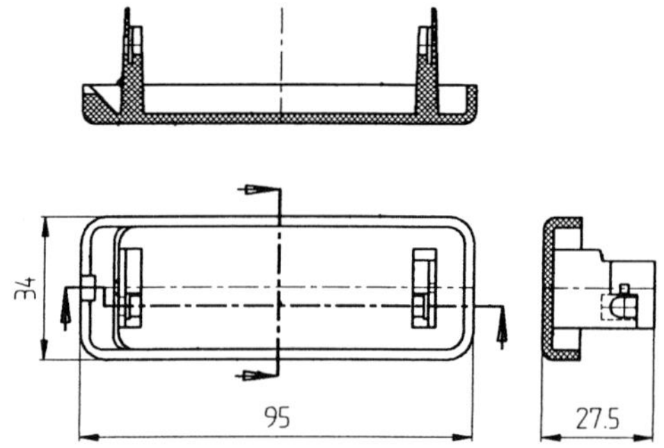

Figure 1 PMMA lighting fixture cover

Gating/Runner System

The mold is filled via a sprue bushing that feeds a short runner leading to a narrow edge gate located midway along each part.

Part Release/Ejection

As soon as the mold opens, the slide (1) begins to move outward. Because of the very thin ejector pins (3), which would bend if subjected to extremely high ejector force, an "ejector delay mechanism" (4) is installed to prevent the ejector plates (5, 6) from advancing until the slide (1) has cleared the part after an opening stroke of 30 mm. Now, the parts and runner can be ejected. The pushback pins (8) ensure that the ejector returns to the correct starting position.

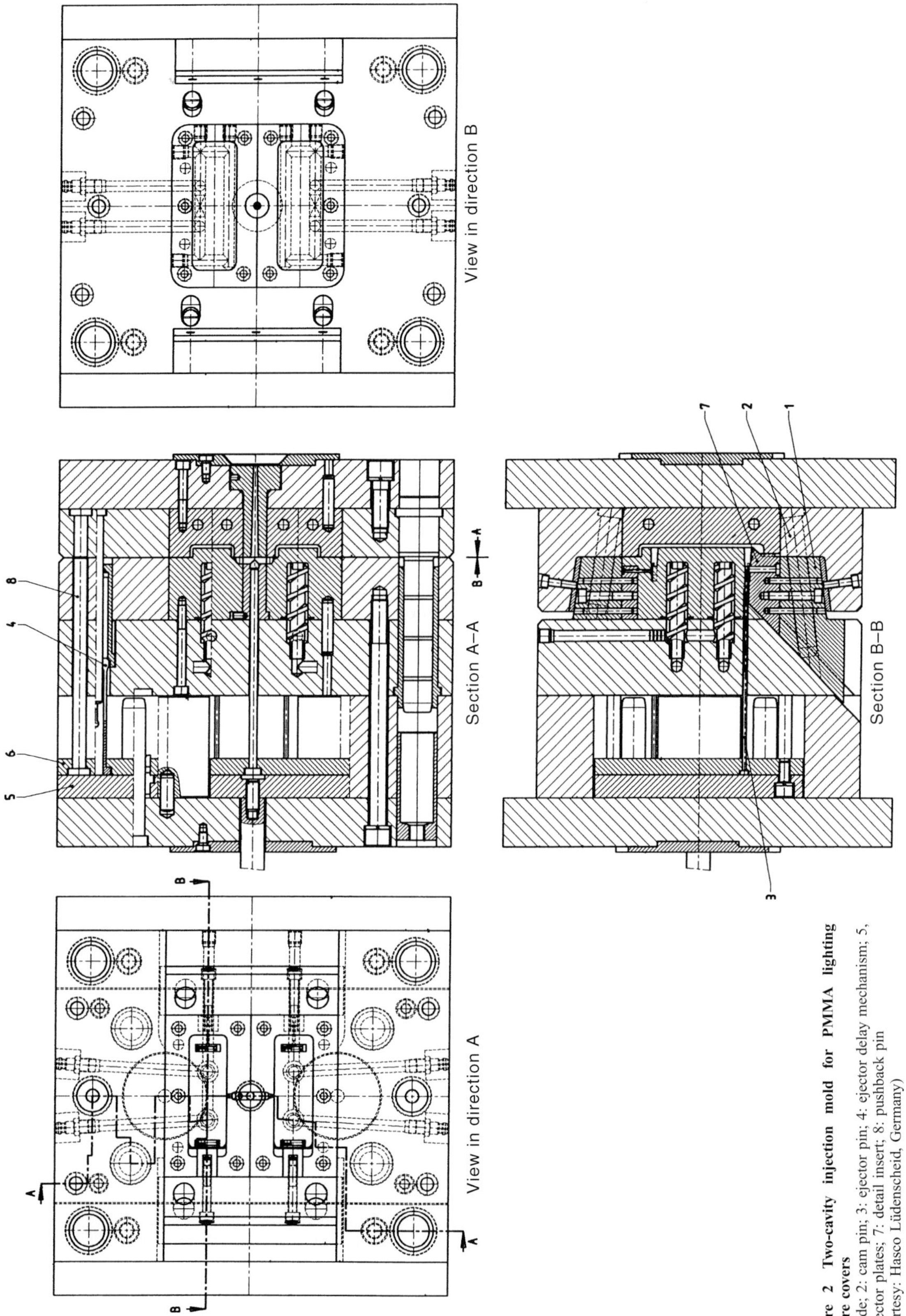

Figure 2 Two-cavity injection mold for PMMA lighting fixture covers
1: slide; 2: cam pin; 3: ejector pin; 4: ejector delay mechanism; 5, 6: ejector plates; 7: detail insert; 8: pushback pin
(Courtesy: Hasco Lüdenscheid, Germany)

Example 118, Two-Cavity Injection Mold for Polyacetal Hinges

The hinge shown in Fig. 1 (dimensions: 86 mm × 61 mm × 33 mm) requires a fairly complex release and eject sequence, because of numerous functional elements.

Mold

The two cavities are oriented in the mold symmetrically (Fig. 2). The bore for the hinge pin is released by means of a core (1) actuated by a cam pin (2) (Section C-C, Fig. 2). The lifter (3) releases the hook on the top surface. An hydraulically actuated side core (4) releases the outer and inner surfaces of the short tube that projects at an angle from the top edge of the hinge.

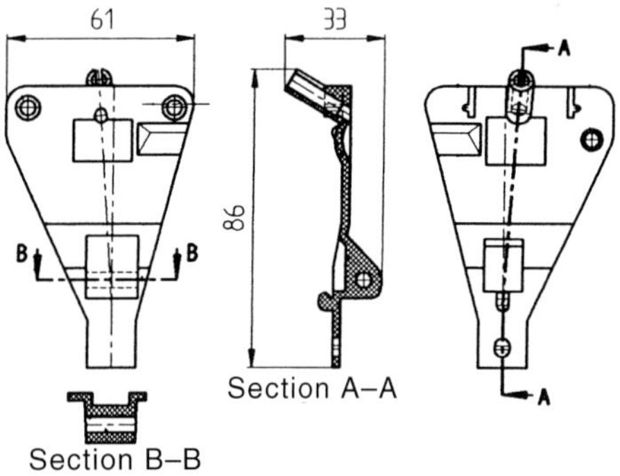

Figure 1 Polyacetal hinge

Gating/Runner System

The two cavities are edge-gated and filled via a sprue and runner cut into the mold parting line.

Part Release/Ejection

Because the latch lock (5) (View x) holds parting line II closed, the mold opens at parting line I. During this motion, the cam pin (2) pulls the core (1) out of the bore for the hinge pin. Opening stroke I is limited by the stop (7). In the meantime, the latch lock (5) has released parting line II. Next, the tapered surface on pin (8) releases the locking pin (9), which is withdrawn from the side core (4) by spring (10), thus permitting the short tube on the top of the hinge to be released. Finally, the ejector rod (11) advances the ejector plates (12, 13), and the lifter (3) releases the hook-shaped undercut it contains, while the ejector pins eject the part and runner.

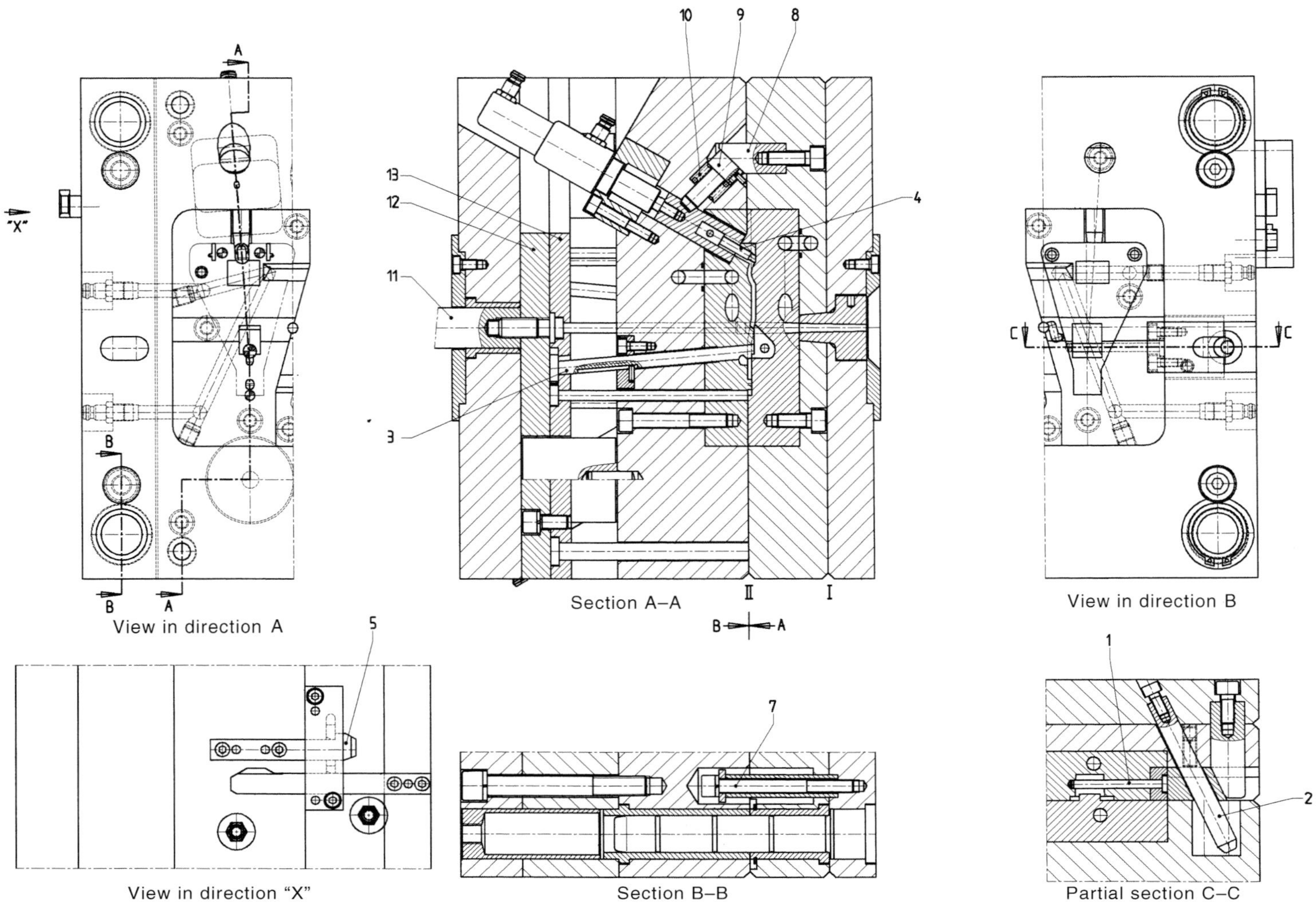

Figure 2 Two-cavity injection mold for polyacetal hinges
1: core; 2: cam pin; 3: lifter; 4: side core; 5: latch lock; 7: stop; 8: pin; 9; locking pin; 10: spring; 11: ejector rod; 12, 13: ejector plates
(Courtesy: Hasco, Lüdenscheid, Germany)

Example 119, Eight-Cavity Injection Mold for PE-HD Threaded Caps

The threaded cap shown in Fig. 1 (dimensions: 30 mm × 30 mm) is part of a spray top for a bottle. At the top end, there is an inner ring with a bead that forms an undercut all the way around.

Mold

The mold is constructed from standard mold plates with dimensions of 296 mm × 196 mm and has a shut height of 356 mm. The eight cavities are arranged in two groups of four cavities each. The internal threads are formed by two-piece threaded cores that have the threads for the molded part at one end (1) and gear teeth at the other end (2). Inside the threaded cores are inner cores (3), the ends of which form the inside surface of the inner ring with its bead and associated undercut. Each set of four threaded cores is driven by a gear (4), which, in turn, is driven by the main drive gear (5). The main drive gear (5) is powered by a motor-driven quill (6), the interior of which houses the ejector rod (7).

Gating/Runner System

The melt reaches mold parting line I via a hot-runner manifold (8) and two hot-runner nozzles (9). There, two spider-shaped runners convey the melt to the submarine gates cut into the sides of the cavity inserts (10). Sprue puller bushings (11) with ejector pins (12) for the runners are located opposite the tips of the hot-runner nozzles (9).

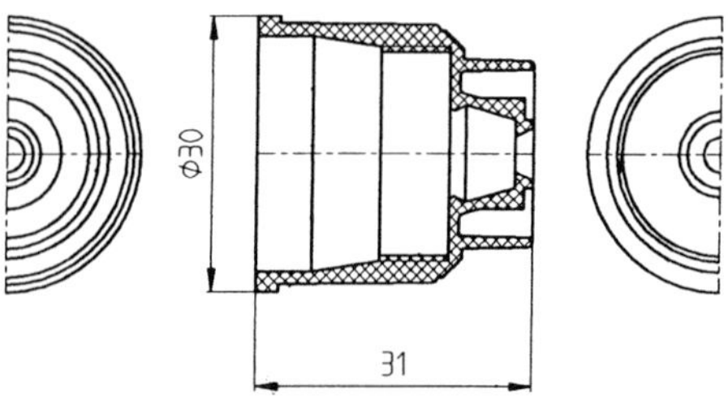

Figure 1 PE-HD threaded caps

Venting

The sleeve inserts (17, 18) provide venting for the region of the cavity that forms the circular rings at the top of the cap.

Temperature Control

The cavity inserts (10) are cooled by circular cooling channels, while the mold plates on each side of mold parting line I are cooled by drilled cooling channels. The inner cores (3) are hollow and contain a bubbler for water cooling.

Part Release/Ejection

The mold opens first at parting line I. The caps are pulled out of the cavities, shearing off the submarine gates. The threaded cores start to rotate and unscrew themselves from the caps. Under the action of the springs (13), the mold opens at parting line III by an amount equal to that by which the cores have unscrewed from the caps. After a distance H (View z), the latch lock (14) releases mold parting line II. The inner cores (3) are now pulled out of the threaded cores as well as the inner rings in the molded caps. The circular undercut is spread apart and released. As soon as the threaded cores are unscrewed completely, the caps are ejected by the remaining ejector stroke available for ejector plate (15). Finally, the ejector rod (7) advances the runner ejector pins (12) to eject the two spider-shaped runners.

The tension on the springs (13) must be adjusted carefully to prevent unacceptable loads or even deformation of the ends of the threads as the result of excessive force having been applied at the end of unscrewing.

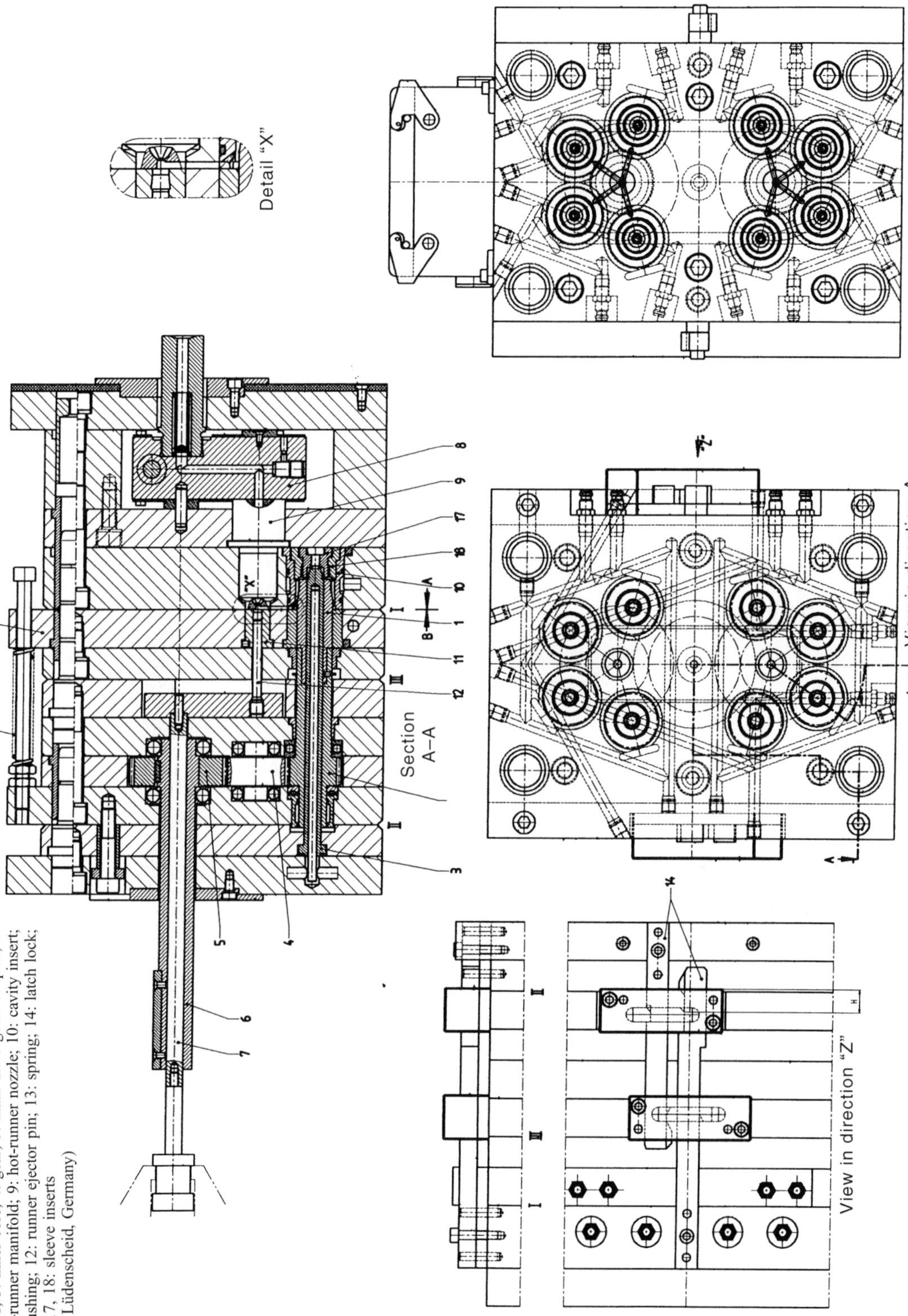

Figure 2 Eight-cavity injection mold for PE-HD threaded caps
1, 2: threaded cores; 3: inner core; 4: gear; 5: main drive gear; 6: quill; 7: ejector rod; 8: hot-runner manifold; 9: hot-runner nozzle; 10: cavity insert; 11: sprue puller bushing; 12: runner ejector pin; 13: spring; 14: latch lock; 15: ejector plate; 17, 18: sleeve inserts
(Courtesy: Hasco, Lüdenscheid, Germany)

Example 120, 4-Cavity Hot-Runner Mold for Connectors Made from Polystyrene

The part, referred to as "connector", is a rotationally symmetrical sleeve with a total length of 134mm. The molded part is divided centrally by a membrane-like intermediate wall and has a collar in the same plane. Its wall thickness all-round is 1/8mm. The demolding incline is 0.2° (Fig. 1). The visible part has a high surface quality, and any gate mark on the surface is unacceptable. The design specifies demolding in the direction of mold opening.

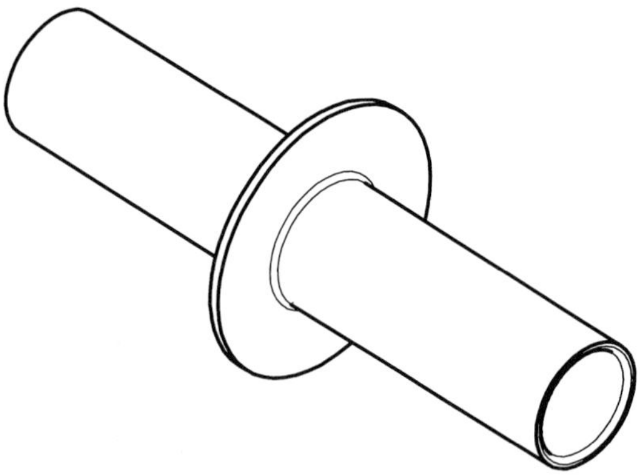

Figure 1 Connector made from polystyrene, diagram

Mold

This mold design with dimensions of 24mm × 246mm is based on a standardized modular system. The mold half on the nozzle side is designed as a three-plate "hot half" and screwed together in blocks (19, 20, 21). The centering and guide elements (31 to 35) are arranged for easy servicing. The hot-runner mold has a high mounting height, dictated by the article. Both cavity plates (5, 6) are equipped with mold inserts (23, 24). Internal support is provided by four support pillars (27), (Fig. 2).

Gating

The direct gating point is located in the middle of the article intermediate wall. It promotes uniform melt flow, thereby producing parts with little warping. The air-insulated melt chamber insert and the gate for the long, slim nozzle (18) are located in the contour insert (24a) made from hardened 1.2343 steel. The nozzle body is screwed to the cavity plates (3). The nozzle and hot-runner manifold are force-fit connected by a sliding seal face.

Thin walls and long flow paths require high injection pressures. The externally heated, fourfold standard hot-runner distributor (17) with shrink-fitted diverters (17a) is naturally balanced. The electric lines from the hot-runner manifold, the nozzle heaters, and thermocouples lead to the connection housing (30) and are connected according to DIN 16765, version B. In order to reduce convective heat loss (so-called chimney effect) and to protect against flashing at the machine nozzle, a flat, form-fit GFK seal ring is mounted over the centering bush (16).

Cooling

Using water as coolant, heat is transferred via the outer surfaces of the mold insert (23, 24) and the cores (25) standing on the ejector side in the core retainer plate (10). Coolant is supplied to separately controlled, parallel configured circuits. Core cooling is done with long diverting elements via the clamping plate on the ejector side (11).

Demolding

The molded articles are demolded by the ejector sleeves guided by the mold core (26). The ejector assembly is set on pillars with fourfold rods (29) and sleeves (28) and connected to the machine ejector via the central ejector rod (13). Four return pins (36) move the ejector plates to start position.

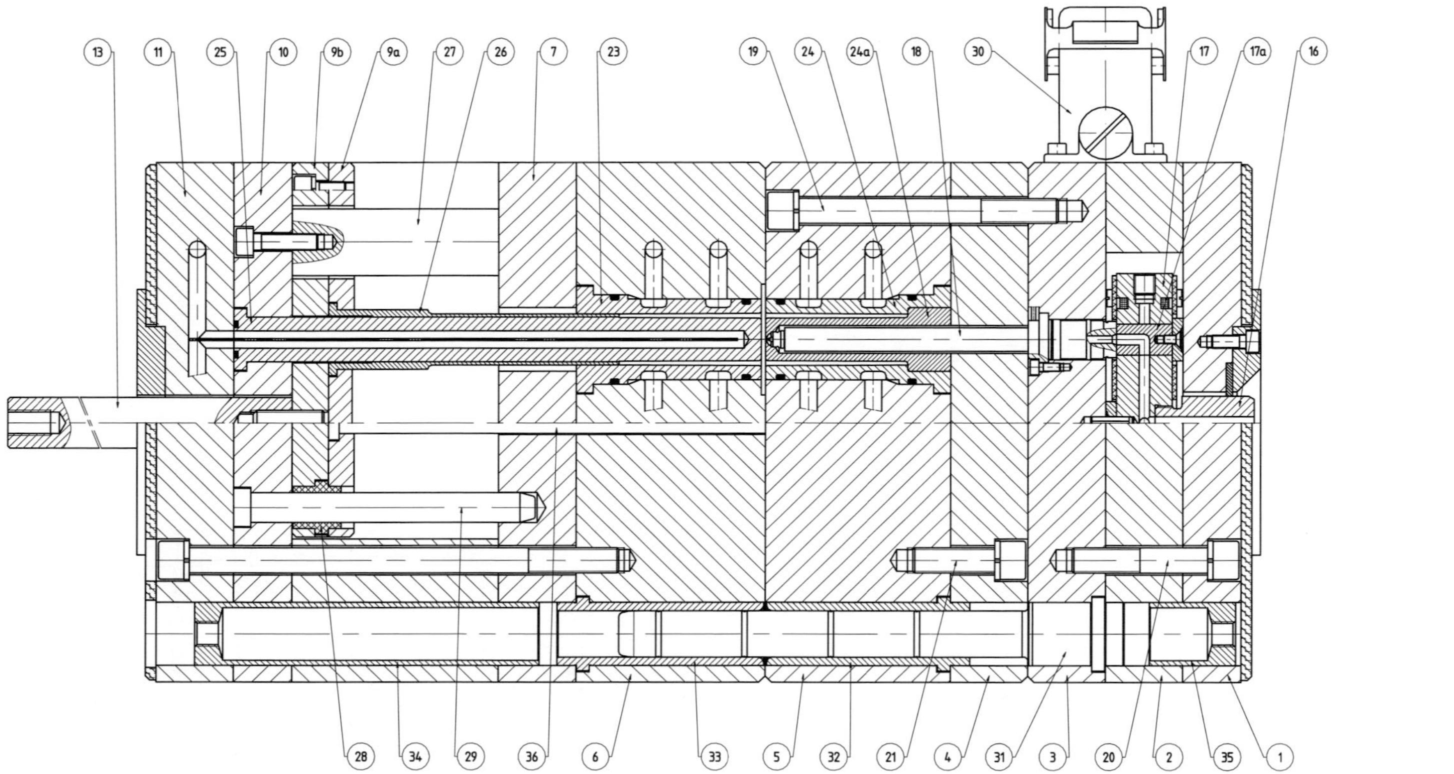

Figure 2 Fourfold hot-runner mold for connecters made from polystyrene
3, 5, 6: cavity plates, 9a, b: ejector assembly, 10: core retainer plate, 11: clamping plate, 13: ejector rod, 16: centering bush, 17: hot-runner manifold block, 17a: diverter, 18: Open sprue bush with tip, 23, 24: mold inserts, 24a: contour insert, 25: core, 26: ejector sleeve, 27: support rollers, 28: sleeve, 29: rods, 30: ancillary housing, 31-35: centering sleeve and guide elements, 36: return pin
(Courtesy: Hasco, Lüdenscheid)

Example 121, Single-Cavity Mold for a Polypropylene Cutlery Basket

The basket (dimensions: 287 mm × 157 mm × 140 mm; Fig. 1) is used to hold cutlery in a dishwasher. It is divided into 16 compartments by three partitions running lengthwise and crosswise. The outer walls and the bottom have a grid-like structure. In addition, two of the partitions have two openings each 10 mm square. The numerous partitions, together with the high shrinkage of polypropylene, pointed toward a high ejection force requirement to strip the molded basket off the mold core. Accordingly, special measures were taken in order to ensure that the part could be ejected without being damaged in spite of its flexible grid-like structure.

Figure 1 Polypropylene cutlery basket

Mold

The mold (dimensions: 596 mm × 496 mm × 687 mm; Figs. 2 to 6) was constructed largely using standardized mold plates from Strack Norma, Wuppertal, Germany. Steel grade 1.2767 (hardened to HRc 54) was used for the part-forming components. The side walls of the basket are formed by four slides (18, 19) (Fig. 2) that move laterally in gibs (33, 34) mounted on the stripper plate (3, Fig. 4). The slides are supported by heel blocks (20, 21, 29) located in mold plate (2) and, when the mold is closed, are held by additional support blocks and rails (30 to 32) in the stripper plate (3). When the mold is in the open position, the slides are held by spring-loaded ball detents to prevent any unintentional movement. The slides are actuated by angled rods (38, 39) set at angles of 15° and 20°. An angle of 20° is used for angled rods (38) for the following reason: support block (30) serves also as a safety stop for the lower slide 18/1. If the angled rod (38) were set at an angle of 15°, a collision with stop (30)

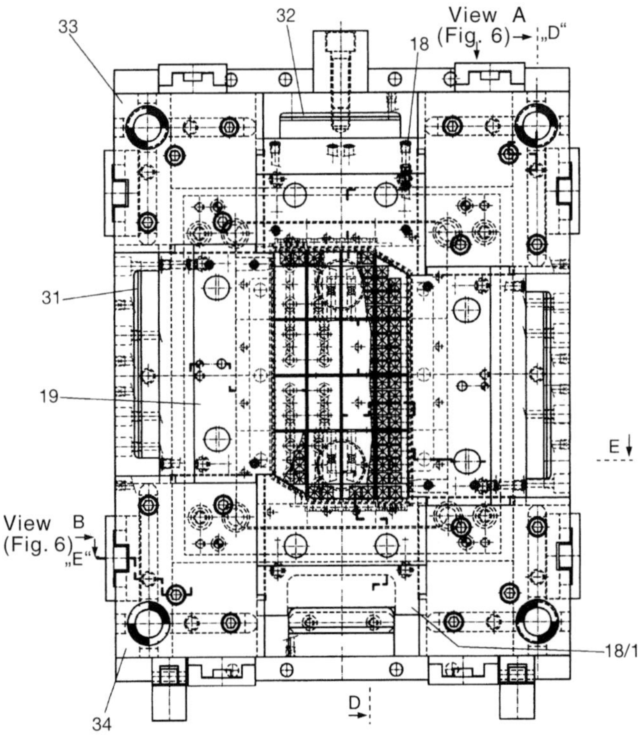

Figure 2 View of the ejector side

Figures 2 to 6 Hot runner mold for polypropylene cutlery basket
2: mold plate; 3: stripper plate; 4: core retaining plate; 6: ejector plates; 7, 8: leader pins; 12: contour-forming insert; 18, 19: slides; 20, 21, 29, 30: support blocks; 23: hot runner manifold; 24: locating rail; 25: core pin; 27: push pin; 31, 32: support rail; 33, 34: gibs; 37: roller guide; 38, 39: angled rods; 40: return pins; 41: stripper bolts; 42: blade ejector; 43: inhibitor pin; 47: compression spring; 54: stripper bar; 57: ball detent
Company illustrations: Friedrichs & Rath, Extertal; Ellersiek & Schaminsky, Bünde, Germany

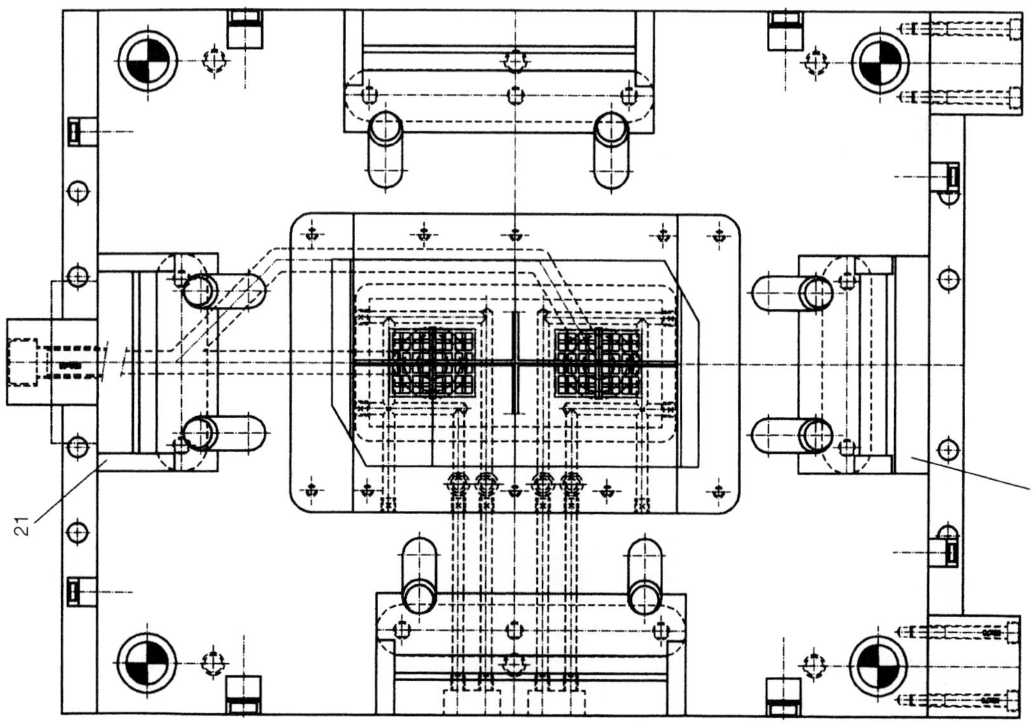

Figure 3 View of the injection side

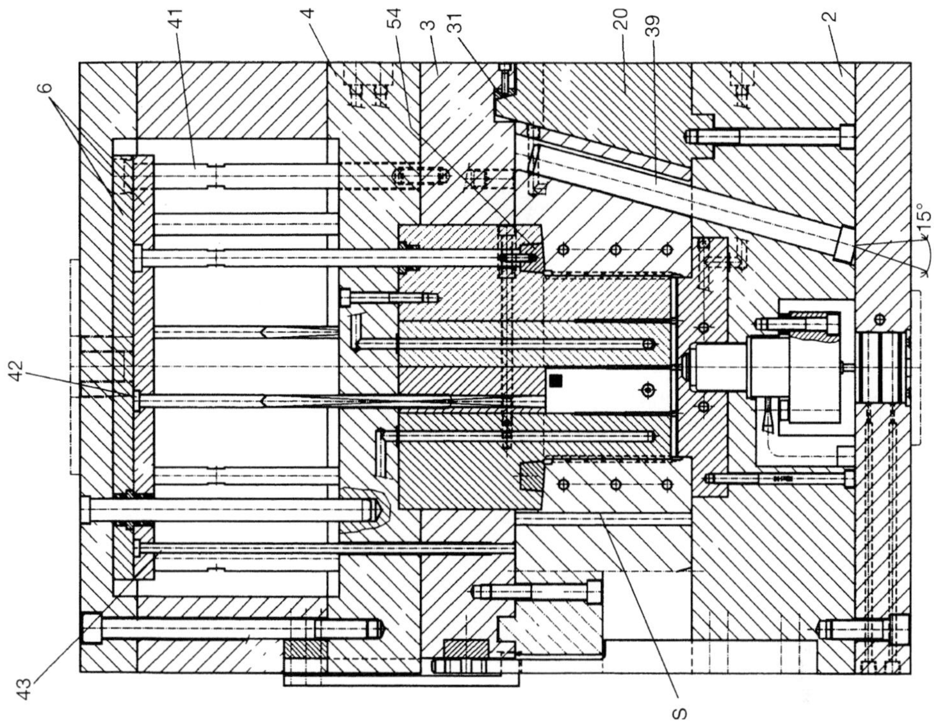

Figure 4 Section E-E

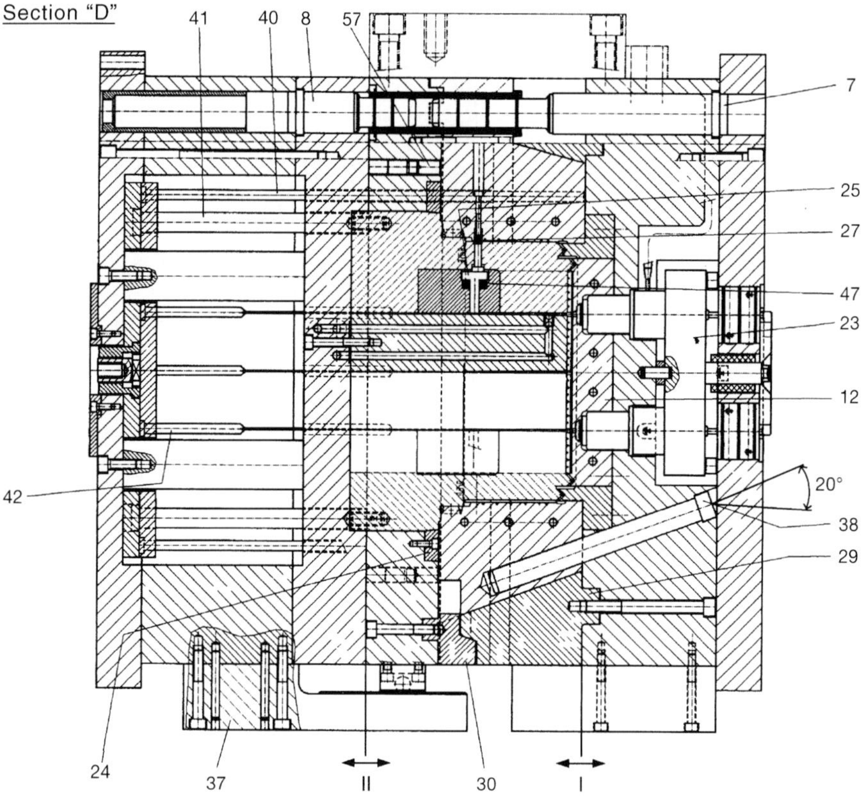

Figure 5 Section D-D

would have resulted. For this reason, an angle of 20° was specified, with the result that, for the same stroke as the other slides, the length of the angled rod could be reduced.

Gating and Temperature Control

The part is filled via a hot runner system (23) using two valve gates located in the bottom of the basket. Eight circuits, comprising cooling channels in the slides, cavity inserts and cores, provide for mold cooling.

Part Release/Ejection

The part release and ejection sequence is controlled by latches (A and B in Fig. 6).

Step 1

The mold opens at the main parting line (I) or a distance of 13 mm, because parting line (II) is initially held closed by latch (A). This releases the part from the bottom of the cavity (12). The slides (19, Fig. 2) that form the short sides of the basket move outward. This releases the core pins (25), which, by the action of the springs (47), retract from the openings in the partitions. The two remaining slides (19) that form the long sides of the basket (Fig. 4) do not move at first, because the bores for the angled rods (39) have 3.5 mm play.

Step 2

Latch (A) releases parting line (II), while latch (B) locks parting line (I) at an open position of 13 mm. As parting line II opens a distance of 10 mm, the slides retain their position with respect to the molded basket. The as-yet unopened slides (19) firmly hold the long walls of the basket, so that it is now stripped 10 mm off the core. Simultaneously, the stripper bolts (41) attached to stripper plate (3, Fig. 4) advance the ejector plates (6) by this same 10 mm, thereby bringing the blade ejectors as well as the stripper bar (54) into contact with the basket and supporting the part release operation.

Step 3

Latch B locks parting line II at the 10 mm position and releases parting line I fully. All four slides separate, releasing the molded part completely.

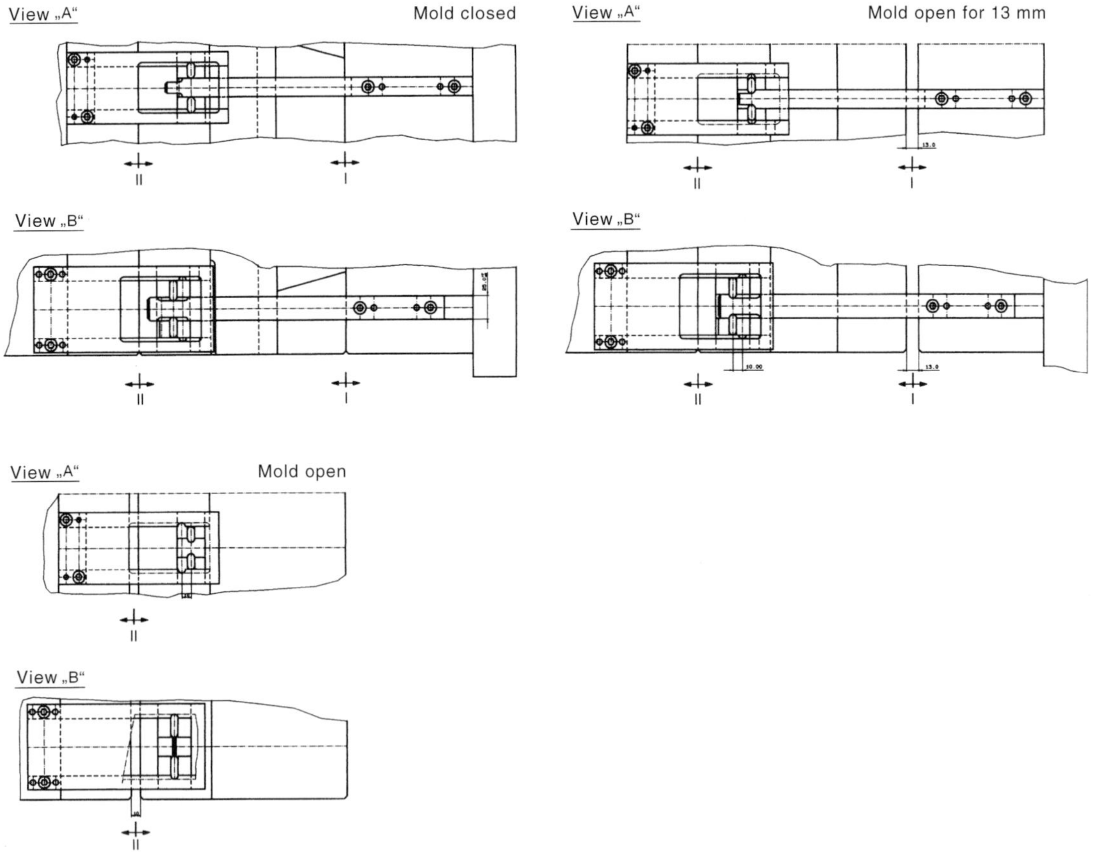

Figure 6 Latch control mechanism

Step 4

The ejector plates (6) are advanced the remaining distance by the machine's ejector. The blade ejectors (42) and the stripper bars (54) strip the basket entirely off the mold core. The positions of the ejector plates, and thus of the blade ejectors and stripper bars, are monitored by proximity switches. To prevent mold damage during set-up or in the event of a malfunction of the ejector mechanism, a mechanical safety is also incorporated:

- The ejector plates (6) are not allowed to be actuated until the mold has opened far enough to permit the part to be stripped off the core completely. To this end, the ejector plates (6) have attached to them inhibitor pins (43, Fig. 4) that prevent movement of the ejector system until the bores S in the slides (19) line up with the pins (43). Once the slides, and thus the mold, have opened sufficiently far, the pins (43) can enter the bores S, and the ejectors can advance.
- In order to prevent the ejectors from damaging the bottom of the cavity and also to avoid jamming the stripper bars into the slides, the mold is not permitted to close until the ejectors are fully retracted. To this end, the inhibitor pins (43) that have entered the bores S in the slides (19) prevent closing of the slides as long as the ejector system is not retracted. As the mold closes, parting line (II) closes before parting line (I) is fully closed. This prevents the core pins (25) from scuffing the contour-forming cores as the slides (18) close completely. Return pins (40) ensure that the ejectors are in the fully back position.

Example 122, Two-Cavity Injection Mold for Cover Plates Made from Polyacetal

The cover plates (dimensions: 25 mm × 30 mm) feature four mounting studs on one side and, in the center of the other side, a pushknob that snaps into a hole (Fig. 1).
Instead of using the usual design based on slides to release the circular undercut formed by the snap-fit head on the pushknob, this mold uses two standard, ready-to-install assemblies (Fig. 2). Each of these (supplier: Strack-Norma, Germany) consists of an outer cylindrical sleeve (1, Fig. 3) that holds two jaws (2) which, when closed, form a truncated square pyramid and move in dovetail guides in the bore of the sleeve. The jaws are actuated by a pneumatically powered piston (3) on the rod end of which the drive washers (4) are attached. The jaws are hardened; the internal shape was produced via EDM. Both the sleeve (1) and the pneumatic cylinder (5) have circular cooling grooves that are sealed by Viton O-rings (6).

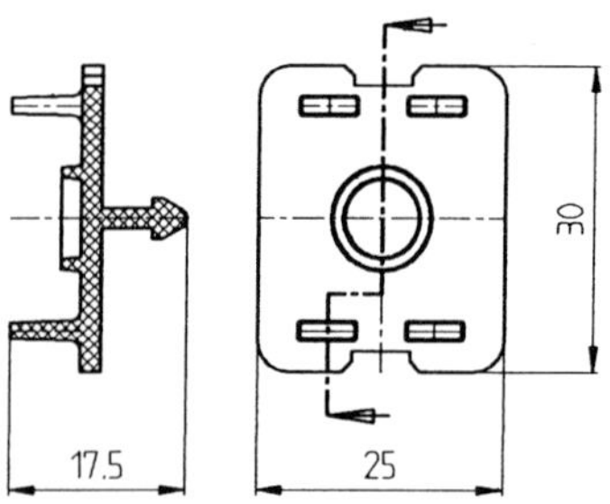

Figure 1 Cover plate

Gating/Runner System

The melt flows through a hot sprue bushing (7) to the runner channel cut into the mold parting line. At each end of the runner, a submarine gate leads to one of the mounting studs on the part.

Part Release/Ejection

As the mold opens, the jaws (2) stay in contact with the face of the moving half of the mold until the undercuts on the molded parts have been released. The molded parts and runner remain on this half of the mold as it continues to open. At the end of the opening stroke, the ejectors advance, separating the runners from the parts and ejecting both.
The collapsible core assemblies are also available as standards with four jaws and – depending on the requirements and conditions – can be actuated by mechanical means, for instance, ejector systems. Elastic ejector pins offer an alternative means of demolding undercuts. Because of their elastic spring properties, these ejector pins release the undercut on mold opening. A spring travel of up to 3 mm is available.

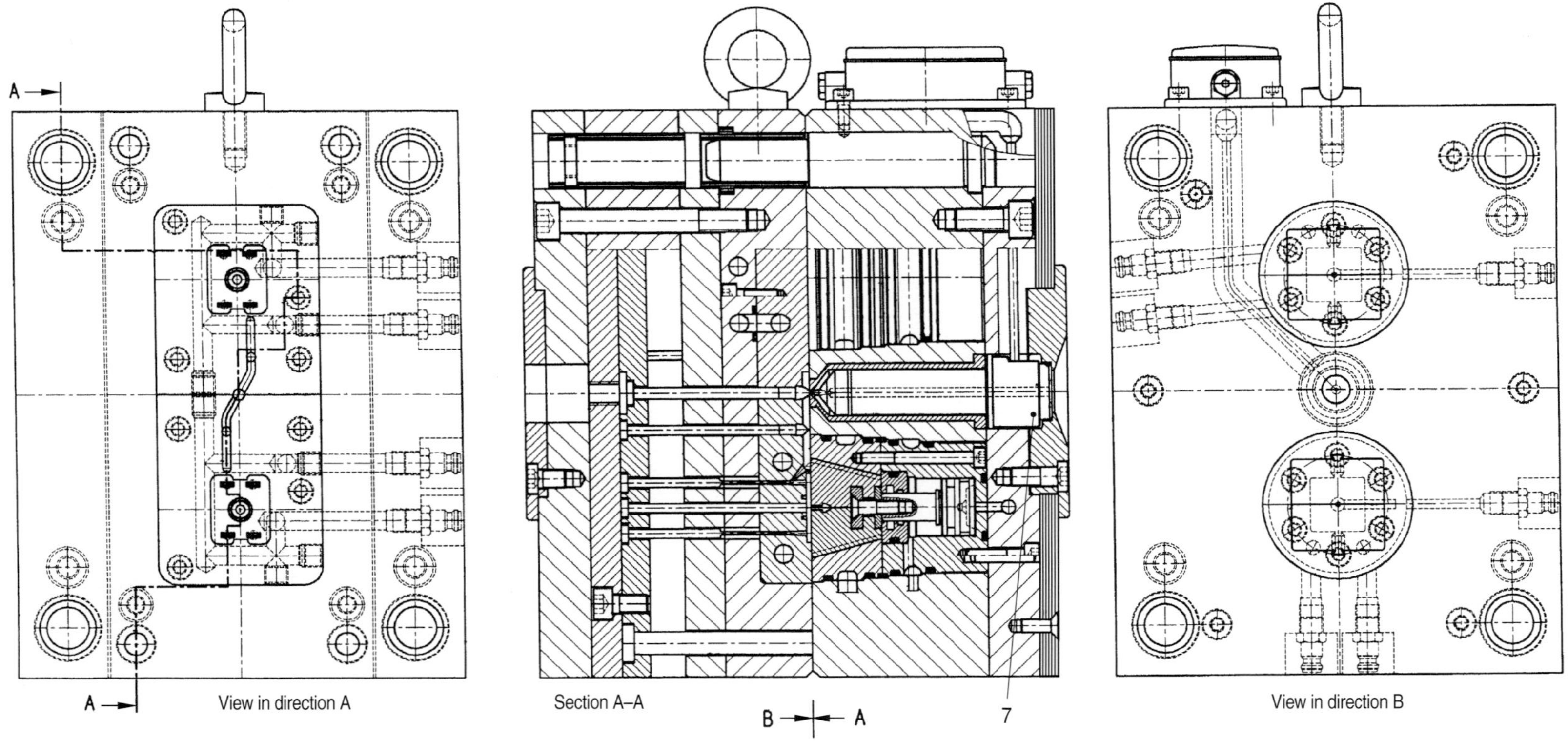

Figure 2 Two-cavity injection mold for cover plates

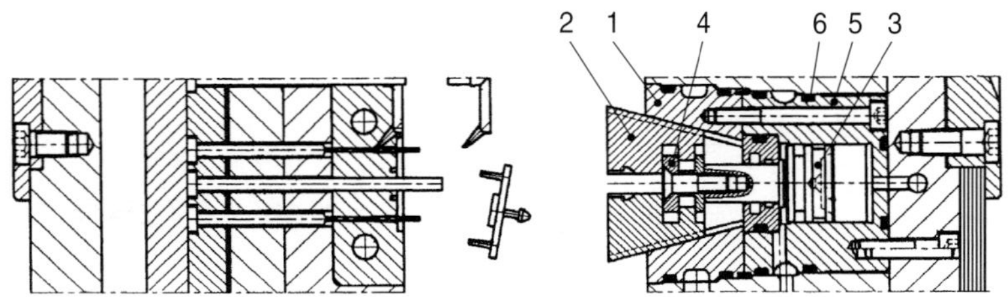

Figure 3 Part release and ejection
1: sleeve; 2: jaws; 3: piston; 4: drive washers; 5: cylinder; 6: O-ring; 7: hot sprue bushing
(Courtesy: Strack Norma GmbH, Wuppertal, Germany)

Example 123, Single-Cavity Injection Mold for a Joystick Baseplate Made from PA 66

In order to satisfy the high functionality demands on the molded part while keeping mold costs under control, the baseplate (Fig. 1) is produced from PA 66 in a single cavity injection mold. The article dimensions are 49 × 49 × 3.4mm. The molded part weighs 2.8g with a sprue weight of 1.6g.

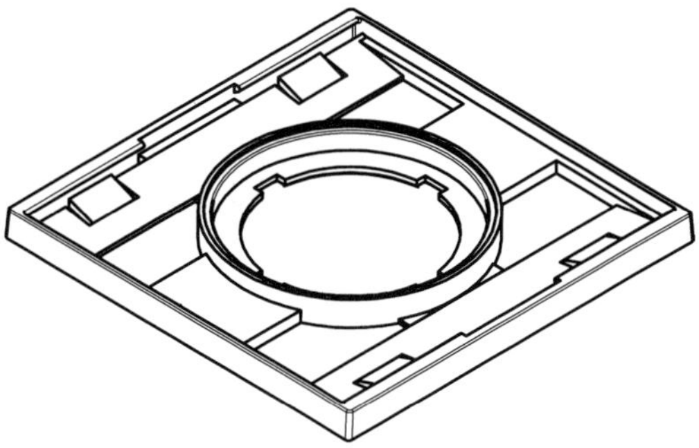

Figure 1 Joystick baseplate made from PA 66, diagram

Mold

This design (Fig. 3) can be characterized as a stripper plate mold with guide pillars (5) mounted on the ejector side and a bush-guided stripper plate (3). The mold is constructed from standardized platens with dimensions 218mm × 246mm and largely prefabricated standardized elements. Due to circumstances in the mounting space for the injection molding machine, longitudinally projecting clamping plates were selected that simplify tooling with threads for lifting and threaded bores. The cavity plates are produced from prehardened steel and have quadratic pockets for installing the hardened inserts.

A central two-stage ejector (6) is located on the closing side of the mold with predefined stroke sequence and variable stroke in order for the sprue to be removed separately upon demolding. The functional connection between the fourfold pillar-guided ejector assemblies (22 and 25) follows the scheme of the two-stage ejector shown in Fig. 2.

The relatively tall ejector box requires two spacer strips (7) that are connected with centering sleeves (8). Support pillars (9) are included to achieve the required bending strength. Precise congruence is provided by three centering units with expansion compensation (10) between the nozzle-side cavity plate (2) and the stripper plate (3).

Gating

A flat machine nozzle is scheduled to feed in the centrally installed standard sprue bush with a long sprue cone (11). The semicircular cross-section of the subrunner is incorporated in the mold insert and demolded in the cycle. Since the sub runner is incorporated in the sprue bush, an anti-twist device (dowel pin, 13) is required. The part is injected laterally via a submarine gate. Internal mold pressure is measured by a pressure transducer (14).

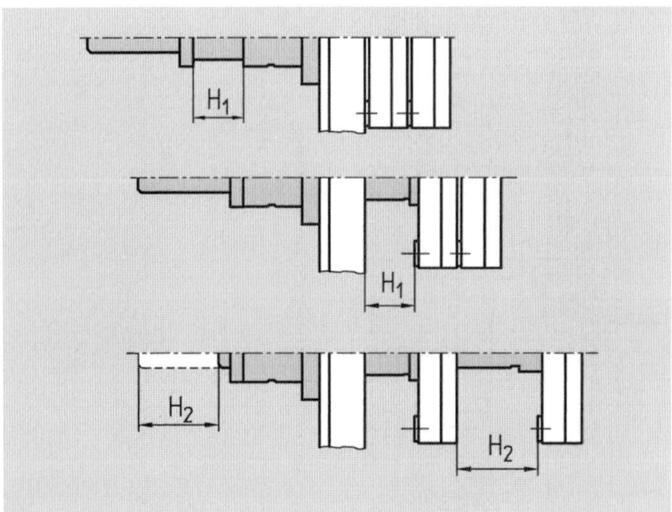

Figure 2 Working principle of a two-stage ejector, H: Stroke

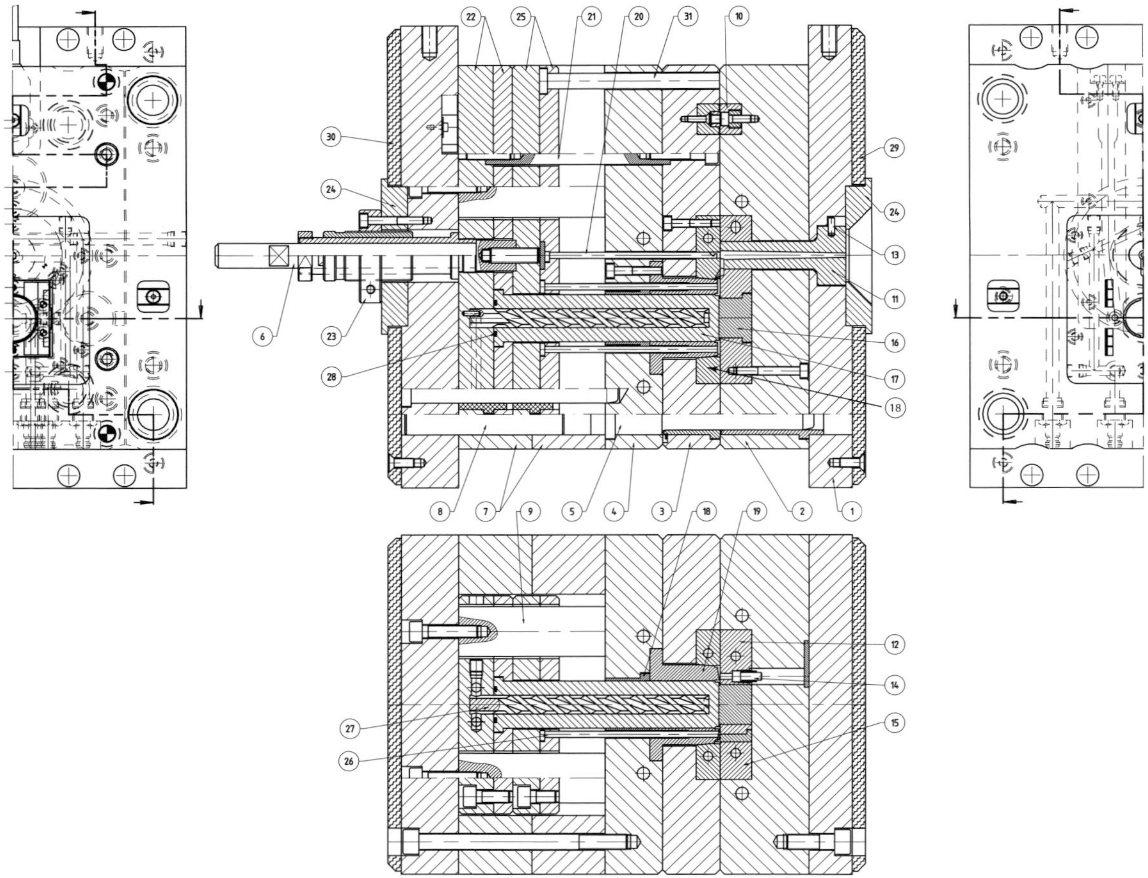

Figure 3 Single injection mold for a joystick baseplate
1: clamping plate FS, 2: cavity plate FS, 3: stripper plate, 4: cavity plate BS, 5: guide pillar, 6: two-stage ejector, 7: spacer strips, 8: centering sleeves, 9: support rollers, 10: centering unit, 11: sprue bush, 12: mold insert FS, 13: dowel pin, 14: pressure transducer, 15: mold insert BS, 16: mold core BS, 17: cooled mold core, 18: BA guide bush, 19: mold core, BS, 20: central, undercut ejector for retaining sprue, 21: ejector rods, 22: rear ejector assembly, 23: flange, 24: centering flange, 25: Front ejector assembly, 26: ejector pin, 27: double-helix core, 28: O-ring, 29, 30: thermal insulation sheet, 31: return pin
(Courtesy: Hasco, Lüdenscheid; Möller, Bad Ems)

Cycle time is approx. 7 s. The cavity is formed by the contour in the inserts incorporated in the nozzle and ejector sides. Each of these consists of several parts that guide and center each other when the mold closes. The quadratic mold insert (12) is positioned off-center in the nozzle plate (2). It holds the contouring mold cores of the top side and is congruent with the mold insert (15) on the ejector side. The cylindrical hole in the center of the article is formed by the core (16). Special precision was required to coordinate the cores (16 and 17) that are stressed by surface pressure and in order to prevent flashing. Precise guidance of the moveable and cooled mold core (17) by means of a guide bush (18) contributes to article quality. The mold core (19) is centered positively in the mold insert (15) of the stripper plate by conical surfaces.

Demolding

When the mold opens, the sprue is pulled from the sprue bush. The central ejector (20) has an undercut for retaining the sprue. The rear ejector assembly (22) is directly connected to the ejector plate (3) via four ejector rods (21). The two-stage ejector (6) is fastened centrally in the ejector-side clamping plate to the centering flange (24) by the stroke-adjusting flange (23).

The two-stage ejector first moves both ejector assemblies with stroke H1 (Fig. 2), moving the stripper plate forward. Thereby, the molded part is released from the cavity and simultaneously shears off the sprue. The secondary stroke H2 from the front ejector assembly (25) uses the ejector pins (26) to finally eject the molded part. At the same time, the sprue and runner retained by the ejector (20) are released for the removal unit. When the mold closes, both ejector assemblies are returned to their start position by return pins (31).

Cooling

In order to optimize cycle time and influence shrinkage and warping, several separate cooling circuits are located in all cavity plates and inserts. The tubing is attached to laterally screwed-in nipples by quick couplings.

The core (17) is cooled with a double-helix core (27). Influx and return are integrated in the pressure plate of the rear ejector assembly (22). The cooled core fastened in this ejector assembly is sealed by an O-Ring (28). Thermal insulation sheets (29 and 30) are screwed onto the mold on both sides.

Example 124, Single-Cavity Injection Compression Mold for Thermoset V-Belt Pulley (Injection Transfer Mold)

The pulley (Fig. 1) for a Poly-V belt in a car engine has a diameter of 91.5 mm. It is located via its bore on a shaft and mounted with three bolts. The mold (Fig. 2), which measures 344 mm in diameter and 350 mm in height, is a core-embossing type.

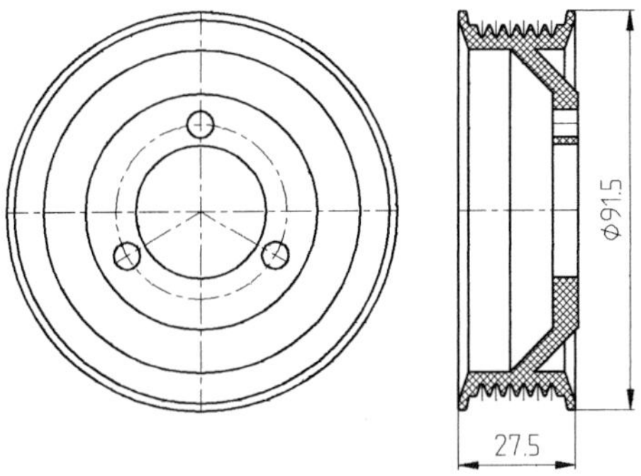

Figure 1 Thermoset V-belt pulley

The unique feature of this mold involves actuation of the compression core (1) by the yoke-shaped slide (2) with the aid of hydraulic cylinder (3). Before molding compound is injected, the core is retracted, so that the transfer chamber is considerably larger (Section A-A, bottom). After the metered amount of molding compound is injected, the core is moved to the right, forcing the molding compound into the actual cavity (Section A-A, top). The flange holes are formed without any flow lines.

After the part cures at a mold wall temperature of about 180°C (356°F), the mold opens. The radially guided splits (7) on the moving mold half are opened by the cam pins (8). The cured gate remnant is pulled out of the sprue bushing (4). After the mold opens, the pulley is pushed off the core by the stripper ring (9); pin (10) ejects any cull remnant.

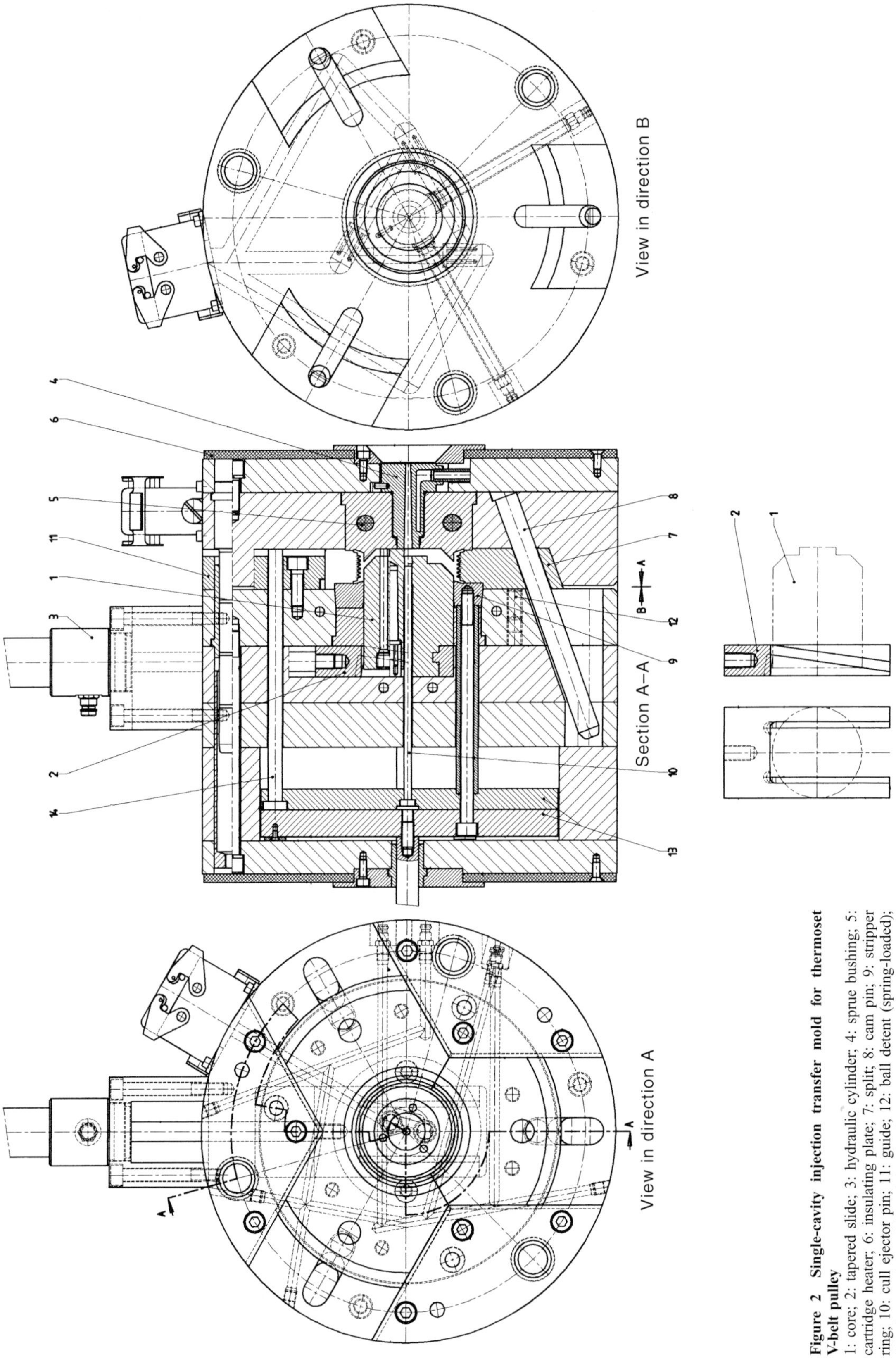

Figure 2 Single-cavity injection transfer mold for thermoset V-belt pulley
1: core; 2: tapered slide; 3: hydraulic cylinder; 4: sprue bushing; 5: cartridge heater; 6: insulating plate; 7: split; 8: cam pin; 9: stripper ring; 10: cull ejector pin; 11: guide; 12: ball detent (spring-loaded); 13: ejector plates; 14: push-back pin
(Courtesy: Hasco Normalien, Lüdenscheid, Germany)

Example 125, 16-Cavity Hot-Runner Mold for Paperclips Made from ABS

Standard mold units from a mini-mold system were utilized to produce these parts on a Babylast injection molding machine. The system has to provide for frequent, trouble-free color changes. The minimal wall thicknesses require high injection pressures. The position of the molded parts in the mold permits relatively low holding pressure. 3D CAD data were available for designing and SLS data (*S*elective *L*aser *S*intering) for producing the prototype. The molded part has dimensions of 27 mm × 8 mm, wall thicknesses from 0.5 to 0.8 mm and weighs 0.2 g (Fig. 1). The clamping effect of the paperclip results from calculated warping. Surface texturing is done by laser treatment of the cavities. Gate diameter is 0.4 mm.

Gating

The parts were direct-feed molded using a standardized mini-hot-runner system with 2 rows of 8 chambered and screwed-in tips (using the principle of indirectly heating by heater cartridges). In the hot-runner manifold, there is a heated pipe (distributive element); flow channels have been cut into its external jacket surface for natural balancing. The gap between the distributor element and the hot-runner manifold block must be kept as small as possible in order to eliminate crevice corrosion. A widening of the gap, moreover, could cause the melt to stagnate in this area, leading to heat degradation which, in turn, can be the cause of melt contamination. In view of the extremely small gate diameters of less than 0.4 mm, the screwed-in tips have to remain precisely centered in the gate under working conditions, i.e., heat expansion has to be calculated in advance; to this extent, it is operation dependent. These circumstances have to be taken into consideration, especially when processing various thermoplastics with extremely different melting points in the same mold. In order to minimize heat loss by conduction, supportive disks were fabricated from titanium alloy. Temperature control for the hot-runner system requires just one control circuit. The thermal sensor is located between the heat source and a thermally conductive nozzle (Fig. 2).

Temperature Control

Temperature control has to be done so as to maintain melt flow in a very narrow space. CO_2cooling was specified for core the temperature.

Ejection

Due to the upright position of the parts in the mold, there are undercuts to be ejected. Profiled ejector racks (19) provide safe ejection. An externally mounted ejector mechanism (5, 6) is moved by a special guide (23, 25) through an ejector bolt (41). Ejection forces are reduced by friction-reducing Dicronite™ through coating on all moving parts and cavities.

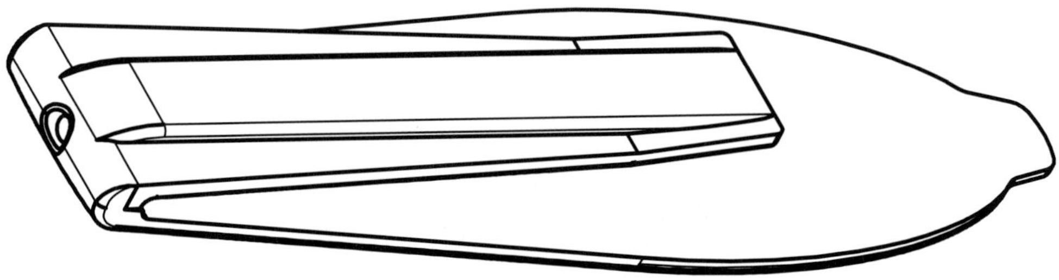

Figure 1 Paper clip made from ABS, diagram

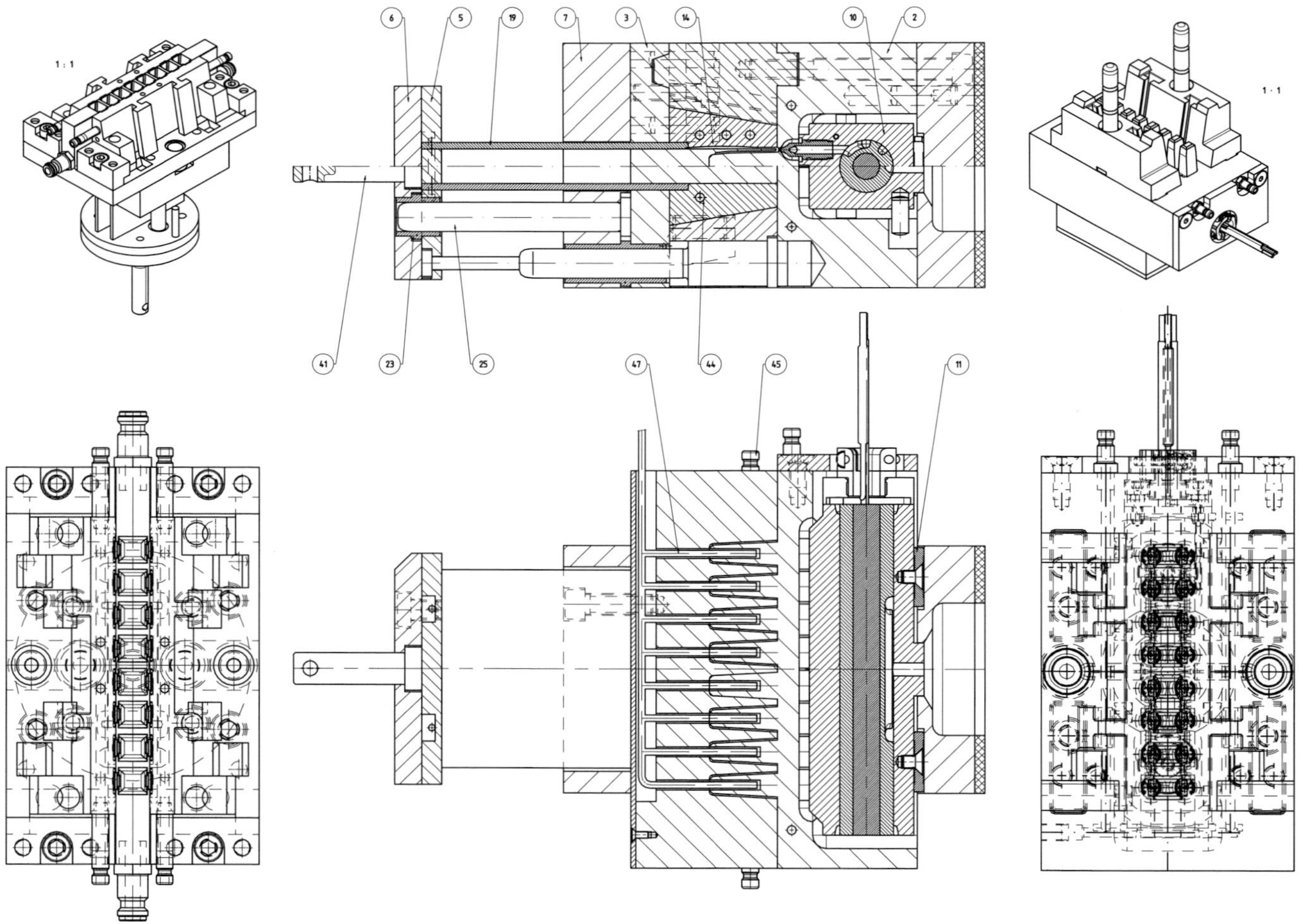

Figure 2 Sixteen-cavity hot-runner mold for paperclips
3: core retainer plate, 5, 6: ejector assembly 7: cavity plate, AS, 10: hot-runner manifold block, 11: support disk, 14: inserts, 19: ejector rack, 23, 25: ejector plate guide, 41: ejector bolt, 44: temperature control, 47: CO_2 cooling
(Courtesy: Hasco, Lüdenscheid)

Example 126, Single-Cavity Injection Mold for a PE-HD Clothes Hanger Produced via Gas-Assisted Injection Molding

The clothes hanger (Fig. 1) is basically a bent round rod 16 mm in diameter with a hook at one end. An I-beam-shaped cross piece with a wall thickness of 2.2 mm connects the two ends of the rod. Two button-shaped projections for hanging women's skirts are located on the bottom cross piece.

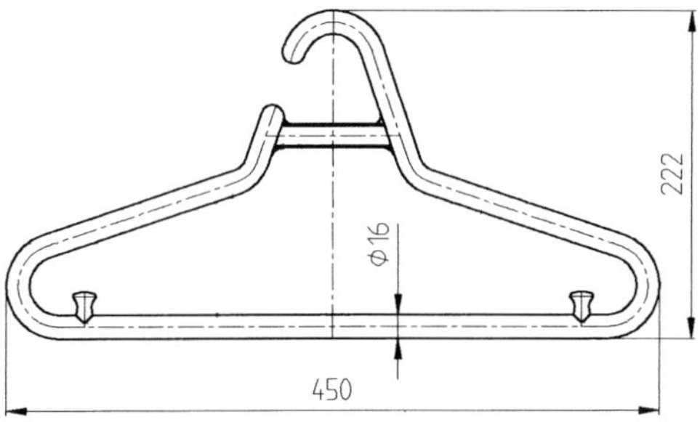

Figure 1 PE-HD clothes hanger

Mold (Fig. 2)

To save material, reduce cycle time and prevent sink marks, the mold (dimensions: 546 mm × 346 mm × 297 mm shut height) was designed for gas-assisted injection molding.

The mold cavity and runner channel, which enters the cavity close to the connecting cross piece, are located between the mold plates (1, 2). The gas injection pin, consisting of injection pin body (4) and gas injection needle (5), is located immediately adjacent to the gate.

Molding Sequence

Molding of the clothes hanger begins with injection of a metered amount of melt. The injection pressure required is relatively low, because the mold is filled only partially and the large cross-section of the part does not create any significant resistance to flow. For the same reason, no melt flows into the air gap between the injection pin body (4) and gas injection needle (5).

Next, nitrogen gas at a pressure of about 150 bar/2175 psi is introduced into the runner, and from there into the melt in the cavity, via the gas injection pin. An ever-larger bubble forms in the melt, while the still-molten core advances via fountain flow to the ends of the flow paths leading away from the gate. The result is a part with a tubular cross-section 16 mm in diameter and a wall thickness of 2.5 mm.

Part Release/Ejection

As the mold opens, the two loosely fitted mold inserts (6) release the bottom cross piece of the hanger from the stationary-side mold plate (2) under the action of springs (7). The sprue is pulled out of the sprue bushing and the hollow runner stem is pulled out of the gas injection pin. Once the mold has opened completely, the profiled ejector pins (8) and the runner ejector pins (9, 10) eject the molded part and runner, respectively. After the molded part has been degated, a hole about 3 mm in diameter that was formed by the nitrogen gas remains at the gate.

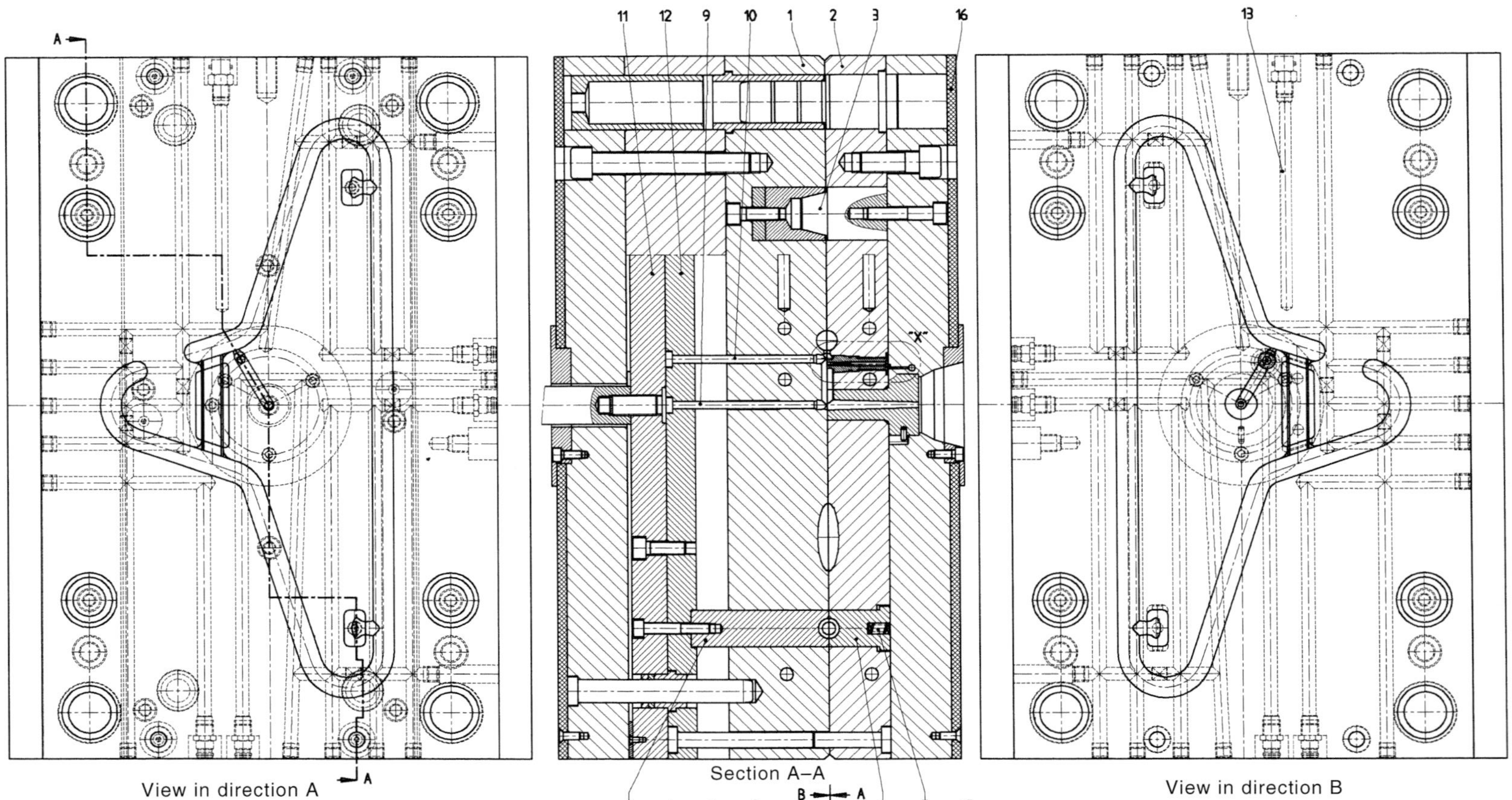

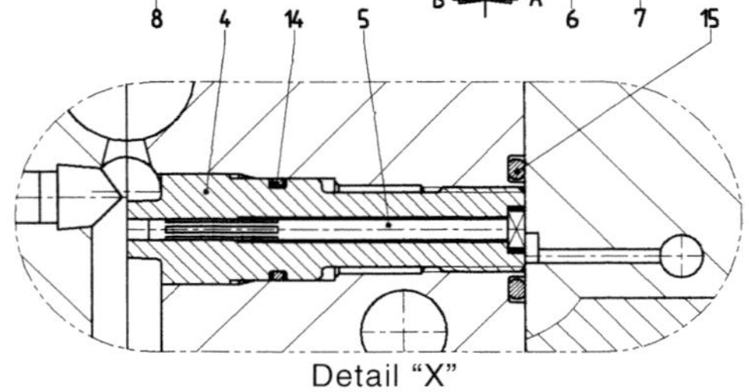

Figure 2 Single-cavity injection mold for an HDPE clothes hanger produced via gas-assisted injection molding
1, 2: mold plates; 3: locators; 4: injection pin body; 5: gas injection needle; 6: mold insert; 8: profiled ejector pin; 9, 10: runner ejector pins; 11, 12: ejector plates; 13: thermocouple; 14, 15: sealing rings 16: insulating plate
(Courtesy: Arburg, Loßburg; FH Aalen, Germany)

Example 127, Single-Cavity Injection Mold for a Syringe Shield Produced via Metal Injection Molding (MIM)

The pellets of molding compound used for metal injection molding (MIM) consist of a mixture of metal powder and a thermoplastic resin. The resin component imparts to the molding compound the flow characteristics of a highly filled thermoplastic. This means that injection molding machines can process the molding compound in molds with complex part-forming geometries.

With the MIM process, a plastic resin-containing part ("green part") is obtained from the initial step of injection molding. In a second step, heat and, if necessary, chemical means are employed to remove the resin binder from the injection molded part. The resulting part ("brown part") is then converted into a dense metal part by means of sintering – as in conventional powder metallurgy. In the course of this latter step, volumetric shrinkage occurs, resulting in a linear shrinkage of from 10 to over 15%. The injection molded item (Fig. 1) is the "green part". The finished item is a syringe shield in the metal alloy IMET N 200 comprising 1.5–2.5% Ni and the remainder Fe.

Mold

As Fig. 2 shows, the two mold inserts (1, 2) form the outer surface of the injection molded part, while the cores (3, 5) and slide (7) form the inner surface. The core (3) is actuated by the hydraulic cylinder (4). The end of the core seats and locates itself in the mating core (5), which is actuated by the cam pin (6). The opening in the side of the molded part has rounded edges on the outside, necessitating the use of slide (7), which is actuated by cam pin (9). When the mold is closed, the core (3) is held in place by the side lock (16), while the wedges (8) and (12) lock the core (5) and slide (7) in position.

Gating/Runner System

The part is filled via a large gate at the lower end. Since the risk of jetting is especially high when processing metal injection molding compounds, the gate is placed so that the incoming melt impinges directly against the core (5).

Part Release/Ejection

As soon as the mold opens, the core (5) and slide (7) begin to withdraw from the molded part. Next, the core (3) is pulled and the molded part is carefully removed and deposited by means of a part handling device. The runner drops and, after being regranulated, is proportioned back into the virgin molding compound. The ball detents (10, 11) secure the core (5) and slide (7) in the retracted position.

Literature

1. H. Eifert, G. Veltl: Metallspritzguss fuer komplizierte Bauteile. Metallhandwerk + Technik. Heft 9/94
2. F. Petzoldt: Advances in Controlling the Critical Process Steps of MIM. PM World Congress 1994/Paris
3. R.M. German: Powder Injection Molding. Metal Powder Industries Federation, Priceton. NJ, MPIF, 1990.

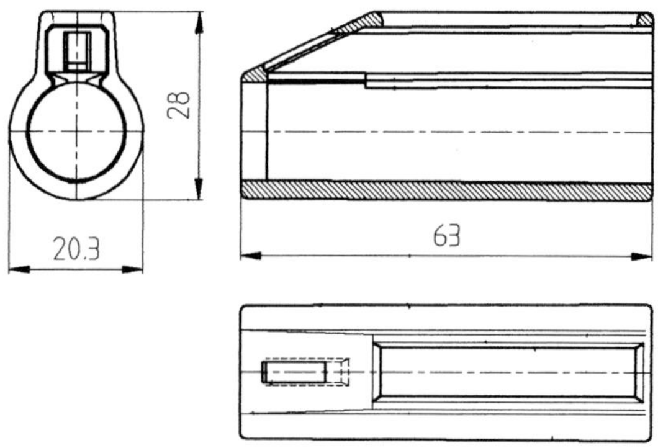

Figure 1 Syringe shield

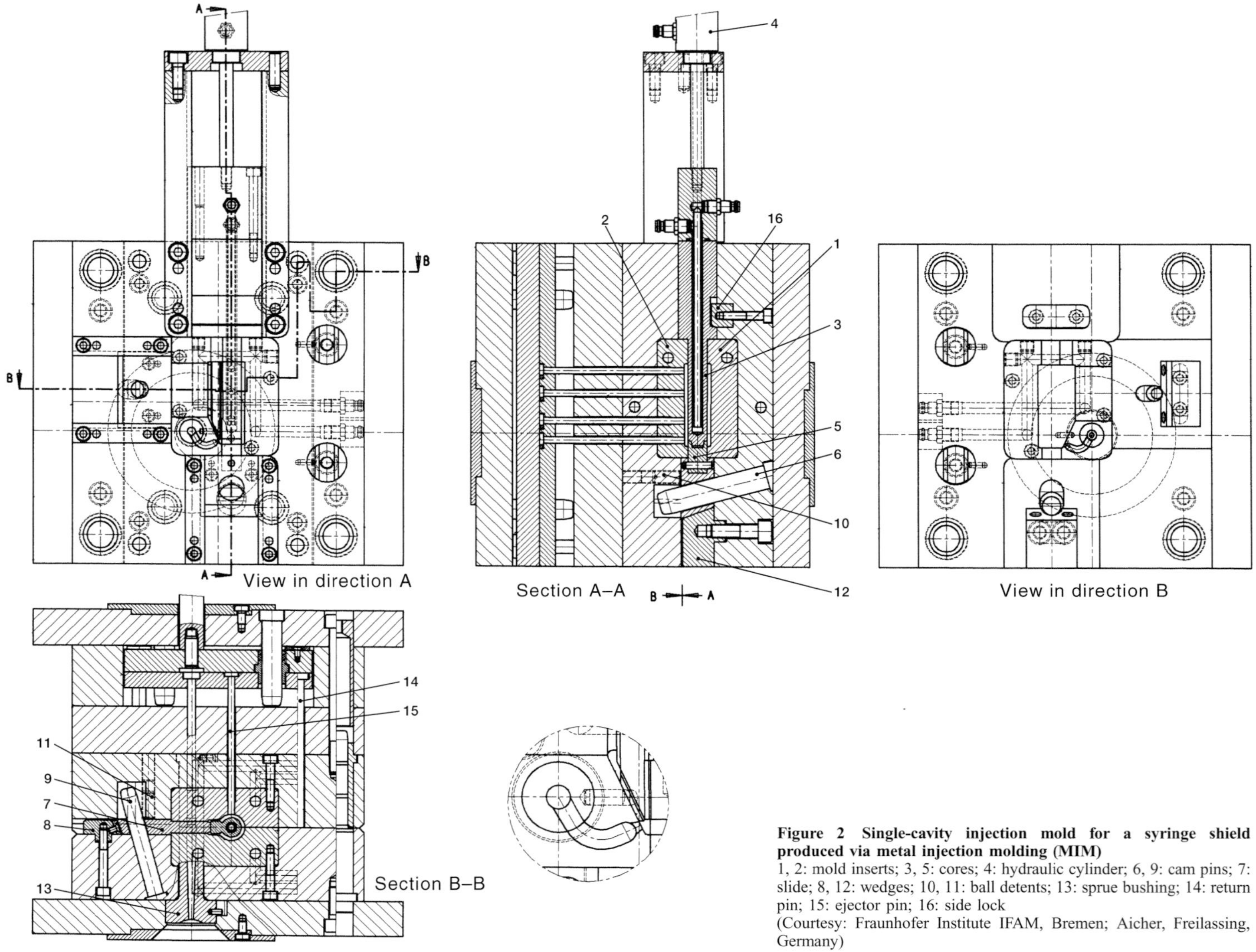

Figure 2 Single-cavity injection mold for a syringe shield produced via metal injection molding (MIM)
1, 2: mold inserts; 3, 5: cores; 4: hydraulic cylinder; 6, 9: cam pins; 7: slide; 8, 12: wedges; 10, 11: ball detents; 13: sprue bushing; 14: return pin; 15: ejector pin; 16: side lock
(Courtesy: Fraunhofer Institute IFAM, Bremen; Aicher, Freilassing, Germany)

Example 128, Three-Station Mold for a Handtool Handle Made from PP/TPE

In the mold, a high-quality handle for a woodworking chisel is produced. The handle weighs 75 g and has an exceptionally heavy cross-section (Fig. 1). The outside is a three-dimensional free-form surface with a non-slip grip section. There is a smooth transition in the non-slip region from a slim, round cross-section to a heavy square cross-section. A polypropylene (PP) is used for the handle body, while a thermoplastic elastomer (TPE) forms the non-slip grip region.

The dimensions of the mold are 696 mm × 646 mm × 596 mm. The shut height dimension of 596 mm does not include the length of the rotating shaft or coupling. Figure 2 shows a simplified lengthwise section through the 4 + 4 + 4-cavity mold. The mold was designed for a molding machine with a clamp force of 2000 kN, the main horizontal injection unit for both hard components, and a second horizontal injection unit in an L-position for the soft component.

The first and second shots for the handle body (PP) and the final soft layer (TPE) are molded successively over the mold cores (3) that form the bore for the tool blade. This requires transferring the first and second shots with the aid of a rotary mechanism, or indexing plate, integral to the mold.

The cores (3) that form the bore for the tool blade are held in the indexing plate (1) by a strip (2). The handle body is molded over these cores. For the transfer step, the indexing plate with the cores and handle shots first advances, then rotates 120°, and finally retracts into the new cavity. The translational motion of the indexing plate is accomplished with the aid of the ejector hydraulics in the machine. A hydraulic motor (4) with gear drive (5) mounted to the mold provides the rotary motion.

Figure 3, a plan view of the moving mold half, shows the arrangement of the cavity inserts. The first station (21) is at the upper right, the second station (24) is at the upper left, and the third station (23) is at the bottom. The four inserts (6) for each station are mounted in their respective plates (7) on both the stationary and moving sides. These, in turn, are fastened to a base plate (8).

The ejector mechanism (10, Fig. 2) is needed to assist part release at the first and second stations by means of an ejector pin (11) and to prevent cocking of the parts during ejection from the cavity. It advances in parallel with the rotating shaft (12) and, through the action of a two-stage system (13), disengages upon completing its stroke, while the indexing plate continues its motion.

The advantage of this mold concept is that handles of different sizes and designs can be produced by a simple changeover of the basic mold. Changeover is limited to the mold inserts (6), the cores (3) used to form the bore for the tool blade, the unit with the ejector pin (11), and the backup plate (14). These components can be changed with little effort while the mold is still installed in the machine.

Sequence of Operation for a Single Cycle

The motions and functions that occur during one cycle are as follows:

- Injection of polypropylene (PP) at the first and second stations by the main horizontal injection unit and injection of the soft component by the second horizontal injection unit in the L-position.
- The mold opens at the primary parting line I.
- The hydraulically actuated part removal robot mounted on the stationary platen advances into the open mold. The end position is sensed by a limit switch.
- Next, the machine ejector advances the entire rotating assembly (rotating shaft, indexing plate, and cores with molded parts). The finished part is now within reach of the part removal robot.
- The part removal robot retracts from the open mold, stripping the finished handle off the core.
- The rotating assembly is indexed 120° by the hydraulic motor and gear drive. The rotating motion is monitored by a limit switch.
- The machine ejector retracts the rotating assembly. The preshot from the first station is now located in the second, the shot from the second station is now in the third, and the empty core is in the first station.
- The mold closes, and a new cycle begins.

Cavities

Because of the very heavy cross-section (approx. 37 mm), the hard component is molded in two stations. This yields a significant cycle time reduction over molding in a single step, and also has a very positive effect on the quality of the molded parts.

Establishing the shrinkage is a very difficult issue. This is very important for the third station, where the soft component is molded, in order to obtain a smooth transition between the hard and soft components. More than a little experience is needed here for an exact calculation and to design the cavities, because shrinkage of both the first and second preshots must be taken into consideration.

Good venting of the cavities is also a major concern, since this can affect various parameters during injection.

Station I

The design of the first preshot largely determines the quality of the handle and the cycle time. It should be

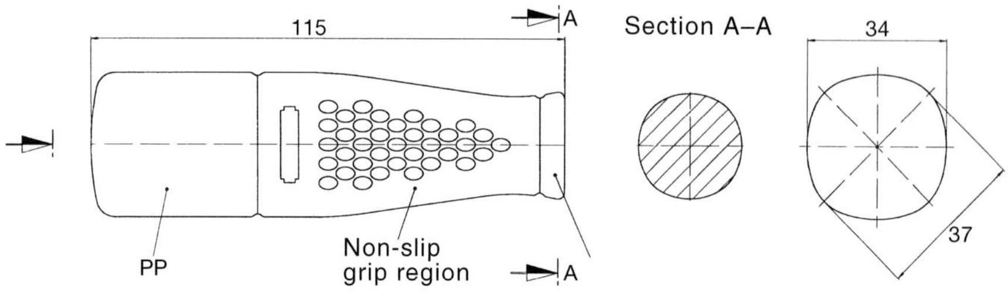

Figure 1 Tool handle of PP with a non-slip grip of TPE

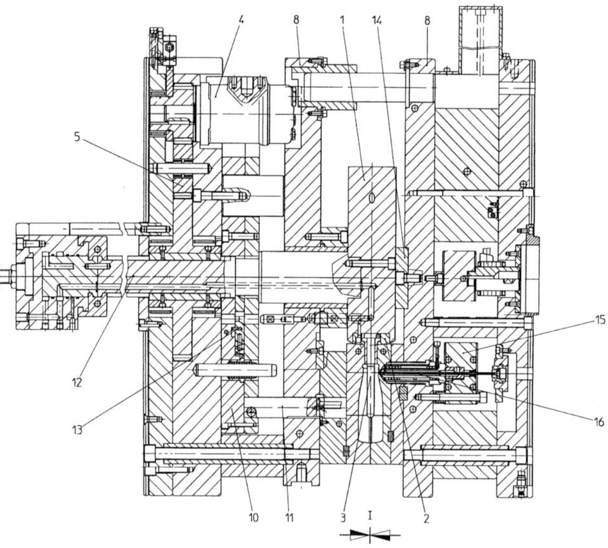

Figure 2 Simplified section through the closed mold
1: indexing plate; 2: retaining strip; 3: core; 4: hydraulic motor; 5: gear drive; 6: base plate; 10: ejector mechanism; 11: ejector pin; 12: rotating shaft; 13: two-stage ejector system; 14: backup plate; 15: hot runner system; 16: needle shutoff nozzle

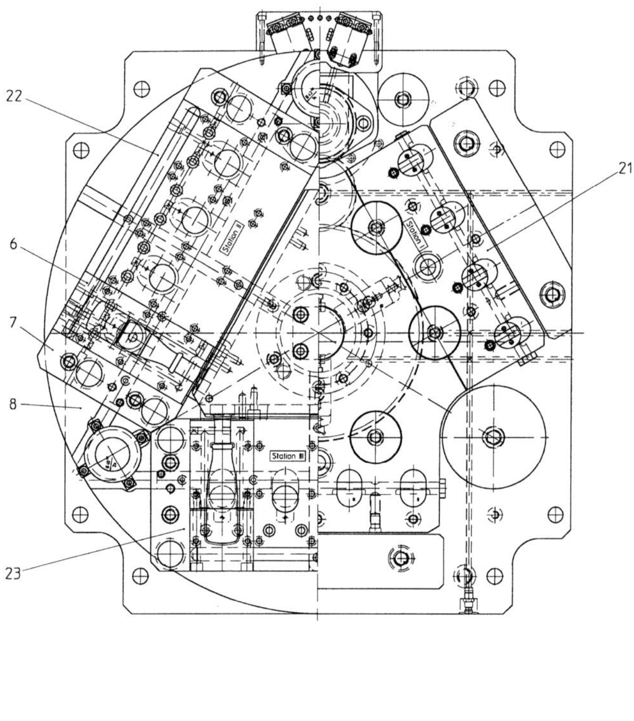

Figure 3 Simplified view of the moving mold half
6: mold insert; 7: mold plate; 8: base plate; 21, 22, 23: mold stations

designed to have as uniform a wall thickness as possible and such that the wall thickness at the second station is uniform and not excessive either. Its heavy ribbing also contributes to enhanced cooling.

Station II

These cavities form the outer surface of the hard component and the relief for the soft component. Absolutely uniform filling is required at this station, e.g. the melt flow must advance around the first preshot uniformly and join precisely at the end of the flow path. Because the part is gated on only one side, this can be achieved only through the geometry of the first preshot.

Station III

At this station, the soft component is molded over a portion of the second preshot. It is extremely important here to predict exactly the correct shrinkage of the first two preshots at the transition from hard to soft.

Gating

Injection at each of the three stations takes place via an externally heated hot runner (15). Feeding of the first and second station takes place via a joint hot-runner manifold with open sprue nozzles. The third station is gated via the second hot-runner manifold with needle shutoff.

Part Release/Ejection

The finished handles are removed by a part removal robot (Fig. 4). It consists of a hydraulic cylinder (17), a guide system (18), a stripper plate (19), and the frame (20). The part removal robot is attached to the stationary mold half, thus forming a single unit with it. A positive stop limits the stroke as the cylinder advances. A limit switch at each end of the stroke monitors the motion sequence. This part removal robot has the additional advantage of being universal. It can be adapted easily to other handle sizes and blade diameters by merely replacing the stripper plate.

Mold Temperature Control

Mold temperature control is provided by ten cooling water circuits

- in the stationary-side clamping plate;
- in the stationary- and moving-side base plates;
- one cooling circuit each for the stationary- and moving-side cavities at each station; and
- in the indexing plate via the rotating shaft.

In this regard, it is important to have uniform temperature control in associated cavities on both the stationary and moving sides in order to ensure a constant cooling rate for the melt and thus avoid warpage of the molded parts. Particular attention must be devoted to the cooling circuit in the indexing plate, since it cools the cores used to form the bores for the tool blades. The effectiveness of this circuit has a significant influence on the cycle time.

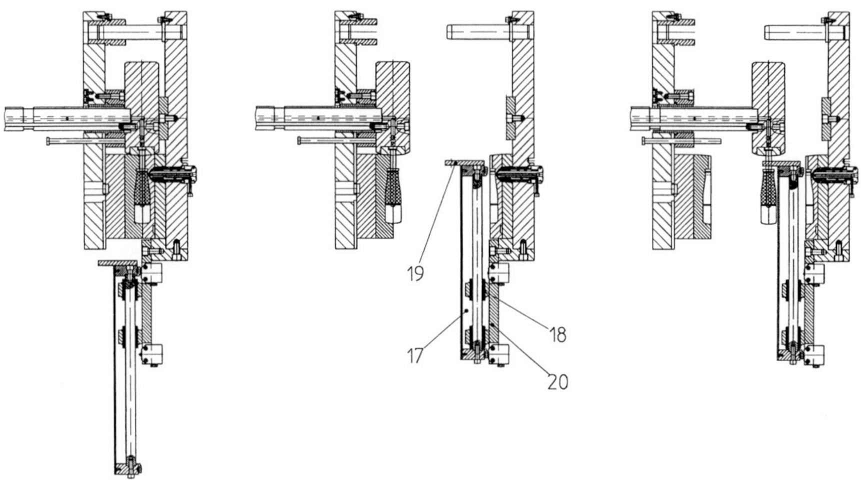

Figure 4 Sequence of operation for the part removal robot
17: hydraulic cylinder; 18: guide system; 19: stripper plate; 20: frame
(Courtesy: Braun Formenbau GmbH, Balingen, Germany)

Example 129, Four-Cavity Injection Mold for Couplings Produced via Metal Injection Molding (MIM)

The coupling (Fig. 1) in the alloy IMET N 200 (1.5–2.5% Ni, remainder Fe) is a component in a small automotive drive assembly.
The mold (dimensions: 156 mm × 156 mm × 225 mm) is constructed from standard mold components. The mold plates (2, 3) are made from the corrosion-resistant steel grade 1.2085 (X33CrS16) with 16% chromium and a strength of about 1000 N/mm^2, while the backup plate (4) and the clamping plate (1) are made from steel grade 1.2312 with a similar strength. The two cavity inserts (26, 27) are fabricated from steel grade 1.2767, hardened to 54 HRC.

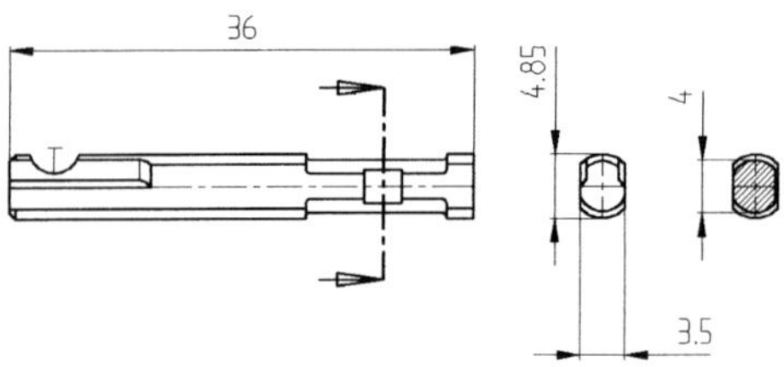

Figure 1 MIM "coupling", ready to use, injection molded and sintered

Gating/Runner System

The four cavities are arranged around an H-shaped runner system with submarine gates (Detail "X"). These are placed in such a manner that the incoming molding compound impinges against the cavity wall. This is intended to counteract the tendency to jetting. The sprue bushing (19) serves as the connection to the injection unit of the molding machine.

Part Release/Ejection

The ejector pins (24) are contoured to match the shape of the molded part and secured against rotation by the dowel pins (30). The molded parts are removed by a part handling device and carefully deposited. At the same time, the ejector pins (23, 25) eject the runner, which simply drops and, after being regranulated, is proportioned back into the virgin molding compound.
The finished sintered part exhibits 17.5% shrinkage as measured with respect to the cavity/green molded part.

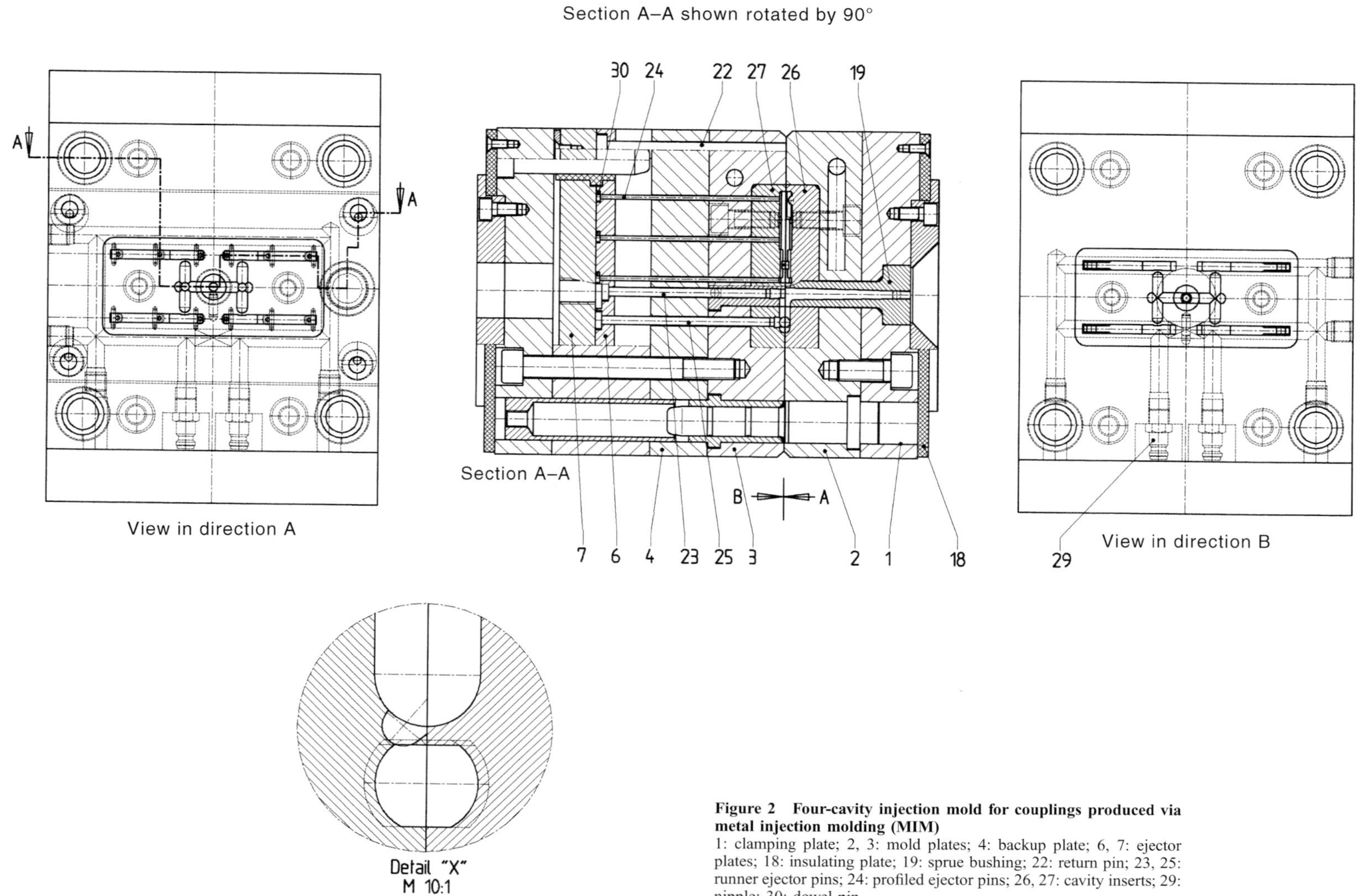

Figure 2 Four-cavity injection mold for couplings produced via metal injection molding (MIM)
1: clamping plate; 2, 3: mold plates; 4: backup plate; 6, 7: ejector plates; 18: insulating plate; 19: sprue bushing; 22: return pin; 23, 25: runner ejector pins; 24: profiled ejector pins; 26, 27: cavity inserts; 29: nipple; 30: dowel pin
(Courtesy: Krebsoege, Bad Langensalza, Germany)

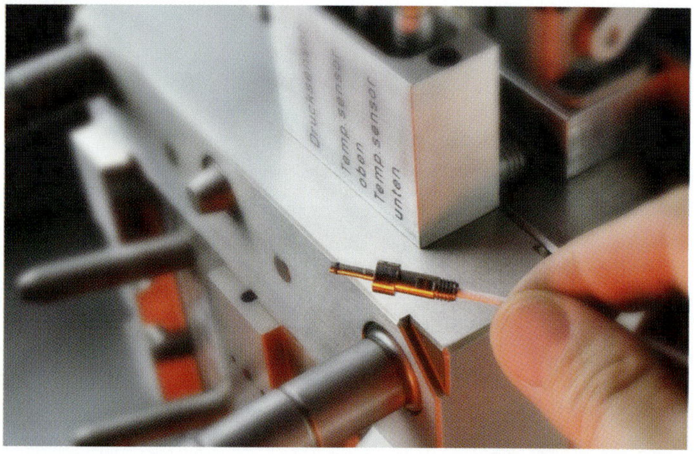